War and Castros

New approaches to the northwestern Portuguese Iron Age

Francisco Manuel Veleda Reimão Queiroga

BAR International Series 1198
2003

This title published by

Archaeopress
Publishers of British Archaeological Reports
Gordon House
276 Banbury Road
Oxford OX2 7ED
England
bar@archaeopress.com
www.archaeopress.com

BAR S1198

War and Castros: New approaches to the northwestern Portuguese Iron Age

ISBN 1 84171 563 8

Printed in England by The Basingstoke Press

All BAR titles are available from:

Hadrian Books Ltd
122 Banbury Road
Oxford
OX2 7BP
England
bar@hadrianbooks.co.uk

The current BAR catalogue with details of all titles in print, prices and means of payment is available free from Hadrian Books or may be downloaded from www.archaeopress.com

Table of contents

Table of contents I
Preface III
1- INTRODUCTION 1
2- PREVIOUS STUDIES 3
1) History of research in the study area 3
2) The growth of regional studies 4
3) The classical sources 7
3- THE REGION 11
3.1) The region defined 11
3.1.1 -Geomorphological structure, hydrology and soils 12
3.1.2 -The climate and vegetation 14
3.1.3 -Mineral resources 18
3.1.4 -Natural routes of communication 19
4- ASPECTS OF CASTROS ARCHITECTURE 21
4.1) -General aspects of dwelling architecture 21
4.1.1) The organization of the space 22
4.1.2) Public areas and utilities 22
4.1.3) The quarters 26
4.1.4) The familial enclosures and their functional areas 26
4.2) The chronological boundaries 31
4.3) Discussion 32
5- THE SITES 35
5.1) Patterns of distribution 35
5.2) Geography, resources and site hierarchy 37
5.3) Discussion 38
6- THE DEFENSIVE STRUCTURES 41
6.1) The emergence of the defensive structures: some model cases 41
6.1.1) The late Bronze Age sites 42
6.1.2) The sites with continuous occupation 43
6.1.3) The early Iron Age sites 43
6.2) The defensive structures 44
6.2.1) The defensive wall: military and social significance 45
6.2.2) The ditch and rampart 46
6.3) Discussion 46
7- THE PRODUCTIVE ACTIVITIES 49
7.1) -Food producing strategies 49
7.1.1) -Evidence from Classical sources and its interpretation 49
7.1.2) -Evidence from archaeology 50
7.1.2.1) -Agriculture and gathering 51
7.1.2.2) -Animal husbandry 55
7.1.2.3) Marine resources: fish, shellfish and salt 56
7.1.2.4) -Evidence from implements 57
7.1.3) -Discussion 58
7.2) -Mineral resources 60
7.2.1) -Evidence from Classical sources 60
7.2.2) -Mineral resources and evidence for their use 60
7.2.3) -Discussion 63
7.3) -Manufacturing 63
7.3.1) -Ceramics 63
7.3.2) -Metalwork 66
7.3.3) Weaving 68
7.4) -Raiding 68
7.4.1) -The evidence 69

7.4.2) -Raiding as a system ... 69
7.5) -Long distance trade and exchange ... 71
8) -Discussion ... 72
8- THE SOCIAL STRUCTURE ... 73
8.1) -Introduction ... 73
8.2) The Celtic 'family' in Europe: some analogies ... 73
8.3) Castros familial and supra-familial institutions ... 75
8.3.1) Family ... 75
8.3.2) *Castellum* ... 76
8.3.3) *Populi* and *Civitates* ... 77
8.4) Discussion ... 78
9 - WARFARE AND SOCIETY ... 81
9.1.1) Lusitanians and Callaicans: who's who? ... 81
9.1.2) The newcomers: ethnic movements in the Northwest ... 82
9.2) The Callaican warrior: a profile ... 84
9.3) The nature of the warrior: the mercenary and political cohesion ... 90
9.3.1) Social hierarchy and institutions ... 90
9.3.2) Indigenous *auxilia* during the Conquest of Hispania ... 92
9.4.1) The military events up to Augustus ... 93
9.4.2) Castros under Augustus: permanence or change? ... 99
9.5) War, economy, prestige and social order ... 100
9.6) Discussion ... 103
10- CONCLUSIONS ... 107
BIBLIOGRAPHY ... 113
APPENDIX I ... 149
1) Introduction ... 149
2) List of Sites ... 150
2.1) District of Braga ... 150
2.2) District of Porto ... 164
2.3) District of Viana do Castelo ... 174
2.4) District of Vila Real ... 192
APPENDIX II ... 203
1) Introduction ... 203
2) Radiocarbon Dataset ... 206
3) Notes ... 273
ILLUSTRATIONS ... 285

a meus pais
e ao Mestre

... y la palabra aunque las letras tengan
iguales transparencias y vocales
ahora es otra y es otra la boca:
la misma boca es otra boca ahora:
cambiaram labios, piel, circulaciones
otro ser ocupó nuestro esqueleto:
aquel que fue en nosotros ya no está:
se fue, pero si llaman, respondemos
"aquí estoy" y se sabe que no estamos
que aquel que estaba, estuvo y se perdió:
se perdió en el pasado y ya no vuelve.
(Pablo Neruda 1963 - Pasado)

Preface

The above thoughts by Pablo Neruda can reflect a very ancient human dilema towards the past. An almost indiscriminate acceptance or rejection of the past conducted particular feelings in the present. A glorious past can always disminish both the social and individual frustrations of an apathethic present. And the past is all Archaeology is about.

The fascination with the past, the nostalgia of origins, the search for a cultural identity for the Northwest, all these have been the major motivations for over one century of a research often conducted by the heart. In fact, Archaeology is a science turned to the study of the human past, and has evolved so fast in the last few decades that it makes every new achievement part of the inexorable destiny of becoming part of the past of archaeological knowledge.

Field research is fortunately evolving very fast in Portugal, due to renewed internal dynamics and wider contacts with European schools of Archaeology. This turnover is undoubtedly related to a healthy attitude of the present older generations of Iberian researchers who lead younger archaeologists towards a scientific evolution in methods, in a context of freedom and commitment that are indispensable for those who need to find their own way. This is why recent achievements in Castros research owes so much to the 'old masters'.

This study is the dissertation submitted in 1992 to the Faculty of Anthropology and Geography of Oxford University for the degree of Doctor of Philosophy. With exception of this block of text and the subsequent careful revision of the text kindly done by Christina Unwin and Richard Hingley for less adequate expressions and spelling mistakes, no other changes were made to the original work. The structure and presentation of this study obey the University regulations for doctoral thesis, in particular the limitation in length to 100 thousand words. I agree the overall presentation of this work might not be the more suitable for publication, and some introductory commonplaces, presented here and there, may hurt the feelings of specialists on the subject. However, some of the generalities may prove to be useful since the English readers had no updated references on Castros Culture for decades[1]. Meanwhile, the continuation of research in this lapse of time, both mine and my colleagues', appears to be substantiating a few of the somewhat shy ideas presented here, namely the ones referring to economical activities and the non-defended farmsteads in valleys.

It is only normal on these occasions to begin by remembering all those who contributed to the fulfillment of such an important work for one's academic life.

My first acknowledgment goes to Oxford University, to its research environment and magnificent libraries, where I first understood the real dimension of a scientific and multidisciplinary research. Colleagues from the Institute of Archaeology and researchers from various departments of the University never denied their help when it was necessary.

My strongest feelings go to my supervisors, for their help, encouragement and kindness.

1 - The articles produced by Hawkes (1958, 1971, 1984) and Savory (1968) were nearly the only written in English up to 1992.

Professor Carlos Alberto Ferreira de Almeida (†), from Porto University, is responsible for my path in Archaeology from the very start. His knowledge and encouragement made me go forward in a project I never thought was feasible.

To Professor Barry Cunliffe, from Oxford University, I must thank his permanent patience and courtesy as well as his attentive and helpful supervision.

The Calouste Gulbenkian Foundation has provided generous financial support during my long stay in Oxford. First as a *recognized student*, to update my limited knowledge in archaeological research, and then during the development of this doctoral thesis. Too much I owe to the generosity of the Gulbenkian Foundation, to the Director of the Fine Arts Department, Professor Nobre de Gusmão, and in particular to the Director of the Portuguese Affairs Department in the UK, Senhora D. Maria Antónia da Silva, whose kindest help and encouragment were important for the course of my research. To the Meyerstein Fund I also owe acknowledgements for the award of one grant for travelling expenses.

The research environment and technical facilities provided by the Vila Nova de Famalicão City Council Department of Archaeology, which I organized and developed in the course of the past ten years, were an important support for my research. Together with the Oxford University Institute of Archaeology and the Donald Baden-Powell Quaternary Research Centre, these were my research bases, and many thanks are due to both the heads and the staff of these institutions for their help and collaboration.

To Pamela Wace, I cannot feel enough obliged for her patient and attentive revision of a particularly unreadable text.

My friends and colleagues, Felisbela Oliveira, Teresa Soeiro, António Dinis, and also Elzira Sá and Chris Unwin, deserve a very particular remark for their help and encouragement.

A word of gratitude is also due to all those Portuguese and Galician companions and colleagues who extended their co-operation beyond patience. They are far too many to be named here, but references are made in the text for every particular case. Their example has shown that research is a team work for which no borderlines can exist, and also how important co-operation, discussion and criticism is among fellow researchers.

And now, looking back to many difficult moments in the course of this work, I feel I also owe an acknowledgement to the Galician oysters and to the Ribeiro. They made me realize that, when going through the bad times, the only way out is to look ahead and carry on.

> "The techniques which produce artefacts are easy to infer, subsistence-economics fairly easy, social and political institutions considerably harder, and religious institutions and spiritual life the hardest of all."
> (Hawkes 1954, 161-2)

1- INTRODUCTION

From obscure beginnings at the end of the Bronze Age to the award of the *Ius Latii* to all the inhabitants of Iberia by the emper or Vespasian (AD 73-74), winds the long and uneven path of the 'Castros Culture' in the Iberian Northwest.

A castro is a fortified village usually located in a prominent geographical position, such as a hill spur, hillock or less frequently, a valley edge. As such a prominent landscape feature, it has given its name to the Iron Age Culture (or group of related cultures) in general, which occupied the region of the Iberian Northwest between the end of the Bronze Age and the beginning of the second century AD. Such sites were characterized by small settlements in their early phases, gradually evolving into urban structures under Roman influence.

A castro represents a model of settlement which has given its name to the 'Culture of the Castros, or the 'Castros Culture'. It should be noted that English writers have already used the singular term 'Castro' in general when describing this culture (Hawkes 1984). However, I consider the correct term to be in the plural, when referring to the culture, and have employed it thus throughout this study.

Much has been written about 'Castros', since the stone ruins of these hillforts first attracted the attention of the *Literati*, and they gave their name to a cultural group. Memory of them has, however, always been a part of the local culture of the inhabitants of the Northwest. Aspects of them survive in the colourful local legends of these northern Portuguese peasant communities, giving occasional glimpses of how life must have been during the Castros´ period. A few themes are common to all these legends. For example, that in which the treasures lie inside the huge rocky hilltop where they are guarded by a woman Moor. Sometimes the guardian is a snake which can change into the shape of a lady. Alternatively this guardian may become an underground passageway, running from the hilltop to the river below (Brandão, AM, 1891, 205).

As a land of mountains and small plateaux, sunshine and plenty of water, the Northwest of Iberia has always shaped the peoples who lived there. To understand this is to appreciate the pride of being *Calaico*, i.e. an inhabitant of *Callaecia*. This designation was often used, both during and after the Roman Conquest[1]. It persists, even today, amongst the communities of the Northwest, where cultural links are still strong.

Until now, the main concerns of archaeologists have been related to cultural identity, the chronological framework, and to the position of the Castros Culture in the broader framework of European Prehistoric Society. According to Maluquer de Motes (1955b, 241), the two major sources of influence in the Iberian Peninsula in Prehistoric times originated in the Mediterranean and Central European (Celtic) areas, and these two lines of research have been actively pursued. There has been considerable open debate as to the importance of these external influences on the Castros Culture. More recently, the significance of local identity and development has come to the fore. What emerges from such discussion is that the Northwest appears to be the point of interaction for these foreign influences. Celtic studies have advanced little beyond this, however, in recent decades.

When I first started this research, Iron Age defensive structures were my main area of interest, together with general questions of typology and social function. Very soon I discovered that the real situation was far more complex. The Castros´ social and cultural interactions were clearly of the utmost importance and could not be studied in isolation. In particular, the social role of war looked like a promising line of enquiry, and I decided to concentrate my attention on that topic. Lively discussion with Portuguese and Galician colleagues confirmed that others were thinking along the same lines. Our knowledge of the subject benefited by being viewed from different perspectives. The outcome of this debate sets the Castros Culture firmly in a European context, where it compares favourably with any other contemporary culture. However, the geographical situation in the northwestern corner of Iberia is a constraint that should not be neglected. Its particular environmental conditions played a determining role in the evolution of social dynamics and cultural strategies within the Castros society.

[1]-Idatius, bishop of Chaves in the Vth century AD, was still calling himself a *Calaico* to designate his origin (Tranoy 1974)

The study period is bounded by the point of emergence of fortified sites during the final stages of the Bronze Age (c.1200-1000 BC) and the Flavian period, at the end of the Ist century AD. The actual beginning of the Castros Culture is however, far from precise and still open to question. By contrast, curtailment by the Flavian epoch, as proposed by Silva (1986a, 61), is well documented archaeologically, and generally accepted.

The principal area of study (Figure 2) lies within the region confined by: the Atlantic coast to the west, the Tâmega-Corgo river line to the east, the Minho river to the north and the Douro river to the south. Data from adjoining regions have been taken into account, however, whenever relevant.

The main theme of this study is warfare in the Castros Culture. A fresh approach has been adopted, to evaluate the importance of warfare at this time and to see how it shaped the development of Castros society as a result. For this, warfare is examined in terms of the total environment in which it occurs, i.e. in terms of the geographical, economic, social and ritual landscapes in which it exists. This study has not been restricted, therefore, to mere studies of weaponry, defences, and their related mythology. To achieve these aims, the considerable body of data gathered across the study area over the years has been re-interpreted. It has further been supplemented by the results of my own recent fieldwork, excavation at several Bronze and Iron Age hillforts and palaeoenvironmental research. As time does not permit an exhaustive study of the subject, selections have had to be made, and some regrettable omissions have inevitably occurred.

One of the limiting factors to further progress in the North of Portugal is the variable quality of the database. For example, the systematic collection of suitable samples for environmental work is still in its infancy. Data from other regions and countries are often used to fill in gaps in local knowledge, so that a clearer picture of the environment under scrutiny may be obtained. Of course, there are a number of difficulties associated with using data 'out of context' in this way, which are fully appreciated here. Again, in an attempt to explain the social and ritual attitudes of Castrossociety towards war, most of the information has been taken from the period immediately preceding the Roman Conquest. In spite of the constraints implied in doing so, these data have been used to reconstruct the emergence of warrior society in the first millennium BC, which the presence of the defended sites suggests. The dangers inherent in equating material culture with 'social culture' (Trigger 1978, 115), or with one particular type of social system, have been taken into account in the following chapters.

Before looking in more detail at the relationship between war and society in Chapter 9, the other aspects which helped to mould this unique society are considered in turn and in more detail. After a brief survey of the classical sources and a history of research in the region (Chapter 2), the actual extent of the region itself is defined in Chapter 3, together with all the relevant environmental factors. Various aspects of the Castros themselves are discussed in Chapters 4-7. Castros architecture is the subject of Chapter 4, an analysis of site distribution follows in Chapter 5, defensive stuctures are analyzed in Chapter 6, and the evidence for the economic, industrial and trading activities which were established within the defences are surveyed in Chapter 7. The subject of the extraction of precious metals and mining was expanded in Chapter 7, because these activities may have influenced patterns of site distribution and regional Romanization. The value of raiding in such societies is a further theme of Chapter 7. Chapter 8 rounds off these integrated discussions with background information to contemporary social and political structure. A site catalogue and a radiocarbon dataset are listed in Appendices I and II respectively.

Finally, it should perhaps be emphasized, that with this study I hope to contribute a fresh and integrated approach to the subject of warfare and its social significance in the Castros Culture of the first millennium BC. A further intention is that this wide-ranging, if rather cursory, treatment of the subject will open up new directions for future research and detailed analyses.

2- PREVIOUS STUDIES

1) History of research in the study area

Studies of the later Prehistoric period in Northwest Iberia have been gradually increasing since the beginning of this century, and a large amount of data have been laboriously gathered by generations of scholars. In this chapter the work on Castros, carried out up to the present time, will be surveyed region by region across the area studied. The regions into which the area studied has been sub-divided for ease of presentation are: Viana do Castelo, Braga, Porto and the adjoining part of Galicia. A survey of some of the more pressing problems in current Castros research then follows. Finally, a summary of the Classical references to the Castros, and to the activities of the Romans in the area is provided. There is clearly much work yet to be done on the later Prehistoric period, classifying and re-interpreting this fine body of assembled data in the light of modern scientific research methods.

Written references to the castros in post-classical times may be found in Medieval documents[1] mentioning the *castrum*, or *castellum*, for the purpose of registering land tenancies. In fact, it can be shown that many of the Medieval castles (Almeida 1978b) were located on the hilltops previously occupied by the Iron Age hillforts, or by even earlier settlements. Settlement of these hilltops continued intermittently into the Medieval period at some of the sites. Most of the Castros however, do not seem to have outlived the Roman period, i.e. beyond the second century AD. The sites chosen for continuing occupation were undoubtedly those which offered strategic advantages for the defence of human and animal populations, resources and the best agricultural and pasturelands. This may imply that skirmishes remained a threat to peaceful existence in the area. Many references to this can also be found in geographical and encyclopaedic works, from the 17th century onwards[2], some of which give fairly accurate geographical descriptions, although field detail is somewhat lacking.

Sarmento was the first scholar to look at the Castros Culture from a scientific point of view. In fact, the discovery of the cultural identity of the so-called 'Culture of the Castros', or 'Castros Culture', and the spreading of these new ideas widely abroad, was entirely due to his pioneering work from the 1870s onwards. Being a wealthy man with an exceptional education in the natural sciences, and an acute perception, he conducted the first ever scientific excavation at the castro called Citânia de Briteiros near Guimarães, between 1875 and 1884. This was immediately followed by a similar study of the Castro de Sabroso near Briteiros[3].

International recognition of the Castros as a distinctive culture in the Northwest of Iberia between the second half of the first millennium BC and the Roman occupation soon became firmly established. This occurred in particular, after some of the delegates to the IXth International Congress of Anthropology and Prehistoric Archaeology (Lisbon 1880), had visited the Citânia de Briteiros, and gained a most favourable impression of it[4]. Among these participants were some of the leading scholars of the time, namely; Cartailhac, Émile Guimet, Henri Martin, and Virchow. The Castros was just one of several major cultural groupings that were beginning to emerge and find a voice in Europe and the Near East at this time, e.g. the Hallstatt in Austria or the Mycenaean in Southeast Europe. As the Castros was clearly a part of the Celtic family of cultures, it came into the forefront of European discussion, and a very profitable interchange of ideas between Sarmento and his foreign colleagues ensued. Amongst these scholars, E. Hübner stands out, and a lengthy correspondence blossomed between the two men[5] .

Around the turn of the century, Leite de Vasconcellos gave some cohesion to the large volume of inadequately studied data through the publication of a *corpus* (Vasconcellos 1899; 1900a; 1902a; 1902c; 1903a; 1934). This work is unique, both for the breadth of its theoretical approach, and for his immense capacity to gather data spread all over a country where, in spite of its relatively small size, overland communications were rather difficult.

Further progress in Castros Culture studies was hindered for a long time by the rarity of any synthesis along similar lines[6]. Other major difficulties concerned the inability to confirm the chronological and cultural phases of the Bronze Age, i.e. at the end of the first millennium BC, with any

1 - Gathered by Alexandre Herculano and published in *Portugaliae Monumenta Histórica.*

2 - Particularly in: Barros 1919 <1549>; Argote 1732-34; Cardozo, L., 1747-51, and 1758; Costa 1868-69; Leal - Ferreira 1873-90; Sampaio 1893-97; Vieira 1886-87; Viterbo 1983-84 <1798-99>.

3 - Sarmento 1884-1909; 1894-5; 1903-5; 1906-9; 1933.

4 - As commented upon by Mário Cardozo (1933) in the introduction to "Homenagem a Martins Sarmento".

5 - The letters between Sarmento and Hübner were later published by Mário Cardozo, in 1947 (Cardozo 1947c).

6 - The first great synthesis focusing on the Iron Age of the Galician region, which included the northern Portuguese data, was produced in the early fifties (Lopez Cuevillas 1953b). Similar work was not undertaken again until the 1980's for the north of Portugal (Silva 1986a). Reference should also be made here to similar studies for the Bronze Age (Coffyn 1985) and for the period of Romanization (Tranoy 1981).

certainty. More recently, thanks to the *radiocarbon revolution* (Renfrew 1979, 53 foll.), the situation has improved considerably, although some inaccuracies do still remain[7] (Figure 22).

2) The growth of regional studies

Regional studies have proved to be an invaluable means of recording the scattered data from all over the country. New approaches enshrined in both general publications and specialist monographs have resulted. Although many of them are based on fieldwork and of a rather descriptive character, such as visits to sites or mere references to the existence of settlements located with respect to administrative boundaries, they form an essential source of information for those times prior to 'national archaeological mapping'.

A rather substantial number of articles has been produced by archaeologists such as Abel Viana, (1926, 1932, 1937; 1955a-b-c, 1960/61, 1963)[8], F.A. Pereira (1895; 1898b; 1898c; 1902; 1904; 1908; 1909; 1910; 1911; 1914; 1915b; 1924/30-31; 1927-29a; 1927-29b; 1930; 1933, 1934; 1938/39/41; 1956)[9] and L.F. Guerra[10]. Such works followed in the footsteps of Sarmento (1884-1909; 1894-5; 1903-05; 1906-9; 1933), who had meticulously reported on his visits to inumerable sites, collecting stray finds as he went, in the region of Entre-Douro-E-Minho especially. This group of resolute local field archaeologists were responsible for the mapping of many sites in the region, some of which have since been totally destroyed, as well as for the systematic recording of stray finds. Although Martins Sarmento covered a wide area and many regions during the course of his research, most of the archaeologists who came after him have tended to restrict their field activities to one particular locality.

For the Viana do Castelo region, significant contributions have been made by: Q. Neves (1942, 1948, 1959, 1965), and L. Alves (1980, 1985) for the area of Caminha, Afonso do Paço and A.P. Quesado (Paço - Quesado 1956, 1957) for a description and location of many of the sites in 'Alto Minho', H.B. Nunes (1973, 1979) and A.A.M. Reis (1978) whose main emphasis was on the Roman sites in the Ponte de Lima area, but where reference was made to Iron Age sites and stray finds in passing.

7 - A list of radiocarbon dates for the Northwest is given in Appendix II.

8 - Also Viana - Neves 1959; and Viana - Oliveira 1954; 1955.

9 - His extensive bibliography was recently compiled by Nunes (1984).

10 - In particular 1900b, 1900c, 1909 and 1982 (=1911), where this last reference is somewhat repetitious.

Staying with the Ponte de Lima area, since the 1970's a systematic and scientifically-based programme of research has been conducted by Almeida (1971, 1978b, 1980a, 1980b)[11], Almeida - Baptista[12] and J.A.T.M. Marques, (1984b), dealing especially with archaeological mapping around Monção.

There are, in addition, relevant excavation reports for the Castros of the area (J.A.T.M. Marques 1984c, 1985a, 1985b, also Marques - Silva 1985), M.A.F. Moreira[13], E.J.L. Silva[14], and T. Soeiro (1981).

Probably the most important research program here was the excavation conducted at Sto. Estevão da Facha (Almeida *et al.* 1981), where this long-lasting site, occupied continuously from the VI century B.C until the I century AD, was recorded and discussed (Almeida *et al.* 1982).

For the region of Braga we have the studies of Sarmento[15], A. Bellino (1909), C. Teixeira (1936; 1939; 1940a; 1940b; 1941; 1943-44; 1955-56)[16], J.L. Vasconcellos (1934), R.F. Ribeiro (1930-34), L. Pina (1940), J.S. Guimarães (1970, 1971)[17], Hawkes (1958), A.C.F. Silva[18], R.M.S. Centeno (Centeno - Silva 1978), T. Soeiro[19], A.R. Cunha (1975)[20], M. Martins (1981, 1983a-b; 1984a-b; 1985a-b-c-d; 1987a-b; 1988a-b)[21], C.A.B. Almeida[22] (1981,

11 - Also Almeida *et al.* 1981. This work is undoubtedly the best attempt so far for an understanding of the cultural and architectural development of the Castros Culture in the area of Ponte do Lima, during its early and middle periods.

12 - Almeida (C.A.B.) and Baptista 1981, for a complete description of the sites and their pattern of distribution in the region of Ponte do Lima, which has one of the largest densities of castros known; Almeida (C.A.B.) 1982b; 1985c; Almeida (C.A.B.) *et al.* 1985.

13 - The author focuses on the coastal area of northern Portugal, Moreira (M.A.F.)1982.

14 - See Silva (E.J.L.) 1982; Silva (E.J.L.) and Marques (J.A.T.M.) 1984; excavation reports of two castros.

15 - See the above references.

16 - Carlos Teixeira has left us some substantial work on the castro Máximo (Braga), and also on the castro of Lanhoso where a significant set of finds was discovered.

17 - On some sparse considerations about the "Picoto de Santo Amaro".

18 - See Silva (A.C.F.) and Centeno (R.M.S.) 1977; Centeno (R.M.S.) and Silva (A.C.F.) 1978; Soeiro (T.) *et al.* 1981. These are reports of trenches dug at the castros of Briteiros and Sabroso (Guimarães), intended to check previous work and establish a chronology.

19 - Soeiro *et al.* 1981; Soeiro 1983.

20 - There are other articles, published in regional newspapers ("Correio do Minho" and "Diário do Minho"), which are listed in Oliveira (E.P.) 1982; 1984, 69-71; 1985, 33.

21 - There is a substantial set of recent articles by M. Martins for the sites in the lower Cávado river valley, mostly dealing with excavation reports. The author is mainly concerned with settlement evolution in this region during the first millennium BC. More recently, theoretical approaches to the subject have become predominant in her work (1987a and b, 1988a).

1982a, 1983, 1985a, 1985b) and A.G. Costa (Costa *et al.* 1980)

Work in the region of Porto was first organized by a group of eminent scholars, most of whose publications may be found in *Portugália*[23]. Other institutions, such as the *Instituto de Antropologia Dr. Mendes Corrêa* and the *Sociedade Portuguesa de Antropologia e Etnologia*[24], also entered the field. The fellows of these societies have left us significant academic work, mainly in the periodicals of these societies.

The concentrated campaigns of fieldwork carried out in the Porto region have given rise to an extensive bibliography. Some of these references are cited below according to subject, otheres are drawn upon in later chapters as appropriate.

Short pithy articles on mining and metallurgy (1927, 1929a-b, 1930b, 1933a-b), and also on the characterization of settlements (1933c), were produced during the prolific career of Rui Serpa Pinto, before an untimely early death ended his flourishing scientific career.

Questions relating to mortuary practice have been raised over the years by: Pinto (1932), based on evidence obtained from Terroso, by Garcia y Bellido (1966a) and more recently by Silva (1986a, 302-3). Although Silva's synthesis incorporated some new data, this tended to confirm the findings of earlier work, rather than open up new avenues of research for the future.

Turning now to the excavations of the castros themselves, the castro of Alvarelhos (1889) and, incidentally, a related collection of metal finds (1902, 1904 and 1905-8), were studied by J. Fortes. He was a member of that older generation of scholars who have already disappeared from the scene. The jewellery theme was taken further by R. Severo (1905-8)c. General studies in these fields, and in the castro locality, were subsequently continued by their contemporaries, namely by J. Pinho (1905-8, 1928, 1929a, 1929b), J. Vasconcellos (1905-8), Peixoto (1905-8), Corrêa (1924; 1928; 1933-4; 1935a; 1935b; 1936; 1940) mainly for the theoretical aspects, and by F.R. Cortez (1946b, 1948a, 1949a, 1949d). Detailed case studies, e.g. of the defences or the domestic architecture, were produced by Almeida (1965, 1984), Cardozo (1946a, 1949a, 1952a), Dias (1946, 1947, 1948, 1949), López Cuevillas (1947a, 1950b), Lopez - Lorenzo (1946), Lorenzo Fernández (1973), Mergelina (1943-44), Paço (1953a) and Pereira (1910, 1911, 1914, 1933). There are, of course, many more excavation reports on castros in varying degrees of detail that could be listed if space permitted, in addition to the major references cited here.

Interest in the castro of Citânia de Sanfins begins with E. Jalhay (1944, 1947, 1949a-b, 1950a, 1950b, 1951). The long history of investigation at this site continued with A. Paço (1952; 1953a; 1953b; 1953c; 1954a; 1954b; 1954c; 1956; 1957; 1959; 1961; 1963; 1965b; 1968a; 1968b; 1968c; 1974)[25], Almeida (1974c) and Silva - Centeno (1980).

The important site of Monte Mozinho has been extensively studied under modern conditions by Almeida (1974b, 1977, 1980c)[26] and Soeiro (1981-82, 1984), and indeed, excavations have continued up to the present time.

Additional regional studies of relevance have been produced by D.P. Brandão (1963b, 1963c, 1965, 1973, 1985), A.C.F. Silva[27] (1980a, 1980b, 1983a, 1983b, 1983-84, 1986b) and M.L. Real (1984)[28]. One very special remark should go to the volume and quality of research done by Mário Cardozo, who has left us an impressive[29] and valuable work covering virtually all subjects relating to the Castros Culture.

This survey cannot pass without some mention of the work of C.F.C. Hawkes who, in co-operation with the Portuguese archaeologist Mário Cardozo, excavated the Castro de Sabroso in 1958 (Hawkes 1958; 1971a; 1974, 192-3) and the Cividade de Âncora and Castro do Cútero in 1959 (Cardozo 1959b; Hawkes 1971a and 1974). Some of these

22 - Also Almeida (C.A.B.) and Baptista (A.J.) 1981; Almeida (C.A.B.) and Neiva (M.A.P.) 1982; Almeida (C.A.B.) and Soeiro (T.) 1980.

23 - This important periodical has now been re-edited after a long absence, thanks to the co-operation of the fellows of the Institute of Archaeology at Oporto University.

24 - This Institute and Society, apart from having good library and archaeological museum, has always gathered to itself a very active group of scholars working in various regions of the country, who mostly deal with anthropological and archaeological subjects.

25 - In addition to some work produced in co-operation with Jalhay. The intense work of Afonso do Paço in the Citânia de Sanfins (Paços de Ferreira), which is one of the largest castros in Iberia, was carried out over a long period of time, thus making this site one of the most important and best known of its kind.

26 - Almeida's excavations at the castros of Mozinho and Citânia de Sanfins (1974c) are undoubtedly the best fieldwork-based examples for showing how the Castros Culture developed under Roman influence during the I century AD. At Mozinho it was possible to record the architectural re-organization of a substantial part of the site under the Flavians. This supports the generally held view that the uniform reforming policy in the provinces, was equally applicable in Entre-Douro-E-Minho.

27 - Also Silva - Centeno 1980 and Silva 1986b. Although these works are wide-ranging in both subject content and area covered, they are important because through a careful interpretation of the data, a picture of the regional cultures could be built up.

28 - A recent urban rescue excavation report, providing interesting data on the occupation of the Oporto city area from the late Bronze Age onwards, showing evidence of trade with the Mediterranean world. See also Silva (A.C.F.) *et al.* 1986c, 107-9.

29 - For an extensive list of his bibliography see Cardozo 1972; with its 168 papers on various subjects, or alternatively Oliveira 1984-85.

results have yet to be published[30]. The importance of these excavations lies in the fact that they were the first castros in the Northwest where stratigraphy was recognized and recorded, and it was to be some while before such methods were applied generally to Castros research in northern Portugal and Galicia[31].

On the Galician side and following in the footsteps of scholarly archaeologists like Lopez Cuevillas, Bouza Brey and Taboada Chivite, the University of Santiago de Compostela has always managed to produce an active group of archaeologists and classicists. They are responsible for the high standard of Castros research in Galicia, and rewarding joint ventures with them take place from time to time.

From this survey, it should have become apparent that the data collected from the area studied are by now extensive, and have been recorded in a bewilderingly diverse range of monographs and periodicals. Any information on Castros that may be required has, therefore, to be gleaned amongst this vast library. Some of these works contain only passing references to the topics of interest. Few attempts at synthesis have been made over the years, although the situation is slowly changing and some monographs on specific themes, or on local geography, have emerged. Short essays by Almeida mark the turning point. These surveys include: a definition of the zones of influence (1973), a typology for castros pottery (1974a), an assessment of the impact of Augustan and Flavian reforms on castros architecture and social growth (1974b, 1977), the definitive interpretation (1974c) of the *monumentos com forno* as thermal bath structures (probably one of the most enigmatic structures ever produced by the Castros Culture), a treatise on the development of the Castros Culture in its most relevant aspects (1983a), an analysis of the types of house structure (1984), a discussion of Castros art (1986) and finally, how the Culture was transformed under Roman influence (1983b).

In order to appreciate how Castros research has developed, what the state of the art is at the present time, and where to direct future efforts, some estimate of the quality and reliability of the data assembled is required. One problem encountered has been the general absence of any stratigraphy at Castros sites. This caused many to doubt the value of applying stratigraphical techniques at these sites, at least until the 1970s (Eiroa 1980, 78-9). The use of modern techniques, in fact, probably began with Almeida's excavations at Mozinho (Almeida 1974b, 1977). The overwhelming majority of northern Portuguese researchers into the Castros Culture now follow his methods and, for example, metallurgical, palaeoenvironmental and phosphate analyses now appear more frequently in archaeological reports.

Another important problem, namely that of establishing chronological boundaries for the Castros Culture, has had an equally erratic history. Until very recently, faith has had to be placed on relative dating methods alone, with the castros being loosely slotted into the space between the end of the Late Bronze Age and the coming of the Romans. The comparatively late date for its start (VIth to IVth centuries) (Jordá Cerdá 1985-86, 263)[32] was based on the assumption that the Northwestern Iron Age was the result of sporadic external influences, and more substantial contacts did not become a part of the life of these local communities until the La Tène period. Early in the seventies, the first radiocarbon dates for castros became available (Eiroa 1973b). This caused the old chronological framework to be questioned, and the 1980s saw some changes being made as the number of radiocarbon dates increased. For example, those sites where a continuous occupation from the Late Bronze Age to the end of the millennium was thought to be secure, on the basis of the occurrence of characteristic finds like carinated ceramics, now produced some surprisingly early radiocarbon dates. Some clarification of the problem has thus been achieved, although many more radiocarbon dates will be required before a solution can be reached. There has been considerable argument over the validity of the concept of a 'Castro Culture' over the years (Höck 1980, 55-70; Martins 1981, 109; 1987a, 724-6; 1988a), but on the whole the topic has met with disinterest.

After a century of scholarly research work, our understanding of the Castros Culture may have increased; but we still know relatively little about the social interactions and cultural development amongst the groups of people that settled in the north of Portugal from the late Bronze Age until the declining years of the Iron Age. Though a reasonable amount of data has been produced, e.g. enabling architectural plans to be drawn or artefact typologies constructed, fundamental questions relating to familial and social organization, or to the economic and political relationships between sites, still remain to be

30 - I am grateful to Professor Christopher Hawkes and Mrs. Sonia Hawkes for kindly letting me go through the unpublished finds and drawings of the excavations directed by the former in Portugal. The conclusions on the stratigraphical sequence and general chronology of these sites has been fully supported by recent work (Silva 1986a).

31 - Hawkes excavated Castro de Sabroso, in 1958, and Cividade de Âncora and Castro do Cútero, in 1959. The yet unpublished plans, sections and observations produced by Hawkes reveal a technique which can be considered outstanding for its time. Hawkes' lessons were generally neglected by his Portuguese colleagues. Only later on Almeida (1974b) was to recognize the value of stratigraphy for an understanding of the archaeological process.

32 - Further references and discussion of this subject will be presented below in chapter 4.

answered. The best known period is, in fact, the later part of the Iron Age, when Classical texts and epigraphy, a by-product of the process of Roman acculturation, could supplement the rich archaeological data collected from excavations at the hillforts. Before the IInd century BC, the absence of additional corroborative sources makes any reconstruction of daily life in Northwestern Portugal extremely difficult, i.e. between the final stages of the Bronze Age and, let us say, the campaigns of Caesar, in 61 BC.

This picture has been gradually changing in the last few years, thanks to the joint efforts of *old* and *new* generations of archaeologists who are starting to pay more attention to an anthropological interpretation of the data, instead of merely constructing chronologies and typological sequences in isolation. There has also been a tendency to overstress the idea of population migrations as the more plausible explanation for every exogenous cultural evidence. There is much still to be done however, and in particular, how to apply research techniques more efficiently, use palaeoenvironmental and spatial approaches more extensively, and integrate these data with the material remains. On the other hand, spatial analysis should be applied with caution. Statistical analyses of this type require that the base parameters, such as topography, economic and other cultural factors, and their relationship to the building of hillforts and other categories of settlement, be well defined and understood. The number of sites in the sample must also be large. At the present time, such pre-conditions cannot be met in the area studied.

The problem of the origins of the Castros Culture is another aspect that has received little interest from archaeologists over the years. First millennium BC studies in this area, until now, have been concentrated into two main blocks, i.e. either the final phases of the Bronze Age or the Iron Age, with little interaction between the two streams.

Pioneering guidelines for the study of the Bronze Age were laid down by Bosch Gimpera (1932) and refined by Martínez Santa-Olalla (1946). These two scholars between them defined and schematized the Northwestern Bronze Age. Later confirmation came from other authors (Hawkes 1952, MacWhite 1951, Savory 1949) and parallels with contemporary cultures in Northern Europe were proposed. The critical overview produced by Maluquer de Motes (1975b) did not change this state of knowledge significantly, although it has somehow affected positively the methodological side of future research, as the guidelines proposed by these studies have, and probably will, survive for a long period.

The greatest limitation to this study of the Northwestern Late Bronze Age is our almost total ignorance of the habitation sites. This occurs at the same time as a substantial body of finds is known which correlates well with that found in several other regions of Europe. Although a good typology has been worked out for Late Bronze Age metalwork, we have very little idea of either the people, or the culture, that produced it, in spite of the suggestions (Calo - Sierra 1983, 32) that 'many castros were occupied from the Bronze Age'. This situation is well reflected in both Monteagudo's and Coffyn's important syntheses (Coffyn 1985, Monteagudo 1977) where a typological analysis of finds is not only dominant, but virtually exclusive. In fact typological studies[33], and the establishment of comparative models with 'European prototypes' (Almagro Basch 1940, 1958; Almagro Gorbea 1973, 1974a, 1974b, 1977; Coffyn 1983), were to remain predominant until the 1980s. More recently, sufficient data have been made available through fieldwork to establish new perspectives, based on a closer correlation between the finds and the sites. The first attempt to relate the Northern Portuguese Late Bronze Age finds with the settlement sites was made by Kalb (1980a), and then slightly re-worked by Coffyn (1985). A difficulty here is that it is not very easy to distinguish Bronze Age and Iron Age sites from one another prior to archaeological excavations. Also, in some cases the stray finds have been allocated to the nearest castro, regardless of the characteristics of the site. Nevertheless, a good contribution was made by Kalb and her work is being developed. Now that an ever-increasing number of finds of ceramics of 'Alpiarça type'[34] are coming to light in the earlier levels at a number of the castros, it may soon be possible to prove that some sites were occupied continuously throughout the millennium, and to confirm that the defended sites of the Northwest emerged during the Late Bronze Age. Finally, it should perhaps be noted that an overview of the situation is available in the Proceedings of the conference held at Santiago (Calo - Sierra 1983, Fariña *et al.* 1983) which, although not entirely realistic, has made a contribution to contemporary research.

3) The classical sources

Reference to the Northwest has been made by quite a few Classical authors. This means that any reconstruction of the ancient history of the area from such disparate sources is bound to be a thankful task. Fortunately, an annotated listing of these references, as found in the Greek and Latin writings on Iberia, is

33 - In addition to the above references, studies, such as those produced by Hencken (1956), Meijide Cameselle (1988) and Monteagudo (1981) for this epoch, are exclusively typological in character.

34 - Although the following reference may already be outdated due to recent unpublished fieldwork, it is known that pottery of this type has been found at Alto da Caldeira, Tapado da Caldeira, and Castelo de Matos (Baião), Castro de Peso (Viana do Castelo) Castelo de Faria (Barcelos) and Monte Padrão (Sto. Tirso).

available in Schulten's *Fontes Hispaniae Antiquae* (Schulten 1922, 1925, 1935a, 1937, 1940 and 1952). Other studies exist and in particular, Schulten (1920 and 1959-63) and Lopez Cuevillas (1933-34) review the same sources.

Of all the Classical authors, Strabo is probably both the most important and the most often quoted on matters concerning the Castros Culture[35]. Some of his remarks have been regarded with suspicion[36], because he never actually visited Iberia but relied on the accounts of previous writers such as Posidonius and Polibius (Lasserre 1966, 3-4), and probably descriptions obtained from sailors and soldiers as well. The fact that most of his books were written during the reign of Augustus (Strabo died in AD 20) makes him a crucial witness for any investigation of the Conquest, and the effects this may have had on the Castros Culture. His commentaries further allow the old and new situations to be compared, from which an understanding of a society undergoing a process of deep change may be gained.

I shall now take this opportunity to justify the credulity I felt, accompanied by the necessary criticism whenever possible, on reading some of Strabo's observations about the Northwest. Archaeological research has now been able to confirm that many of his observations were true, even those which seemed to be fantastic. This has provided us with rare and precious insights into the Castros Culture. Two brief examples will be developed below, namely his references (III, 3, 6) to the *Callaican* steam baths, from which a new interpretation of the 'monumentos com forno' resulted (Almeida 1974c), and the movements of the Turdulians (III, 3, 5), for which the archaeological evidence (Silva 1983c, 1986a) of their presence along the southern shore of the Durius River is irrefutable. There is no doubt, however, that some of Strabo's descriptions of the Northwest are far too general and even misleading, although on the whole he is worthy of trust. The more credible references will be described later, along with some critical reviews of Strabo (footnote 35). The way in which his references fit with the available archaeological data will also be elaborated upon.

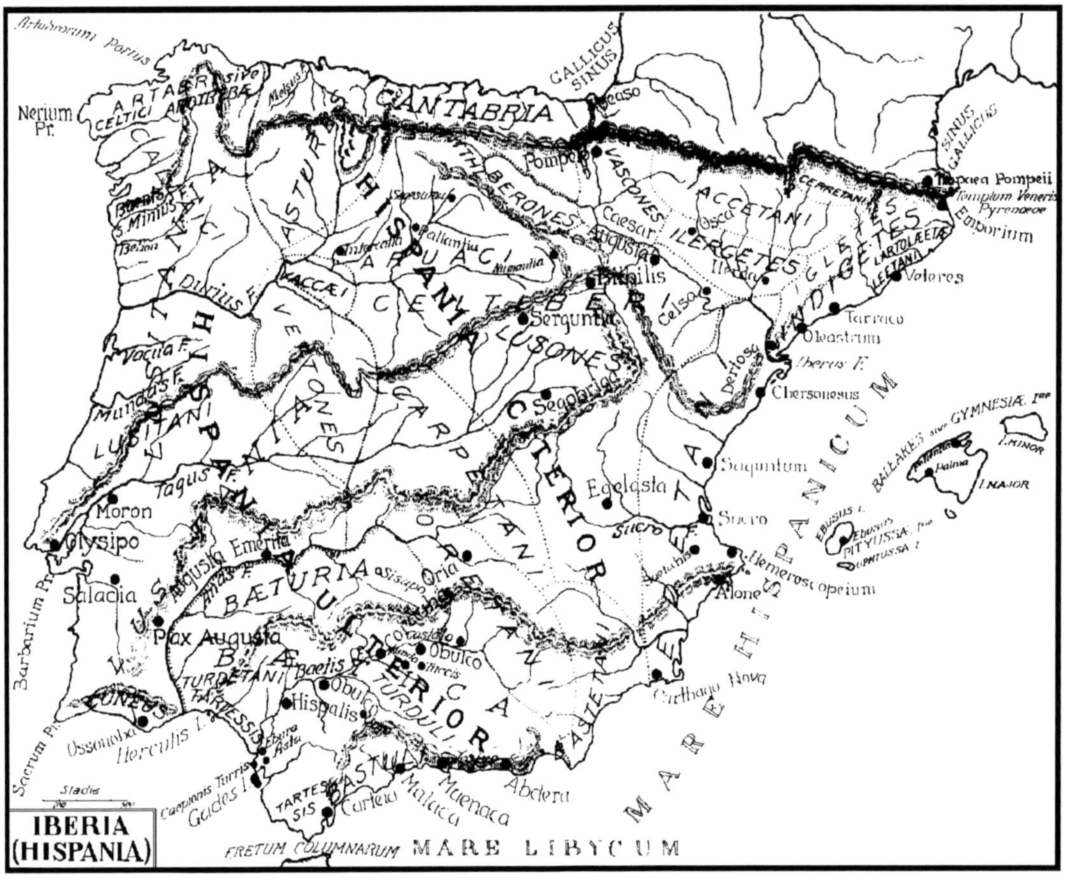

35 - There are many articles that explore specific topics referred to by Strabo. For example, see Alonso Nuñes 1975, Bermejo Barrera 1979, Blazquez 1962a and Garcia y Bellido 1945a.

36 - See Alonso del Real (1979) for a critical review of sources and interpretation and, for particular aspects, some studies by Bermejo Barrera (1981a and 1981b, 1983).

At this stage it may be wise to note that some of the Classical sources can be rather general and apparently inaccurate on occasion. In a few accounts, their particular philosophical, political, or chronological standpoints may distort the facts, as they are perceived by us on reading these observations today (Bermejo Barrera 1978; Lasserre 1966, 3; Nash 1976). This and the well-known distortions that can arise from subjective historical comment should always be borne in mind. What is more difficult to condone however, is the acceptance or rejection of Classical comment by modern scholars on purely subjective grounds and, in particular, the conscious selection of information to fit in with current research ideas!

Transcriptions are often used in the course of this study, both as a means of illustrating a point, and to outline any conflict with the archaeological data. Most of the references are taken from Jones' reliable translation of Strabo, which is regarded as preferable to either Lasserre's (1966), Schulten's (1952), or to the one produced by Veloso - Cardoso (1965), especially for discussions of the more controversial questions.

Apart from Strabo, other Classical writers are also quoted, namely, Diodorus Siculus who lived in the mid I[st] century BC and Livy who was a contemporary of Strabo (he died in AD 17). Again, Livy is an ideal commentator of the Conquest period but, unfortunately, his references to the Northwest are few. As a contemporary of Florus, Appian writes in the second century A.D (he died circa 160). Some of his sources are reliable, namely the references to the *Iunius Brutus* campaigns in Iberia (cf. Posidonius and Polibius), which date to the second century BC. Similarities in some facts between Appian and Strabo suggest that the same sources were used.

Now we shall take a look at some descriptive references from book III of Strabo's *Geography*, in order to illustrate the view of this geographer concerning the peoples and the region of the Northwest.

Strabo's Geography, book III

3, 3 -"And yet the country north of the Tagus, Lusitania, is the greatest of the Iberian nations, and is the nation against which the Romans waged war for the longest times. The boundaries of this country are: on the southern side, the Tagus; on the western and northern, the ocean; and on the eastern, the countries of the Carpetanians, Vettonians, Vaccaeans, and Callaicans, the well-known tribes; it is not worth while to name the rest, because of their smallness and lack of repute. Contrary to the men of to-day, however, some call also these peoples Lusitanians. These four peoples, in the eastern part of their countries, have common boundaries, thus: the Callaicans, with the tribe of the Asturians and with the Celtiberians, but the others with only the Celtiberians.

3, 5 -"Last of all come the Artabrians, who live in the neighbourhood of the cape called Nerium, which is the end of both the western and the northern side of Iberia. But the country round about the cape itself is inhabited by Celtic people, kinsmen of those on the Anas; for these people and the Turdulians made an expedition thither and then had a quarrel, it is said, after they had crossed the Limaeas River; and when, in addition to the quarrel, the Celtic peoples also suffered the loss of their chieftain, they scattered and stayed there;"...."Now about thirty different tribes occupy the country between the Tagus and the Artabrians, and although the country was blest in fruits, in cattle, and in the abundance of its gold and silver and similar metals, still, most of the people had ceased to gain their livelihood from the earth, and were spending their time in brigandage and in continuous warfare both with each other and with their neighbours across the Tagus, until they were stopped by the Romans, who humbled them and reduced most of their cities to mere villages, though they improved some of their cities by adding colonies thereto. It was the mountaineers who began this lawlessness, as was likely to be the case; for, since they occupied sorry land and possessed but little property, they coveted what belonged to the others. And the latter, in defending themselves against the mountaineers, were necessarily rendered powerless over their private estates, so that they, too, began to engage in war instead of farming; and the result was that the country, neglected because it was barren of planted products, became the home only of brigands."

3, 6 -"...Now some of the peoples that dwell next to the Durius River live, it is said, after the manner of the Laconians -using anointing-rooms twice a day and taking baths in vapours that rise from heated stones, bathing in cold water, and eating only one meal a day; and that in a cleanly and simple way. The Lusitanians are given to offering sacrifices, and they inspect the vitals, without cutting them out. Besides, they also inspect the veins on the side of the victim; and they divine by the tokens of touch, too. They prophesy through means of the vitals of human beings also, prisoners of war, whom they first cover with coarse cloaks, and then, when the victim has been struck beneath the vitals by the diviner, they draw their first auguries from the fall of the victim. And they cut off the right hands of their captives and set them up as an offering to the gods.

3, 7 -"All the mountaineers lead a simple life, are water-drinkers, sleep on the ground, and let their hair stream down in thick masses after the manner of women, though before going into battle they bind their hair about the forehead. They eat goat's-meat mostly, and to Ares they sacrifice a he-goat and also the prisoners and horses; and they also offer hecatombs of each kind, after the Greek fashion -as Pindar himself says "to sacrifice a hundred of every kind". And the mountaineers, for two-thirds of the year, eat acorns, which they have first dried and crushed, and then ground up and made into a bread that may be stored away for a long time. They also drink beer; but they are scarce of wine, and what wine they have made they speedily drink up in merry feastings with their kinsfolk; and instead of olive-oil they use butter. Again, they dine sitting down, for they have stationary seats builded around the walls of the room, though they seat themselves forward according to age and rank. The dinner is passed round, and amid their cups they dance to flute and trumpet, dancing in chorus, but also leaping up and crouching low. (...) All the men dress in black, for the most part in coarse cloaks, in which they sleep, on their beds of litter. And they use waxen[37] vessels, just as the Celts do. But the women always go clad in long mantles and gay-coloured gowns. Instead of coined money the people, at least those who live deep in the interior, employ barter, or else they cut off pieces from beaten silver metal and pass them as money. Those who are condemned to death they hurl from precipices; and the parricides they stone to death out beyond their mountains or their rivers. They marry in the same way as the Greeks. Their sick they expose upon the streets, in the same way as the Egyptians did in ancient times, for the sake of their getting suggestions from those who have experienced the disease. Again, up to the time of Brutus they used boats of tanned leather on account of the flood-tides and the shoal-waters, but now, already, even the dug-out canoes are rare. Their rock-salt is red, but when crushed it is white. Now this, as I was saying, is the mode of life of the mountaineers, I mean those whose boundaries mark off the northern side of Iberia, namely the Callaicans, the Asturians, and the Cantabrians, as far as the Vasconians and the Pyrenees; for the modes of life of all them are of like character." ...

4, 15 -"The Iberians were once, virtually all of them, peltasts, and wore light armour on account of their brigand life (as I said of the Lusitanians), using javelin, sling, and dirk. And intermingled with their forces of infantry was a force of cavalry, for their horses were trained to climb mountains, and, whenever there was need for it, to kneel down promptly at the word of command."...

4, 18 -"... for instance, it is the custom among the Cantabrians for the husbands to give dowries to their wives, for the daughters to be left as heirs, and the brothers to be married off by their sisters. The custom involves, in fact, a sort of woman-rule -but this is not at all a mark of civilization...."

As one can see from this long quote from Strabo, his references to the peoples of the Northwest often confuse and mix the Callaicans with other peoples such as the Lusitanians and the Asturians. The reasons for this inaccuracy are not yet completely clear, but it seems that lack of precise information would be the major argument. It has been suggested that this mixing of different ethnic groups could be explained by the use of sources based on Roman administrative divisions (Santos Yanguas 1985, XXVI), where *Callaecia* was a part of *Lusitania* in the first Augustean administrative organization of Iberia, being later integrated into *Tarraconensis* in the second re-organization. On a theoretical basis, it is not possible to know to what extent it would be legitimate to apply some of his descriptions of the characteristics of the Lusitanians to the Callaicans, without any supplementary evidence. It has been argued that the name Lusitanians represents a group of peoples, rather than one single people, who inhabited the region between the Tejo and the Douro rivers (Alarcão 1988, 16), and a similar concept can be applied to the Callaicans north of the Douro. Perhaps the correlation between the archaeological data and the literary sources does not always permit such comparison, but it will be attempted in the following chapters nevertheless.

---

37 - Most probably meaning *wooden*, as referred to in Shulten's edition.

3- THE REGION

In this chapter, a definition of the area studied, as well as the extent of its boundaries, will be the first topic discussed. The likely geomorphology, hydrology, distribution of soil types, climate and degree of vegetation cover in this region during the first millennium BC will then be surveyed, so that we may better understand the reasons for the rise in hillfort construction at this time, as well as some of the choices relevant to their location. Other features in the region which helped to shape its cultural identity, such as the importance of mineral resources and good navigable waterways to facilitate communication and trade, complete this section.

The boundaries of this study lie within the region repeatedly defined by many authors as that demarcating the main distribution area of the Castros Culture. In fact, after López Cuevillas' initial proposal (1933, 105-6) for a distribution area of the Castros Culture, he later suggested a revision (López Cuevillas 1933; 1953b, 55-6; 1973, 513-5), which has now become the working basis for many archaeologists.

For the defined area, the relevant boundaries established by Lopez Cuevillas (cf. above) are: the Vouga river to the south, and the line between the Tâmega and Tua rivers to the east, together with its hinterland, or cultural interface. Any discussion of the details of the various proposals for the limits of the Castros Culture cannot be presented here as they are too extensive. Although most of the studies on castros archaeology somehow start by referring to this subject[1], little advance has yet been made on the earlier work of Lopez Cuevillas. Further, the region defined for this study also lies within the limits of the *Conventus Bracaraugustanus*, but again, these boundaries are not entirely coincident. The limits of the *Conventus* itself have been the subject of discussion between various authors. An updated survey is presented by Tranoy (1981, 160-2), with references. However, if the discussion of the precise limits of this Roman juridical territory seems to be of little relevance to this study, the fact that these limits may have originated in a region where the previous indigenous ethnic and cultural personality was recognized by the Roman administration is of the utmost importance to us.

The relationship between man and environment is a two-way process and it has to be understood as such when determining the effects of cultural development on the landscape. This relationship may also affect, to a certain extent, both productive strategies and the concomitant patterns of geographical distribution and social organization. However, it is also true that the human influence on the landscape has been increasingly dramatic, from Prehistoric times until the present day. In fact, since the application of palaeoenvironmental sciences, it has become more and more difficult to see ecology as a natural process in Europe. Human pressure on the environment has been significant since the Mesolithic period (Harding 1982, 8), largely due to anthropogenic deforestation.

The approach expressed in this chapter, and also below in Chapter 7, owes much to the theoretical directions of those who have made major contributions, mainly in methods of information retrieval and analysis, to the development of Palaeoecological studies on a scientific basis during the last few decades (Biek 1963, Brothwell-Higgs 1963, Cornwall 1958, Dimbleby 1967, Hodges 1964, Rosenfeld 1965). Such pioneering studies allowed the interaction between culture and environment to be determined. More recent generations of environmentalists have substantiated and made progress in this kind of work (Brothwell - Higgs 1970, Butzer 1971, 1975, 1982; Jarman *et al.* 1972, Renfrew, J., 1973; Shackley 1981 and 1985), building a solid scientific discipline, and promoting broad contacts and an exchange of ideas through the organization of conferences[2] and publications (Hastorf - Popper 1988, Higgs 1972, Jones 1983, Sheridan - Bailey 1981).

3.1) The region defined

Due to its geographical characteristics, the northwest of Iberia was considered a very rough and inhospitable region[3] by the Classical authors. Landscape is an important factor which probably affected, and even conditioned, the cultural development of the peoples that settled there during later Prehistory. However, it cannot be used as a monocausal argument to explain unrelated cultural factors, as has been repeatedly implied from Strabo (III 3, 8) until more recent times.

1 - The most significant references on this subject are Acuña Castroviejo 1973, 28; 1977, 7-8; Almeida 1974a, 1975492; Cardozo 1973b, 262; Fariña Busto *et al.* 1983, 98; Maluquer de Motes 1975a, 271; Romero Masiá 1976, 3, and probably also Blanco Freijeiro 1960, 179, as based on aspects of the architectural plan. See also the updated review in Calo Lourido 1991, 25-33.

2 - C.f. the work and dynamics produced by several associations for environmental studies

3 - "...And Northern Iberia, in addition to its ruggedness, not only is extremely cold, but lies next to the ocean, and thus has acquired its characteristic of inhospitality and aversion to intercourse with other countries; consequently it is a wretched place to live in..." (Strabo III, 1, 2).

The area this study refers to (Figure 2) is part of the region usually called Entre-Douro-E-Minho[4], or the northwest of Portugal[5] and it is limited by the Atlantic coast, the Tâmega-Corgo[6] line, the Minho and the Douro rivers. Expanding this definition further, the area studied is bounded by the Minho river to the north and the massif Peneda-Gerês, which runs north-south and then bends eastwards towards mount Larouco. To the west there is the short stretch of Atlantic coastline, cut by several river estuaries. The southern limit is the Douro valley as far as its small tributary on the northern side, with the Corgo river demarcating its western end. This latter runs N-NW in the direction of the 'high' Tâmega river valley, thus following the massifs of Marão, Alvão and Barroso-Cabreira.

The Atlantic ocean is an obvious natural boundary. The Minho river was chosen as the northern geographical border because it shows very different and distinctive cultural aspects[7] on its northern margins (e.g. Santa Tecla). This suggests the presence of a cultural sub-group of the Castros Culture.

The Douro river to the south does not show, in fact, any such clear cultural distinctiveness, but then it was the southern boundary for both the *Conventus Bracaraugustanus* and the *Gallaecia.* Furthermore, the use of this river as a boundary between peoples stands out, both from an archaeological standpoint, and by reference to the Classical sources. Moreover, the importance of the boundary created by this river is reflected in the fact that it divided the territories of two important peoples, the Callaicans and Lusitanians, according to Pliny (IV, 21) and Strabo (III, 4, 20), although the hypothesis has also been put forward that Lusitanian territory eventually extended to the northern side of this river (Alarcão 1988, 34).

Similar geographical and climatic conditions prevailed north of the Minho river, in Galicia[8]. There is thus no geographical reason for the existence of any cultural differences on the two sides of this river.

Hills and rivers are the dominating features in the northwest of Portugal. Indeed, there is no place where mountains or hills are out of sight. Water from springs, streams and rivers flowed down hill slopes and along valley floors, and a lush vegetation of evergreen trees, bushes and grasses was supported as a result. The Atlantic climate dominated the region and moulded the cultural lives of the local communities. A small strip of mountainous hinterland has also been included in this study, on the eastern side of the region. The geographical characteristics here are again different, as the altitude and distance from the coast produced a distinctive environment.

3.1.1 -Geomorphological structure, hydrology and soils

The relief is basically oriented in the direction northeast-southwest, towards the coast, as defined by the river valleys (Figure 3). The mountains of Peneda and Gerês-Larouco, roughly aligned in the direction northwest-southeast, continue as the Barroso-Alvão and Marão-Aboboreira massifs (Figures 4 and 5). These latter, running southwestwards towards the Douro river, demarcate the zone of Atlantic influence from the hinterland by forming a barrier. Their role in the definition of the geographical boundaries (Ribeiro 1955, 48) is important, as they regulate rainfall by obstructing the masses of humid air as they pass over, thus reducing drastically the precipitation in the hinterland (Girão 1958, 7), and therefore conditioning its climate, orography and ecology.

Between these mountains there are small river valleys, although in many cases (sections of the Tâmega and the Douro) the rivers make their way through steep slopes and gorges carved in the bedrock. Some river valleys are relatively broad and flat, as a result of their geological formation, but the only truly flat zone is the coastal fringe (Figure 3 and 5), which forms a strip of variable width parallel to the Atlantic.

Some of the more compact granite massifs, such as Peneda, Barroso, and Marão-Aboboreira have developed small plateaux. This feature is responsible for both the characteristic local scenery, and the typical settlement plans found in such areas[9].

---

4 -In fact, the region defined for this study goes a little further west than the Entre-Douro-E-Minho, as it includes a substantial part of the Vila Real district area, which belongs to the Trás-Os-Montes province. However, we will keep this expression as more representative than any other one could eventually use for this purpose.

5 -"North Atlantic", one of the three major geographical divisions of Portugal (Ribeiro 1955, 235) is, perhaps, the expression that best defines the identity of this region. When using the expression Northwest alone, we will be referring to the northwestern corner of Iberia, the region defined for the Castros Culture.

6 -This eastern limit is not very coherent in terms of geographical and regional identity but, as explained below, it has some relevant archaeological implications for the purpose of this study.

7 - Such as a rather particular architecture, as well as a different style of Iron Age pottery, i.e. that with plastic decoration.

8 -Atlas Geografico de España ill. nº 38, 44, 48.

9 - It is precisely in such areas that most of the known megalithic tombs groups are located (*in genere* Jorge, VO, 1982).

The eastern side of this region is limited by the important geological fault of Chaves, which separates the Trás-Os-Montes plateau to the east, from the Barroso and Alvão plateaux to the west. As this fault coincides with the Tâmega river bed at Chaves, it produces a small and fertile alluvial plain (circa 3x7 km on the Portuguese side) only 3 to 4 meters above the river level, called 'Veiga de Chaves'.

The terrain then steepens towards the western massifs, where heights of 1000 meters are common. There is also a sequence of hill terraces (Teixeira *et al.* 1974, 8), and many of these spurs were occupied by hillforts[10].

Hydrology

Rivers and streams make a very complex hydrological network, extending water supplies to nearly everywhere, and contributing to the erosion by slowly carving out channels through the mountains and depositing the debris in the valleys, where alluvial soils are formed.

The rivers of the Northwest played a very important role in the cultural development of the region. Since Prehistoric times, they have served as natural routes of communication, had a regulatory influence on the climate and orography, and been a source of valuable metal resources like gold and tin. Human settlement has correspondingly been attracted to them. However, it would be a mistake to overstress their importance to the detriment of other geographical factors, when looking for the reasons for the development of the unique cultural identity of this region (Ribeiro 1955, 234).

There are five major river basins leading directly to the Atlantic: the Minho, Lima, Cávado, Ave-Este, and Douro, although the Douro had little influence on the study area. Due to the characteristics of the terrain, all these rivers show a great difference of level (Figures 7 and 8), i.e. between the spring in the mountains and the estuary (Girão 1958 and 1960, 147, graphs), although their unevenness reduces considerably as they get closer to the coast.

The river Minho runs for most of its course through Galician territory, where it is joined by its most important tributary, the Sil river, as well as by many other Galician streams. Its valley is surrounded by a large granite massif. The lower reaches of its course form the administrative borderline between Portugal and Spain. At this point it loses most of its previous gradient, due to the flatter configuration of the landscape (Figure 8.1). Since the beginning of the Quaternary, a build-up of alluvium and sand banks has taken place here, in a river-bed now wider as it approaches the estuary, due to the slower flow. These Holocene alluvial deposits produced large clay-sand-pebble banks of different height within the valley zone. Such deposits are composed of various levels of terrace (Teixeira 1956, 7-8), the height of which can range up to 100 metres above sea level. The whole area corresponds to a tectonic depression (Ribeiro - Moreira 1986, 22), that spread in the direction ENE-WSW, and which is responsible for the existence of various springs, both of carbonated and thermal waters. Aplite-pegmatite mineral veins are numerous in the rock within this valley and its surrounding area (Teixeira - Assunção 1961, 33), occurring in both schists and the periphery of the granite formations. Accordingly, some evidence of tin mining may be found in the region, dating from ancient to more recent times. This probably reflects an intense ore extraction that is clearly corroborated by the numerous finds of bronze metal and metallurgical moulds (Coffyn 1985) of Bronze Age date. The real value of this region lies in the fact that it has one of the few known sources of tin ore in Europe.

The Minho and Lima basins are separated by the line of mountains Peneda-Arga (Figure 5 C-D), whose altitude decreases towards the sea shore. The Arga massif is cut by the valley of a small river, the Âncora.

The upper course of the Lima, which rises in Galicia, is narrow and steep (Figure 8.1). The gradual modification of this course in the lower reaches[11] produced a wider bed, and some well known irregularities in water flow.

The Lima river valley (Figure 5 A-B) has the massifs of Padela and Nora on its southern side, which lie perpendicular to the coast. These mountain ranges are separated by a slight depression, which forms a natural route. Along most of its course, the flanking mountains are intensely cut by deep valleys (Teixeira *et al.* 1972, 7) containing small rivers and streams. It is an alluvial valley, with a wide floodplain that tapers progressively up-river, as the altitude rises and the relief steepens (Figure 3).

The separation between the Lima and Cávado rivers is not quite as dramatic as that between the Minho and Lima, especially in the middle course of the Cávado. The Gerês massif, which divides these two rivers and dominates their upper courses, gives way to a lower hill terrain dotted by small enclosed valleys in the lower reaches.

The waters of the Cávado run fast due to the steep terrain, from the area of Larouco mountain to the estuary (Figure 8.1). However, in its middle and

[10] - Probably the most impressive example is Castro do Lezenho, Boticas, at 1074 metres a.s.l., being a relatively large and well-defended hillfort.

[11] - This change can be noticed in the course of this river between the cities of Ponte da Barca and Ponte de Lima.

lower basin, after it receives the waters of the Homem (Figure 8.1), the land is very fertile and geographically well-balanced (Figure 3 and 5 A-B), similar to that of the Ave-Este rivers. This is an area of low hills and well-watered small valleys. The course of the Cávado is cut by many geological faults, thus dividing it into 'compartments' (Teixeira *et al.* 1973, 54), where alluvial planes and rocky gorges alternate on the river bed. This characteristic is also shared by the Ave river (Figure 8.2), which flows along a geological fissure with its line of thermal springs, very much like the Cávado.

For the purposes of this study, the Douro river is only notable for the navigability of its lower course (Figure 8.2), and thus for its value as a means of communication. Its margins are arid (Girão 1960, 142, 148) and its basin is very wide in the centre of Spain and northeastern Portugal, but it tapers increasingly in its lower course, between the Spanish border and Portugal, towards the estuary.

The Tâmega is the only river basin, within the region defined, that does not flow directly to the Atlantic, since this river is a tributary of the Douro. It is characterized by a small area of valley in its upper basin (Figure 8.2), near the border, which is embedded in a geological step fault. This is important from an archaeological point of view because human activity has been concentrated around it since Prehistoric times.

The small basins of the Leça and Neiva rivers[12] will not be considered here as they are relatively insignificant in the surrounding landscape of low hills, and for the purposes of this particular study.

Soils

The types of soil found in this region (Figure 6) are fairly uniform, and the well-defined pattern may thus be determined. In general terms, this area is dominated by humic cambisols derived from igneous rocks and schists[13]. Luvisols occur in small areas of the coast and are indicative of the Atlantic influence. In the mountainous areas of Peneda, Gerês, Larouco, Barroso, Cabreira, Marão and Alvão, rankers are predominant. The soils of the Valongo strip (east of Porto) are composed from Ordovician schists and quartzites.

Being able to determine soil type is important for our understanding of economic strategies and site distributions. Erosion[14] is a major element in assessing the agricultural potential of a region, as it greatly displaces the soils and their organic components. Throughout Prehistory, erosion has often been associated with deforestation of the hills (Daveau 1988, 196), and also with particular types of settlement like the hillforts, that became common in the Iberian northwest from the Late Bronze Age onwards.

Although a wave of anthropogenic deforestation may have occurred across Europe from the Neolithic onwards, the great volume of soil deposition in the alluvial valleys in antiquity corresponded with the Roman period (Bintliff 1982, 156). This was due to a noticeable increase in agricultural and mining[15] activities. The available data are still very scarce, but a picture can tentatively be sketched for the Northwest. As it will be argued below, associated factors, such as deforestation, indicate that the erosion of the hills may have increased dramatically during the Late Bronze Age. Thereafter it progressed further in several steps. This assessment is necessarily conditioned by the absence of data for previous epochs. It is not yet known if this phenomenon predates the Late Bronze Age. Relevant data will be presented in Chapter 7, adding to the impact that human communities may have had on both soil erosion and general environmental change through deforestation (Figure 10) during the course of the millennium.

3.1.2 -The climate and vegetation

A knowledge of climatic conditions is probably essential to our understanding of how productive strategies developed, and why particular species of plant were chosen for use, and how they spread as a result. Thus, a proper combination of factors, that we call climate, make agricultural practices possible. However, as Harding (1982, 8) points out, extreme caution is necessary when applying climatic results to archaeological data, always bearing in mind that they should not be overstressed as the determining factor in the development of productive activities.

The major disciplines for the study of climate in the Northwest during the first millennium BC are palaeoethnobotany, palynology and sedimentology. Palaeoenvironmental studies for the Northwest are still in their infancy, in comparison with the large body of data available for many other regions of Europe. Being the result of an application of various sciences, only a coherent and large body of data can reflect both Prehistoric ecology and the way it has changed due to human activity.

12 -Referred to in Girão 1958, nº 4.

13 - *Carta dos Solos de Portugal* 1971.

14 - The concept of erosion, on a regional scale, must be understood as a two-way process of erosion-deposition. The erosion of the hills implies a certain degree of accumulation in the alluvial valley and/or a change in profile of the river's estuary.

15 - The best example for the case of Iberia is undoubtedly the extensive Roman mining, particularly in León and Asturias, where the landscape has changed very dramatically.

Palynological studies have been the major support for palaeoclimatic studies in Europe (Frenzel 1966) as well as in the Northwest, although one can argue over the reliability of the results, at a given level, as being affected by climatic and sedimentation conditions which may blur the real environmental picture.

Charcoal analysis can provide precious information on the associations of forest species, which can also be used for determining climatic characteristics (Beug 1982, 88-90), although the anthropogenic selection of the varieties present in the samples (Vernet 1988, 46) should always be taken into account. At present, there are some recent data available for this study area[16], so that some conclusions as to overall climatic conditions in the northwest of Portugal during the first millennium BC can be drawn. The combination of these two research methods may prove to be both useful and accurate in the future.

Other than its widespread use for dating, dendrochronology can also be of the utmost importance in determining climatic conditions (Aitken 1990, 36-46), since tree-ring growth is influenced by the ratio of rainfall to temperature. However, there are some limitations to reconstructing palaeoclimates by this method (Pilcher-Hughes 1982, 82-3), and its application to agricultural cycles also requires caution. As a matter of example, let us consider an unusually rainy Spring, which is characteristic in the Northwest. This would be represented by a thicker tree-ring growth, meaning a good cycle for vegetation growth but also, and inevitably, a bad harvest of Spring cereals.

The analysis of insect and mollusc remains is also a good source of climatic data (Lamb 1982, 24). Most species migrate little and adapt slowly to climatic change, making them ideal subjects for the method. They can also be found in contexts which are representative of certain ecological environments (Osborne 1982, 66), which may be human-related. However, such studies require a preliminary broad characterization of the regional climate and environmental conditions, as well as more precise determinations of the absolute and relative chronologies of the archaeological contexts, a state of affairs not yet possible for the majority of cases in the Iberian northwest. Remains of insects were found, in fact, in substantial quantities at the Castro de Penices, amongst the many soil samples which, unfortunately, have not yet been analyzed. This example demonstrates that insect remains are one source of information which is available in castros. However, conclusions on this subject cannot be drawn from the information available from just one site. The method does offer considerable possibilities for future work in this region.

The first millennium BC may be divided into two major climatic periods which are observable on a European scale: namely the sub-boreal and the sub-atlantic.

Sub-boreal period

The sub-boreal period is characterized by a relatively high temperature and dry weather. Magny's (1982) studies for the Jura region show that there was a progressive climatic deterioration during this period, which continued through the following phase. The long-term trend was thus a lowering of annual temperature, accompanied by an increase in rainfall. However, these data should not be overstressed, as temperatures at the end of the Bronze Age (1300-900 BC) in the middle latitudes of the northern hemisphere (Lamb 1982, 30) seem to have fallen by only about 2° centigrade, and the variation in precipitation did not exceed 10-15% (Magny 1982, 37). This also means that by the end of the Bronze Age, average temperatures were slightly higher than either during the Iron Age or at the present time. These effects can, of course, be under or over estimated according to the conditions prevailing at sub-regional level. This period, from 2500 to 600 BC, is not particularly important for this study, on the whole. However, its later phase may be of relevance for a full understanding of the contemporary environmental conditions, which may be one reason for the emergence of hillforts in the Northwest at this time.

Sub-atlantic period

The next period, the sub-atlantic, runs from 600 BC to the present day. It is marked by a considerable decrease in average temperature, and some of its sub-phases seem to have had a higher rainfall.

The VIII-VIIth centuries BC in Central Europe witnessed a rapid climatic deterioration (Bouzek 1982), with a change towards cool Atlantic conditions. Two later, but less severe sub-periods, corresponded with the Hallstatt and La Tène periods of the Iron Age respectively; the former being of the VI-V^th^ , and the latter of the II-I^st^ centuries BC. It is not yet known to what extent these climatic changes may have affected Iberia, but it is a known fact that they roughly coincide with the broad archaeological phases of Northwestern archaeology, namely the decline of Bronze Age cultural patterns and the establishment of the castros on one hand, and the development of the Castros system, prior to and during the Roman Conquest on the other hand.

16 -I am particularly indebted to Isabel Figueiral for the kind suggestions and advice provided on the subject of charcoal analysis.

General climatic conditions in Iberia

The general climatic patterns established for Prehistoric Europe as a whole can be applied to the northwest of Iberia on a preliminary basis, i.e. after the corroborative conclusions of Margalef (1956, 7) have been taken into account. This author does recognize a certain lack of statistical quantification in the data upon which his conclusions are based. For that matter, there is no doubt that the sedimentological studies of river beds such as the Minho, Lima or Cávado rivers will provide some very interesting data on regional climatic change during the first millennium in the near future.

Use of the various environmental studies in association with the related Classical references to Mediterranean climate (*in genere* Pliny and Strabo), for Iberia in particular, show that the climate in the Iron Age was broadly similar to that of the present century.

Today, the climate of the Entre-Douro-E-Minho, as a whole, is characterized by cool Springs and Autumns, with alternate rainy and bright days. Summer average temperatures are moderate, with little rain but a high relative humidity. Winter is cold, humid and very wet. The proximity of Atlantic air masses moderate temperatures slightly but there is copious snow in the mountains and frost in the more exposed valleys. In the highest massifs, summers are short (Ribeiro 1955, 48), cold and dry, and winters long and cold, with copious rain and snow.

The region under study is relatively uniform in its average annual rainfall values (Figure 9), and it forms the area designated as 'North Atlantic' (Daveau 1977, 9; Ribeiro 1955, 236), which is characterized by a climate whose humidity (Daveau *et al.* 1985) is controlled by its proximity to the Atlantic. The relief is oriented towards the sea like an amphitheatre, and the river valleys are aligned in the same way. This allows the mass of humid air from the Atlantic[17] to penetrate the valleys and condense out over the mountains. Annual rainfall is correspondingly very high (Figure 9). Precipitation is present virtually throughout the year, leaving only a short dry period during the summer.

The eastern side of the area studied overlaps another area, namely north *Trasmontano*[18] (c.f. Ribeiro above), which is characterized by a more continental climate. A very different culture may be found in this hinterland area, the result of climatic and geomorphological factors. In fact, apart from determining either the whole, or some aspects, of the ecology of a region, climate can also be responsible for various social effects and tensions (Harding 1982, 3), as well as changes in the dynamics of material culture. There are also types of disease that are typical of, or commoner in, certain climates.

Degree of vegetation cover

For the study of vegetation we rely principally on anthracological, carpological and palynological studies, bearing in mind their complementarity for this kind of approach. Unfortunately, very little work has been done for the first two millennia BC in the north of Portugal, and therefore any coherent approach suffers from lack of data.

Amongst the sciences referred to above, anthracology stands out for our purposes, particularly [19] because the recent and valuable work of Figueiral (1990), on charcoal analysis from three Late Bronze Age settlements and four castros in the Northwest of Portugal, based on dated contexts, allows us to connect the results with a chronological framework closely related to cultural phases.

The sample sites are Bouça do Frade (Jorge 1988a, 1988b), Lavra (Sanches 1988), and casteo de Matos (Figueiral-Queiroga 1988, Queiroga 1984), all three being Late Bronze Age settlements in the Baião council area; as well as the castros at Crastoeiro near Mondim de Basto, (Dinis, AP 1993), Cruito at Baião (Pereira-González 1988), and Ermidas (Pautreau-Queiroga 1989, Queiroga 1985a, 1985b) and Penices (Queiroga 1987) at Vila Nova de Famalicão. The results for the three studied Late Bronze Age sites are consistent, and in accordance with the characteristics of an Atlantic climate, although the proximity of these sites should be borne in mind to avoid excessive generalization.

The analyses of the carbonized wood remains from the above sites (Figueiral 1990) are shown in Figure 10. The following points should be noted:

(1) This is a chart representing the variation of species throughout the first millennium BC. It is based on Figueiral's (1990) study of the carbonized wood remains from sites in northwestern Portugal.

(2) The percentage values presented here have been taken from the absolute values (Figueiral *ibid.*) of the following samples:

Castelo de Matos: 29A, 30A.

Lavra: 31A.

Bouça do Frade: 28A, 32.

Castro das Ermidas: 18A, 19A, 21B, 22A, 24A, 26A, 35A, 35C.

Castro de Penices: 3A, 4A, 6, 9A, 10A, 11A, 38A.

Castro do Crastoeiro: 15A, 20B, 21C.

Castro do Cruito: 12A, 13A, 21A, 34A, 35E.

The results were then organized into three major phases, i.e. phase 1 (sub-divided into two) is

---

[17] -The Atlantic here is a very important factor as a regulator of the temperature (Girão 1960, 166).

[18] - Trás-Os-Montes, literally meaning "behind the mountains" is the hinterland region to the north of the Douro river and to the east of the line of mountains Marão-Alvão-Barroso.

[19] - An interesting development of methodological guidelines for this kind of study was done by Badal Garcia (1987, 1987-88, 1988); and in particular a study for the statistical reliability of sampling and analysis strategies.

Late Bronze Age and refers to both open and defended sites, with phases 2 and 3 relating to the pre-Augustan and Augustan-Flavian periods respectively.

Any conclusions drawn from this chart must take the following points into consideration:

(1) The Late Bronze Age sites are relatively close to one another.

(2) The Late Bronze Age and pre-Augustan samples are fewer in number than those of the Augustan-Flavian period, which should be taken into account for statistical purposes.

(3) Castro do Crastoeiro is a hinterland site, and therefore a different geological, topographical and ecological environment. The chart is not sensitive enough to show up the main differences observable at this site.

(4) Samples 20B and 32 are from pits, and 21A, 21B, 21C are from hearths.

The reason for presenting the data as a chart in this manner is to outline the major lines of interaction between the human and ecological landscapes in each of the phases. Charcoal analysis can only reflect man's particular selection of species in the environment in a limited way (Vernet 1988, 46), considering the cultural factors involved in this process. On the other hand, and from an archaeological point of view, samples from pits and debris areas may be considered as roughly equivalent, as the fillings of storage pits do not reflect the primary function necessarily, and they normally contained perishable materials such as cereals anyway, but rather their gradual or immediate filling with refuse after they had been abandoned for primary use.

Bearing all these points in mind, it becomes fairly clear that the progressive denudation of the deciduous forest during the millennium seems to have slowed down by the first century AD.

The distinction made between open and defended sites in the Late Bronze Age appears to be supported by the larger diversity of species present in samples of the open sites. This reflects a more dynamic interaction with the environment, indicating that this group of sites, when contemporary with the defended sites, was probably more concerned with productive activities.

A surprising diversity of species occurred in the Augustan-Flavian period, and notably a consistent set of edible plants was found from site to site. No doubt this reflects a dramatic increase in agricultural practices, as well as a more controlled selection of those plants with some economic value.

Two species of oak (*Quercus pirenaica* and *Quercus robur*) are clearly dominating this picture of extensive deciduous forest. Other species associated with this type of forest, such as hazel, mountain-ash, hawthorn, holly, chestnut and yew, are also present. Scots pine has also been identified. Local people

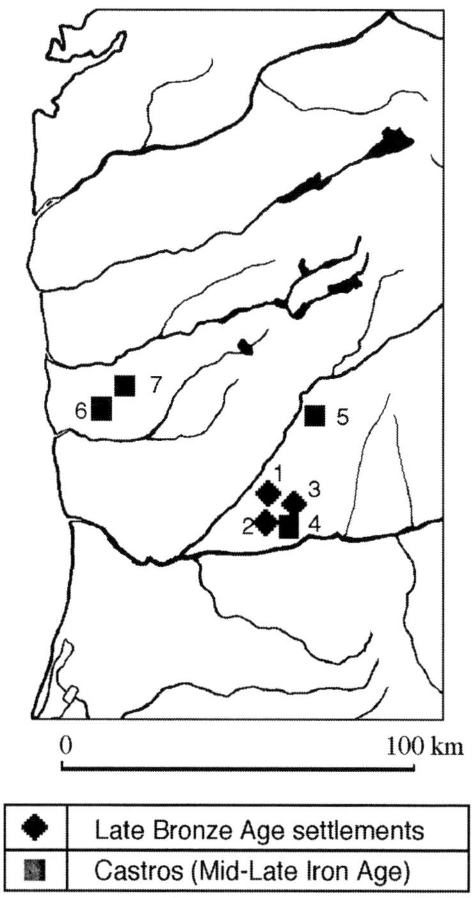

1 - Bouça do Frade; 2 - Lavra; 3 - Castelo de Matos
4 - Cruito; 5 - Castroeiro; 6 - Penices; 7 - Ermidas

also seem to have exploited the woodland alongside watercourses, as those species which are characteristic of river and stream margins, such as alder, willow, ash, poplar, alder, ivy and walnut, have also been identified among the carbonized remains from the studied sites.

This list of trees is supplemented by the low bushes of the *Leguminosae* and *Ericaceae* families, which are of the utmost importance as indicators of the ecological degradation of the original woodland (Figueiral 1990, 102-3) due to human action. In fact, broom, fern, heather, strawberry tree and box are good colonizers of cleared ground, and are able to spread immediately after the woodland has been cut or burnt, or fields abandoned. Bramble in particular is a good indicator of the abandonment of cultivated fields, due to its preference for moist and fertile soils.

From the Late Bronze Age onwards, there is evidence for increasing denudation of the natural vegetation as a result of human activity. Agricultural practices played a major role in this (Morzadec-Kefourn 1983). Palynological studies in Galicia (Menendez Amor 1969, Torras Troncoso *et al.* 1980) and an increasing amount of evidence for seeds (Peña

Santos 1987, 131)[20] indicate how grasses spread within the man-controlled environment of this period.

If the ecological situation in the early part of the first millennium BC looks fairly static (Figueiral 1990, 113), this may be due to a low population density in the Northwest, particularly in comparison with later periods. In fact, closer to the end of the millennium, a very different picture emerges[21]. The evidence available for this period reveals a flora more conditioned by human presence, with deforestation encouraging the spread of *Leguminosae*, which are in competition with oak. Different types of interaction between man and the environment may be demonstrated in the various zones of the area studied. For example, field-rotation (Figueiral 1990, 118) was probably practised at Ermidas, and a pastoral economy at Crastoeiro. Such differences probably reflect specifically local geographical conditions.

This picture, built-up from charcoal analysis, is supported by a palynological analysis from sites in neighbouring Galicia (Aira Rodriguez *et al.* 1989, Margalef 1956, Aira-Vazquez 1985), and by one sedimentary deposit from Serra da Estrela (Janssen-Woldringh 1981) in the centre of Portugal.

In summary then, the late Bronze Age is marked by a noticeable human denudation of the woodland, but the ecological balance suggests that population density was low. The significant reduction of the woodland implies that erosion may have increased substantially at this time, especially on the steeper slopes where the layer of humus is thinner, and so may be rapidly washed downhill.

Under such conditions, colonization by *Leguminosae* and other bushes takes longer, leaving the soil exposed and subject to further erosion. By the end of the millennium, the population had increased somewhat and there was an apparently intense and uniform deforestation, resulting in the spread of a more open landscape which was colonized mainly by *Leguminosae*, *Ericaceae* and grasses.

Such a landscape was more subject to erosion than had previously been the case. The vegetation composition is indicative of a temperate, humid, Atlantic climate, whose annual temperature variations are very much in accordance with those of the present day. It is likely that a combination of climate, deforestation, erosion and a substantial increase in population contributed to the multiplication of hillforts, so much in evidence by the turn of the millennium. These hillforts are spread however, over a range of ecological environments.

20 - Other than the large quantities of cereals found at the Castro de Torroso, cereals and pulses are also present in samples from Castelo de Matos.

21 - It is indeed a fortunate fact that three of the four sites studied by Figueiral (1990) are located in different geographical areas, which accounts for the reliability of the application of this study to the northwest of Portugal.

3.1.3 -Mineral resources

The geology of the Northwest is relatively simple . It is basically formed of granites[22] and their derivatives, and schists (Castro, LFM, 1968), the latter being associated frequently with quartzites and marbles.

The granites have the widest occurrence. A complex of migmathitic and metamorphic schists spreads through the upper and central basins of the Tâmega river. It extends to the Barroso mountain and Montalegre region, appearing again in a strip parallel to the coast, which runs from the margin of the Minho river to the zone around Braga and Barcelos. Starting in the estuary of the Cávado river and extending to the southeast, i.e. between Porto and the confluence of the Tâmega and Douro rivers, there is a very intricate pattern of bands of schists, greywackes and quartzites, occurring both alone and in association. These elements also appear on the Marão mountain ridge, running southeastwards between the towns of Vila Real and Lamego. In parts of the central and lower basins of the Minho and Lima rivers, and also by the estuaries of the Douro and Ave, there are small areas of quaternary deposits, composed of sands, pebbles, arenites and clays. The *Veiga de Chaves*, in the upper basin of the Tâmega, is another significant example of this type of mixing.

This region is relatively rich in metal ores (Figure 11), the most important ones being found in the quartzite veins (Thadeu 1965, 23). The ores of copper, tin, and zinc are frequent in some areas. For example, cassiterite is often found in alluvial deposits, particularly in those of the Tâmega and Beça rivers (Costa, HM, 1965-66).

The occurrence of tin in the alluvial deposits of Iberia was a well-known fact to the Classical writers (Diodorus 5, 38, 4; Pliny 34, 156; Strabo III, 2, 9). Iron ores may be found in sedimentary deposits (Thadeu 1965, 10), and veins of tin and gold also occur in the granites and their derivatives.

Metal production was no doubt a very important factor for the cultural development of this area. This will be discussed further in Chapter 7, as here our main concern is the relationship between mineral sources and settlement distribution.

The rich mineral resources of the Iberian peninsula caught the attention of various Classical authors, ranging from a more or less sober geographical description (Strabo III, 2, 9-10), through a tale of mythical adventure (Strabo III, 2, 8, quoting Posidonius; Justinus 44, 3; Florus II, 33), to a full account of the technical details (Pliny 33, 67-77).

22 -Among these the characteristic and very common (Girão 1960, 49) porphiroid granite stands out.

3.1.4 -Natural routes of communication

The natural routes of communication in the region, have, until now, received a mainly geographical interpretation. This can be supported by some evidence suggesting that these routes were used by the ancient communities. This evidence is still scarce for the Iron Age. However, recent regional studies (Almeida, CAB, 1990a; Martins 1987b) have outlined the relationship between watercourses and their valleys as natural routes, and the spatial distribution of hillforts, in particular along the Lima and Cávado river valleys.

The most conclusive evidence can be found in Strabo's Geography, which refers to the existence of tracks (Strabo III, 3, 7)[23] as well as river routes (Strabo III, 3, 7)[24], although archaeological studies have not yet corroborated these suggestions.

The major rivers of the region offer an easy means of communication between the coast and its hinterland (Figure 4), both on account of their good navigability and of the topographical configuration of their valleys.

Both the Minho and Douro rivers were navigable (Strabo III, 3, 4-5)[25] by small-sized boats far into Spain. The Lima and Cávado rivers cannot be considered navigable for more than two thirds of their total length but they might have been used more extensively in Antiquity. This is probably dependent on the types of boat used (Garcia y Bellido 1945b, Filgueiras 1974), for which, unfortunately, very little evidence is currently available.

The use of leather boats is referred to by Avienus (V, 101-107), Strabo (III, 3, 7), and Pliny (XXXIV, 156), although the extent of their use in the Northwest remains unknown. Strabo's reference (*ibidem*) sounds particularly coherent, for making a distinction between the times prior to Brutus' campaign when only leather boats were used by the Northwestern peoples, and the time at which he wrote the Geography. Dug-out canoes were rare at this time and presented as a novelty in the region.

23 -"Their sick they expose upon the streets..."

24 -"Again, up to the time of Brutus they used boats of tanned leather on account of the flood-tides and the shoal-waters, but now, already, even the dug-out canoes are rare."

25 - "After these two is the Durius, which, coming from afar, flows by Numantia and many other settlements of the Celtiberians and Vaccaeans, and is navigable for large boats for a distance of about eight hundred stadia inland. Then come other rivers. And after these the River of Lethe, which by some persons is called Limaeas, but by others Belion; and this river, too, rises in the country of the Celtiberians and the Vaccaeans, as also does the river that comes after it, namely the Baenis (others say "Minius"), which is by far the greatest of the rivers in Lusitania -itself, also, being navigable inland for eight hundred stadia."..."Off its mouth lies an island, and two breakwaters which afford anchorage for vessels."..."Now this river was the limit of Brutus' campaign, though farther on there are several other rivers, parallel to those mentioned."

The ease with which the Lusitanian *auxiliary* crossed the rivers using their skin-floaters, during the last episodes of Caesar's campaigns of the Civil War in the south, is indeed an indication that small personal floaters were used, although they are referred to as 'waxen vessels'[26] by the sources.

However, this image of the methods and techniques of navigation provided by the Classical sources, and somewhat justified by the lack of archaeological evidence, has been questioned (Alonso Romero 1976b). Comparisons have been made, with the Galician rock carvings, with the material evidence of earlier contact around the Western Seaways between Iberia and the British Isles (Alonso Romero 1976b and 1976c, Hawkes 1952, Lopez Cuevillas 1958a, MacWhite 1951), and also with the experimental use of reconstructed ancient boats (Alonso Romero 1976a)[27].

The technical feasibility of long-distance transport in the Atlantic with such boats during the Bronze Age has now been demonstrated as possible, thanks to the experiences of Webb (Alonso Romero 1976b, 73).

As the state of research stands at present, it is virtually impossible to trace Iron Age roads or paths in a region so densely populated since that epoch. Ancient roads do however, seem to have been re-used and renewed throughout time, rather than abandoned.

A road network has the function of connecting various points within an area and will either continue in use, or be abandoned, according to the development of human occupation in that zone. The ancient roads of the Northwest show a remarkable adaptation to topography, as one might expect in a mountainous region such as this. Roman engineers, who greatly improved the lines of communication in the Northwest, very frequently adapted their advanced techniques of road construction, either to local topography (Almeida C.A.B. 1979, 88-9; 1987, 277) or to previous routes. This adaptation to previous paths may explain some of the irregularities (Caamaño Gesto 1979b, 284) found in Roman road construction in Gallaecia.

A route connecting two castros prior to the Conquest would inevitably be kept if these two sites also had Callaico-Roman settlement phases. Obviously these arguments are more difficult to apply to local tracks or paths with a more utilitarian function, such as the circulation of products within the site catchment area.

26 - In his edition of Strabo's *Geography*, Jones (pp. 74-5, foot. 1) remarks upon the various translations and interpretations of this aspect, which does not seem to have been focused on by Athenaeus.

27 - An interesting experiment has been done at Santiago de Compostela University, reproducing an Iron Age type of boat made out of leather, wood and fibrous plants, in an attempt to prove that navigation towards the North, following the "tin route", was possible using ancient boat technology.

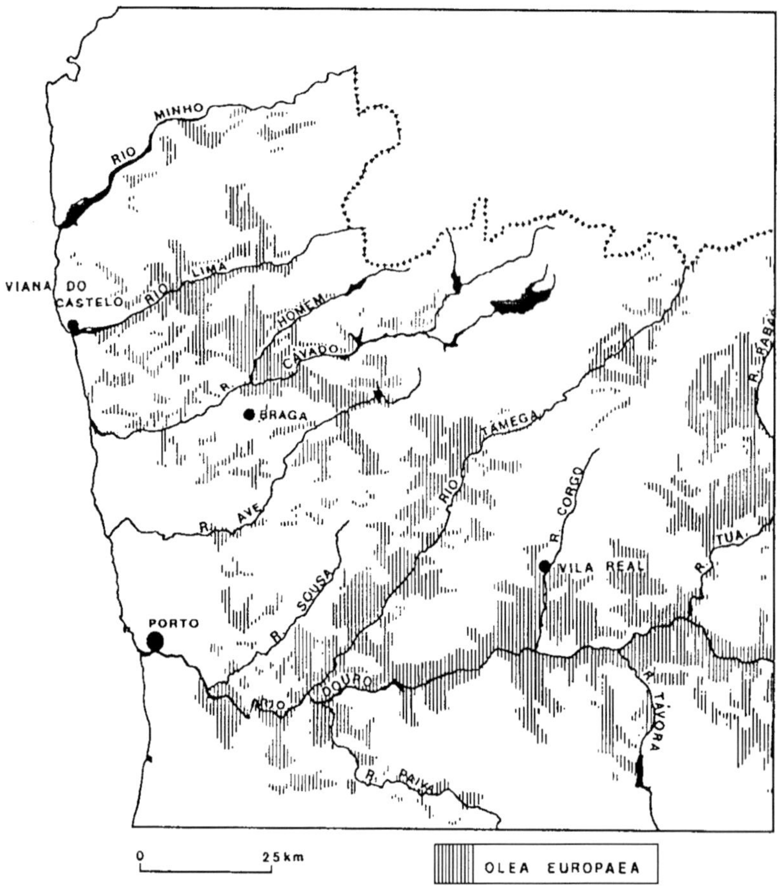

Present day distribution of the olive tree in northwestern Portugal

The geographical environment of the Entre-Douro-E-Minho provides good conditions for natural routes of communication, either by following the river valleys, or by taking lines parallel to the mountains, where some small plateau tops were suitable for such a use.

The rivers are undoubtedly a good means of communication, used since very early times, and much quoted by the Classical authors. Their importance as routes of cultural diffusion may have been overstressed in some case studies (Calo - Sierra 1983, 56), especially when an overview of the surrounding geographical features would make the more feasible routes obvious[28]. However, in many cases, what was written by Charlesworth (1924, 154) still rings true: "... *Spain -like Gaul- was marvellously supplied with the great natural highways of rivers.*"

The suggestions expressed in Figure 4 represent a model of the possible natural routes of communication. Since any correlation between natural routes and the pathways actually used has to be supported by archaeological evidence, further information can be obtained by comparison with the distribution of Late Bronze Age finds (Figure 61), the course of Roman roads (Figure 50), and site distribution in the region (Figures 24.1 to 24.5). This subject will be discussed further and corroborated with the distribution of imported finds (Figure 51) which, above all, outline the importance of the Atlantic as a route. However, the lack of evidence for wheeled transport in the Northwestern Bronze and Iron Ages, leads one to suppose that man-made roads were narrow and only suitable for foot-transport, as Roman armies might indeed have found out the hard way during the Cantabrian wars (Torres Rodriguez 1976, 25).

The northwest of Portugal is a region where natural resources are plentiful and the climate mild. Some valleys and slopes may have had their own particular microclimates and environments, being protected from the dominant winds and blessed by plenty of sunshine. These milder climatic conditions greatly improved agricultural productivity. An indication of this fact today may be seen in the relatively even distribution of Mediterranean species, such as olive, in this region, thanks to the existence of these *micro climates*. Some variation across the area studied may have resulted in environments with different economic potential.

Human ecological activity has undoubtedly had dramatic effect on some geographical aspects of the Northwest.

Deforestation is probably the most significant factor in a region characterized by mountains and alluvial valleys, as this would inevitably result in increased erosion. This denudation of the woodland was associated with an increase in agricultural and mining activities, and this probably first became noticeable in the Bronze Age. The loss of good farming land through erosion would have had far-reaching effects on the economy and the distribution of settlement. It may also have been a contributory factor in the rise in the number of hillforts during the later part of the millennium.

28 -Many examples of old tracks are known in the mountainous Basto region, mostly medieval and modern. They run through the deepest slopes, choosing the middle-slope or sometimes a line close to the top, but they are very rarely found along the base where the river carves its most inaccessible passes in the rock and provides water for a luxuriant vegetational cover. Normally, these paths only descend to the valley when crossing the river becomes necessary.

4- ASPECTS OF CASTROS ARCHITECTURE

4.1) -General aspects of dwelling architecture

Social organization is reflected in the way the settlement is structured. The relationship between material culture and the social structure of the communities who lived there has already been outlined (Brun 1987, Champion - Megaw 1985, Collis 1984c, Cunliffe 1984b and 1986, Härke 1979, Pauli 1980, Wells 1980 and 1984) for some areas of Iron Age Europe. In this chapter I will attempt to sketch an image of the society and its interactions with the environment by examining the structure and characteristics of the dwelling architecture and its related features.

An analysis of the dwelling structures is used to help us understand better those aspects which may not be reflected directly in the archaeological record. In fact, the dwelling unit in the castros is a remarkable indicator of social organization, economic production and related everyday activities. However, theoretical models and cultural affiliations should not be taken too far, running a risk of misusing the data.

The distinctiveness of the architecture in the castros of the Entre-Douro-E-Minho region is one of the major cultural characteristics of the Northwest. Both the architecture and the organization of internal space within the settlement express more than just material culture. They may be considered as the most characteristic element for any definition of the 'southern area' (Almeida 1973, 1983a, 1983b) of the Castros Culture.

If one can consider the castro itself as the determinant element, and symbol, of the Castros Culture, then the stone built architecture is undoubtedly the most typical feature of the castro. Indeed this aspect is so characteristic, that the Castros Culture has been described as a *civilization of stone* (Almeida 1984). A castro is a fortified village usually located in a prominent geographical position such as a hill spur, hillock or, less frequently, valley edge. The culture which gave rise to this type of settlement is broadly related to the Iron Age Complex of Europe, and it flourished in the Iberian Northwest between the end of the Bronze Age and the beginning of the second century AD. Being characterized by small settlements in its early phases, urban sites developed under Roman influence. This development was accompanied by a change in regional organization and intrasite hierarchy, as will be discussed below.

The term '*castro*' may have been used differently in other regions[1] so it is stressed that the expression is used here only for the Iron Age hillforts of the peninsular Northwest.

Although it has been suggested for a long time[2], only recent research (Figueiral-Queiroga 1988; Martins 1985a; Peña Santos 1987; Queiroga 1984; Silva 1986a, 33-7) has been able to show that castros, as typical stone-built settlements, derive directly from the Late Bronze Age fortified settlements of the Northwest. In fact, the places where these first defended sites were built continued, in some cases, to be occupied throughout the Iron Age, although architectural and cultural patterns changed within them. Most of our information, however, derives from later phases. From the period of *petrification* (Almeida 1983 and 1984) to the end of the first millennium, the overall aspect of the castros is clearly dominated by the presence of massive defensive structures and the apparently haphazard distribution of the dwelling structures within the defended area. This fact may be related directly to local topographical conditions. The hilltops are usually rocky, and therefore the organization of the functional areas is conditioned by, and must be adapted to, e.g. by the use of terracing. However, the effort required in many castros, from the earlier phases onwards (Martins 1986b; Peña Santos 1987, 113), to level the slopes by building terrace walls has been little recognized to date (Masiá 1976, 18-9).

A recent study has shown that the castro may not be the only type of settlement in the northern Portuguese Iron Age. Other than the various castros, the existence of non-defended, i.e. open, settlements located at the valley edge should also be considered, as clearly demonstrated by Almeida, CAB (1990a, 53, 59) after an exhaustive survey of the Lima valley. These important data raise crucial questions that are likely to change our general approach to patterns of occupation in the later phases of the Castros Culture, but first let us review the evidence.

In the small valley of Facha (Ponte de Lima), the remains of occupation spanning the period from circa 20 BC to 20 AD (Almeida, CAB, 1990a, 58-62) were found by chance in a field at low altitude (70 metres a.s.l.), on the edge of the valley. A circular structure was found here, 6.3 metres in diameter,

---

1 -This expression has been incorrectly used in the course of recent research in the south of Portugal to designate defended sites dating from the Neolithic or Copper Age, in addition to those sites which are characteristic of the Castros Culture.

2 - C.f. the numerous references to finds of the Late Bronze Age, namely metals and the 'Alpiarça type' pottery, found in castros of the Portuguese Northwest. Also Calo - Sierra 1983 and Fariña *et al.* 1983.

internally paved with a floor made of fired clay and containing three hearths, which has been interpreted as a typical Castros dwelling structure. The associated finds include local wheel-made cooking and storage vessels. Geographically isolated from the hillforts that exist in the vicinity, this dwelling structure appears to be directly connected with agricultural activities. This also indicates that, by this stage, the immediate surroundings and catchement area of the castros had been extended towards the valley. This feature may have been either a permanent dwelling structure, or a temporary farm-house used seasonally during the more important periods of the agricultural cycle (e.g. for the surveillance of crops). Nevertheless, the finding of this structure in such circumstances may eventually lead to a better understanding of the social and economic dynamics of the Castros populations.

The apparent lack of organization within the early sites, such as Terroso (Figure 13), was replaced by a clearly organized urban system of quadrangular plan from the Augustan period, in which public areas had a higher profile.

The architectural re-organization of many sites under Augustus changed the plan of the castros, and displayed a new magnificence in the public and dwelling areas, all organized in quarters. The increasing importance of symbols of communal prestige is reflected in this reconstruction and reinforcement of the defensive structures. The enlargement of both individual dwelling space and the site area as a whole undoubtedly reflects a substantial increase in population. Simultaneously, many new sites were created, virtually doubling previous numbers (Almeida 1984, 37), and new types of settlement patterns arose, such as the so-called specialist *agricultural castros* (types 2b and 3, Figure 23).

Architectural adaptations to the environment and the available resources appear not to have produced major differences[3] between coast and hinterland, which supports the idea of cultural uniformity within the Entre-Douro-E-Minho region. Only items more susceptible to change, such as pottery, could be indicators of any discrepancies.

In this section I will try to analyze the most relevant features of the Castros habitat, as a portrait of the society that built and reshaped the site according to its particular needs and living strategies.

4.1.1) The organization of the space

The organization of space within a settlement reflects both its function and the various activities performed there. The oldest phases of the Castros Culture have not been excavated intensely enough to make possible any understanding of their characteristic architectural forms and functions.

The internal organization of the sites prior to the turn of the millennium is still barely known. Under Augustus there was considerable architectural re-organization in many castros (Alarcão 1988, 35; Almeida 1974b, 1977), and this fact has to be understood in a context wider than that of Imperial policy in Hispania. In fact, it probably reflects an important change in social structure, politics, material culture and economy throughout the Roman world.

The Augustan and Claudian architectural improvements (Almeida 1983a and 1983b) have greatly destroyed the remains of previous habitations and urbanism, of which only fragments are now represented in the archaeological record. Archaeologists who excavated sites prior to the last two decades normally did not go any further down than the last phase of occupation[4]. The inner area of a *traditional*, or Roman-contemporary, castro is organized into functional areas. Amongst these, one can recognize areas of public domain, and areas connected with familial life.

4.1.2) Public areas and utilities

An examination of spatial and architectural organization within the better excavated castros reveals the existence of areas and structures for communal use, which became more distinct by the turn of the millennium. These features are important for our understanding of the social aspects, but their usefulness is much wider. The public areas of some castros also reflect daily habits and priorities, and the level of social investment in equipment and facilities. It is just unfortunate that evidence for public equipment in the earlier phases is not available and generalization would not be wise, considering the substantial difference in social change during the phase the available evidence relates to.

Herding was a major element of the productive system but it has left little trace in the architectural structure and even if it had, no consistent approach has been produced so far which could unravel such

---

3 - Obviously one has to bear in mind the existence of small architectural differences, such as the use of pebbles for paving in the coastal sites, as can be seen at Cividade de Âncora (Silva 1986a), or the apparently dressed masonry stones from Castro de Penices.

4 - Probably the best example of this fact comes from the recent excavations directed by A.C.F. Silva and R. Centeno at Cividade de Terroso, which uncovered important pre-Augustan evidence beneath the area excavated by R.S Pinto in the early thirties, which had not gone beyond the uncovering of remains before the later occupational phase. A further example can be seen in Silva - Centeno 1977.

evidence. Small domestic animals could have been kept in the familial enclosure, although there is no significant evidence for this assumption at this stage of research[5]. A very limited number of sheep/goats could have been kept in such an enclosure[6], but there is no doubt that herds and the larger animals would have been kept somewhere in the site outside the dwelling area[7] but, probably, inside the defensive structure and in a communal areas. However, the existence of watering places for small animals within the dwelling area (c.f. below), in addition to the proliferation of small enclosed areas, could indicate that some domestic animals were kept within the familial enclosure.

Some sites, such as Carvalhelhos, have special areas, surrounded by defensive walls and with stone-built structures inside, which have been interpreted as corrals (Taboada Chivite 1977, 82). It looks as though there was a general separation of human and animal dwellings within the site. Visits made to numerous unexcavated sites have revealed the existence of small terraces, outside the external defensive wall on many sites, which could well have been enclosures for animals. Only future studies can substantiate this suggestion. The Castro de Lagarelhos (Nogueira da Montanha, Chaves) has a defensive wall embracing an elongated hilltop. Its interior is divided by one wall of constant thickness, creating two separate enclosures which, in my opinion, represent one for residence and the other for the herds, rather than for dividing the human community (Silva 1981-82, 88) into two in the same castro.

Water tanks and wells are known in some sites of northern Portugal and Galicia (Briteiros, Sanfins, Santa Luzia, San Cibrán das Lás, Borneiro). They would supply water for both the inhabitants and the domestic animals[8].

Water for human consumption could be obtained from other sources, such as the springs sometimes located inside the site and the water tanks which supplied water for the 'monumentos com forno' (Almeida 1974c, 159-60).

The structures known as 'monumentos com forno' are a remarkable example of a communal facility. They are usually located at the periphery of the sites, outside the main defence but still enclosed by a wall (Figure 45). Although local variations are known, a 'monumento com forno' generally consists of a large cuppola-covered furnace, the main chamber, a vestibule, and a main yard with a water tank. The whole structure is below ground level (Figure 18) and buried, with the exception to the main yard which is not covered.

The recognition of these most peculiar structures started with the discovery of the 'monumento com forno' at Citânia de Briteiros (Cardozo 1931), which was integrated with another known element (Cardozo 1985, 129-30) at the same site (Cardozo 1935b); namely the much acclaimed 'pedra formosa', found in the early XVIIIth century and later taken by Martins Sarmento to the Museum at Guimarães.

In the early thirties, a succession of studies and explanatory theories for the function of this kind of structure was produced by various authors. Amongst these[9], the proposed functions funerary cremation[10] has been widely accepted (Cardozo 1932, Garcia Bellido 1968a Lorenzo Fernandez 1948, Uria Riu 1941), as have uses as shrine connected to the cult of water or of water deities (Azevedo 1946, Jorda Cerda 1969, Sarmento 1933, 139), a thermal bath structure for medicinal purposes (Chamoso Lamas 1955, Conde Valvis 1953, 1955), a pottery kiln (Fernandez Fuster 1953, 1954, 1955), a communal baking oven (Gomez Tabanera 1973-74, 1980, Santos Junior 1966a), and finally as a furnace for metal smelting (Monteagudo 1952a), thus reflecting the enigmatic character of these structures.

This problem was conclusively solved after the intentional discovery[11] and excavation of the

5 - Considering the absence of structures or enclosures which might indicate the presence of domestic animals within the dwelling structures, one of the possible ways is to determine the amount of phosphate in the occupational layers of different areas. Such a study has already started at the Castro de Penices, although significant results have not yet been achieved due to insufficient sampling.

6 - In fact, their status as "domestic animals" is claimed by Silva (1986a, 113) due to various footprints of sheep/goat found in tiles and ceramics from many castros. However, the presence of a relatively large number of these animals would not fit with the architectural evidence of the known castros. On the other hand, considering the economic dependence of the castros on agriculture, and the assumption that fields lay in the surroundings of the site, then the presence of goat herds would have required particular care. Being able to breed in poor conditions, these animals have, though, always been considered very damaging to the fields, to the extent of being forbidden to roam near the villages and fields in regions where goat herding was a significant activity.

7 - There is little evidence for the existence of enclosures for keeping herds within the defended area of sites, and a project to measure the amount of phosphates in some occupation layers of the areas which have enclosures and no dwelling structures is necessary to provide some fresh data on this subject.

8 - Although not exclusively, as at Briteiros, for example, where the water crosses half of the site through a channel before it reaches the tank, a careless attitude towards the possible contamination of the water is suggested.

9 - It is not my intention to develop this subject here: see the abstracts of Garcia Bellido (1968), Masiá (1976, 150-7), Queiroga (1992), Silva (1986a, 53-60).

10 - This was the more widely accepted explanation, which has survived until very recently.

11 - Due to their location, at the edge of the hillforts, these structures are very difficult to find. Until recently, the "monumento com forno" of Sanfins was the only one to be intentionally discovered, the others being found by chance, or in

'monumento com forno' at Citânia de Sanfins (Almeida 1974c), and was corroborated by a new interpretation of its practical function based on a re-assessment of a *forgotten* reference from Strabo (III, 3, 6), i.e. that the Castros people from the Douro region had the habit of taking steam baths. Definitively settled by the subsequent excavation of three more examples, at Galegos (Silva 1986a, 56-8), Freixo (*ibidem* pp. 60), and Eiras (Queiroga *et al.* 1992), the social practice of steam bathing[12] is now an established fact for the peoples of the Northwest, taking into account the distribution (Figure 15) of these structures.

The entrance to the main chamber of the 'monumentos com forno' of the Northwest of Portugal consists of a large granite slab. Some slabs have a degree of decoration, which varies from site to site. The decoration is, as a rule, more carefully executed and profuse on the front surface of the slab, which seals the main chamber, suggesting a symbolic significance which has yet to be deciphered. The decorative care put into these structures produced a magnificent feature, which indicates that baths must have had an important functional and symbolic role in society. It certainly was a communal structure which also reflected the prestige and wealth of the site and its inhabitants.

The typology of these structures suggests that there were two main types, which I have called 1 and 2, with type one having two variants, "a" and "b".

The better studied form is type 1. Type 1 has a more significant distribution and better preserved structures, and is concentrated in the north of Portugal between the Douro and Lima rivers (Figure 15), with the exception of the bath at Santa Mariña de Aguas Santas (Galicia) which lies to the north of the main concentration. However, and rather surprisingly, this latter find is still within the boundaries of the *Conventus Bracaraugustanus*. The distribution of Type 1 is thus very similar to that of the statues of warriors (Figure 58), supporting the existence of distinct indigenous cultural and ethnic group which was later to inspire the establishment of the boundaries[13] of this Roman administrative unit. Type 1, distributed through the Entre-Douro-E-Minho, can be divided into two sub-types:

-Type '1a' includes most of the excavated examples and has the more significant characteristics for the purpose of this approach, with a relatively late date (I-II centuries AD), massive stone architecture, and a house-shaped profile to the front slab and main chamber.

-Type '1b' is little known. The only examples of this type are two front slabs, one from Sardoura, and the other found very recently at Arcos de Valdevez[14] without any archaeological context. Until now there have been no diagnostic elements to date these two finds. The bath from Sardoura was destroyed during agricultural work and its associated material (Cardozo 1949c, 496) suggests a period of abandonment in the early to mid Ist century AD.

The 'pedra formosa' from Arcos de Valdevez, being a stray find, has no chronological context. The architectural elements associated with the two front slabs are not known, and therefore it is not possible to define their morphology. However, the similarity between these two slabs suggests that they are representative of one type of 'monumentos com forno'. The question is to know whether they are contemporary with type '1a' and characteristic of small and less populated settlements, where there was little need and capacity for building monumental structures for social use. Alternatively, they may correspond to an earlier, pre-Augustan phase. Their rounded profile and decoration gives the appearance of archaism, possibly recalling some types of wooden hut, and no hint of similarity with the bath structures of the Classical world (Queiroga 1992, 149)[15] can now be defended.

Further north on the coastal fringe of Galicia, and separated by a large zone devoid of finds, a different type of bath is known. The main differences are the lack of a 'pedra formosa' and the use of locally available schist, which alters both general appearance and architectural features. These structures (Garcia Bellido 1942b, 225; 1942c, 297; 1968a, 41-2; Ramil González 1990, 39) are contemporary with the decline of the Iron Age and rise of Romanization in the region.

The date range for the baths in the Entre-Douro-E-Minho is between the Ist and IInd centuries AD[16], and their masonry work certainly reflects a mastery of quarrying and stone carving. However, the date given by Strabo's reference to this well established and widespread cultural habit, in addition to some archaeological evidence, shows that these structures and the habit of steam bathing are clearly indigenous (Queiroga 1992, 148-9), and therefore not associated with any adoption of Roman patterns of thermal bathing. Of course one cannot deny that most of the 'monumentos com forno' are the

the course of extensive excavation, but without any prior knowledge of their existence.

12 - This explanation did not have an immediate and overwhelming acceptance (Tranoy 1981, 341-6), although no new arguments were brought up in defence of the former theories.

13 - See Rodriguez Colmenero (1972) and Tranoy (1981) for discussion of the proposed limits.

14 - This information and permission to use the drawing (Figure 16.1) were kindly granted by António Martinho Baptista, to whom I am most grateful.

15 - I am pleased to be able to correct my own conclusions from this paper written in early 1986.

16 - Although some might have been reused (Almeida 1974c) for secondary functions, such as a water supply.

archaeological materialization of a post-Augustan feature, and as such are associated with certain characteristics which are typical of the architecture of this phase, such as monumentality and profuse decoration. But let us assume that baths already existed in the IV-IIIrd century BC, located at the edges of the hillforts, which adopted the architectural characteristics of this time; namely the use of perishable materials for their construction. Would Northwestern archaeology, as we know it, be able to locate and identify such structures under these circumstances? It would probably not. So taking into account the available body of evidence, future research should note the probable existence of indigenous baths in earlier phases of the Castros Culture.

The cultural and social significance of the bath structures should not be evaluated as an isolated phenomenon, confined to the Northwest. In fact, a quick review of the subject, such as the one attempted below, reveals the existence of similar patterns in other parts of Europe.

Acompained by an abundance of old interpretations and arguments from various authors[17], the so-called *burnt mounds, fulach fiadh, cooking places*, or whatever other local designation has been given to these archaeological finds, have long been a matter for discussion. A recent and refreshing approach by Barfield - Hodder (1987), along the lines of previous studies (Barfield - Hodder 1981, 200), suggests a broad connection, on a scale surpassing the boundaries of Prehistoric Europe, between different types of archaeological evidence for the steam bath, sauna[18] or sweat bathing. The emergence of bathing appears to have little connection with the Classical world, and the different strategies adopted in the various parts of Prehistoric Europe, now supported by recent ethnological parallels from North America (Barfield - Hodder 1987, fig 2), look like autonomous technical developments, both in the structures and forms of bathing. Also, the rather early dating for some of these structures in Ireland (Brindley - Lanting 1990, 56), Scotland (Barber 1990, 102), Wales (Kelly 1990, 123; Williams 1990, 138) and England (Hodder 1990, 108), mainly based on radiocarbon dating, is surprising given the scarce and undiagnostic character of the finds associated (Cherry 1990) with these structures. Early Greek baths, the *laconia*, and presumably similar bathing structures in northern Italy dating to the VI-V^th centuries BC (Barfield - Hodder 1987, 375)[19], indicate that this phenomenon did reach Mediterranean shores.

There is a wide range of radiocarbon dates to show a concentration of bath structures between the Bronze and Iron Ages (Hedges 1974-5), although this phenomenon survived into the Middle Ages. Structurally similar features are known in Sweden (Larsson 1990), the *Skärvstenshögar*, which date to the beginning of the first millennium (*ibidem* pp.145).

The corroborative references from the Classical sources added to the same evidence from other parts of Prehistoric Europe (Herodotus IV, 73; Strabo III, 3, 6) also imply that the archaeological meaning of this type of feature has to be understood on a large scale.

At Citânia de Briteiros there is a large house, about 11 metres in diameter and situated by the upper defensive wall, on the periphery of the dwelling area which has been referred to as the 'house of the council'. In fact, being far larger than any other known Castros house, equipped with a stone bench along the interior wall, and clearly disociated from the other architectural elements which are characteristic of the 'familial enclosures', this structure has been interpreted as the archaeological evidence for the existence of a site council (Silva 1981-82, 85; 1986a, 53, 285), as can be inferred from Strabo (III, 3, 7), and substantiated by the corroborative interpretations of other authors (Bermejo Barrera 1978, 13-22; Caro Baroja 1970, 28-9; Silva 1986a, 268; Tranoy 1981, 140).

Some interesting examples of enclosures, located on the highest points of many sites, are further indicators of communal activities, although they cannot be related exclusively to a political purpose. These are relatively small areas enclosed by a massive wall, with very few or no structures[20] inside. The best available date for such a feature is undoubtedly that from Mozinho (Almeida 1977, 24), which suggests that these enclosures are late, and probably connected with Flavian architectural and social improvements which opened up the way for a wider participation of the individual in local political and social dynamics.

The Augustan and post-Augustan architectural re-organizations also indicate the adoption of a new social discipline, reflected in the organization of the quarters and streets. Although this has much to do

17 - Curiously, a very similar path to the one of the Iberian *monumentos com forno*.

18 - Although differences between these types cannot be clearly distinguished, one could suggest a difference in the amount of steam as a basis of separation of the types.

19 - Quoting De Min, M. 1984:Adria Antica. in Aspes, A. (Ed.) *Il Veneto nell'antichità*. Verona, pp 809-30.

20 - At Mozinho, no remains of structures and very few finds were found (Almeida 1977, 23-4). At Santa Luzia there is, in fact, some evidence for structures, although not necessarily contemporary with the enclosure, and inside a similar feature at Castro das Eiras there is evidence of elaborate structures carved in the bedrock.

with Roman concepts of urban organization, it also implies the subjugation of former patterns, and of individual rights, to communal (or, perhaps one could more precisely say at this stage, political) rules. This can be clearly seen at Briteiros and Sanfins, where the sites were re-shaped to a quadrangular plan, with parallel streets which intersected one another at right angles, creating quarters. Such plans were now more independent of the topography. Such change obviously involved a great deal more manpower and specialized labour, which is also the result of increasing complexity in the social system, and it is widely known in the larger *oppida* where, as seen at Sanfins (Almeida 1974c, 171), private rights were sacrificed on behalf of the community.

The habitation area is crossed by streets which define the dwelling zones, or quarters. These quarters will be examined in the next section.

4.1.3) The quarters

The concept of 'quarters' in Castros urban organization is only perceptible with the Augustan, and subsequent, architectural modifications. In the Entre-Douro-E-Minho this is exceptionally well expressed at Briteiros, Sanfins (Figure 13, 45), Mozinho (Figure 14) and Santa Luzia. In these sites, a network of paved streets, which are straight and virtually always intersect at right angles, creates areas of quadrangular configuration. Therefore, the familial enclosures are distributed and modelled according to the available space defined by these major *streets* (Figure 13). This is yet another new characteristic of the later phase: namely that the dwelling units, being parts of the urban structure, will have to fit into a previously arranged maze, whereas formerly the structure of the maze was defined by the apparently disorganized association of these units. Although this may be true at the Entre-Douro-E-Minho, it cannot be taken as a rule. The extensive and accurate planning done at Santa Tecla (Mergelina 1943-44) in contemporary architectural phases shows, in fact, how an equally organized urban planning can develop according to rules which appear to be more based on topographic setting than on geometric guidelines, even if a similar layout is the result for the purpose of social analysis. This should be noted for future research, when attempts to define the patterns of Castros indigenous urban development are made.

4.1.4) The familial enclosures and their functional areas

The Castros inner space contains dwelling modules organized in areas with groups of associated similar structures, which represent the familial dwelling units or 'composite houses' (Almeida 1984), in the Mediterranean sense of this expression.

These composite houses were inhabited by a family, or part of a family, and this is why they have been referred to as *barrios* (quarters) (Masiá 1976, 103) or, more accurately (Almeida 1984), *familial enclosures*. An estimate of the average size of a family living in such a familial enclosure, together with some explanation for the function of the various inner areas, has not been provided by recent archaeology. However, the rather uneven dimensions of the known units reveal that, in most of the cases, the group that inhabited these units could not be as large as that suggested for a family (Tranoy 1981, 106-8). Therefore, one should bear in mind that the concept of *family*, as a larger group based on blood relationship, should be differentiated from its component parts, which are represented archaeologically by the dwelling units, and which are commonly also designated as families.

The dwelling architecture is a reflection of the social organization (Silva 1986a, 268), but it also shows a very efficient use of the available space, often incorporating the uneven topographical

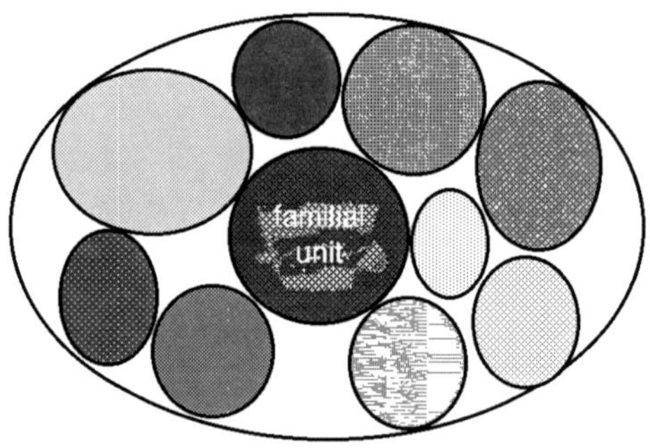

FAMILY

conditions to good effect.

The 'familial enclosure' is basically a composite house, containing several component parts arranged in the area and enclosed by a wall. These component parts usually have their entrances oriented towards the centre of the unit, where there is either a yard or a central house.

Prior to the period of Romanization, the shape of the familial unit was circular and irregular, and fashioned according to the dimensions of site terraces or the topography. It may also have been the result of earlier architectural style, when circular plans and perishable building materials prevailed. By the turn of the millennium, concepts of urban planning emerged, which were undoubtedly connected with the increasingly available resources of labour and skill in masonry, developments in familial structure and habits, and a more unequal distribution of wealth.

The structure of the dwelling units has been hotly disputed in the past. The Castros house of the Augustan period developed directly out of previous local dwelling structures, but it also resembled other types of dwelling unit (Garcia Y Bellido 1971, Hogg 1966) found either in the Mediterranean, or in northern Europe (Figure 20).

Both types are composite, containing areas with different functions, which are enclosed and separated from the outside by a wall. However, it looks as though this is a case of functional convergence, dictated by local conditions rather than by the diffusion of any architectural and functional model. When observing the *domus* at the castro of Monte Padrão (Santo Tirso), we see the same traits recurring, like the yard paved with granite and surrounded by rectangular granite structures of local masonry, without mosaics and located on hilltop, and cannot help thinking how, after Augustan urban re-organization, the habitation patterns of the Castros Culture slowly adapted to the Roman model whilst retaining a distinct cultural identity reminiscent of earlier times.

Other relevant activities associated with the familial enclosure are the religious practices, and specifically those related to burial and domestic ritual.

Although apparently strange in a region closely connected to the Mediterranean sphere, burial within the dwelling units has been suggested by the evidence since the beginning of this century (Peixoto 1905-08c). This subject was later reviewed for the Northwest (Garcia y Bellido 1966a). The available evidence for burial within the settlements was neglected for a long time[21], because it contradicted the concept of *Mediterraneanization*[22] in the Castros Culture prior to the turn of the millennium. More recent field work carried out at Terroso (Silva 1980a) has resulted in a proper evaluation of the previous data, to which corroborative evidence for burial inside the familial unit at Cividade de Âncora has been added (Silva 1986a, 49-50).

The finds at Terroso (Garcia y Bellido 1966a, 9-10) were three pits (Figure 21.1) which lay under the pavement of a house, surrounded by granite stones and divided by one slab. Recent excavations have uncovered a small rectangular pit (Figure 21.2) surrounded by granite stones and paved with brick (Silva 1980a, 311-12), located within the dwelling unit and against the outer face of a circular wall, at the level of the pavement. The other evidence from the Entre-Douro-E-Minho comes from Cividade de Âncora, where a set of eleven structures (Silva 1986a, 49-50, 303), similar to those from Terroso, was uncovered. From the list of associated finds, mainly indigenous pottery and one fragment of Arretine

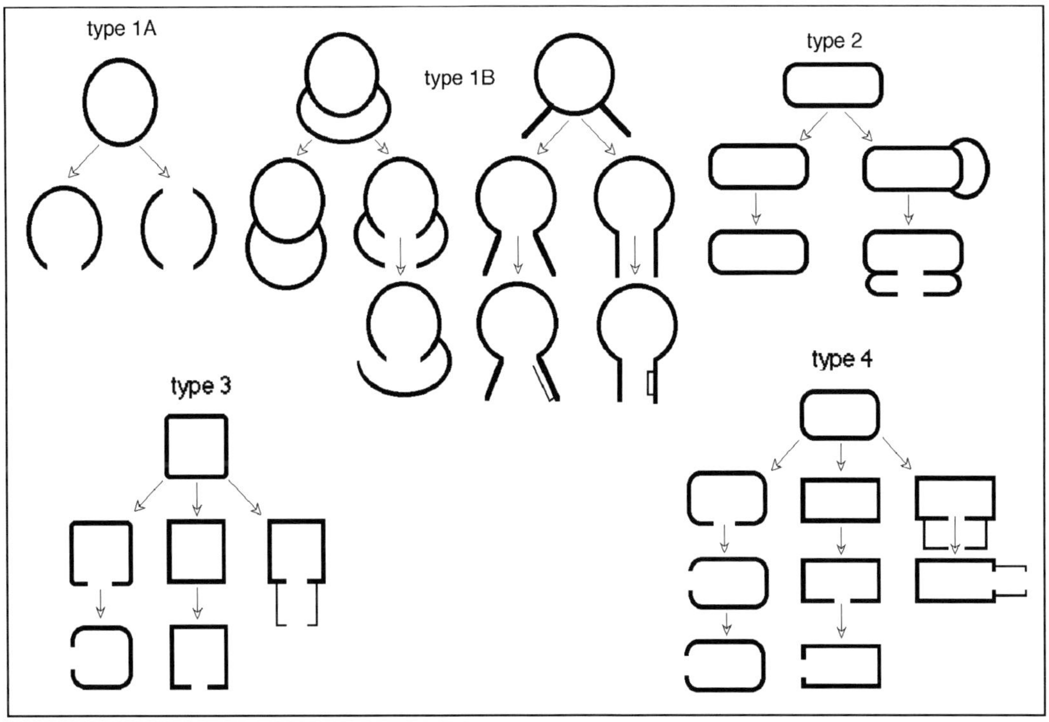

Some types of stone dwelling structures and their probable development
(after Eiroa 1975, and Romero Masiá 1976)

21 - With the exception of Alarcão's overview (1983, 193) and Almeida's monograph (1972, 17).

22 - In particular, the widespread rule for inhumation outside the settlement, as a result of the "Law of the Twelve Tables".

sigillata[23] (Silva 1986a, 303), a carinated beaker stands out because it is a rare form for the Castros Culture[24], although some significant examples have been found in connection with jewellery. The presence of these pots seems to indicate a funerary, votive, or hoarding function, as proposed by Silva (1986a, 128), and the form survived for a relatively long time.

Some of the small spaces confined by the walls of the structures of a familial enclosure can be designated as annexes, as their function has not yet been defined precisely. However, as Garcia y Bellido outlined (1966a, 11) and Silva confirmed (1986a, 302-3 *in genere*), one of the possible functions for these spaces was funerary or, in more general terms, religious practice.

Taking the above data as a coherent set, we have new and important references which stress both the strength of blood relationships within the Castros family and the functional multiplicity of the dwelling units.

However, in spite of this new evidence the problem is still not completely clear. If burial within settlements was a common practice in Celtic Europe (Bartell 1973)[25] and in the central Iberian Iron Age (Garcia y Bellido 1966a, 7-8; Maluquer de Motes 1954, 8; Tarradel 1965, 174-5), it is indeed surprising to find this custom at sites like castros where Roman influence was strong and particularly as Roman law was strict about funerary practices, dictating that burial had to be outside the settlement. Alternatively, this may be just another reminder of the cultural resistance which is so characteristic of the Castros Culture. The lack of suitable data does not help us in this matter. Why is evidence for funerary practices so rare? The last twenty years have witnessed the excavation of a substantial body of dwelling structures in the Entre-Douro-E-Minho[26], and yet the scarceness of these data is striking. Could one assume, under these circumstances, that funerary structures existed in all the familial units, or perhaps exclusively in the *patriarch* unit, which belonged to the head of the extended family? Available evidence favours the second suggestion. In fact, the enclosure in which the remains of funerary practices have been found, i.e. at Cividade de Âncora, shows undeniable evidence of communal structures and ritual, which may have been restricted to the familial world. The large house equipped with a bench (Silva 1986a, fig. XXVII-vi) may have been used for meetings and later feasting (Silva 1986a, 49, 122)[27], as suggested by Strabo's reference (III, 3, 7) to Castros familial and communal meals and feasts.

The connection referred to above, between archaeological evidence, its interpretation, behavior and belief, is still ambiguous in the light of the available data, but it has necessarily to be considered if any understanding of the role of the family within the community is to be achieved.

Inside the dwelling enclosures there are also structures to meet the basic needs of a family: these are the houses and the annexes.

The earlier houses, known in castros prior to the end of the IV century BC (Almeida 1984, 35-6), were built with perishable materials and clay, but this cannot be taken as a rule without considering local particularities. This type of construction is known in defended sites of the oldest phases of occupation, such as the castros at Neixón, Castromao, Cameixa and Villadonga in Galicia, and at Facha, Santiago and Penices in northern Portugal, as preceding the stone-built structures. Evidence from Castelo de Matos and S. Julião (Figueiral-Queiroga 1988, Martins1985a) shows that the circular hut, made with clay and perishable materials and with a floor of fired clay, is characteristic of Late Bronze Age patterns of building. The few available radiocarbon dates place this type of construction in the X-VIIIth centuries (Martins 1988, 126-7; Queiroga - Figueiral 1991) but it may have survived for some time where suitable local resources were plentiful. The huts from Torroso (Galicia) seem to be in perfect accordance with the model proposed earlier by Almeida (1984, Almeida *et al.* 1981), suggesting development of composite structures at an early stage. The Torroso huts have low walls of rustic masonry (Peña Santos 1987, 118), and small stones and a great deal of clay mortar were used in the structures. The upper part of the wall is not of stone, but rather built with clay mortar and the branches of trees and bushes. The walls are relatively thin and the hut diameter large (circa 7 metres), and the two superimposed pavements in the interior are made out of fired clay. The dating for the Torroso type of hut (circa 605-565 BC, Peña Santos 1987, 132), obtained from the associated material, suggests that it stands at the point where Iron Age dwelling patterns were beginning to emerge.

Remains of hut pavements made out of fired clay with post-holes are also present in the oldest phase at Castro de Penices (Famalicão)[28], showing a reasonably high occupation density and dating to the VIIIth - Vth centuries BC. The stage at which stone-

23 - Goudineau 27 form, manufactured around the last decade (10-8 BC) of the millennium (Goudineau 1968, 376).

24 - Rare examples were found at Briteiros, Faria, Laundos, Terroso (Almeida 1974a, 25-6, type 7).

25 - *In genere* Wait 1985, with various references.

26 - The most extensive excavations were at Mozinho (Almeida 1974b, 1977, 1980c; continued by Soeiro 1981-82, 1984) and Sanfins (Almeida 1974c), following previous excavations, but smaller-scale excavations were conducted at many other sites.

27 - In particular, the significant fact that the only pottery found inside this structure was a drinking vessel.

28 - The data referring to Castro de Penices are the result of my own excavations in co-operation with Felisbela Oliveira and António Dinis, under the supervision of Professor Ferreira de Almeida.

built architecture was adopted at this is not yet clear, as there may be a chronological and typological gap between this phase and the following one.

The fact that the remains of these oldest phases of occupation were nearly always damaged or seriously destroyed by subsequent building, together with the reduced size of the areas excavated to date, hampers any attempt to study the organization of the dwelling space. Therefore, it becomes difficult to understand the function of the utilitarian areas, as well as to produce inferences on the size and structure of the group. These are important aspects to be dealt with by future research.

This earlier type of house was still being used when the first stone-built defensive structures were introduced, and was to leave reminders (Almeida 1984) up to the end of the millennium. Unfortunately, study of these constructions is difficult, as their presence has been barely identified. No relevant data are available to determine the structure, internal organization, or associations of these huts within the site, with which to enlighten this obscure phase of hillfort history.

Evidence suggests that from the third century BC onwards, the house becomes a stone structure, whose function is normally identified by the presence of a hearth. At Facha (Almeida 1984, 36-7; Almeida *et al.* 1981, 64-5), the first stone houses were circular, 3-4 m. in diameter, but with thin (± 30 cm) and presumably low walls. The stones used for building the wall do not bear any signs of the mason's pickaxe and are generally small. Following an old tradition, the interior is paved with a layer of clay, which was smoothed while wet and then hardened with fire. This *modus operandi* for building floors was to survive as long as the Castros Culture itself. The oldest clay floors (Almeida 1984, 37) are relatively thin. From circa 100 BC to 100 AD these floors were frequently rebuilt, and it is not rare to find houses with a sequence of several floors, each one built directly on top of the previous one[29]. During the course of my own excavations, at Vermoim, Ermidas, Penices,and several others, I noticed that damaged clay pavements were covered by new ones, instead of being repaired, which resulted in a rise in the level of the interior of the house. It is possible that a higher circulation and level of activity inside the house was responsible for rapid deterioration of the floors. The addition of these thick layers of clay would inevitably reduce the inner height of the house, so that the wall and door-step might also have to be raised in some cases. As the damaged floor was not removed before adding a new one, this gradual raising of the level of the house could also have been intentional, to combat dampness, for example, or for some other reason.

[29] - One house at Mozinho had six consecutive clay floors.

The Roman Conquest, and especially after the beginning of the first century BC, started a trend towards urban development that only ended with the Flavians. These buildings are constructed exclusively in stone cemented with clay and the outer side of the walls is normally more carefully executed using larger stones, in contrast to for the smaller and less well dressed stones chosen to face the interior. As I argue below, this could indicate the use of plaster on the internal walls, but the archaeological evidence is not secure on this point.

At some stage, the space inside the house became too small for the family's needs, and some houses were enlarged[30] by adding a *vestibule*. Vestibules appear from the Augustan period onwards[31] in this region. These structures undoubtedly reflect the deep cultural changes of the Castros Culture at this time. It has been suggested that vestibules are the architectural indicator of population growth, as they are the result of a sudden desire to enlarge the inner space of the house. It is not yet clear whether the vestibules had low walls, and columns to support the roof, as suggested by Pereira (1933, 28), or were merely at the same level as the house walls. However, it is certain that they housed utilitarian activities, as some of them contain a baking oven or a bench, although their function probably differed from that of the main house. It is important to note that vestibules are considered to be an addition to the original plan of the house, and are understood as such. Architecturally, the wall of the vestibule always fits against the house wall, thus clearly indicating that it is of later date. They probably reflect an unforeseen need for enlarging the common familial space, otherwise the house could have been bigger from the start. This could be explained either by a growth of the family, or by an increase in the complexity of the household activities.

The roofing of these units was a thatch made out of timber and plants, such as straw (Azevedo 1945) or broom. Romanization introduced the use of *tegulae* and *imbrex*, although they were used in conjunction with plant material for a long time, on account of their weight.

Almeida (1984, 39) has suggested a model, establishing that the huts where there is a hearth and the entrance is at ground level are houses, and that the structures with no hearth where the entrance is above the ground level are granaries. In this model, the house with a hearth was a domestic dwelling, for living, eating and sleeping in. However, one should also consider that these houses may have been too

[30] - Eiroa (1975, 159) considers the vestibule as a variation of the circular house, when in fact it is neither a variation nor a restraint to the circular houses.

[31] - From Caesar's campaign onwards, in Almeida's (1984, 37) opinion.

small to lodge all the household activities, and that the so-called granaries (Almeida 1984, 39) could have been used as *houses for sleeping*, which would account for the absence of artefacts within them.

The hearth is an important element in the dwelling structure, and reflects various aspects of the *familial world* (Taborin 1989, 80). It is a utilitarian feature, a source of heat and a means of processing food in its various stages[32], but its social and symbolic functions (Audouze 1989, 333) were no less important. The central location of the fireplace in Castros houses was fundamental to an equal distribution of the members of the familial unit around it (Almeida 1977, 31-2), and *meals were taken around the fire*, according to Strabo (III, 3, 7). Also,

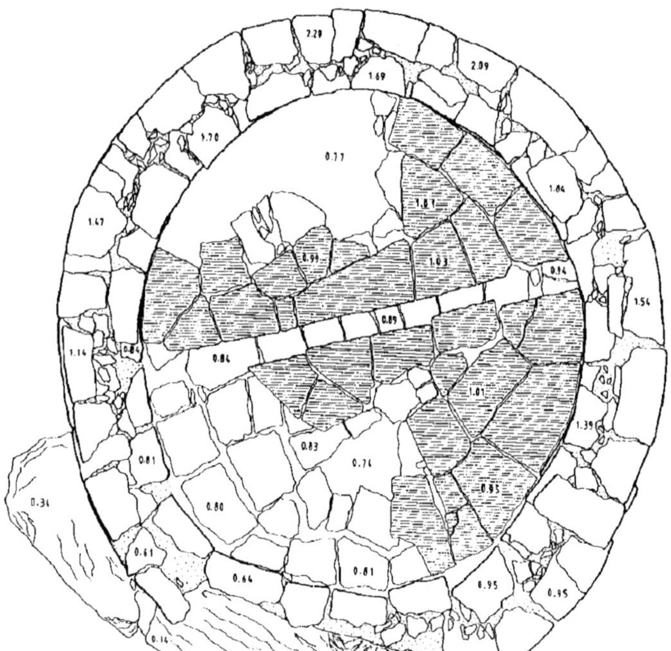

Castro do Cabeço - Granja, Boticas

it was noticed at Penices, and has been well documented at Santa Tecla (Mergelina 1943-44), that the central fireplace is positioned with its rear side facing the door, so that the person who sits by the fire or cooks is always facing the entrance. The hearth and its related elements can provide an important insight into familial relations and domestic behaviour, but decoding the evidence is difficult. Moreover, any attempt to interpreting the data in terms of ethno-archaeological models will have to attend to the fact that evidence reflecting attitudes which can be considered universal, can often mask the existence of distribution patterns of very short range (Gallay 1989, 110), and allowances will have to be made in any reconstruction for those undiagnostic elements associated with habitation.

Some architectural solutions are of such a local character that to try and analyze them in terms of cultural representativeness would probably be a mistake. Recent evidence on this matter has been provided by the houses excavated in past years at Castro de Penices. Here, from the turn of the millennium to the Claudian period, the local masonry bears some signs of the use of the pickaxe, but this does not allow the establishment of a regional pattern. In fact, this site is located at a geographical interface, between granite and schist, and the locally available stone splits easily into 'stone bricks' with reasonably smooth surfaces and of suitable proportions,without any need for reshaping with iron tools. On the other hand, using stones which cleave along natural fissures produces straight surfaces that do not fit very well into circular walls, so that corner stones are rough and gaps resulting from poor jointing tend to let in the damp. This set of circumstances probably led to the custom of plastering the walls, as they were at Penices with a poor mixture of clay plaster made out of local decomposed bedrock. At other sites, such as the Cividade de Âncora (Cardozo 1959a, Hawkes 1971a, Silva 1986a), plaster was still used (although here the stones show a great deal more pick axe work) for insulation and for hiding rough stone joints and walling. In addition, it has been suggested that these plastered houses were painted (Almeida 1977, 33), which has been confirmed at Castro de Penices.

According to Taboada Chivite (1977, 73) the size and arrangement of the huts reflect a scarce daytime use. A composite house should have covered the basic functional needs of an agricultural community, with the exception of tasks carried out in communal areas.

As the first century AD progressed, mostly in its second half, quadrangular house plans emerged, connected with developments in building technique, such as disposing the stones in horizontal courses.

Many examples indicate that a great deal of attention was paid to drainage (Romero Masiá 1976), both inside and outside of the dwelling structures. This should not surprise us, given the significant levels of average rainfall in this region (Figure 9), in addition to a lower rate of rain retention on the granite slopes. It may also be an indication of inadequacies in the roofing system. At Castro do Cabeço, a gutter channel lies inside the houses.

The quadrangular house plans seem to belong to a later stage of the Castros Culture (Almeida 1984,

32 - Food processing in agricultural societies is far more than food preparation prior to consumption. A wide range of activities should also be included in this concept.

39; Eiroa 1975, 163), becoming commoner after the Flavian period. Yet, the circular house shows a remarkable survival in the Castros architecture, as such houses were still being built at the end of the first century AD (Almeida 1974b, 1977) at Mozinho.

Those structures with no door step at pavement level and no earth floor are interpreted as familial granaries (Almeida 1984, 39; Masiá 1976; Taboada Chivite 1977, 75), although other types of granary (Appian 75) probably occur. It is, however, conceivable that timber-built off-the-ground granaries were built, but these would leave little trace in the archaeological record. In fact, the present distribution of the traditional off-the-ground granary in the northwest of Iberia is significantly coincident with the region defined for the Castros Culture. Furthermore, as Frankowski points out (1918, 5), it is a surprising fact that granaries are systematically built in wood in those regions where the remaining architecture is stone-built.

At the Castro de Penices, one circular structure dating to the Julius-Claudian period was excavated, which had no hearth and no remains of fire, on the single floor, and in which the remains of paint were still visible on the plastered inside wall. Careful flotation of soil samples taken immediately above the clay floor did not produce any seeds, and no other finds, such as ceramics or metals, were found in the interior. The single layer of clay floor in this structure is unusual. This could represent either a very short term use for this hut, which is not acceptable due to other cross-evidence, or indicate that there was very little circulation and activity in the interior of the *huts without hearth.*

Considering the average size of these huts, it is possible that the houses with hearths were for daytime activities, whereas sleeping took place in the structures without fireplaces. Apart from being too small for the whole of a family's activity, some of the houses in the castros have a bench along the internal wall (Silva 1986a, 49, 52) which reduces the area available for work, and appears to be unsuitable for sleeping in. It is possible, however, that the houses with hearths could have had a timber-built loft for storage[33].

The yard in the middle of the familial enclosure, roughly paved with beaten clay and the early phases of occupation, and stone-paved closer to the turn of the millennium, was undoubtedly an inner space that greatly assisted household activities. Processing of the cereals could have been done in this space (Almeida 1984, 39), which is often so carefully paved. However, analysis of carbonized cereal samples taken from the soil between the stones of the paved yard at Castro das Ermidas (Vila Nova de Famalicão), and from various other contexts (c.f. Chapter 7), showed that they contained very little chaff[34]. This implies that cereal processing took place outside the familial enclosure, probably in a communal area. Although a useful indication, this evidence is still too scarce to be considered as part of a general pattern.

As it has been argued above, the yard could also have been used to keep small domestic animals, considering the presence of possible containers for feeding or watering small animals (Pericot-Lopez 1931, 16; Taboada Chivite 1977, 82)[35] in some very small enclosed areas, which were normally made simply by the addition of a wall in a corner or space between two structures.

4.2) The chronological boundaries

Attempts at producing a sequence of chronological phases for the Castros Culture have been made by various authors, mostly from the seventies onwards. Expressions, such as Hallstatt and La Tène have occasionally been applied to Iron Age phases in the Northwest on a purely comparative basis with the European chronological and cultural framework, instead of trying to establish a direct correlation with these cultures. Indeed both the chronological limits and internal phasing within the Castros Culture have long been the subject of controversy.

A survey of the most relevant chronological schemes for the Castros Culture is presented in Figure 22. Apart from the establishment of periods and phases, the most important consideration is to define chronological boundaries or, in other words, the beginning and the end of the Castros Culture, in the light of recent research.

In spite of the volume that has been written to date on this subject, the beginning of what has been called the Castros Culture is still not clearly defined. The development of the following discussion is based not on new proposals for chronology, which might be redundant given the large number of suggestions and data available to date, but on questions over the identity of the Castros Culture (Martins 1987a). This will certainly have the effect of perpetuating the discussion and creating a new set of hypothesis.

33 - But not necessarily built with timbers. A circular house excavated at Ermidas revealed several relatively small granite columns spread amongst the rubble around the house, and many similar pieces can still be seen in rubble piles at Briteiros. Although a function for these elements has not been suggested, they certainly were positioned on the upper parts of the house wall, and might have been a support for the roof timbers or an upper floor.

34 -According to Joy Ede's report on carbonized seeds from Ermidas 1986, see figure in page 51.

35 - The reference to small containers or carved stones located by the door of the Castros house suggests that small animals might have been housed within the familial dwelling area.

The major obstacles to any establishment of a periodization or chronological boundaries, especially for the earlier phases, are our elementary knowledge of the material culture and habitats, and the relatively small number of radiocarbon dates suitable for dating sequences of occupation, as can be seen in Appendix II. In fact, a quick review of the bibliographical references presented with each of the radiocarbon dates shows just how recent most of these dates are, and how some of the phasing proposals were established without the aid of absolute chronologies[36]. In fact, the radiocarbon calibration curves (see Appendix II) are least accurate between the eighth and fifth centuries BC which, unfortunately, coincides with the least well known phases of the Castros Culture. Therefore, the result from the conversion of radiocarbon years into calendar years is greatly affected by the relative flatness of the calibration curve, as the number of intercepts increases thus enlarging the possible date range in calendar years.

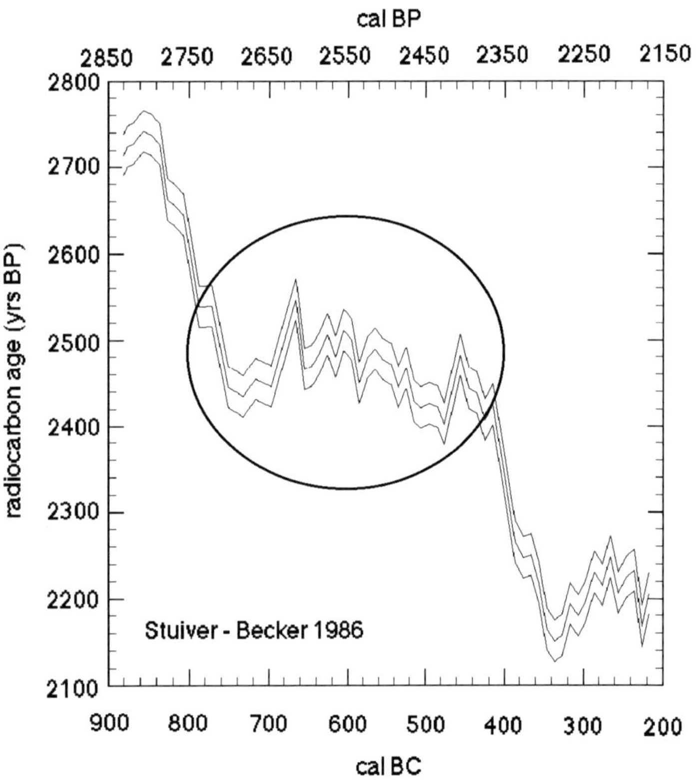

Radiocarbon calibration curve
(outlined for the VIIIth-Vth centuries BC)

36 - Exception shall be made to the more recent ones, such as Silva's (1986a) and in particular to the well balanced and critical use of a few radiocarbon dates by Martins (1987a) for the presentation of regional phasing.

The end of the Castros Culture has not been subject to any particular controversy, and two main proposals have been put forward: either the point of total integration caused by the Roman Conquest, i.e. the period of Augustus; or at the end of the I century AD during the Flavian period. A compromise solution suggests the Augustan-Flavian boundary as a marker for the end of the Castros Culture.

4.3) Discussion

In the body of evidence presented above, some aspects of habitation in the Castros Culture have been discussed, which may lead to a better understanding of patterns of social organization before the onset of Romanization. Such patterns become more perceptible in the archaeological record from the turn of the millennium onwards, showing how dramatic the changes in this domain were.

The impression gained on the use made of habitational space before the Conquest, based on relatively little data, indicates that the management of the space was so well organized into functional areas, that the temptation to believe in the existence of a pre-urban phase from the *petrification* of the sites until the Augustan period is great, followed by a full urban phase, defined by the development of highly structured central places, at later sites like Briteiros, Mozinho and Sanfins. This concept of early urbanization depends on two factors: acceptance of the suggestion that what we call *central sites* existed prior to the Conquest as a result of internal dynamics (Collis 1982, Cunliffe 1986, Martins 1990), and an understanding of indigenous patterns of complexity in space management (Romero Masiá 1976) within a castro.

If future research on a regional scale proves these assumptions to be true, then the major difference between the pre- and post-Augustan systems can be understood in terms of two concepts of urbanism, namely the indigenous and the Mediterranean. It should be stressed, however, that these two types do not differ in all aspects, and indeed appear to have common features, such as the *composite house*. A diachronic view of the various aspects the *composite house* can assume is presented in Figure 20.

This urban phase, established with Augustus only to decline under the Flavians, was greatly influenced by Roman patterns of space management, although many architectural features either retained, or developed, their indigenous traits.

Social areas and equipment are also of the utmost importance for an understanding of the complexity of social organization. In the case of the Castros Culture, the evidence only allows us to make cautious suggestions, rather than well supported statements. Social differentiation according to the characterization of the space within a site has only been briefly referred to by Almeida (1977, 10) when suggesting the existence of poorer quarters at Mozinho. We are thus confined to a general picture of an *egalitarian society*, which is supported by other studies on the subject.

However, some features might be diagnostic, not only for a better understanding of the social system but also for an appropriate correlation with contemporary European cultures. This is the case for the equipment, such as bath structures, which indicates social practices that necessarily have to be related to a particular social organization. The above evidence suggests that this might be an autogenous element in various parts of post-Bronze Age Europe. Although attempts at a correlation between these structures and the social and cultural environment have not yet been successful, their proposed date in northern Europe appears to be contemporary with the emergence of complex metal-producing warrior societies there (*in genere* Coles - Harding 1979), which generated an increasingly intricate social and political system in which rituals, such as feasting and the use of alcohol and other stimulants (Sherratt 1987), became an important item for the establishment and reinforcement of hierarchical links[37]. If broad parallels such as these can be related on the basis of a mere suggestion, then one could argue that the habit of steam bathing and fumigation, for cleanliness or medicinal purposes (Almeida 1974c) had an equally important social and ritual meaning, such as that described by Herodotus (IV, 73) for the Scythians. Now that one significant step has been achieved, the ritual function of the practices related to the *monumentos com forno*[38] is now, more than ever, open to discussion.

The structure of the Castros house is the best indicator for some of the most important aspects in this study. It reflects a structure which is representative of an agrarian society: the basic dwelling unit is consistent with the model of a community based on both agricultural production and the balanced use of the available resources. In fact, this model should not be considered any longer as a *Mediterranean* type but rather as *agricultural*. The existence of contemporary enclosures of similar form (Figure 20) in Britain (Hogg 1966), where they are described as *farmsteads*, or agricultural production units, suggests that a revision of present concepts of the function of the Castros familial enclosures would be in order, so that they reflect the social and productive systems that gave rise to them.

37 - Later transformed into a more accessible public utility, baths have long been connected with social status, throughout the Roman empire and in Medieval times (Vigarello1988, 29).

38 - Future studies on the possible rituals carried out in these and other places, should note the possible use of either stimulants or narcotics (Sherratt 1987). Our natural fears at penetrating the necessarily imaginative world of ritual interpretation can now be relieved by the emergence of diagnostic data. The possible use of narcotics in the IIIrd millennium BC can now be substantiated by the discovery of several heads of *Papaver somniferum* recently found in Trás-Os-Montes, as reported by Figueiral (1991, 24).

5- THE SITES

5.1) Patterns of distribution

The particular distribution pattern of sites within a territory may be the result of the degree of interaction between communities, the availability of local resources and strategies adopted for their use and control, and may even reflect hierarchical relationships between sites. The study of inter-site relationships is particularly important for this study because the underlying assumption is that a permanent state of warfare existed between the castros.

This kind of approach must begin by mapping all the sites in the region studied. Site distribution maps are presented in Figures 24.1 to 24.5, which refer to the lists in Appendix I. In this chapter, the different categories of site will be defined, and some factors affecting patterns of settlement distribution discussed.

An attempt to estimate the number of castros in the area defined for the Castros Culture is a difficult task which runs the risk of error, either by assessing the figure too high (Castillo 1908; Lopez Cuevillas-Bouza Brey 1929, 95; Lopez Cuevillas 1953b, 255-6) or too low (Bouiher 1979, 12-3). However, recent work on a local scale (Almeida, CAB, 1985b, 1986, 1987, 1990a and 1990b; Almeida - Baptista 1981; Marques, JATM, 1984b; Martins 1987=1990; Martins, JB, 1984; Moita 1962, 1966, 1971; Montalvão 1971; Regalo 1986; Soeiro 1984, Silva 1986b), in addition to other unpublished fieldwork of relatively wide coverage which has been designated as a 'National Archaeological Mapping'[1] project, undoubtedly provides us with a more realistic picture of the number of castros in the Entre-Douro-E-Minho region.

Any study of patterns of settlement distribution in the north of Portugal faces a major difficulty in the lack of accurate definition for the chronological boundaries of a site. This lack of definition is the result of less intensive fieldwork in recent years. An overall view of patterns of settlement distribution in the Castros Culture cannot be obtained at present because of gaps in the evidence, sometimes over large areas. Other types of information, such as surface finds or architectural remains, can supplement the existing data to some extent, but there is really no substitute for a well organized and detailed field survey.

The course of the past decade has witnessed an increase in regional studies regarding settlement distribution (c.f. above), which is an important step forward. However, the extent to which these studies can be used to obtain a general pattern for the region, especially when considering the diversity of small-scale cultural adaptations that exist within it, has yet to be evaluated The subject has to be dealt with very cautiously, and the regional studies with long listings of insufficiently calibrated data (Silva 1986a, 69-108)[2] are of little value for constructing coherent models of development.

Various factors should be borne in mind for a proper understanding of settlement distribution. Amongst these, geographical factors such as climate (Bouzek 1982, 185), hydrology, geology and soil types, can all affect site location and explain occupation strategy in a region. The interaction between such factors is complex and often poorly understood, so that great care is required in the application of statistical models in such circumstances.

Another important aspect is the subsistence economy. One cannot assume, even for the same area, that all the sites had the same range of productive activities, or made use of shared resources. Two sites with the same structure and in identical environments may have practiced very different economic strategies. Analysis of the site catchment territory must be based on the actual use of resources and the patterns of material culture, preferably to any geometrical construction drawn around the site at random (Martins 1990, 211-16).

The answer to these methodological doubts lies not only in a more extensive study of the sites at regional level but also on more excavation. The study of material culture, such as the finds and architecture, provides less fine detail than environmental or economic analyses.

A typology of distribution according to site structure will necessarily be general, due to the lack of relevant data in this region. A secure knowledge of site structure can only be provided by extensive excavation, as has occurred at Briteiros, Mozinho and Sanfins, for example. Very few hillforts show evidence of defensive structures at surface level prior to excavation. It is, therefore, necessary to analyze and interpret the topographical features in order to determine the defensive complexity of a site. But

[1] - I regret that this study does not refer to any sites discovered and listed during the course of this project, as several requests to the "Regional Service of Archaeology" for the north of Portugal for consulting the recent work done on Castros mapping have never been authorized.

[2] - In fact, this catalogue should be viewed with suspicion. During the course of this study, and although the catalogue in Appendix I does not claim to be particularly exhaustive, many discrepancies were detected in Silva's work, such as sites that do not exist at all (being just a reference), chronological discrepancies (either Bronze Age defended sites or Medieval castles referred to as castros) or, less frequently, sites with a different location.

this process can be misleading, as typological differences may not always be related to function and chronology. An example to illustrate this point can be the difference between terrace-supporting walls and defensive walls, as we will see bellow. For example, at first glance and without further investigation there may appear to be little difference between walls built to support terraces and those constructed for defence. On the other hand, any reinforcement or enlargement of a defensive enclosure might well be mistaken for a terrace intended to hold a dwelling. Accordingly, any attempt to establish the defensive complexity of a site should be made only after a careful interpretative analysis of the survey of the whole area has been carried out, taking into account the relationship between topography and both the size and form of the features in question.

The collection and interpretation of data also includes a subjective factor. For example, is the site with two defensive ditches really typologically different from the site with three ditches, or does it merely represent a defensive reinforcement where natural protection is less strong? Because of such doubts in methodology, models constructed in this study will be based on observation.

The chronology of occupation is also lacking in precision. In fact, the only assumption supported by recent research is that all the castros were occupied by the turn of the millennium. Changing this into historical boundaries, it broadly corresponds to, let us say, the reign of Augustus[3]. Thus, this period is the only common point of occupation for sites in the region, and therefore any study regarding site distribution would have to be confined to it. Based on my own personal knowledge of the region, some sites were abandoned in the middle of the first century AD, probably around the reign of Claudius, rather than at the turn of the millennium itself. This has been observed at the Castro de Penices, Castro do Lago (Martins 1990, 209) and also at the Castro de Sabroso (Soeiro *et al.* 1981).

Little is known concerning the occupation of sites in other periods and generalizations should thus be avoided, even on a regional scale. A comparison may be made between the model suggested by the best regional study produced to date (Martins 1990, 198-211) and the data presented in the next chapter[4], just to show how basic and fragmentary our knowledge still is about occupational phases and regional variability. However, one must not forget that several regions have not yet been extensively surveyed, and many sites await discovery in the future[5].

A great advance was made with the recognition of a new type of castro by Almeida, CAB (1990a), i.e. the so-called *agricultural castro*. This has brought us new and diagnostic data on settlement patterns but, at the same time, caused debate as the validity of the concept *agricultural* and even of the *traditional* (Almeida, CAB, 1990a, 261) castro model, as both are determined by the patterns of geographical distribution and complexity of their defensive structures. Although the assumptions underlying both models[6] can be considered correct, the use of these terms may be misleading, since there is no evidence to make us believe that the productive economy of the small defended sites on the alluvial plain was any more agriculturally-based than that of the defended castros on the higher spurs. In fact, examples of agricultural sites with a similar structure and size occur only a few hundred metres from the sea shore. The most diagnostic example of this is the Castro de Labruge. Are sites in such locations really more dependent on agriculture than on the exploitation of marine resources? Probably they are not, but the final word for a functional explanation of sites of this kind must come from a study of both the artefacts and the remains taken for environmental analyses.

As it is difficult to date the length of occupation at many sites, I will propose a model of site distribution according to their geographical context. Size is a further site characteristic that cannot be determined at present. On this basis, the following types of site are defined:

Type 1A- Located on hilltops well detached from the massif, these sites have little need for defensive reinforcement by man-made ditches and ramparts since natural steepness is relatively even all around the site. Although the best example of this kind of site in the region is Outeiro de Lezenho, the most significant and well known site is probably Santa Tecla, on the northern bank of the Minho

[3] - From 27 BC to 14 AD, although this is just a broad point of reference for this model.

[4] - Let us take just one example to illustrate the contradictions found in data from the same region. It is argued (Martins 1990, 209) that sites like Castro do Lago (Martins' C type) were created in the late Ist millennium BC and abandoned by the mid Ist century AD. Now, Castro de Penices is in a nearby area and has both the same structure, size and geographical characteristics. Yet, it was continuously occupied from the beginning of the Iron Age (see Appendix II) to the mid Ist century AD. Of course one can argue that data from one site (Penices) may not be sufficient to dismiss a model, but then evidence from one site (Lago) would also have been insufficient to build it.

[5] - The list presented by Silva (1986a) indicates four castros in the area of Vila Nova de Famalicão. One year later due to intensive fieldwork, ten sites were known in the same area. On the other hand, the study produced by Martins (1990), being more intensive and on a smaller scale, will be less affected by future research. These examples show how studies based on occupation density can be affected by the development of research.

[6] - One point of disagreement, however, is the proposed chronology for this type of site, which Almeida, CAB, suggests begun at the turn of the millennium. Yet, ceramics from much older phases have been found at one site of this kind: the Suvidade at Recezinhos (Soeiro 1985-86a, 22-6).

estuary. Citânia de Sanfins could also be included in this group, even though the site expanded southwards and a small ditch was built on the north side.

Type 1B- These sites are located on the edges of small plateau, sometimes on spurs, defined by the granite bedrock. Such sites overlook both the valley and the plateau and the lie, therefore, at the interface between different ecological zones, with varying resource potential. A peculiar association of three sites of this kind can be found on the hill of Sta. Cristina (Figure 26): the castros of Vermoim, Sta. Cristina and Eiras.

Type 1C- The sites included in this group are located on spurs. Their structure is relatively homogeneous, since the spur location requires the existence of ditches and ramparts in addition to defensive walls. Rather than the altitude above sea level, which may vary, their major characteristic is that they lie some distance away from either the valley or the plateau. This suggests that a proportionately large area of slope is part of the exploitation territory and, therefore, that the range of subsistence activities is wide. Sites in this geographical position are Faria, Senhora da Assunção and Crastoeiro, for example.

Type 1D- This type is essentially the same as 1C, with a slight variation: the site appears to be concerned with the control of natural routes. Besides, according to the more dramatic examples, these sites cannot be agricultural due to the steepness of the terrain. Such settlements probably depended on pastoralism and the extraction of metal ores for their existence. However, their location, normally overlooking a river confluence, implies that territorial control was the chief concern. The best examples in this group are the Outeiro dos Mouros (Ribeira de Pena) and Castro de Álvora (Arcos de Valdevez).

Type 2A- These sites are located on a spur overlooking either the sea shore or a river valley. Their structure comprises the traditional association of defensive ditches, walls and ramparts. The geographical situation of these sites reflects the range of environmental riches that could be exploited, i.e. valley bottom, woodland on hill slopes, and river and marine resources. Castros such as Côto da Pena, Penices and Cividade de Âncora are a few examples.

Type 2B- This type can be differentiated from the previous groups by a closer attachment to the sea or to the valley. Although these sites may be structurally similar to type 2A, their position is determined by proximity to the subsistence resources, to the detriment of a suitable topographical location. Castro de Labruge, Castro de Carreço and Castro de Sto. António are good illustrations of this type.

Type 3- Normally located in alluvial valleys, these sites are very small and structurally simple. A small mound (Figure 30) is flattened and its detachment from the level of the plain increased by building ditches and ramparts. These sites have been referred to as *agricultural castros* (Almeida, CAB, 1990a, 261). However, their function is undoubtedly connected with the exploitation of valley resources, in which the role of agriculture has yet to be established.

5.2) Geography, resources and site hierarchy

The available evidence suggests that changes in settlement distribution in the Castros period, from the VIth to the beginning of the Ist centuries BC, were probably more concerned with the exploitation of regionally available resources than with any (re)establishment of hierarchy within the system. The wide coverage of different ecological and geographical environments, such as the coast, river valley or mountainous hinterland, is typical of the whole period and cannot be related to any particular phase. A few known cases indicate that sites located adjacent to a range of different environments show evidence for early occupation.

The Late Bronze Age is characterized by the existence of both open and defended sites. The latter were undoubtedly connected with the control of production, and a trade in metal resources. The transition from Late Bronze Age defended sites to castros is associated with an increase in the number of defended sites, and the end of open settlement. This does not necessarily imply an increase in population. Such a widespread phenomenon as the fortification of the settlements may, however, be the result of competition for the metal resources. The *recession* in the material culture of sites around the VIth century BC does not necessarily mean that metal sources had been worked out by that time, with a corresponding collapse in the output of bronze (Martins 1990, 204-5). In fact, metallurgical practices continued, even in *agriculturally inclined* sites such as Penices, from the VIth-Vth centuries until the end of the millennium, according to the remains found in soil samples. The reduction in metallurgical production could also result from a decline in long distance trade. Besides, this decline in production is based on a marked drop in the number of hoards, which may not be a reliable indicator (*in genere* Bradley 1985, 1988, 1990), particularly if long distance trade variability is involved, and manufacture is carried on outside the settlement itself.

Site structure and patterns of distribution remained static until the Ist century BC. Thereafter, a process of rapid change began which affected the whole picture.

Contact with Roman culture brought many changes to the Castros Culture. For example, distinctive regional development, which can be traced in all phases, became more marked as a result of a differential response to the pressure and influence of Romanization (Almeida 1983b, 187). Amongst these changes, recognition of the existence of site hierarchy, and of different strategies of occupation in the available space is a recent achievement in the archaeology of northwestern Iberia. In fact, as already explained in previous chapters, the end of the first millennium brought a general re-organization of economic, cultural and social patterns. This is also reflected in the regional distributions, functions and hierarchical interactions (Figure 25) of the hillforts. This has produced a much clearer definition of central and secondary sites and their relationship.

Although regional variability in the distribution of castros according to occupation, phasing and hierarchy, is still not well known in the area studied, it is apparent that the time around the turn of the millennium was a crucial period for the establishment of what could be called a hierarchical re-organization of the territory. In fact, at this time some sites were abandoned, but new sites were built, many of which were to grow in both size and population density. A similar and contemporary trend towards centralization has also been noticed for other areas of Europe (Collis 1982, Cunliffe 1986).

The use of exclusively theoretical models, when studying the territory of both settlements and ethnic groups, can obscure the importance of geographical factors such as rivers[7] (Almeida 1959; Bouza Brey 1942; Cardozo 1926; Cardozo - Santos 1953; Garcia Fernandez 1986; Lopez Cuevillas 1935, 1955a; Vasconcellos 1905, 227 foll.) and mountains as natural boundaries (Albertos Firmat 1974; Fontes 1980; Penas Truque 1986; Rodriguez - Fontes 1980) and other aspects of the Castros Culture, such as religion in Callaico-Roman society. Within the catchment area of a site located on a spur, the river may act as the boundary of the lower and more fertile zones suitable for cultivation, whereas the hilltops and higher slopes define the woodland (slope or plateau) areas for cattle grazing and for gathering. A site with a balanced productive economy would require access to all these resources, and any spatial analysis should take such aspects into consideration (Cunliffe 1973). However, when it comes to defining site catchment zones, one has to take into account not only whether all the hillforts of a region were inhabited at the same time, but also if the landscape we are *assigning* to each site is in accordance with the range of resources we presume, or know in advance, might have been exploited (Butzer 1971, 1982).

Evidence suggests that until the end of the Ist millennium the economy of the hillforts was based very much on subsistence activities, with trade as a lesser sideline. In fact, whether any growth in trading activities occurred at those sites in suitable locations, such as the castros on river estuaries, has not been recorded. Of course any discussion of the mechanisms of trade in the Northwest is beyond the scope of this study.

We have seen that most of the castros prior to the turn of the millennium had a degree of self-identity and economic independence, although they were certainly linked by some kind of hierarchy, as yet undefined. This raises questions as to the particular role of these hierarchical links between the communities, how they were maintained, and how they developed throughout this period. Shall we take as our underlying assumption that political dependence was not uniform, that it affected the economic development of the communities little and that it was connected with an independent sphere of interactions, such as the warrior system? Although some indicators may lead us to believe this, the general level of social investment in items related to the display of communal prestige at these sites seems to contradict this suggestion. On the other hand, a permanent state of competition for resources cannot be neglected as a factor in the development of site individuality. The archaeological data now appear to be at odds with the Classical sources. It is unlikely that large-scale and continuous warfare was extended to neighbouring communities, at least within the same *populus*. Rank within the warrior community might, therefore, be established more through actions outside the territory than by internal dynamics, which would create greater instability within the system.

5.3) Discussion

According to the available data, the distribution of sites in the landscape is relatively uniform throughout the Iron Age. The communities of the Castros Culture built their settlements in both high and low positions, in the hinterland and on the sea shore. Any inferences drawn on the relationship between site structure, geographical location and chronology should be avoided at present due to lack of evidence, especially for the period between the VIth and Ist centuries BC.

Sites were generally located with aspect to the locally available resources, and particularly where a range of different environments could be exploited, and contribute towards the broad spectrum subsistence economy.

7 - The importance of rivers and mountains as boundaries is expressed by Strabo (III, 3, 7) -"Those who are condemned to death they hurl from precipices; and the parricides they stone to death out beyond their mountains or their rivers". The execution of a member of the community guilty of such a serious crime would dishonour both the executioners and the territory, and this is the reason for solutions such as hurling from precipices or stoning, which were widespread in Antiquity (Bermejo Barrera 1978, 24-6). This social attitude of supporting the punishment of certain kinds of crimes by death but, on the other hand, refusing any personal or territorial complicity in the act, led to both the professionalism and social segregation of the executioner (*in genere* Delarue 1979).

An analysis of site distribution (Figure 24.1-5) using preliminary estimates of site duration, shows that there was a large increase in the number of hillforts towards the end of the millennium, if not slightly earlier. This phenomenon has to be more accurately defined on a regional scale, before further generalizations can be made. The majority of sites shows some signs of contact with Roman culture in the middle phases of the Iron Age, in a few sherds of Campanian ware or Arretine *sigillata*, for example. Contact increased in the later phases to include both imported Roman finds and local imitations of them. Thus, what happened to the Castros Culture of the Entre-Douro-E-Minho region regarding populational growth, site hierarchy and urban complexity may have been greatly influenced by Rome, although not necessarily in a direct and straightforward manner.

In fact, the growth of some sites by the turn of the millennium cannot be explained solely by Roman pressure, since the indicators of acculturation in material culture are insufficient to justify the dimension of the changes, as Martins has already suggested (1990, 205). It is possible that the changes in the system had already started prior to the Conquest, as the result of internal dynamics and a trend towards centralization. Romanization may have just accelerated this process.

Bracara Augusta was the centre for the diffusion of Romanization in the region. It did not have the status of *municipium* prior to the Flavians (Tranoy 1983, 197), in spite of its significant role in the Romanization of a vast and important region. Was this fact due to a simple neglect in Imperial policy of the urban and cultural importance of *Bracara* in the region, or was this city allowed a slower development in the interests of harmony and balance between the two opposing cultures? The suggestion that during the early part of Roman occupation *Bracara*, as an *oppidum*, may have been administrated by the indigenous aristocracy supports the latter.

The territory administred by this city was large, according to facts presented by Alarcão (1988, 55). However, the mechanism of this administration, in both the city and its dependent territory during the first century AD, is not yet clear. From the whole process stands out the idea that the castros of the northwest of Portugal enjoyed an effective administrative freedom, and yet were bound to *Bracara* by some kind of hierarchical link. In fact, those castros which developed into large cities during the first century AD, and particularly those which showed substantial growth after the turn of the millennium, may have played an important role in the re-organization of intrasite hierarchy (Figure 25-C) in relation to Roman political and social domination, being positioned immediately under the major centres of Roman influence in the hierarchy. By reinforcing centres of indigenous leadership instead of creating new ones, Rome was probably able to control the whole system of the Castros Culture. This was achieved by re-educating local leaders and by following a strategy of support where resistance to Roman law and culture was minimal. Undoubtedly enormous savings in administration were made as a result. This slow but effective political and cultural pressure was probably the main reason for the emergence of one peculiar feature of Roman provincial life, namely the survival of castros at the end of the first century AD.

6- THE DEFENSIVE STRUCTURES

6.1) The emergence of the defensive structures: some model cases

Until the 1980s, very little was known concerning the earlier defensive structures of the Castros Culture, or of their chronological and cultural framework. Indeed, our lack of knowledge on the material culture, between the Bronze Age and the first century BC, has delayed a proper understanding of this problem. The publication of the first few radiocarbon dates (Jorge 1985; Kalb 1974-77; Martins 1985a, 1990; Peña Santos 1987) from well defined archaeological contexts brought new insight to a problem that had been based exclusively on a typological analyse of the artefacts typology[1].

The separation of the different aspects of the material culture, with no attempt to link them together, especially through the social system, has probably long delayed any development in a subject as important as the emergence of defensive structures in the Northwest.

Scientific excavations carried out in recent years on sites where a proper correlation between early defensive structures, late Bronze Age and/or early Iron Age ceramics, and evidence of metallurgical activities can be seen, all supported by radiocarbon dates, have led to a revision of the subject. A substantial contribution to the problem and summary has been produced by Martins (1988a).

A defensive structure is understood here as the presence of a ditch, rampart, wall or palisade (or any combination of these features), which completely or partially encloses the settlement. Its role is seen as that of a fortification, to protect the settlement and its activities from external attack during periods of warfare.

Bronze Age settlement is characterized by open sites, in which perishable materials were used exclusively for building. By the end of this period, sites were beginning to be located on prominent hills surrounded by a very simple defence. Defended sites co-existed with open sites, but the latter disappeared at the end of the Bronze Age. Some defended sites were occupied continuously until the Iron Age, when a change in material culture occurred, but others were abandoned at the end of the Bronze Age. New sites were also built at the beginning of the Iron Age. An understanding of this phenomenon is of the utmost importance for the establishment of the mechanisms which led to the need for site fortification and to the emergence of the castros.

[1] - Approaches to this subject can be found in Kalb (1980a and 1980b) and Calo - Sierra (1983).

Those sites which have been extensively studied in recent years, were selected to illustrate three different mechanisms of change, which may account for the transformation of the Late Bronze Age type of settlement into the castro.

The examples chosen as representative of these three major patterns of change are as follows:

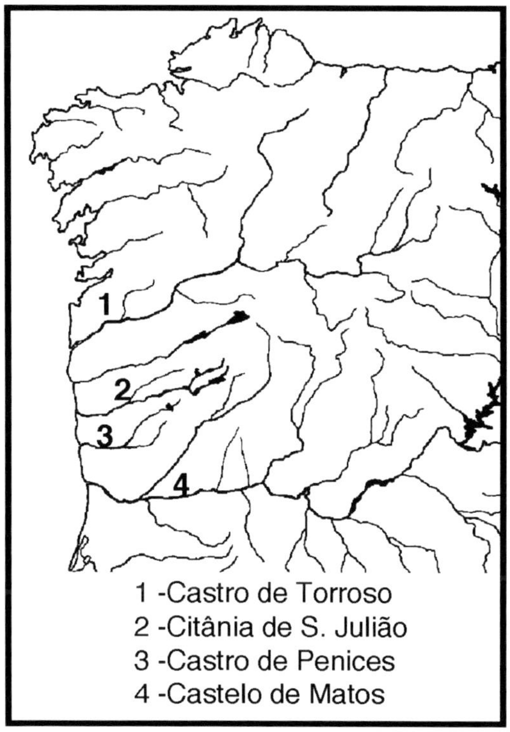

a) -Castelo de Matos is an example of a Late Bronze Age fortified site that was abandoned by the end the period. It cohexisted with a number of open settlements. The results of excavations conducted on this site are available (Queiroga 1984, Figueiral - Queiroga 1988, Queiroga - Figueiral 1992), although some of the information cited here is still unpublished.

b) -Citânia de S. Julião is a defended site which was occupied continuously from the Late Bronze Age into the early Iron Age. It has been extensively studied by Martins (1988b) in recent years.

c) -Castro de Penices and Castro de Torroso[2] represent complementary examples of a typical early Iron Age defended site.

[2] - Although Castro de Torroso is located outside the study area, it is still in the Minho river valley. Moreover, it has been included here to illustrate a model as there are similarities with other sites in this region.

Attention should be drawn to the fact that the examples illustrating these three models overlap chronological to some extent. Both cultural and radiocarbon dating are still inaccurate for this period. For example some types of artifact such as fibulae, glass beads or pottery, can remain in use for a long time, and radiocarbon dates are still too few for each site to provide any definite clarification of this matter.

The question of the uniform development in the transition between the Late Bronze Age and the Castros Culture has been raised (Almeida 1983a, Eiroa 1980, Maluquer de Motes 1975a) and the cultural and chronological mechanism reviewed (Calo - Sierra 1983). New data are emerging from excavations at sites where populations had settled and built defensive structures during the final phases of the Bronze Age. Although many sites have produced pottery typical of this period, such as the *Alpiarça* and *Baiões* types[3], in substantial quantities, such vessels have rarely been found in connection with contemporary defensive structures[4]. Such factors must be taken into account when formulating models, either in a regional context (Martins 1987=1990) or as part of a wider approach (Martins 1988a, Silva 1986a), which might explain the emergence of fortifications and their continuity with the types of castro. Theoretical discrepancies, arising from the very different approaches to this subject by Northwestern archaeologists, have been reviewed and summarized by Jorge (1988b), although the burning question as to whether or not there is cultural continuity from the Late Bronze in to the early Iron Age is still under discussion.

In the following sections, I will attempt an archaeological characterization of the three types of defended sites, i.e. late Bronze Age, transitional and early Iron Age, and define any pattern of development between these stages using the evidence currently available.

6.1.1) The late Bronze Age sites

Although the existence of earlier defended sites has been suggested, on the evidence from just one site (Jorge - Sousa 1980, 129-30), the first known defensive structures in the Northwest date to the final phase of the Bronze Age. These defended sites are usually located on hilltops, which have good natural defences and a commanding view over the surrounding landscape.

[3] - Almeida-Soeiro 1980; Almeida *et al.* 1981; Almeida, CAB, 1985a; Martins 1988b; Queiroga 1984; Silva 1986a; Soeiro 1981.

[4] - Figueiral-Queiroga 1988; Martins 1988b; Peña Santos 1987; Silva 1986a.

Castelo de Matos is a small site (Figure 27.1 and 27.2) located on the top of a hill spur, 894 metres aabove sea level and overlooking the Aboboreira massif and the Ovil river valley. People clearly settled in such an inhospitable, yet magnificent, place for its strategic potential. Excavations were conducted here between 1982 and 1986 (Queiroga 1984, Figueiral - Queiroga 1988) when a small area at the eastern edge of the platform was uncovered.

Few structures dating to the Late Bronze Age now remain because the construction of a Medieval castle on this hilltop destroyed most of the earlier evidence. However, we can see that the edge of the natural platform was strengthened in the Bronze Age with a compact mass of stones which formed the base for a palisade, whose course may be traced in the circles of charcoal along this edge. Nearby are the remains of a hut floor made with clay, which was evenly fired to increase its resistance. The diameter of this hut, which was built in perishable materials, was about 2 metres.

Apart from the evidence for fortification, the finds constitute the most interesting feature of this site. Among the pottery, the existence of the local and imported wares is suggested (Figueiral - Queiroga 1988). The Alpiarça and Baiões types the are most diagnostic of the important wares. Evidence for metallurgical practices occurs in the form of clay moulds, the fragment of a sword or daggers (Figueiral - Queiroga 1988, Fig. 10), and two bronze arrowheads. The extent of long distance contacts is reflected by the presence of one amber bead (Figueiral - Queiroga 1988, Fig. 11) which is a rare product (Ferreira 1966) on the western Iberian fringe. This unique find may have come from Scandinavia along the western sea ways, in part exchange for metals.

The stratigraphy and the characteristics of the sediments suggest that the period of occupation was relatively short. Radiocarbon dates[5] were obtained from samples of charred seeds and wood. From this time, it seems that the site (*c.f.* illustration below) was probably occupied from circa 1187 to 633 cal BC.

Contemporary open settlements which are located on a nearby plateau (Aboboreira) are now reasonably well known (*in genere* Jorge 1988a). The main differences, apart from the defences, include the fact that defended late Bronze Age sites systematically show huts paved with hardened clay[6], whereas the huts in contemporary open sites are either entirely post-built or pits-granaries[7] with no solid pavements.

[5] - More detailed discussion of these dates is presented in the introduction to Appendix II.

[6] - Cf. S. Julião (Martins 1988b) and Castelo de Matos (Figueiral - Queiroga 1988), in addition to some other sites from the period of transition.

[7] - Such are the references to Bouça do Frade (Jorge 1988a) and Lavra II (Sanches 1988).

This illustrates the transitory nature of the open sites, although such sites may have been interded for temporary or seasonal occupation.

6.1.2) The sites with continuous occupation

Citânia de S. Julião is a castro with evidence of occupation from the Late Bronze Age onwards. This site was studied by Martins (in particular 1985a, 1990=1987, 1988b), who conducted excavations there for over five years. It is located on an hilltop (Martins 1985a, 218; 1988b), 297 metres above sea level, with a good potential for natural defence (Figure 28.1) and privileged position to control the adjacent Cávado river valley.

The occupation dating to the Late Bronze Age was concentrated on a terrace (circa 30x20 metres in plan) at the very top of the hill. The edge of this terrace (Figure 28.2) was bounded by a line of stones, in addition to one ditch and one rampart (Martins 1988b, 133), and probably also by a palisade. The dwelling structures were confined to the interior of this enclosure. The huts were circular in shape, about 5 metres in diameter (Martins 1988b, 135), and built with perishable materials, although some were reinforced with a line of stones along the base. The huts had pavement floors made with fired clay, and there was probably a fireplace in the interior.

The wide range of ceramics included both local and imported wares. Metal implements were found (two bronze daggers), but no evidence for metallurgical activities.

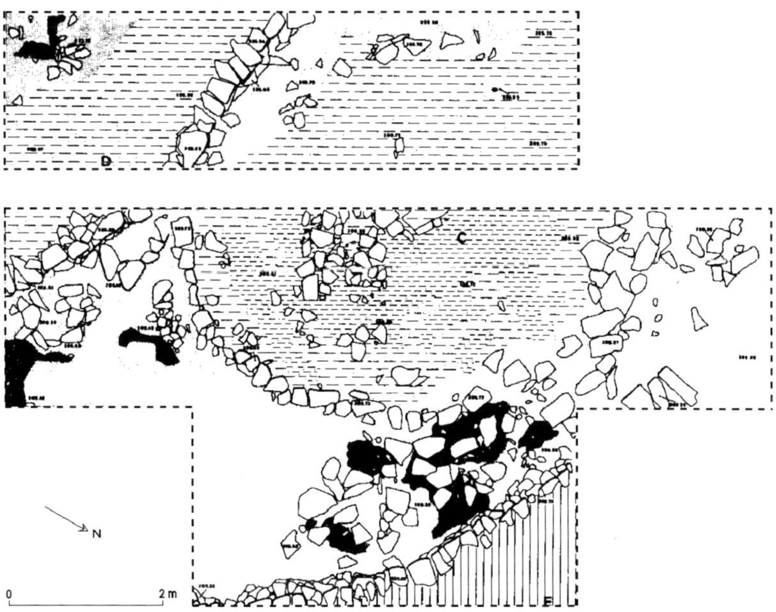

Circular dwelling structure dating to the first phase of Citânia de S. Julião (source: Martins 1990)

The material culture reflects extensive contacts with other regions (Martins 1985a, 219), such as the centre and south of Portugal.

During the next phase (phase IIA; Martins 1988a, 158-78) of occupation two simple defensive walls of granite were built. This indicates a growth of the settlement area in addition to increasing complexity in site structure. This phase also witnessed the emergence of the first stone-built dwelling structures.

The most interesting new aspect of the material culture is the widespread production of micaceous pottery at this time.

The date range proposed for this phase lies between the VII/VIth and the end of the IInd centuries BC (Martins 1990, 112). In spite of being a rather long period, it is notable that the site was continuously occupied from the Late Bronze Age to the end of the millennium.

6.1.3) The early Iron Age sites

Two examples can be used to illustrate a pattern of development at the beginning of the Iron Age in sites with no previous evidence of occupation. The best example is Castro de Penices, although Castro de Torroso will also be cited where appropriate.

The data collected from Castro de Penices are the result of recent excavations which have not yet to be published. A brief summary of the results is given here, since a full presentation of the available evidence lies beyond the scope of this study.

Penices is a small site (Figure 32) located on a spur overlooking the Este river and its alluvial valley. Although very well defended by natural and artificial means, this site is close to the fertile valley.

Evidence dating to the first phase of occupation shows that the site was enclosed by a granite-built defensive wall (circa 3 metres wide), consisting of two block-built faces filled with rubble in between. Three ditches and three ramparts reinforce this fortification on the accessible side of the spur, although they may not all be contemporary.

Dwelling structures of this phase were built in perishable materials and their floors were paved with fired clay. These huts were circular (Figure 33) and some contained hearths. Very few sherds of the initial occupational phase had mica inclusions in the sandy fabric, although the fabric became increasingly micaceous as the period progressed.

Metallurgical production is suggested by the presence of smelting residues in soil samples, by one bronze bar and several bronze ornaments.

Extensive soil sampling has revealed the carbonized remains of seeds, which indicate that a wide range of species was cultivated, especially cereals and pulses.

According to the few radiocarbon dates available for the earlier layers (Appendix II), this phase could have lasted from circa 968 to 380 BC.

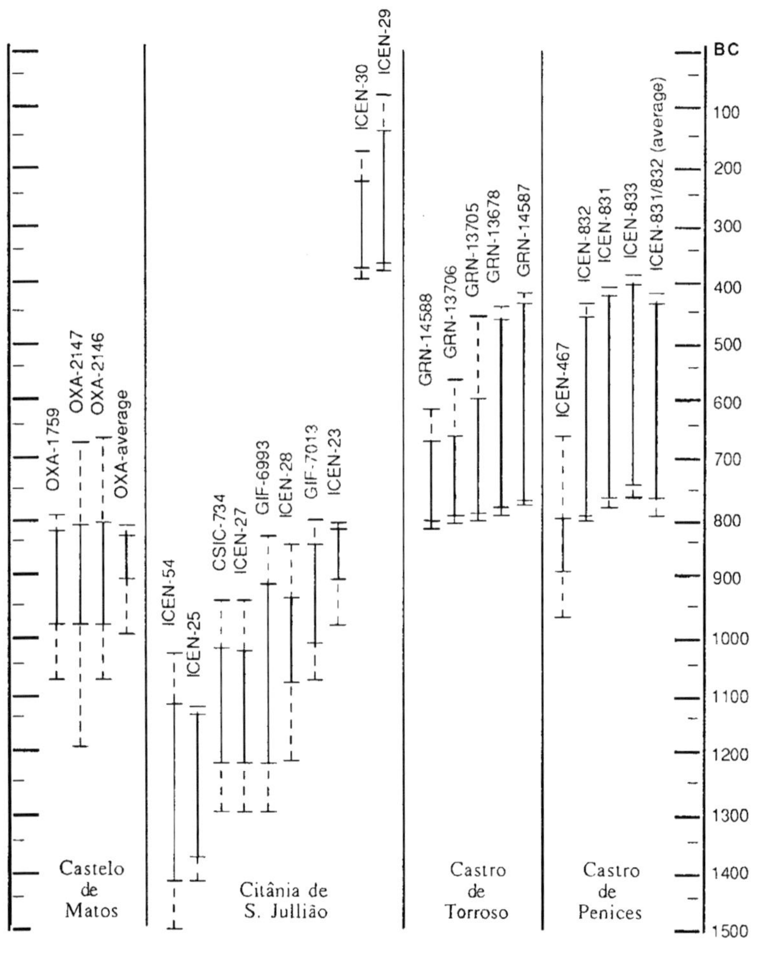

An interesting point concerning occupation at this site is that the settlement area was not enlarged throughout the millennium, although some repairs were done on the particularly unstable defensive wall.

The Castro de Torroso lies in Galicia, not far from the north bank of the Minho river, and has been studied by Peña Santos (1987, 1988). This site is located in a strategic and naturally defended position. The settlement area has an oval plan, of circa 130 x 160 metres (Figure 29), inside which two terraces can be seen. The more vulnerable side is reinforced by a defence of three ditches, and the edge of the dwelling area was probably originally encircled by a palisade (Peña Santos 1987, 120). Dwelling structures of circular plan were defined by stones. Inside these huts, which were built with perishable materials, the remains of hearths are still visible. An examination of the stratigraphy indicates an increasing use of stone in the construction of domestic dwellings.

The pottery is in keeping with the Late Bronze Age wares of the region (Peña Santos 1987, 123). Among the metal finds (Peña Santos 1988), bronze utilitarian and decorative artefacts and an iron sickle are the most relevant.

The available radiocarbon dates[8] indicate a period of occupation between the VIIIth and Vth centuries BC.

These examples illustrate both an increasing trend towards the defence of a settlement through the period studied, and a greater strength and complexity in the fortifications. The investments of time and labour necessary to effect such changes imply that these settlements and their contents were worth protecting. The emphasis of the following Chapter is thus on a detailed examination of the subsistence economy and the activities either practised in, or controlled by, these settlements, which gave rise to such defensive structures.

6.2) The defensive structures

The defensive structures can be considered to be the most obvious archaeological indicator of warfare amongst the castros communities. The defences can be divided into two major groups: the walls and palisades, and the earthworks. The palisades cannot be considered typical, since they can only be found in the Late Bronze Age, in a phase of the formation of Castros Culture, as discussed in previous sections.

An analysis of the defensive structures should start by clear definition of the methods and concepts of analysis. The sites where defensive walls have been uncovered and dated are still very few. The ditches and ramparts are virtually unknown, and the best example is still Castro de Carvalhelhos (Santos Júnior 1973a), although their chronology remains uncertain. The surface analysis is the only method of

8 - Excluding the GAK dates, as discussed in Appendix II.

study for the overwhelming majority of the sites which were not excavated.

The description of the defensive structures, when based on surface topographical analysis, has raised a substantial number of ambiguous descriptive concepts which may prove to be misleading when it comes to correlating the surface observation with the actual archaeological layout. Therefore many references made in the past to the number of defensive walls of a given site do normally include terrace walls.

6.2.1) The defensive wall: military and social significance

The currently available data indicates that the regional and diachronic variability on the structure of castros' defensive walls prevents any attempt to establish a typology, which may be premature, given the present state of research. An exception could probably be made for the structures built from the I[st] century BC onwards, which are better known to date. However, some tentative conclusions can be made, bearing in mind that they probably cannot be generalized.

As it was argued above, the beginning of the Castros' period is characterized by the first stone built defensive structures. These have been found at Neixón (Acuña Castroviejo 1976), Castro de Penices and Côto da Pena (Silva 1986a), all showing a relatively uniform typology (Figures 32 and 34.1) and chronology.

The evolution of this structure in the second half of the millennium is still unclear, in spite of the recent attempt (Silva 1986a, 32-3) to establish a typological model as based on evidence from the Entre-Douro-E-Minho.

One most common building pattern appears to survive throughout the millennium. It is characterized by two parallel walls, the space between them being filled with rubble[9] (Figures 32 and 39) or dry stone[10] (Figures 31.2, 34.1, 35, 37). A more rare variation within this pattern is provided by Castro de Terroso (Figure 34.2) where the walls are separate and the inner space filled with soil (Pinto 1932; Silva 1986a, 30). The most significant development within this pattern, between the middle of the millennium and the end of the I[st] century AD is the increasing use of more elaborate masonry. Therefore, the stones are more carefully cut and arranged, in order to produce a better appearance of the surface and a more solid wall.

Only a limited number of defensive walls were excavated, and therefore a large majority of the studies are based on surface evidence, namely the archaeological mapping studies, where the continuous lines of slopes are normally interpreted as surface evidence for defensive walls. The experience arising from extensive field work has shown that this kind of analysis may be misleading. The designation 'defensive wall' in surface analysis should only be used when it actually encloses the whole settlement, or a substantial part of it, as sometimes it may have worn out from the slope due to erosion or quarrying, or even may never have existed due to sufficient natural defences. Similar enclosures, which define terraces inside the main defensive walls, should not be interpreted as defensive structures but rather as terrace walls. They can be sometimes very extensive, and their function is to create a terrace.

The existence of massive stone-built defensive walls raises the question of quarrying sources and iron tools (Hawkes 1984, 188), manpower for their construction, and earthwork techniques. In fact, the Northwestern Atlantic fringe is characterised by a granitic relief where an immense mass of rock emerges from the ground, in particular in the zones more exposed to erosion such as the hills and mountains. This obviously creates a good accessibility to the stone sources, and it is often necessary to clear the stone in order to build a castro, and therefore plenty of stone will be available. Furthermore, the amount of local raw material is so impressive in an overwhelming majority of the cases, that the castros' interior areas still show massive quantities of stone in the form of large round monoliths above the ground level. Other than the clearance of the surface rock during the building of terraces (Almeida, CAB, 1990a, 50), the digging for the ditches could also provide stone and soil for both defensive and terrace walls, as well as for the ramparts.

The defensive walls, other than strictly for defence, as it has been currently assumed, also produced the terraces which enabled a better organization of the dwelling units. This, so to say, double architectural function has been outlined already for surface surveys (Almeida - Neiva 1982, 8; Almeida - Baptista 1984, 105) and cumulatively demonstrated (Queiroga 1983, 10, fig. 5) by the excavation of a defensive wall (Figure 31.1).

The other important function of the defensive wall is the one of restraining populational growth. In fact, there is a balance between the available resources within the site catchment area and the size of the community. An abnormal growth of the latter may cause economic and social disfunction unless it is accompanied by substantial changes of the system. Any population growth has to fit within the

9 - A mixture of stone and soil, among which remains of previous occupations can be found, as seen at Penices.

10 - Such is the case of Castro de Sabroso (Cardozo 1958c and 1959a; Hawkes 1958; 1971; 1984, 192-3; Soeiro et al. 1981), Briteiros (Silva 1986a, 32), S. Julião (Martins 1988c; 1990, 140), Barbudo (Martins 1989; 1990, 151-4), Lago (Martins 1986b) and Ermidas (Queiroga 1985), among other known in the region.

enclosure, by the reorganization of the inner space, or be split off and a new site created. Significant support for these suggestions can be found at Castro de Penices, where the same perimeter is maintained from the foundation of the site, in the early Iron Age, to its abandonment in the mid Ist century AD.

However, this model can be true for many settlements, it cannot be generalized, since many castros have expanded their area and built new defensive walls by the end of the millennium. Yet this growth was possibly due to increasing centralization and the use of new economic resources, such as mass production, regional trade and hierarchy. The development of these castros is only possible by their role as central places, and their growth will in some cases cause the abandonment of smaller sites by the mid Ist century AD.

Explanatory model for the construction of the defensive structures of castros. Examples from the excavated sites indicate that this system has been used from the oldest phases to the end of Castros Culture.

The defensive walls, being a communal structure, were also an indicator of the prestige of the inhabitants of the castro. Their construction demanded such a substantial manpower and economic surplus that it is difficult to accept that they were built by exclusively using the resources of the community who lived in the site. This impression is even more striking when taking the small sites as examples, in that a collaboration between sites for the construction of the defensive structures, even for the low-status sites, should be considered.

The defensive wall is complemented by other structures, namely the ditch and the rampart.

6.2.2) The ditch and rampart

The ditch and rampart are the best indicators of the defensive preoccupations of the Castros communities. Unlike the defensive walls, they do not represent boundaries, and usually are carefully designed according to the topography of the hill. The ditches and ramparts of castros are a reinforcement of the defences as a result of the unfavourable conditions due to the configuration of the spur. They are always built in parallel and adjacent to the exterior of the defensive wall, thus forming a barrier between the site spur and the main massif.

The ditch and the rampart are defensive earthworks which are normally associated. This association does not exclusively relate to defensive techniques, but also to the particular way these structures are built. The ditch is a large trench dug out in the ground, from where earth and stone are taken. The stone is preferably used for the construction of the defensive wall, and the earth is disposed alongside the edge of the ditch, thus forming the rampart.

Evidence from the many sites visited in the course of this study indicates that the number of ditches and ramparts varies much according to the type of site, with its chronology, and particularly with the topography. Castros of type 1A (Figure 23) do not have defensive earthworks, since the hill presents an even steepness on all the sides. The defences of sites located on spurs are, however, almost always reinforced by earthworks. Yet, these seem to have been neglected in the large castros of the later phase, such as Mozinho, Sanfins, Briteiros and Eiras, to the detriment of the carefully built and massive defensive walls.

6.3) Discussion

It is a generally accepted fact that the final phases of the Bronze Age in the Atlantic fringe, and in particular in the Iberian Northwest, bear the dominant characteristic of a massive production of bronze artefacts. These have been found in variable quantities and types, mostly in land hoards and river deposits, and more rarely in the sea and uncovered from sites during the course of archaeological studies[11]. This occurrence leaves a gap between a highly important cultural fact, the late Bronze Age metallurgical productions, and the peoples who have executed it, the Northwestern late Bronze Age communities. The existing data may allow some thought on this subject, as recent finds of metallurgical activities in a few sites do now allow preliminary correlations, in order to establish a proper integration of a rather substantial amount of stray finds.

Work done in recent years is indicating that the late Bronze Age metallurgy was produced, and certainly also distributed, by the communities established in defended sites[12] located in hilltops from the dawn of the Ist millennium onwards. These, however, did not necessarily constitute the only type of settlement at this time, at least not in the first centuries of the development of the system. Communities living in non-defended settlements, such as those of of the type of Bouça do Frade, Baião (Jorge 1988a and 1988c), might have coexisted for a long period within the system, as producers of food and metals.

The end of the Bronze Age brings a multiplication of hillforts, which is accompanied by the disappearance of the open sites. By the final

11 - As far as this latter situation is concerned, the deposit from Senhora da Guia, Baiões (Silva *et al.* 1984; Silva 1986a 165, foll.), is undoubtedly a model case for its typological importance and conditions of unearthing.

12 - The significance of Late Bronze Age hoards bearing evidence for metallurgical activities, such as the one from Viatodos (Fortes 1905-08b) is not yet clear, although the suggestion that they correspond to itinerant metallurgists has been widely defended.

stage of the Bronze Age only defended sites existed. They were built with perishable materials, but later stone begins to be used for the building of the most important communitarian structures: the defensive walls. By this time, around the VII/VIth centuries BC, this important architectural development is accompanied by the first productions of intentionally micaceous pottery, and this interface indicates the beginning of the Castros Culture.

From this period on, all the constructions are gradually built with the locally available stone, and the internal organization of the settlement becomes clearer.

Closer to the change of millennium, probably by the III/IInd centuries BC, there is a renewal of the defensive walls of some sites, and they are equipped with more careful masonrytechniques, nevertheless, the available evidence does not suggest any generalized enlargement of the sites. The reinforcement and reconstruction of the defences will continue from this period until the end of the Castros Culture.

A particular remark should be made regarding the turn of millennium, when the first large nucleated settlements emerge.

The enlargement of the utilitarian space of the house for cooking[13] through the construction of a vestibule, which seems to be widespread during and after the change of millennium, is a clear indicator of the population growth, or at least of familial growth.

From this period onwards, some castros seem to develop towards major political and trade centres, presumably working as local branches of the capital of the *conventus*. One of the main doors at San Cibrán de Lás (Eiroa 1988, 111; Pérez Outeiriño 1987)[14], by its very particular structure and the existence of benches in the interior (Figure 38), could have been a toll gate (Almeida pers. comm.) for the city market. This increasing centralisation reflects not only a economic reorganization and renewed strategies of occupation of the space, but principally a growth of the population which is also shown by the new enlargement of the dwelling spaces.

The defensive walls certainly had an important role in the defence of the sites, although this may not have been their exclusive function, since they also served as a means of display of the community's prestige. In addition, the defensive walls were also important to confine the size of the castro's population within a certain limit, thus creating a balance between the dwelling space, the number of inhabitants, and the locally available subsistence resources.

---

13 -The terminology "house for cooking" intends to outline the difference between this and the house-dormitory, as explained above.

14 - More precisely the eastern gate of the inner defensive wall.

7- THE PRODUCTIVE ACTIVITIES

The body of data necessary to produce a coherent picture of the environment and the productive economy in the region is still scarce and fragmentary for the first millennium, since most of the available data refers to the end of this period. In the present study, I will attempt to correlate and evaluate the environmental evidence, the use of resources and territorial occupation within a cultural background. I will show that the Classical sources, although often sketchy and open to criticism, can often be highly informative and as able to give us a wealth of other detail. Among more recent studies, I will cite the important work of Vázquez Varela (cf. bibl.) with reference to the Castros Culture, alongside the articles of various other authors. Finally, I will draw on my own research in the area of Vila Nova de Famalicão, where a great deal of attention has been paid to environmental studies.

The main objective of this chapter is to establish the major characteristics of the Castros productive system by using the available data. A further goal is to present a new approach to the subject by re-interpreting both the Classical sources and earlier fieldwork, and incorporating the most recent data. In this way it is hoped that a realistic view of the Castros productive system, its use of resources and inter-site relationships, will be obtained. The results of this approach will be used to determine whether or not raiding, and warfare in general, can be explained in terms of the productive system.

7.1) -Food producing strategies

Production is one of the essential components of an economic system. However, the concept of '*productive activities*' or the '*productive economy*', should not be mistaken here for an economic system. The latter is far too complex, and such a study would be a research subject in its own right. On the other hand, the separate elements which contribute to a productive system, referred to here as productive activities, may well be relevant to the understanding of social interactions in the Castros Culture of northwestern Portugal.

7.1.1) -Evidence from Classical sources and its interpretation

References to the productive activities of the peoples of the Iberian Northwest are, as a rule, rather vague and scattered throughout the works of various Classical authors. Most of these early authors merely offer us a personal account of what had been previously reported by their predecessors. These reports have often been used indiscriminately, and at other times have been reduced either to literary lyricism, or to political statement. Recent data, however, appear to have restored some credibility to these Classical sources. Whether or not they reliabily reflect the times and places they represent has yet to be confirmed by the archaeological evidence.

The establishment of links between the archaeological evidence and Classical sources, and criticism in general, will be attempted in this chapter, and transcriptions frequently used to outline the importance of the various sources.

Strabo's Geography, book III

3,5 -..."*Now about thirty different tribes occupy the country between the Tagus and the Artabrians, and although the country was blest in fruits, in cattle, and in the abundance of its gold and silver and similar metals...*"

3, 7 -"*All the mountaineers lead a simple life, are water-drinkers.... They eat goat's-meat mostly, and to Ares they sacrifice a he-goat .... And the mountaineers, for two-thirds of the year, eat acorns, which they have first dried and crushed, and then ground up and made into a bread that may be stored away for a long time. They also drink beer; but they are scarce of wine, and what wine they have made they speedily drink up in merry feastings with their kinsfolk; and instead of olive-oil they use butter.... Their rock-salt is red, but when crushed it is white.*" ...

4, 17 -"*...For example, these women till the soil, ... and while at work in the fields....*"

4, 18 -..."*in Cantabria...there was a scarcity, not only of other stuffs, but of grain too; and only with difficulty could they get supplies out of Aquitania on account of the rough roads.....for the daughters to be left as heirs, and the brothers to be married off by their sisters. The custom involves, in fact, a sort of woman-rule -but this is not at all a mark of civilization....*"

Justinus

XLIV, 3, 7, 10 "*feminae res domesticas agrorumque culturas administrant, ipsi (homines) armis et rapinis serviunt*".

Pliny, *Naturalis Historia*

XVI, 15 - "*...Acorns at this very day constitute the wealth of many races, even when they are enjoying peace. Moreover also when there is a scarcity of corn they are dried and ground into flour which is kneaded to make bread; besides this, at the present day also in the Spanish provinces a place is found for acorns in the second course at table. Acorns have a sweeter flavour when roasted in the ashes...*".

On the whole, most Classical sources present the idea of an economy based on the herding of goats, sheep and horses, with grain cultivation supplemented by acorn gathering. Rich veins of metal ores, particularly gold and tin, were exploited for local metallurgy and trade.

Several descriptions (Strabo *e.g.* above, Dio Casius 37, 52-3), refer to the existence of widespread herding activities from very early times. Iberian horses, especially the Gallaican, Asturian and Lusitanian breeds, were famous in the Roman world (Appian, Iber., 75, Pliny VII, 166, Silius Italicus 3, 335-7 and 16, 583, Strabo III, 4, 15). The use of horses in this region has also been documented archaeologically. They can be seen, along with horsemen, on the rock carving of Sanfins (Jalhay 1947), the gold diadem of San Martin de Oscos[1] (Lopez Monteagudo 1977), the decorated pottery sherd from the castro of Faria (Barcelos) (Martinez Santa Olalla 1948) and on the decorated vessel from the castro of Santo Ovídio (Fafe) (Martins 1981).

According to Strabo and Pliny (*c.f.* above), acorns were gathered as a very important source of food, although other fruits might also have been used. Strabo suggests that these perishable foodstuffs were preserved and stored, because their consumption took place over at least two-thirds of the year. A reference to beer (undoubtedly of local production) implies the cultivation of some cereals (Dio 53, 29). The use of acorns in making beer in the Castros Culture, thus following a habit well-documented elsewhere (Renfrew, J., 1973, 154; quoting Howes 1948, 172), supports the archaeological evidence for the widespread use of this nut over a long time span. The consumption of wine was clearly not common and was a luxury. Mention of this product could thus be an indication that it was imported from the Mediterranean.

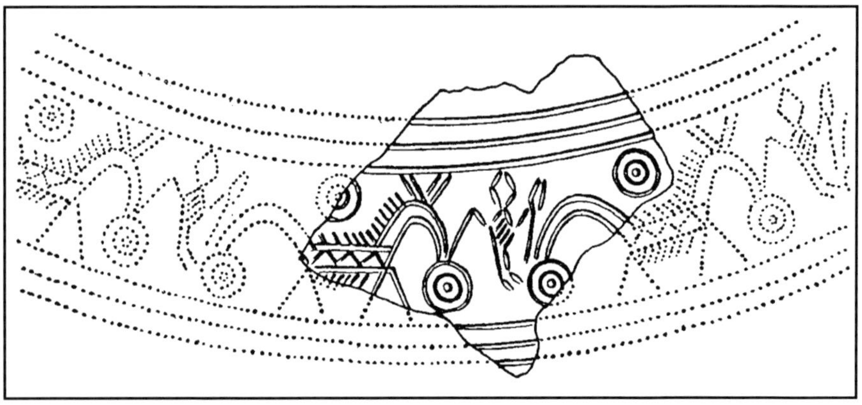

Decorated pot sherd from Castelo de Faria (after Martínez Santa-Olalla 1948)

At present, there is no evidence for rock-salt in the Northwest, and perhaps Strabo's reference applies to sea-salt instead. This is indeed supported by Strabo’s mention of the red colour. This colour is due toa micro-organism which lives on the surface of man-made salt pans.

The fact that agricultural tasks were carried out by women (Strabo *e.g.* above, Justinus 41, 3, 7; Silius Italicus 3, 346-53) is supported by the contemporary belief that Castros agriculture was relatively underdeveloped, and contributed to the peculiar social organization of the culture. The last reference from Strabo (III, 4, 18) is of the utmost importance as it implies that, apart from agriculture and probably gathering, the house and its fields were owned by women, indicating their almost complete control of the productive economy.

7.1.2) -Evidence from archaeology

The Castros food producing system has been the subject of several recent accounts (Blanco Freijeiro 1960, Blázquez 1968, Caro Baroja 1974, 1944 and 1976, Gomez Tabanera 1975 and 1980, López Cuevillas 1953b and 1966, Maluquer de Motes 1968, Taboada Chivite, 1977[2]), which are based mostly on the accounts in Strabo's Geography, and demonstrate a palaeoethnographic approach to the economy. At the same time they deal with sparse and all too often unreliable[3] archaeological data.

The northwest of Portugal belongs to the communitarian-agrarian region of the western Douro valley. It was first defined by Caro Baroja (1943 and 1944) and further described by Blázquez (1968) and other historians (c.f. above), who studied the economy of pre-Roman Iberia. These authors built a model in which the economy was based on goat-herding and a *hoe-agriculture*, with no use of the plough (Almeida 1983b, 194), i.e. a type of garden horticulture rather than field cultivation.

More important is the acceptance by all the authors that this economy was inefficient, leading to poverty and a need for male brigandage in neighbouring regions. This too is fully supported by the Classical references.

[1] - Also known as the diadem of Ribadeo.

[2] - This author (op. cit. 72) already points to the subsidiary role of gathering in the Castros economy.

[3] - The data these studies rely on were produced at a time when archaeological methods did not appreciate the relevance of stratigraphy. Excavations were mostly done by ‘spade and pickaxe’, with little reference to any verification of soil type, context, or disturbance.

Recent palaeoeconomic and palaeoenvironmental work[4], based on archaeological data, has given us further relevant contributions. However, these remain few and restricted to only a small number of sites, so that no quantification is possible. These new data have been used for a complete revision of the economic system (Bintliff 1981), alongside other evidence. A hitherto unsuspected economic complexity has resulted, which was unobserved in the Classical sources alone.

7.1.2.1) -Agriculture and gathering

Recent archaeological research has changed the view about the productive economy of the peoples from the Iberian Northwest expressed in the Classical sources, which stresses the primitive farming techniques (Blanco Freijeiro 1960, 183; Taboada Chivite 1977, 73; Tranoy 1981, 93)[5] and economic underdevelopment. The emergence of Palaeo-environmental studies in northern Portugal, following the example set elsewhere in Europe (Ganderton 1981), was crucial in changing this traditional picture. In fact, until very recently, no soil sampling for environmental and economic analysis (Jarman *et al.* 1972) had been carried out. Although seeds were kept from excavation, a quantitative approach to the reconstruction of agricultural activity (Ede 1988, 71) was obviously impossible, in spite of Pinto da Silva's arguments (Silva A.R.P 1988, 17).

The evidence indicates a very neat involvement, and probably control, of women in both the productive systems and familial and social interactions. In fact, according to the Classical sources, they did 'till the soil' (Strabo III, 4, 17) and certainly carried out the domestic tasks, but so far there is no reason to assume that they were involved in pastoralism.

Acorn gathering seems to have been a very important subsistence activity from early times until the turn of the millennium and, as Pinto da Silva (Silva, ARP, 1988, 18) suggests, edible plants can be kept in the wild state to produce substantial regular crops, without any alteration of their morphological characteristics. Accordingly, the existence of gathering activities does not necessarily mean that we are in the presence of a undeveloped economy, or that we are looking at times of famine, but rather that this was a pre-cereal, or pre-Mediterranean[6] , economy. Besides, gathering should be understood her as a supplementary activity[7], given the various economic resources listed below.

The association of acorn use with castros has been pointed out by the Classical authors[8] and confirmed by archaeology[9], although no systematic research has been done so far. In a recent study (Oliveira *et al.* 1991), an explanation of the process of acorn use for human consumption during the Iron Age was attempted, in order to understand its role and significance in an economy such as the Castros Culture.

The reason for the disappearance of acorns from sites as wheat becomes more common, around the turn of the millennium, has been attributed to Roman influence in the administrative, social, and agricultural areas (Almeida 1986, 161). Reliable archaeological evidence for this change lies in the abandonment of saddle querns at this time (Almeida 1984, 39; Oliveira *et al.* 1991; Silva 1986a, 112), and in the fact that circular querns are not suitable for acorn grinding (Oliveira *et al.* 1991)[10]. This argument could easily be substantiated by an extensive study of seed remains from dated contexts, which would make a useful project for the future. On the other hand, the lack of gluten in acorns makes it necessary to mix acorn flour with some cereal flour for bread-making (Renfrew, J, 1973, 98, 191). However, as Cardozo (1962a, 413) suggested, could the Castros acorn bread have been different to our idea of bread? Besides it may have been difficult to keep acorns for a long time (Renfrew, J, 1973, 194), as

---

4 - *In genere*, the work of Guitian Ojea, Silva ARP, Vazquez Varela, Santos Yanguas, and, to a lesser extent, some suggestions from Silva (1986a, 111-15) awaiting proper publication of the archaeological data they are based on. We owe a special remark to Vazquez Varela's admirable studies on Castros palaeoeconomy, not so much for the depth of the conclusions, as for his use of appropriate methodology.

5 - *In genere* all the studies referred to above are based on the economic analysis of the Classical sources.

6 - Bearing in mind that the typical elements of the Mediterranean economy, namely bread, wine and olive oil, only began to appear after Augustus.

7 - Taboada Chivite (1977, 72) stressed the 'subsidiary role' of the 'predatory economy' elements such as 'hunting, fishing and gathering'. More recently, this position has been reviewed by Vázquez -Pombo (1991a, 124-5; 1991b, 197), drawing upon the complementarity of both 'productive' and 'predatory' activities.

8 - Strabo III, 3, 7 - "*And the mountaineers, for two-thirds of the year, eat acorns, which they have first dried and crushed, and then ground up and made into a bread that may be stored away for a long time.*"; Pliny XVI, 15

9 - Many castros have produced carbonized acorns, but there has been no serious attempt either at quantification or at relating the find to a specific archaeological phase. For the evidence of carbonized acorns at sites: see Cardozo 1959e; Silva A.R.P. (in Santos - Isidoro 1963) and 1976; Vazquez Varela 1974-75, Taboada Chivite 1977, 74; Silva C.T. 1979, 17; Silva 1986a, 111 (footnote nº3 for a list of sites where acorns were found). Over five hundred acorns were also found on the fireplace of an Iron Age house (Almeida, CAB, 1990a, 59) at Facha, Ponte de Lima.

10 - Evidence suggests that the replacement of acorns by wheat and millet, in connection with an increase in masonry work, was responsible for the spread of round querns in castros by Augustan times, and it was not a side effect of the technological spread of the use of the wheel, which Silva (1986a) interprets in close connection with the vulgarization of the potter's wheel at this time. In fact, the former is the result of a new diet, and the latter corresponds to technical developments and mass production in pottery manufacturing.

conditions in the round *granaries* of the castros would have facilited the oxidation and subsequent rotting of acorns. Also, pits with good biochemical storage conditions for large quantities of grain are not known in castros[11] (*in* Cunliffe 1984).

Pollen diagrams from Galicia show an expansion of chestnut early in the first millennium BC (Aira Rodriguez*et al.* 1989; Torras Troncoso *et al.* 1980, 54). Thus, chestnuts may have become an increasing part of the human diet during the Iron Age. Carbonized wood from this species (Figueiral 1990, 98) has been found at three sites from the late Bronze Age to the Iron Age. However, the almost complete absence of carbonized chestnuts in the archaeological record is rather enigmatic. The woods of walnut, fig, pear, cherry and plum have also been identified in castros (Figueiral *ibidem*), and the use of their fruits in the diet of these communities should be considered.

By contrast, during the course of research at Castro das Ermidas and Castro de Penices, both at Famalicão, a noticeable increase in cereal production by the turn of millennium was found. In fact, the quantity of wheat in samples from contexts dating from Augustan times onwards seems to have increased considerably, and this is matched by a corresponding rise in the number of granite round querns found in the archaeological record.

Agricultural activities included the cultivation of a variety of species. The lack of any evidence for fields in the Castros Culture of northwest Portugal is perhaps not surprising, as there have been dramatic changes in soil erosion and land division in this densely populated area during the last two millennia.

The geographical configuration of the landscape does not permit the existence of large agricultural units, or fields, except perhaps in the alluvial valleys. On the other hand, the sites studied are mainly on hilltops and at inter fluves. It is probable that the fields were close to the site, i.e. would be located on plateaux, hill slopes and beside small streams. This idea is in accordance with other factors, and might even account for the emergence of a later widespread and peculiar adaptation to the type of orography found in the Northwest, namely the growth of cultivation on small terraces.

Terracing is a very effective strategy against erosion, and also a means of collecting the fertile soil washed down from the upper slopes. Soil on these terraces would, therefore, be fertilized naturally and built up in depth, and rainfall and drainage would provide plenty of water for the crops. This agricultural system would allow an intensive exploitation of the land, and more than one crop per year may have been produced, particularly under local climatic conditions. Furthermore, it offers ideal conditions for the cultivation of species such as pulses, which require fertile soils. Terracing also makes it easier to protect the crops from wild animals.

Evidence gathered in the course of my own research[12] from sites dating between the late Bronze Age and the Ist century AD indicates that a substantial percentage of *garden crops,* in comparison with cereals, were grown during this period. Incidentally, the percentage of garden crops grown was probably higher in the earlier phases, only to be progressively supplanted by cereals, which implies a more extensive use of land.

Of further interest is the spread, around the turn of the millennium, of the bramble (*Rubus* sp.), represented in the archaeological record by carbonized thorns and seeds from Castro de Penices, for example. The spread of this weed may be an indicator of the abandonment of fertile agricultural fields (Silva, A.R.P, pers. comm.), as it requires moist soils with a high humic content to thrive. The fast growth and spread of this species might also be the result of the temporary abandonment of intensively farmed fields, deemed necessary so that productive capacity could be regained. A more consistent body of evidence is necessary for a proper evaluation of this point, especially palynological data. It is unlikely that the carbonized remains of bramble found at dwelling sites will accurately reflect the bramble population in the agricultural fields, as this species was seldom burnt in domestic fires.

Castros fields lay close to the settlement, probably on the hillslopes or at the base of the hill. It is possible that some of the terraced fields found in castros were used as gardens (Queiroga 1985, 2)[13], rather than as enclosures for animals (Silva 1986, 114). One argument I believe to be of relevance for the existence of gardens, either inside or very close to the site, is based on the fact that certain species require careful nurturing and protection from wild

---

11 - A couple of pits dug in to the bedrock at Tecla (Galicia) have been interpreted, as wells (Mergelina 1943-44) and the large pits at Santo Ovídio (Fafe, Portugal) are viewed as rubbish pits (Martins 1981).

12 - Particular importance is given here to my own work on this subject, as studies done to date in the north of Portugal have not taken the necessary quantification of sampling into account. Under these circumstances, such data are unreliable for determining both the range of species cultivated and the type of agricultural system developed. Therefore, an incidental reference to the finding of a pulse at a site can be used as an indication that these species were cultivated, but it can not reveal its importance within the productive system. It should also be borne in mind that the carbonization of seeds is, as a rule, accidental.

13 - At Ermidas there are several terraces, retained by Iron Age walls, where no remains of habitation were found during the 1984 excavation. Moreover, these terraces seem far too large to be explained as animal pens. At Citânia de Briteiros there are several rectangular enclosures with no remains of houses in them.

animals damage[14]. This was in fact one of the greatest concerns in ancient Iberian farming, as much agricultural activity was directed towards pest control, and some references can be quoted. For example, the Romans offered prizes to those who caught mice, according to Strabo (III, 4, 18). Also, in the tale of the miracle of *St. Antony of Padua*, the saint jailed the sparrows in a cage while he was away so that they would not destroy the crops. In addition, Medieval peasants had obligations to hunt a certain number of bird pests for the manor house, and it was forbidden to kill birds of prey (Almeida p.c.), such as goshawk.

sample number		1773	5	7	9	1768	1770	13	15	18	16	17	22	23	25	1766	26	1769	
cereals	**English name**																		**total**
Panicum miliaceum	common millet	18	26	33	49	41	30	30	34	42	49	10	12	151	99	85	278	200	1187
Setaria sp.	millet	2	4	4	5		3	5	9	5	3	2		4	10	25	51	25	157
Setaria/Panicum		9	20	26	10	38	43	22	9	20	15	15	4	109	25	81	136	108	690
Triticum cf aestivum gr.	cf. bread wheat			2	7	3	4	3	3	6	7	4	12	14	1	15	36	21	138
fr.														3			11	4	18
Triticum sp. indet.	wheat														2				2
Triticum sp. spikelet fork								1								1			2
Hordeum vulgare, HS	hulled barley			1															1
cereal indet. grain				1					1				3	1					6
frag.		+		+	+	+	+	+	+	+	+	+	+	+	+	+	+	+	
rachis														1					1
total cereal grain		29	50	67	71	82	80	60	56	73	74	31	31	279	137	206	501	354	2202
other plants																			
Cruciferae																			
Brassica sp.	mustard															cf. 1			1
Caryophyllaceae																			
Stellaria sp.	stitchwort															2			2
Spergula arvensis	corn spurrey						1												1
indet.							2							1					3
Leguminosae	legumes																		
large,round																		1	1
small,round				cf. 1	cf. 1											1	1	1	5
squarish																		1	1
Rosaceae																			
Rubus fruticosus agg.	bramble		14		2		2												18
Polygonaceae																			
Polygonum sp. *flat*	bistort															1	1		2
Polygonum sp. *round*	"						1										1		2
Rumex sp.	dock						1										1		2
Oleaceae																			
Olea europaea	olive															2			2
Rubiaceae																			
Galium sp.	bedstraw								1										1
Compositae																			
indet.	daisy	1											9						10
Cyperaceae																			
Scirpus sp. *round*	club-rush																1	cf. 1	2
Gramineae																			
ssp bulbosum	grass											1							1
indet.	grass					1	4	1	2					1		4	6		19
total non-cereal		1	14	1	3	1	11	1	3			1	9	2		11	11	4	73

Chart referring to seed analysis of some soil samples from Castro das Ermidas, dating to Augustus-Claudius (Ede, report on seed analysis)

Our knowledge about the number of species of edible and cultivated plants used for human consumption during the first millennium has increased substantially over the last few decades.

14 - It is opportune to recall here information given by Teresa Soeiro about the present cultivation of millet (*Panicum miliaceum*) in the Penafiel region. As harvest approaches, it is necessary to tend the fields from very sunrise to sunset, in order to scare the wild birds away. If no proper attention is paid to this task the crop can be totally lost. This is usually the children's task, as their eldess are doing other work. This ethnographic aspect could be of some relevance when arguing about work division in Castros society.

The cultivated species of cereal so far known to be present in castros[15] are barley (*Hordeum vulgare L.*)[16], wheat (varieties *Triticum compactum Host.* and *Triticum aestivum L.*), millet (*Panicum miliaceum*)[17], and rye (*Secale cereale L.*)[18]. Varieties of pulse are fewer, and are represented by the broad-bean (*Vicia faba L.*) and the pea (*Pisum sativum*).

Barley cultivation might have been connected with the brewing of beer (Blázquez Martínez 1968, 195), which was widely consumed by the Castros peoples according to Strabo (III, 3, 7).

So far there seems to be no direct archaeological evidence for the cultivation of flax (*Linum usitatissimum*), although it is mentioned by the Classical sources (Strabo III, 3, 6 and Pliny XIX, 10). Linen is referred to as being widely used for clothing and warrior garments. The *Zoelae*, known to be located around the Trás-Os-Montes province, were famous for their flax production. During the course of the conservation of bronze finds from Castro das Ermidas, which date from the pre-Augustan period, special care was taken to look for organic remains (Cameron 1988, 81-8, Queiroga - Silva 1991, 169), and some evidence for mineralized plant fibres was found (Anon 1965, Ryder - Gabra - Sanders 1987), although particular species could not be identified.

It has been claimed that the olive, one of the features of the Classical Mediterranean diet, did not reach southern Portugal until the second century AD (Silva A.R.P. 1988, 15-6), although this view may soon change dramatically in the light of recent research. In fact, some evidence for the olive has been found at Castro das Ermidas[19], so that it is possible that the cultivation of this species began with Romanization. From what we know about the present cultivation pattern of this species[20], the north of Portugal has a number of micro climates within it, which would have provided suitable conditions for the olive.

Grape (*Vitis vinifera*) does not occur in the archaeological record before the second century AD (Silva A.R.P. 1988, 16). They were probably not cultivated in the Castros Culture (Silva 1986a, 112), although Taboada Chivite (1977, 75) suggests that they could have reached *Callaecia* before the Roman Conquest.

Recent work (Figueiral 1990, 124) has revealed the carbonized remains of vines in three castros, in contexts dating from the first century BC to the first century AD. It is not possible, however, to determine whether these are wild or cultivated varieties. Considering the sociological importance of wine (Strabo III, 3, 7), and the numerous finds of amphorae in the castros[21] indicating the existence of a wine trade since Republican times, this evidence would not be difficult to accept.

Grape seeds were recently found at Castelo de Matos, in a Late Bronze Age context[22]. Grape seeds have, of course, been recovered from Castro das Ermidas and Castro de Penices, in Augustan-Claudian contexts. It is, therefore, plausible that some wine production took place before the coming of the Romans, although there is not enough data to support this hypothesis. Moreover, evidence such as that of Rianxo (Romaní Martinez 1974-75)[23], showing the penetration of wine amphorae into the hinterland via the small river valleys by the turn of the millennium, reflects a continuing use of earlier routes of distribution. This might be expected when considering the consumption of goods of such wide-spread social importance.

---

15 - Most of the references below are quoted from the synthesis of Pinto da Silva (Silva, ARP, 1988, 5-36, an account of all the previous work, with references and bibliography), and Silva (1986a, 112), in addition to the results of our excavations at "Castelo de Matos" (Baião), "Castro de Vermoim", "Castro das Ermidas" and "Castro de Penices", these last three at Vila Nova de Famalicão.

16 - Barley was found in samples from "Castelo de Matos", demonstrating the early cultivation of this cereal in the region (v.g. radiocarbon dates). Other than barley, broad beans, millet, and wheat were find at this site in context dating to the late Bronze Age.

17 - Millet appears in a surprisingly large number of our samples dating from the Julio-Claudian phase, which was prove to be of some significance when assessing the degree of "agricultural Romanization" at some castros. This cereal would have been used mainly in bread, making as its latin name "panicum" suggests, although another cereal with "gluten" would have had to be added to assist the process of fermentation (Oliveira *et al.* 1991).

18 - Carbonized remains of rye have been found at "Castelo de Faria" (Vasconcellos, J.C., 1950, 37-8), although no chronological reference is available. This site was continuously occupied from the late Paleolithic until the Middle Ages. However, the association of wheat (var. *antiquorum* Heer), rye, broad-bean and millet might be consistent with Iron Age agricultural patterns, in spite of the doubts expressed by this author (pp 37).

19 - Two olive stones were found at "Castro das Ermidas" (Joy Ede, field report on seeds analysis) in 1986, in a context dating to the mid first century AD. However, it would be more cautious to wait for further evidence from charcoal analysis in order to establish whether this fact represents cultivation or just consumption of olives which may have been imported. I believe there is much to be achieved in this matter, and especially in clarifying the surprising scarcity of Dressel 20 amphorae in the Northwest, which can by no means be explained as a persistent taste for animal fats, such as butter, by the Callaicans (Strabo III, 3, 7). Carbonized olive wood, dating to the second half of the I century AD, was found at Castro of Cruito (Figueiral 1990).

20 - Cf. Comissão Nacional do Ambiente, Atlas Nacional do Ambiente, Carta de Distribuição da Oliveira, presented in Chapter 3.

21 - Silva (1986a, 112). The results of my own excavations, and a general overview of others, show that wine amphorae are found in castros, frequently in pre-Augustan contexts. The known forms are mostly Haltern 70 and Dressel 1.

22 - These seeds were found in a Late Bronze Age context at "Castelo de Matos". Radiocarbon dating (≈990-800BC) fully supports this chronology (see Appendix II for radiocarbon dates).

23 - See also González Fernandez (1983), Hidalgo Cuñarro (1982a, 1982b, 1983) and Naveiro López (1991) for evidence of coastal trade in Galicia.

The very early presence of the wild grape has been recorded in various areas of southern Europe and Iberia. Buxó i Capdevilla (1988, 42) refers to its presence on Catalonian sites from the Copper to the Iron Ages, and at El Prado (Murcia). Evidence for a cultivated variety has been found in the Iberian South (Walker 1985, 44-5) as early as the third millennium BC. Vine cultivation seems to have spread through the Aegean during the Bronze Age (Renfrew, J, 1973, 205), reaching the western Mediterranean with the Phoenicians, probably around the VIIIth century, although its intensive exploitation (Marinval 1988b, 92-3) can not be earlier than Greek influence in this area. The early evidence for distribution of the cultivated vine in France (Marinval 1988b, fig. 30) shows a clear concentration around the southern coast. This seems to support the theory of a Mediterranean diffusion for the cultivation of the vine. However, a very early use of the wild vine, i.e. a semi-cultivation from the Copper Age (Marinval c.f. above p. 93), has to be considered as a distinct possibility from the evidence. The wild variety of grape found at Castro de Penices (Figueiral 1990, 118)[24], dating to the Augustan period, suggests that this species was known and used throughout the millennium without the major morphological changes that would be indicative of its domestication. This may mean that this species was not in regular use, or that it was in a wild state; that it was either part of the diet, or used for wine making, similarly to the above arguments for the acorn.

Strabo (III, 3, 7) says that wine was a rare commodity but well appreciated by the inhabitants of the castros. This suggests that it was probably imported from the South. Also, there is sufficient literary (Strabo III, 3, 7; Orosius 5, 7, 13-14) and archaeological evidence to indicate the fact that beer was also consumed. Furthermore, the sociological importance of alcohol in male drinking rituals (Sherratt 1987), which are particularly common to peoples where warfare plays a significant social role, seems to explain in part the strangely unruly consumption amongst the Callaicans referred by to Strabo. Parallels can be drawn with Celtic feasting. Alonso del Real suggested (1979, 66) that the custom of 'to keep at hand a poison[25]? (Strabo III, 4, 18) could well refer to any type of stimulant bearing a similar function. The similarity between some habits of the Scythians and the Iberians (Strabo III, 4, 17), e.g. on the aspect of fidelity "*to whomever they attach themselves, even to the point of dying for them*", is perhaps surprising. As I argue elsewhere in the course of this study, this may be necessary for any understanding of social links and rituals, and particularly for those connected with hierarchical institutions such as the *fides* and *cliens*.

[24] - After Buxó i Capdevila personal communication.

[25] - Still, it should be pointed out that an interesting remark from Jones accompanies the translation (III, 4, 18, footnote), in support of the interpretation as poison, with some convincing Classical and ethnobotanical references. Otherwise, this poison could be extracted from the yew (Silius Italicus 3, 329; Florus 2, 33, 50), of which carbonized evidence (Figueiral 1990) was found in one castro.

7.1.2.2) -Animal husbandry

Small domestic animals were certainly present in the castros. There is little evidence for this assumption, although consideration of the archaeological details[26] raises some questions in this matter. For example, footprints of pigs, goats, dogs and cats (Taboada Chivite 1977, 81) occasionally occur in ceramics, mostly on *tegulae* of the later period. It must be stressed, however, that this type of evidence is contemporary with the process of Romanization. Besides, the domestic cat is believed to have been imported from the Near East, and is not related to the European wild variety.

The fact that the Classical writers refer to the Castros Culture as a herding-based economy leads us to believe that domestic animals may have been present in reasonably large numbers. Relying on this evidence, most of the studies (Caro Baroja 1976, López Cuevillas 1953b, Blázquez martínez 1968, Taboada Chivite 1977, 77) picture an economy based on sheep and goat herding.

In any analysis, one should be cautious in interpreting the osteological remains as an indicator of the importance of a given species without some sort of previous Ethnoarchaeological filtering of the data. In fact, the presence of an equal number of pig, sheep and ox bones in the archaeological record might signify that each of these species had a particular economic and ethnoarchaeological meaning.

Stone-rings, normally interpreted as holding pens for animals[27], are relatively common in castros, especially in post-Augustan contexts.

Sheep (*Ovis aries*) and goats (*Capra hircus*) have been relatively well documented by palaeontological studies, particularly in Galicia where more evidence is available (Penedo Romero 1988, 329; Taboada Chivite 1977, 81; Vázquez Varela 1973; 1977), and by iconography (Conde Valvis 1952; Lopez Cuevillas 1951b and 1951c; López

[26] - During the course of the 1989 excavation at castro de Penices (Famalicão, Portugal), the lower half of a broken storage pot of reasonable size was found buried, outside the entrance of a house vestibule, at the occupation level of the yard. When digging the interior of the vestibule, in a context corresponding to an earlier occupation of the house, without vestibule, a similar piece was found by the entrance, which at this time would have been outside the house. Bearing in mind the many archaeological and ethnographic parallels (Taboada Chivite 1977, 82), it is possible that these broken pots could have been re-used for feeding or watering small domestic animals.

[27] - In fact, further suggestions have been made regarding the function of these pieces. See Calo Lourido (1991, 763-6) for an updated summary on the subject.

Monteagudo 1973 and 1977). According to the study of Vazquez Varela (1977, 643), the known remains of *ovicapridae* indicate that only young animals were used for consumption. This suggests that a certain value was placed on the productive capacities (for reproduction and milking purposes) of the adult animals, as it is unlikely that these species were used exclusively for meat, as suggested by Vazquez Varela (*c.f.* above). The size of sheep appears to have diminished from the Bronze to the Iron Ages in Asturias (Altuna 1980, 65), although goat size remained constant.

Factors such as the large number of cattle (*Bos taurus*) found on some of the Galician sites, a comparatively low rate of consumption of young specimens (Vazquez Varela 1973, 314-15; 1977, 643), and a general reduction in size (Altuna 1980, 56, 81), can be interpreted as a clear indicator of the use of this species for traction purposes (Taboada Chivite 1977, 76). There is little archaeological evidence, however, supporting the domestic use of cattle, and the entrances and paths of the castros are not apparently wide enough to allow access to wheeled vehicles. The above quote from Strabo suggests the relatively intensive use of milking animals, although not exclusively of cattle.

Domestic pig (*Sus scrofa domesticus*) has been found in archaeological contexts at Galician coastal sites (Vazquez Varela 1973, 314), in contrast to the scarcity of references from Classical sources[28] about its presence in the Castros productive economy. The Iron Age brought an increase in pig breeding (Altuna 1980, 81) and, according to Strabo (III, 4, 11), Asturian hams were soon to become valued highly by the Romans. It is likely that this increase in breeding and consumption was related to the pig's adaptability in its dietary preferences e.g., to acorns. The inclusion of pork in rituals and feasts looks plausible, either because the taste of this meat was much appreciated in Celtic feasting, according to Poseidonius, or because the pig was sacred, as reflected in the statues of boars[29] found in the northeast of Portugal and in the west of the Spanish Meseta.

The various Classical references to the famous Callaican horses do not seem to have much support in archaeology (Altuna 1980, 51; Jalhay 1947; López Monteagudo 1977; Penedo Romero 1999, 330; Taboada Chivite 1977, 78-80; Vazquez Varela 1977, 643-4; 1986, 233). This is perhaps surprising, when we are probably dealing with a warlike society, in which the horse would have been an asset and an exclusive warrior item. Yet, at Coto da Pena (Silva 1986a, 113) there is evidence for the consumption of horse, although this fact 'should not have been frequent' (Vázquez Varela 1977, 643). Iron Age horse gear is generally absent in the region, however. There is thus a marked discrepancy between data from Classical and archaeological sources.

7.1.2.3) Marine resources: fish, shellfish and salt

Archaeological evidence for the exploitation of marine resources comes from research carried out in Galicia by Vazquez Varela (1975a, 1975b, 1975c, 1976, 1986), and at coastal sites in northern Portugal (Silva 1986a, 114-5). It is now clear that fishing and shellfish collecting was relatively intensive[30] and constituted an important economic supplement for these coastal communities.

The fish remains found in pre-Roman contexts (Vazquez Varela 1976, 85) indicated that species such as *Labrus berggylta L.*, *Sparus auratus L.* (gilt-head bream), *Crenilabrus melops L.*, *Tranchurus tranchurus L.*, *Merluccius merluccius L.* (hake) and *Morone labrax L.* (haddock), were consumed. The presence of species typical of deeper waters, like haddock, implies the use of boats and *off-shore* fishing[31]. Net weights or sinkers are relatively common at these coastal sites. Fish hooks in bronze (Queiroga 1983 ill.) and iron (Silva - Centeno 1980, 63) are known on inland sites, in contexts dating to the first century BC, suggesting that line-fishing was practised prior to the Conquest.

Shellfish remains are numerous in coastal castros. Work done on this subject (Silva 1986a, 114; Vazquez Varela 1975a), and especially on the samples taken from the debris areas (*concheiros*), suggests that shellfish consumption was related to availability and to ease of collections, rather than to any particular gastronomic taste, or nutritional value. Coastal resources became available in the hinterland (Almeida 1969, 38) as Roman acculturation progressed, although the exact mechanisms involved in this process are unknown as yet.

The exchange of marine resources, at least within the distance that could be traveled in one day[32] from the coast, certainly existed before the coming of the Romans. Salt, which is essential to existence as well as a useful commodity, would certainly have been available to the peoples of the hinterland throughout the first millennium, and it was brought

28 - Excepting Pliny's reference to the quality of the Cantabrian smoked hams (Schulten 1920)

29 - Approach to the religious aspects of the so-called "boar culture" in Silva (1986a, 299), and Tranoy (1981, 118-9).

30 - Recalling here Silva's comment on the lack of fish remains on the sites he studied (1986a, 115), we will have to take it as a fact that this is due to soil erosion and deficient sampling methods, as the Galician sample is uniform (Vázquez Varela 1976). Adequate soil sampling might possibly change this.

31 - The use of sea resources, and in particular the capture of species of deeper waters will increase with Romanization (Cano - Vázquez 1991, 74-5).

32 - In the sense of the Portuguese expression "*jornada*", which is roughly estimated as 20-25 km in this region.

in along with other marine products. This may be presumed by the relatively short distance between the coast and hinterland in the Entre-Douro-E-Minho. In addition to a desire for fish and shellfish, the existence of a trade in processed goods, such as salted fish, should also be considered as a distinct possibility.

A likely preference for the consumption of pork in this region has already been discussed above. It is stressed than salt is necessary for the preservation of pork and especially of ham, which is known to have been widely consumed by the inhabitants of the mountain areas (Strabo III, 4, 11) before the coming of the Romans.

There is little archaeological evidence at present for an Iron Age salt industry (Almeida, CAB, 1990a, 29), although some recent unpublished data[33] seems to confirm our previous assumptions. Salt was exploited on the coast[34], probably for regional and long-distance trade throughout the millennium, and made accessible to the hinterland. The salt and *garum* industries were higly developed and (Strabo III, 4, 2) in the Iberian south and were connected with Roman trade; there is also evidence for such activities in the coastal Northwest (Almeida 1969, 38-9; Naveiro López 1991) later in the IVth century AD.

Wooden ard of 'radial' type (after Dias 1982). This type of ard is still in use in some areas of northern Portugal.

7.1.2.4) -Evidence from implements

The study of the implements related to productive activities promises to be rewarding, even though a wide variety of tools made out of perishable materials will not have survived in the record. In addition, there is sometimes an ambiguous relationship between the shape and function of those tools whose function is not known, particularly when there are no suitable ethnographic examples for comparison. Indeed it is always hazardous to try and determine the range of human activities associated with one tool type, particularly when the analysis is based exclusively on typological aspects. A relatively small number of tool types were, in fact, used in agricultural practice. The majority of these belong to the later phases of the Castros Culture, when iron becomes more common.

A number of iron mattocks and pickaxes have been found (Romero Masiá 1977, 232; Taboada Chivite 1977, ill.) during the course of excavation at the castros (Figure 46). This implies that builders in stone in the castros were busier from the I[st] century BC onwards. Many iron agricultural and crafts tools have been found in recent decades. Reaping and pruning hooks are some of the commonest types of iron tools. Several of these tools have been found at castros, e.g. at Mozinho (Soeiro 1984), Ermidas (Queiroga-Pautreau forth.), Borneiro (Romero Masiá 1987), Cidá do Castro (Lopez Cuevillas 1958), documenting not only that the cultivation of fruit trees took place, but also the management of the surrounding woodland, (as suggested above). Most of these iron implements come from contexts dating from the time of Augustus, i.e. a period of advanced Romanization. These tools may have been produced in response to new cultivation techniques, as the larger range of cultivated plants known (Figure 10) suggests. Such tools are rare in the archaeological record before the Augustan phase.

The use of the plough in the Castros Culture is suggested by an increase in cereal production, and the more frequent occurrence of cattle bones[35]. The cattle may have been used for traction purposes. Ploughs are referred to by Silius Italicus (3, 351) and Justinus (44, 3, 5), and the cart of Costa-Figueira (Cardozo 1946b) shows two bovids attached by a yoke. The overall palaeoenvironmental picture indicates that substantial areas of the territory may have been cultivated, as the increase in cereal pollen implies (Aira-Vazquez 1985). Such a dramatic ecological change in a varied landscape such as this,

33 - Recent research carried out by Almeida (CAB) in the area of Esposende, and in particular the chronological confirmation of data provided by the excavation of Castro de S. Lourenço, establishes that salt was being exploited by coastal communities by the turn of the millennium. Small schist containers previously found on the sea shore, and assumed to be for salt production by the evaporation of brine, have been found re-used, for roofing houses at the above site. I am grateful to this colleague for his kind permission to use this unpublished information.

34 - Although Strabo (III, 3, 7) refers to the existence of rock-salt, there has been no confirmatory evidence for it so far.

35 - Evidence for ox is substantial in the Asturian settlements studied by Altuna (1980, 56), and appears to have increased (*op. cit.* pp. 81) during the Iron Age.

even after taking the high site density into account, is difficult to accept without considering the use of the plough. It is true that there is little evidence for the plough in the archaeological record, but there are well-documented European post-Neolithic examples of ard and plough (Rees 1979), mostly made out of perishable materials. Modern examples still exist in northern Portugal (c.f. above figure) where only a small piece of iron is used for the share[36], so that remains might easily be overlooked, or have been badly corroded in the acidic soil of the region.

The importance of animal *secondary products* (Sherratt 1981) in the subsistence economy Prehistoric communities has been widely recognized. The use of secondary products in the Castros Culture has been clearly documented in several ways, for example in Strabo's reference to the use of butter[37], and the presence of possible ceramic cheese strainers (Silva 1986a, 114, graph 3, type F1). Milk from cattle and goats, as well as its by-products of cheese and butter, was therefore an essential nutritional supplement. In addition, leather was widely used (Cardozo 1965f) for clothing, utensils and implements, shields and floaters to cross rivers, just to quote a few examples from Classical sources. The importance of wool is suggested by the numerous spindle whorls found in all the excavated castros, from the earlier contexts to the late Ist century AD. Although weaving tools are rarely found, technical developments in cloth manufacture, and therefore the extensive use of raw material, are suggested by the decoration and details of dress on statues of Callaican warriors from S. Julião and Outeiro de Lezenho (Figure 56.2 and 56.3).

Hunting, other than as a *sport*, added supplementary protein and minerals to provided a useful source of food during times of harvest failure and stand outs within the main productive system (Vázquez Varela 1991). Remains of wild species of mammal have been recorded in castros, particularly in Galicia.

Fish hooks may not be numerous, but they have been found on a wide variety of sites and in many phases. This implement gives us direct evidence of fishing activities (Ferreira 1968, Figueiredo 1898), and it is referred to by Pliny (N.H. IX,XVII). It occurs on coastal sites and also inland, even in mountainous regions where there is a nearby river or stream. Bronze hooks were recovered from my own excavations at Vermoim and Ermidas (Vila Nova de Famalicão), in contexts dating from the mid-late Ist century BC and the mid Ist century AD. respectively. The use of this implement in river fishing (Garcia Alonso 1981-82, 322) is now fully supported by archaeological finds. Although bronze hooks are relatively common in the Northwest, their manufacture in iron did not begin until the later second century AD (Garcia Alonso 1981-82, 317), as evidence from the Mediterranean coast of Iberia implies.

7.1.3) -Discussion

According to Strabo (III, 4, 17) and Justinus (XLIV, 3, 7), agricultural tasks amongst the Northwestern peoples were done by women. Several authors have analyzed this reference, and they all seem to agree[38] that land and agricultural practices belonged to women.

The Castros productive economy might well be described as a developed system, and self-sufficient rather than market-oriented. However, there is no doubt that this productive system incorporated a wide range of activities and may thus have been able to accumulate a considerable surplus, which can be detected in the material culture, thanks to a well-balanced ecological adaptability and exploitation of local resources. Productive activities were supplemented by gathering.

Evidence from some areas showing the relative independence of the Castros economy in that particular geographical and ecological environment suggests a well-balanced and controlled use of resources. An illustration may be seen in the presence of sheep/goat, cattle and domestic pig on the small Cíes islands (Vazquez Varela 1973 and 1975), at the entrance to the Vigo's *ría*, where a more marine based economy might be expected. This shows how animal husbandry (Vazquez Varela *c.f.* above) has developed from the early phases of the Castros Culture until the first century AD.

The pastoral component of the economy varied from area to area and, as might be expected, depended on local conditions. The richer valleys and coastal fringe were preferred for cereal cultivation. The vegetation composition reconstructed at Crastoeiro (Figueiral 1990, 103) is indicative of over-grazing, and the scarceness of charred seeds in the charcoal-rich samples (Dinis, AP, pers.com.)

[36] - Many examples are known in Europe of ard-shares made out of stone, normally polished, from the Neolithic onwards until the Middle Ages. This function has been suggested for some early Spanish stone implements (Mélida 1942, 57, 59), although there has been some criticism of this interpretation on technical grounds (Dias 1982, 138) European parallels (*in genere* Rees 1979) might easily demonstrate the feasibility of the use of stone ploughshares in ards of primitive type. A similar type of wooden ard using an iron ploughshare was used at Danebury (Cunliffe 1984-b, 354-7, Fig. 7.14) dating to the Iron Age, but this type could still be found in the north of Portugal in the XI century AD (Barroca 1988, 168, Fig. 6).

[37] - (Strabo III, 3, 7) "...and instead of olive-oil they use butter...", also Pliny XXXIV, 156.

[38] - See Almeida (1983a, 1983b), Alonso del Real (1979) and Bermejo Barrera (1979) for a social and anthropological analysis; Blázquez Martínez (1972; 1974, 367-9; 1975, 18) Caro Baroja (1973), Maluquer de Motes (1968b, Garcia Bellido (1968b) for an economic approach.

retrieved from the occupation levels of this site is notable. It is, therefore, plausible that pastoralism was the dominant activity here. In any case, particular attention should be paid to differences of this kind, especially between sites in the hinterland and on the coastal fringe[39], where a much wider range of productive activities was possible.

By the turn of the millennium, both deforestation in advance of cultivation and continued ecological degradation caused by the pressure of goat/sheep grazing were responsible for the faster pace of erosion on the slopes and hilltops around these sites. This erosion reduced soil fertility and led to a decline in economic production at the castros. The resulted in yet more deforestation, and further away from site. These are important factors to consider when examining the social and cultural causes for the multiplication and reorganization of the castros during this period.

Iron Age archaeology in Galicia is more advanced than in northern Portugal. On Galician sites sheep/goat bones (44.3%) slightly outnumber those of cattle (41.2%) (Penedo Romero 1990, 339, figure 10). On average, these figures suggest a slight preference for sheep/goat consumption. When size differences between these two species, and the fact that small domestic animals are more frequently consumed than the larger ones are taken into account, then it becomes clears that the pastoral component of the Castros economy was dominated by cattle.

If future research corroborates this, then many other aspects of the economy may have to be reviewed. One would have to consider a Castros agricultural system in which the ratio of cereals and pulses in the fields was controlled, and the fields were located according to the constraints of local topography. Romanization appears to have had little effect on the productive system. This is suggested by evidence from Facha (Almeida, CAB, 1990a, 59), where at an open agricultural settlement, a rise in the importance of gathering (acorn in this case) was the response to change in the productive system. This is yet another point to consider, in support of the idea that the pre-Augustan economy was more concerned with the basic needs of the community, than with the intensive exploitation of the potential resources of the region. In some cases, technical development in agricultural practices appears to have been neglected.

Recent work on the distribution of castros and Roman settlements in Galicia (Carballo Arceo 1986, 143-4) shows a marked coincidence between the two. This indicates that the same territories continued to be occupied in the Roman period, and identical explotation strategies practised.

The apparently low emphasis on hunting (Lopez Cuevillas 1953d, 363-4; Silva 1986a, 114; Vazquez Varela 1986, 233) is surprising. However, the acidic soil of the region and erosion may have played a part in destroying the evidence. Game was certainly plentiful, and hunting might well have been a sport or prestige activity, as the rock engraving (Jalhay 1947) of a horseman hunting a deer with a spear at Sanfins suggests. Indeed, as has been stressed above, there are very few representations of horsemen in this region, and therefore the association of this example with hunting emphasizes the likely social importance of this activity.

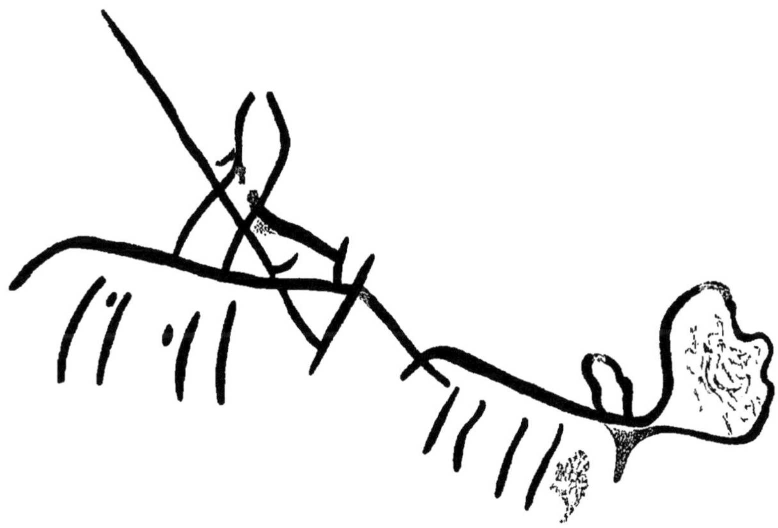

Deer hunting on horseback. Rock engraving from Sanfins.

The interpretation by the Classical authors of gathering as a sign of economic underdevelopment should be viewed cautiously in the light of new archaeological evidence which presents gathering as a rather productive activity. In fact Clark (1986, 167) in his work on the Cantabrian Prehistoric economy[40], estimated that the Cantabrian forests were capable of producing 0.5 ton/ha of hazelnuts, 3 ton/ha of chestnuts and 500-1000 kg of acorns from a mature oak in any year. This estimate is impressive

[39] - C.f. the above references to Vázquez Varela, in addition to the results from the castros of Penices and Ermidas.

[40] - However, one should be more cautious about serious discrepancies in some other aspects of the Iron Age productive economy suggested by this author (Utrilla Miranda 1987, 337).

if one considers the distribution of castros before the Roman Conquest, and the open nature of the deciduous forest in the region. Yet all this has to be understood in terms of cultural attitudes. A decline in gathering does not necessarily accompany agricultural intensification, although it may reflect a change in cultural attitudes towards the Mediterranean economy and its pattern of life.

In fact, from the turn of the millennium onwards diet became more wheat-based, and the value of trade gained in importance in a now market-oriented economy. In this case, the spread of agriculture does not exclusively mean that even bigger surpluses were accumulated. Evidence rather suggests that an intentional increase in overall production was used to obtain a surplus for long-distance trade, and to facilitate entry into the pre-monetary and monetary economy, where prestigious imported goods became essential for social display in a Romanized society.

The dating and circumstances of the establishment of Roman *villae* on the coastal fringe to the north of the Douro river are not clear. If some of them were already in existence in the mid first century AD, which would agree with the industrial growth of *Bracara Augusta* (Alarcão 1988), these *villae* could well have contributed to the agricultural development of the region by opening up routes of exchange and diseminating technological knowledge. The number of 'liberated' noticed by Almeida (1969, 29) during the study of inscriptions in the Maia region adds support to the idea, although it might alternatively refers to the mining activities of the nearby Valongo gold mines.

7.2) -Mineral resources

7.2.1) -Evidence from Classical sources

There are many references from Classical authors to the variety of metals which could be extracted in the Northwest. For the Entre-Douro-E-Minho the number of references is particularly large. For example, Strabo and Pliny may be quoted as an illustration of the quantity of metal available at the time of the Conquest.

Strabo, Geography, book III.

2, 8 -"...For the whole country of the Iberians is full of metals...". "...Up to the present moment, in fact, neither gold nor silver, nor yet copper, nor iron, has been found anywhere in the world, in a natural state, either in such quantity or of such good quality. And the gold is not only mined, but is also washed down by the rivers and the torrents...".

2, 9 -(Tin) "...is produced both in the country of the barbarians who live beyond Lusitania, and in the Cassiterides Islands; and tin is brought to Massilia from the British Islands also. But among the Artabrians, who live farthest on the north-west of Lusitania, the soil 'effloresces', he says[41], with silver, tin, and 'white gold' (for it is mixed with silver). This soil, however, he adds, is brought by the streams; and the women scrape it up with shovels and wash it in sieves woven basket-like."

Pliny, *Naturalis Historia*

XXXIV, 47 "It is now known that it (tin) is a product of Lusitania and Callaecia found in the surface-strata of the ground which is sandy and of black colour. It is only detected by its weight, and also tiny pebbles of it occasionally appear, especially in dry beds of torrents."...

7.2.2) -Mineral resources and evidence for their use

Amongst the metals available in the region, gold, tin, copper and iron were particularly desirable. Their exploitation had repercussions for the development of the Castros Culture throughout the millennium.

There are two major periods when metals were extracted in the Northwest, namely the period prior to the Conquest and the period of Romanization. Archaeological evidence for the former period is slight in comparison with that available for the latter. Also, an attempt should be made to distinguish the indigenous exploitation of these resources under the influence of external contacts from that conducted exclusively by the Romans.

Large-scale mining was practised before the arrival of the Romans. In any case, all traces of any substantial pre-Roman industry, would have been destroyed by the later massive earthworks and systems of the Romans. Therefore, all reference to early mining activities in this chapter is largely supposition.

There is no doubt, however, that mining began early, in the Copper Age (Pinto 1933a, 1933b; Uria Riu 1956). It should thus have been a well-established activity in the Northwest during later Prehistory.

Mining activities in the Northwest from the Bronze Age until the coming of the Romans would probably have been small-scale and employed relatively few people. Accessible sources of metal ores, such as the veins of native copper, copper oxides and carbonates, or alluvial tin, were exploited initially, surface deposits would have naturally been preferred. The process of erosion resulting from increasing deforestation, from the Bronze Age onwards, may have uncovered many new ore deposits during the millennium, further promoting the development of a metal-based industry.

---

41 - Strabo is quoting Poseidonius.

Again, it has been suggested that much of the metal extraction, such as sand-washing for gold, silver and tin (Strabo III, 2, 9), was done by women. Open-cast mining could also have been carried out seasonally (Joleaud 1929, 135[42]; Monteagudo 1954, 83) using non-specialist labour.

In some cases, the Romans continued to exploit ore sources used in the Bronze and Iron Ages[43]. In spite of technical advances[44], the extraction of gold, from the Bronze Age until the coming of the Romans, was probably done by sand-washing[45] (Strabo III, 2, 8-9), and by women (Alonso del Real 1979, 67), a pattern which was still in use even a few decades ago (Gomez Moreno 1941, 461; Cabré de Morán 1931, 59; Lopez Cuevillas 1951c, 7) in the Minho, Sil, and Tormes river areas.

In the Lima valley, evidence for pre-Roman mining has been recorded (Almeida, CAB, 1990, 210-11), the dating depending on the find of a single bronze axe. Mining may even have been a well-developed industry before the Romans arrived, so that a source of labour was already in place (López Cuevillas 1951c, 16, for the gold).

It can be argued, that before Romanization a hierarchy able to control the means of production already existed[46]. The intensive exploitation of gold sources by the Romans probably started immediately after the Augustan Conquest (Domergue 1990) in all the mining zones of the Northwest. This rapid re-organization was probably assisted by the existence of mines that had been worked by the indigenous populations in the I century BC, which the Romans merely took over. However, this postulated centralization of manpower and production does not fit entirely with the archaeological evidence.

There is no reason to believe that the exploitation of primary sources of gold, involving the removal of a considerable volume of sediment followed by a degree of processing, (Sánchez - Pérez 1989, 20) was done in the Northwest in pre-Roman times. On the other hand, the extraction and refinement of gold from alluvial deposits, although not documented archaeologically, is a simple process[47]. It is also referred to by Strabo (III, 2, 9).

The substantial number of finds of utilitarian goods and jewellery of local manufacture, datable to the first millennium BC, is probably the best indicator of the use of local metal resources.

A decrease in the purity of gold with Romanization, as suggest by the analyses of jewellery from the castros (Silva 1986a, 237), may be related to a change to the exploitation of primary sources of gold (Pérez Outeiriño 1986, 109). Therefore, one can assume that most of the pre-Roman gold from the castros came from river deposits, and that sand-washing was the commonest way of obtaining it.

Roman gold mining seems to have been a highly organized and intensive activity throughout the Northwest, leaving the impression that the sources were exhausted by them, (Allan 1965, 154-5, 171), although not all authors take this view (Sánchez - Pérez 1989, 16).

There may have been a decline in metallurgical production with the coming of the Iron Age. This is demonstrated by differences in both the quantity and variety of bronze work in sites and hoards[48], and by the scarcity of ironwork before Romanization. This may be due to a substantial reduction in the availability of the metals to make bronze, in addition to increasing competition. On the other hand, gold metallurgy does not appear to have suffered a similar 'crisis'. Normal levels of production were maintained in the Iron Age (Lopez Cuevillas 1951c, 107-8), although the simplicity of the techniques[49] used should be borne in mind here.

The availability of gold resources in the Northwest appears to have determined, to some extent, the distribution of Iron Age settlements (Sanchez-Palencia 1983a, 82-3) after the Conquest. This suggestion can now be corroborated in some mining areas in Entre-Douro-E-Minho, particularly in those areas where detailed regional studies have been carried out (Almeida, CAB, 1990a; Soeiro 1984, 24-5).

---

42 - It is worth while mentioning Joleaud's reference on this subjecthere concerning, archaic mining methods in the small parish of Muradas (Beariz, Orense, Galicia) earlier this century. In this parish mining was a seasonal activity carried out mostly by women.

43 - Such is the model defended by Mañanes (1976, 42).

44 -At Jales (Trêsminas) a bronze palstave was found inside the mine (Almeida, F., 1970, 296). This tool is characteristic of the peninsular Atlantic Late Bronze Age, and so far there is no evidence to demonstrate its continued use in subsequent times. Therefore, it is possible that the exploitation of this source dates from that. Early copper mining, is known in others regions (Ferreira-Viana 1956, 526-8, fig. 1; Gossé 1942, 47, 54; Saunier 1910; Serra Ráfols 1924, 159-84; Uria Riu 1956, 631, 634). Similarly, in other parts of Europe there are a few examples of early mining with extensive underground galleries, what might allow us to suppose that a certain development of mining techniques, along with a social capacity for the concentration of manpower had occurred by the that time.

45 -The major accounts are Domergue 1970a and 1970b; Sanchez-Palencia 1983; Sanchez-Perez 1983, although they do not agree with each other. This is a very controversial subject which will be reviewed later.

46 -Mainly the epigraphic sources (eg. the question around *Camalus* and *Coroneri*, from Briteiros).

47 - According to the estimates made (Sánchez - Pérez 1989, 23), this activity could prove to be very productive, in particular when considering the non-specialized labour involved. In fact, taking as a model the out put of the Sil river "sand-washers" in the last century, which was 12-14 gm per person per season, many of the Iron Age gold jewels could have been made with comparatively low production costs, in comparison with their social and intrinsic value.

48 -C.f. the numerous finds reported by Coffyn (1985) and Monteagudo (1977) for the region, and the magnificence of the hoard at Senhora da Guia, Baiões (Silva *et al.* 1984).

49 -It can be extracted by sieving the sand from various rivers in the northwest of Iberia , in reasonable quantities(Schulten 1959, 396) without mining.

It is unlikely that silver was a primary product of the Northwest during later Prehistory, but it may well have been a by-product of the gold extraction process.

Silver normally occurs in association with gold, both in primary and secondary mineral sources, and can be obtained during the refinement of gold. The amount of silver in native gold varies according to the geological composition of the deposits. Gold bearing over 25% of silver was called *electrum* in Antiquity (Pliny, NH, 33; Strabo III 2, 9). Silver jewellery is also a rare commodity prior to the Conquest, in comparison with the relative profusion of goldsmithing.

The availability of tin ores in the Iberian Northwest from the Bronze Age[50], and indeed throughout Antiquity, is a well known fact that has been the subject of various studies. Some of these studies were concerned with founding the location of the mythical Cassiterites Islands (Blázquez - Aguilera 1915; Garcia Bellido 1945a, 247; Monteagudo 1968, 247[51]; Schulten 1959-63 (1), 377; Tranoy 1981, 27-9), which are believed to be around the Galician coastline, although this has yet to be defined with precision. Amongst those who have studied this problem, Dion's (1953 24-5) objective approach is noteworthy. This author dissociates the expression *Cassiterites* from any precise geographical place, and proposes instead that it was the broad designation for a place where metal was exchanged rather than extracted.

For the moment, any further discussion of geographical discrepancies and descriptions, in Herodotus (III 5), Diodorus (V 38), Strabo (II 5-15 and III 5-11), Pliny (IV 119; VI 56; XXXIV 47), Ptolemy (II 6-73) and Avienus (94-116), is irrelevant. However, evidence for early mining and metallurgy around the Atlantic coast of Iberia, and its relation to the trade routes of the western seaways, will be considered further. An Atlantic and Mediterranean trade in metal goods was probably already well established (Ruiz-Gálvez Priego 1986) during the early phases of Castros Culture[52] in the Northwest, and this may have included both raw material and artefacts. Metals, which could be easily extracted from surface deposits and then processed in the castros, probably formed the basis of this trade. This implies that exchange outlets in the form of strategically located coastal entrepôts[53], with shore zones set aside as the easily accessible termini of penetration routes into the hinterland, developed.

For this, it is necessary to make a connection between zones with mining potential, or where ancient mining remains are visible, and the coast, by following natural routes in order to determine which zones were more likely to have dealt with long distance trade. In fact, some evidence is already available[54] to help us understand the vitality of these coastal sites, as shown by their imported finds (Figure 51), whose presence can only be explained in terms of the raw materials having been sent from the hinterland to the coast for exchange.

The northwestern corner of the Iberian Peninsula is very rich in tin sources (Monteagudo 1954, 82) which occur in association with quartz in granites and schists (Allan 1965 156), sometimes in cassiterite ores in massive concentrations (Figure 11). The extraction of tin ores did not necessarily require intensive mining techniques (Pliny XXXIV, 47), as most deposits in this area (Allan 1965; Neiva 1944) can be reached easily from the surface.

The northwestern corner of Iberia is not particularly rich in copper ores, although its exploitation is referred to by Justinus (44, 3). Copper occurs in the north of Portugal, generally associated with quartz veins in granite, or in mineral veins with lead and zinc (Noronha - Abrunhosa 1987, 140-1), some of which occur as outcrops on the surface. Copper in the form of oxides, sulphides and carbonates at ground level might have been subject to economic exploitation in later Prehistory, without any need for extensive earthworks. This may be one of the reasons why a bronze industry flourished throughout the millennium.

The subject of iron metallurgy in the Castros Culture has been a particularly neglected one. This has not been due to any lack of interest amongst archaeologists, but rather to the lack of application of scientific methods of analysis in this particular field.

The use made of the many sources for iron oxide can only be traced archaeologically by an analysis of the slag. The extraction methods used when mining iron ores may be less demanding than for copper (Cavalheiro 1989, 126), but the smelting process requires a much higher level of skill. This will be dealt with further below, in the discussion on metal manufacturing.

50 - A fact supported by the volume and variety of metallurgical production in this phase, which seems to be archaeologically reinforced by the bronze hoards of this region (see Coffyn 1985), which i.e. mostly in the coastal area.

51 - Reviewing, in this work, the ideas of others on the subject.

52 - See Hawkes (1984, 212, fig. 1) for an account of the sailing voyage of Pytheas around the Atlantic coast of Iberia, probably in the year 325 BC. (c.f. above page 211)

53 - There is no doubt that sites such as Coto da Pena (Silva 1986a), considering the chronological variety of its finds, could have been used as an entrepôt for long distance trade.

54 - See also Silva 1986a for reference to various Mediterranean imports in coastal areas of the northern Portugal.

7.2.3) -Discussion

Consideration of the whole process of exploitation of the mineral resources in the Northwest reveals that, to some extent, it was technically unsophisticated. This level of technical development was probably related directly to how accessible the resources were, and how easy the ores were to extract. Evidence indicates that mining was a non-specialist activity carried out seasonally as shown above, for example, in the extraction of gold by sand-washing. This was a widespread process, in use throughout the millennium and a scientific analysis of the pre-Roman gold jewellery (Hartmann 1971, 130) indicates that the metal was obtained from secondary deposits. It can, therefore, be assumed that alluvial deposits were the most intensely exploited sources of gold, until the Romans began extracting gold from mineral veins.[55].

Recent ethnographic evidence (Monteagudo 1954, 82-3) from Galicia has shown that mining, earlier this century, was still a seasonal activity carried out mostly by women.

Intensive deforestation from the Late Bronze Age onwards, as discussed above, far exceeded normal agricultural needs. As Schulten (1925, 69-70) points out, the forest fires referred to by Poseidonius, through Strabo and Diodorus Siculus[56], could have been started intentionally by shepherds to enlarge their pasture lands. In any case, the result of this degradation of the forest was a marked increase in erosion, which incidentally revealed new mineral veins of cassiterite and pyrites, and caused more gold to be flushed into the rivers. Maluquer de Motes (1977, 9) assumed a similar however based ona different arguments[57].

We have then a picture of a region relatively rich in metal sources, particularly tin, iron, gold and silver. Each site carried out its own prospecting, mining and subsequent processing of the metal ores. The high level of craftsmanship in the Castros Culture, both in jewellery and bronze tools, suggests either that manufacture was the specialism of some sites, or that it was the work of itinerant craftsmen. Increasing competition between sites may also have had a significant role to play in the regional development of metal production and the distribution of artefacts.

Much of the production activity and the quality of artefacts imply that manufacture was carried out by highly skilled craftsmen. Not all production may have been all the same standard, however, and there may have been differences in method between goods for home use and those destined for export. This topic will be discussed in the next section.

7.3) -Manufacturing

7.3.1) -Ceramics

Archaeological ceramics, as Almeida (1983b, 197) has justly pointed out, are an important indicator of *manufacturing techniques , table fashion*, and *cooking and dietary habits*. The use of proper strategies in their study can provide invaluable information for the understanding of both daily routines and economic activities. However, the particular stratigraphical and architectural conditions of the castros virtually disabled these important means of interpretation. Being sites with a high level of architectural activity and a lengthy span of occupation, the continual phases of construction tended to increase the disturbance of the soil to cause deterioration of the deposited finds, in addition to the fact that pottery is only abandoned on a site when broken and therefore no longer of service[58]. Indeed, as Lopez Cuevillas (1953b, 248) has noticed, the extremely fragmentary state of Castros pottery makes any attempt to study it a very difficult task. The Castros funerary practices do not provide any complete vessels[59], and the enormous variety encountered in local production, together with the lack of fairly accurate chronological boundaries, makes any attempt at establishing pottery typologies and sequences[60] a very unrewarding task.

The pottery production of the north Portuguese Iron Age is an important means of study for the establishment of cultural phasing and variability. Pottery, when properly studied, as Lopez Cuevillas has stressed (1953b, 247), can provide a large body of evidence .

On a first approach, there is homogeneity within the ceramic production of this region during the second half of the first millennium BC, which is suggested by the resemblance provided by the predominance of mica in the fabrics. The range of typological variations appears to be due to local or

[55] - However, it should be left open to discussion whether or not the exploitation of the primary sources by indigenous communities preceeded the Conquest, as has been suggested by Domergue (1970a, 264; 1970b, 166-7) and Sánchez Palencia (1983a, 79), and indicated by the different metal compositions of the indigenous jewellery by that time.

[56] - V, 35, 3, quoted above.

[57] - This author suggests that the process of erosion which uncovered surface mineral veins is more related to a drier Sub-Boreal climatic phase, between 2500-500BC, than to any anthropogenic erosion.

[58] - However, examples of the utilization of damaged pottery for purposes which are different from the original are various: strainers, spindle works containers, polishers.

[59] - An overview of the funerary practices is attempted in Chapter 4.

[60] - Other than the essays generally presented in excavation reports, which are necessarily limited to one site, attempts at building a typology of Castros pottery within a chronological framework were made by Almeida (1974a), Martins (1987a=1990, 1987b) and Silva (1986a).

regional feature of the products. Some of these first impressions, however, may prove to be misleading on closer inspection.

Regional characteristics in the production of ceramics are, at the present stage of research, more noticeable at the level of products and materials (Little 1990, 122, 154), than on a typological basis. They indicate the use of local raw materials and production, but they could also be grouped according to ethnic traits, as recent work (Silva 1986a, 126-7) has attempted to demonstrate.

Castros pottery can be differentiated from other ceramic products, such as those of the Bronze Age or Roman period, by the intentional inclusion of mica in the temper, which is responsible for the shiny appearance of these vessels.

Studies produced to date suggest that the pre-Augustan period of the Castros Culture is characterized by a rather static situation in forms (Figure 48) and manufacturing techniques. The range of pottery forms is limited. It is characteristic of an agrarian-pastoral society with a basic degree of nutritional sophistication and table habits. However, the existence of many other containers which were manufactured from different raw materials should also be considered. Horn and wooden objects (Strabo III, 3, 7), precious metal vessels, or imported Greek, Punic and Campanian wares (Figure 51), are rarely represented in the archaeological record.

The shapes of Castros pottery developed from the late Bronze Age types of the region (Martins 1987b). Forms with a round base, which are characteristic of that phase and more suitable for positioning the pot over the embers for cooking, were soon abandoned. Throughout the Castros period there was a wide adoption of forms with a flat base, with the exception of some types of pot with interior handles. These flat-based cooking vessels now sat over the base of the hearth surrounded by the embers, and the clay fabric of the hearth made a better heat transfer medium. This fact is suggested by the absence of any residues of charcoal on the base and at the base-edge, in comparison with their frequent presence all over the external surfaces of the round-based vessels (Figure 48)

By the I^{st} century BC, or perhaps slightly earlier, there was a dramatic increase in pottery production at both local and regional level and it became more standard, implying the use of a potter's wheel. Some forms of imported ware were copied in local products (Almeida 1984a), and the households became better supplied with a wider range of imported ceramics as the I^{st} century AD progressed.

An overview of the Castros pottery, summarized in Figure 48, provides some indication of the economy and social habits of the Castros inhabitants. The pottery vessels fall broadly into the following major groups, according to criteria based on morphological, technical and archaeological analysis: cooking, drinking, processing of food products and storage.

The vessels for cooking can be divided into two groups: the ones used on the surface of the fireplace and the ones suspended over the fire. The first type is shown in the "S"-shaped profile of the pot illustrated. The number of handles can vary from none to two. Suspended vessels include a pot with an *ear-type* handle which is relatively short lived, and a *pot with interior handles*. Such pots remained in use from the mid I^{st} century BC to the mid I^{st} century AD.

Covering virtually all the Iron Age (Silva 1986a, 130), the 'pot with interior handles' (Figure 48-B) is a most interesting type amongst the Castros ceramics, so some special remarks will be made here. This form is present in the castros from the end of the Bronze Age onwards (Peña Santos 1987, 123; Silva 1986a, 120) and its peculiar form and frequency suggest a special importance and particular function in household cooking activities. These pots always show the effects of long-term exposure to fire on their external surfaces (Almeida 1974a, 19), and their mode of utilization has been suggested as suspension over the fireplace, in the same manner as cauldrons.

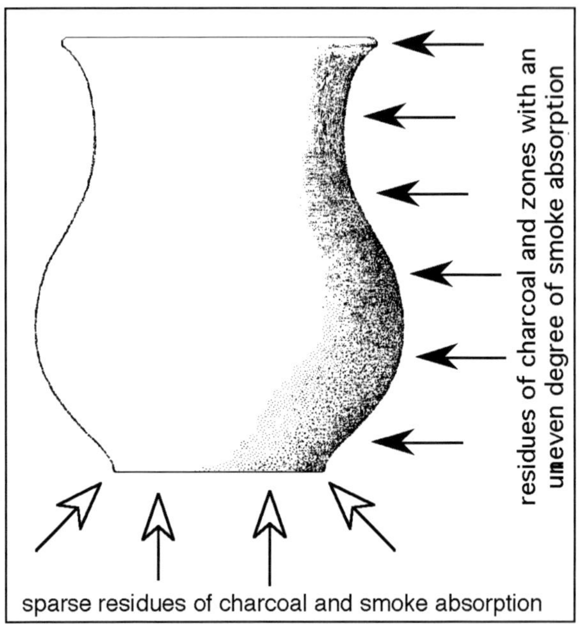

This function seems perfectly clear, according to the evidence. These forms would also have been suited to constant boiling and less delicate cooking, such as that suggested for acorns (Oliveira *et al.* 1991, 260) or meat, or simply for keeping water permanently hot. As there is usually no abrasion on the inside handles of the pots, one could suggest that leather strips were used to hang them over the fire. Other cooking took place on the hearth itself, for which different pottery forms were used, as referred to above.

A small cup, locally called *copa*, was used for drinking (Almeida 1974a). The internal surface and rim were usually carefully polished, which is consistent with the idea of being comfortable to drink from.

The sieve, or strainer, used for making cheese is the commonest example of a type related to the processing of food products. There are, however, other more enigmatic examples, such as the pot illustrated in Figure 47. This pot was clearly used for cooking, yet it has one hole in the base which was intentionally made after firing.

The type connected with storage is represented by larger "S"-shaped vessels in the oldest phases, being gradually replaced by the *dolium* with Romanization. The suggestion that this vessel had a lid is substantiated by the carinated profile of the neck. The lid of this container may have been made out of wood, since pottery lids are very rare.

The use of wooden vessels by the Callaicans is mentioned by the Classical sources (Strabo III, 3, 7)[61], but there is only scarce evidence for this, which seems only natural considering the acid nature of the soil. However, a surprising find, in the oldest occupation layers at Facha (Almeida *et al.* 1981, 70 - fig. 32, 3-5)[62], of two fragments from the rim and neck of an oak vessel, proves the veracity of Strabo's reference and shows that wooden vessels probably existed. Besides, a quick review of the Castros ceramics (Almeida 1974a, Silva 1986a) calls attention to the scarcity of containers suitable for water transport, as in many cases the water supply was probably at some distance from the living quarters. It is, therefore, plausible that wooden containers were used for tasks involving carriage and storage, as well as for agricultural work, due to their higher resistance to breakage in comparison with pottery.

A general impression gathered from the pottery is that there was little interaction, or contact, between castro sites from the end of the Bronze Age until the I[st] century BC, i.e., during the period of the Castros Culture. Ceramic forms and manufacturing technology show no significant development during that time and long-distance imports are few and restricted to Greek and sand-core glass wares (Figure 51). The regional and local individuality of Castros indigenous pottery suggests that each site produced enough pottery to supply its own needs until the first decades of the first century AD. The relative simplicity of the techniques utilized, such as low firing temperatures (Guitian-Vazquez 1979) and a visible dependence on local raw materials (Little 1990, 152-3) again show that pottery was produced on site and probably by the site's specialist potter (Almeida et al. 1981).

Fresh data come from three recently studied sites at Famalicão, the Castros of Ermidas, Penices and Vermoim, whose geological context is similar and which lie only a few kilometers apart. Here, a noticeable difference in design and material between ceramics from contemporary periods at each site can be seen, although the existence of morphological similarities should also be noted. One particular fabric characteristic of the older phases at these sites is a light colour of the surface and a black mica temper. At the same time, there are vessels which have a contrasting white mica tempering material.

The habit of mending broken pottery is known to be reasonably widespread in the mid and late phases of the Castros Culture. Mending is usually done by joining the parts of a broken vessel with a metal staple. It can, alternatively, be done by using fresh clay, but this means that the pot has to be re-fired. The surprising number of examples exhibiting this technique, even on coarse undecorated wares[63], during the last centuries of the I[st] millennium, constitutes interesting and new evidence. It seems that pottery, even if very coarse, was valued for its utility ever after repair.

Another relevant aspect is the unusually high content of mica which, more than any great variation in typology of form, is the most striking characteristic of Castros pottery in the northwest of Portugal. The presence of mica in such generous proportions in the fabric, from the beginning of the Iron Age until the turn of the millennium, cannot be explained on technical grounds because it reduces both the breakage strength and the thermal resistence (Little 1990, 156) of the vessels. On this subject, the work recently produced by Little (1990) is of outstanding value for the understanding of many important aspects of Castros pottery manufacture, and will certainly guide directions for future research on this subject. However, some of her results and considerations are limited by the absence of a chronological framework for the pottery samples used. The samples selected (Little 1990, 53-3) cover a relatively long period and there is no attempt at a reduction of the time span. The absence of closer chronological definition for these samples does not allow either the establishment of a chronological relation between material and shape, which could be important, nor a connection to historical events (Little 1990, 73, 155). Perhaps a more accurate characterization and dating might illuminate statements such as the 'micaceous pottery appeared more standardized on the regional level than did the granitic' (Little 1990, 121), since function might

61 - As mentioned elsewhere, we will neglect Jones' translation (waxen vessels) in favour of Shulten's (wooden vessels).

62 - According to Lopez Cuevillas (1953b, 215), fragments of a decorated wooden vessel were found at Castro de Cameixa.

63 - This was recorded in the course of my own field research carried in castros of the Vila Nova de Famalicão, region, such as Vermoim and Ermidas, in contexts dating to the I[st] century BC, but other references (Almeida *et al.* 1981, 58, fig. 30-1; Silva 1986a, 123, fig. 48-4) indicate a wider use of this technique in earlier stages of the Castros Culture.

also have had a role to play in the selection of such inclusions.

Macroscopic observation of several groups of pottery from Castro de Vermoim, dating to the Ist century BC, suggests that the use of quartz and feldspar may be associated with a particular vessel of rounded profile and with interior handles. This is the 'pot with interior handles', a local type of 'ceramic cauldron' (Figure 48-B) which was suspended over the fire for long periods. This vessel was the only one with a fabric composition quantitatively dominated by quartz-sand. Although tempering with quartz might have detrimental effects if the vessel were to be used at temperatures higher than its firing temperature (Rye 1981, 34), evidence indicates that this would rarely be the case. Sherds of this particular material also show a consistent firing, and are more resistant to intentional breakage than those with a micaceous temper. Conclusions might be drawn from this fact that a relatively unusual fabric is connected to a particular shape and function where particular thermo-plastic properties are needed. One must assume, therefore, that it was known by the Castros potters that the inclusion of mica would diminish the resistance of the vessels to heat and breakage (Anderson 1984), but mica was probably used for decorative purposes and for a pleasing effect. The absolute chronology and cultural phasing for this technical variability is one important aim for future research.

Discussion is also needed concerning the chronology and circumstances of the spread of the potter's-wheel in the region. A close relationship between the spread of wheel-made pottery and the rotary quern has been suggested (Silva 1986a, 122; 1990, 314) but a purely technical explanation for this such as the use of the round quern being responsible for the spread of wheel-made pottery, does not seem adequate. Indeed one has to bear in mind that they are related to very different production functions, and are not necessarily related during became of this technical novelty in the region.

The proposal made by Snodgrass (1980, 368) for the spread of iron metallurgy may be applicable in this region. According to it, one should make a clear distinction between exclusively technical factors and economic, or cultural ones. In fact, a large body of evidence now indicates that ideas and techniques traveled surprisingly fast in first-millennium BC Europe. Wheel-made pottery was produced in central and southern Iberia as early as the VIIth-Vth centuries (Aubet Semmler *et al.* 1983, 108-16; González - Pina 1983), and ideas certainly traveled along with the Vth-century BC Greek pottery found in some castros. In the present state of research, it cannot be accepted that the late adoption of the wheel in castros, applied either to pottery manufacture or to cereal grinding, is exclusively due to technical ignorance. Besides, the data indicate that use of the slow potter's wheel had already spread throughout the region by the IInd century BC (Silva 1986a, 122), whereas the rotary quern did not displace the saddle quern until Augustan times (according to my own field research). Any explanation should take in to account contemporary changes in the social system, and the related economic and demographic development, which could be pointed to as the motor for the start of the mass production of pottery in the Augustan period, when the use, and probably spread, of the fast potter's wheel becomes indisputable in the archaeological record. The round quern, is connected principally with the intensification of agriculture and the rising importance of cereals in the diet, and its relationship to the mastery of granite masonry and the availability of iron tools is necessary to any understanding of its diffusion.

7.3.2) -Metalwork

Metallurgical activities occurred across the Northwest during the first millennium BC. The research of recent years has greatly contributed to the building of a coherent picture, not particularly by the addition of new finds, but rather by the acquisition of contextual and technological information which has permitted a more complete understanding of the data produced by past generations.

Evidence for a thriving bronze industry can be found in the Northwest throughout the Ist millennium BC, in the form of both metal artefacts and residues that reflect their manufacture, such as slag and moulds.

The importance of Northwestern metallurgy in the Bronze Age has been fairly evaluated in the past, and properly studied and correlated with other regional European cultures (Almagro Gorbea 1977; Hawkes 1952; MacWhite 1951; Martinez Santa-Olalla 1946; Monteagudo 1977; Savory 1949), although only more recently has a successful attempt to insert metallurgy into the broader cultural and social aspects been made (Calo - Sierra 1983; Coffyn 1983 and 1985; Jorge 1988b; Kalb 1980a and 1980b; Silva 1986a) been made.

The most remarkable bronze metalwork in the whole millennium is undoubtedly that of the Late Bronze Age, if we consider both the quantity and variety of the finds (Coffyn 1983, 1985; Monteagudo 1977; Silva *et al.* 1984). Finds occur in hoards, river deposits, more rarely in settlements, and their distribution has an Atlantic bias. The numerous Northwestern bronze finds dating to this epoch reflect not only a massive production of metal and the control of the mineral sources but more importantly, a mastering of metallurgical technique, as in other regions of Europe. This noticeable control of metal production and the trade networks would also have become possible through an increasingly elaborate social system. The connection of many finds with their site of manufacture was at first hesitantly attempted (Coffyn 1983 and 1985; Kalb 1980a), but soon new data[64] (Figure 66 and 67) enabled this point to be clarified. Cumulatively supported by this evidence, it became an established inference that many of the hilltops where Northwestern castros were built may have had a earlier occupation during the last phases of the Bronze Age. This has tempted some archaeologists to date castros as far back as the Late Bronze Age, and allowed others to question the validity of continuing to use the expression 'Castros for an all-embracing and chronologically-wide cultural phenomenon.

Excavation in recent years has substantiated Monteagudo's early suggestion (1954, 81) that from the late Bronze Age onwards every site, or group of sites, may have had its own metallurgical production centre.

How metallurgical activities developed from the Bronze to the Iron Ages in the Northwest of Portugal is by no means clear, in spite of many new detailed studies. There is still a gap in our knowledge of technological changes between the two periods.

In spite of the Castros Culture being of Iron Age date, there is a notable sparseness of iron artefacts at these sites until the last centuries of the millennium. For the Augustan period, however, the quantity of iron tools increases and slags have been found in castros during excavation. Complementary evidence, such as the noticeable spread of granite masonry (Almeida 1984) during this epoch, explains the spread of iron. The Castros masons of the Ist century BC/AD would have needed large quantities of tools made out of carbon-containing iron in order to produce their massive stone architecture, because granite is a hard rock which requires a considerable investment of labour in this quarrying (Hawkes 1984, 188) and for its cutting and dressing.

Our knowledge about iron metallurgical activities in the Castros Culture is still very limited. Evidence concerning the extraction of iron ores is non-existent, although other materials implying similar mining processes, such as cassiterite and the presumed remains of its furnace (Costa, HM, 1965-66; Santos Júnior 1957, 48; 1964, 361-5; 1966b, 183, 186), are known in castros. The processing of ores cannot be deduced from the archaeological evidence, but taking into consideration the case of Muro da Pastoria (Soeiro 1985-86b, 25) where the fragments of a presumed furnace were found in a context dating to the Ist century BC, this lack of data may be due to limited field research. Slags and over 35kg of iron ore were also recovered from this site. This evidence is sufficient to indicate the presence of metallurgical activities at Muro da Pastoria.

Evidence from other areas also reveals that iron working was a common practice in castros by the turn of the millennium. A reasonable quantity of iron and bronze tools, as well as weapons, slag, crucibles, casting moulds and blacksmith's tongs, was found inside a circular house (Pericot - Lopez 1931, 17-37) at Castro de Troña (Galicia). Blacksmith's tongs were also found at Meirás, Faria (Luengo y Martínez 1950, 106) and Sanfins (Silva 1986a, Fig. 88-1). Whereas the reduction of iron ores was most probably done outside the hillfort, the production of metal tools could generally proceed within the settlement, probably in a peripheral area (Pleiner 1980, 396).

It is not yet clear whether the raw materials were transformed into metal close to the mining site, or transported to central sites for smelting. Widespread evidence for metallurgical activities in the castros (Almeida *et al.* 1981, 70; Almeida, CAB, 1982b, 24; Carballo Arceo 1983; Cardozo 1980, Fig. 31-1; Costa, HM, 1965-66; Eiroa 1988, 111; Silva 1983a, 134; 1986a, 168-82; Santos Júnior 1957, 48; 1964, 361-5; 1966b, 183, 186; Soeiro 1985-86b, 25) suggests that even if the smelting of ores was conducted outside these sites, then at least the final removal of the slag impurities and tool making was done by resident blacksmiths in the castros. This is no surprise if one considers the fact that iron smelting techniques prior to the turn of the millennium were relatively primitive, and therefore the resulting metal carried an excessive amount of impurities (Cavalheiro 1989, 131) which would have to be expelled by subsequent hammering and quenching.

These activities may be limited to a relatively small number of centrally-placed sites, such as the castros of Penices and Ermidas, where sorting of soil samples under the binocular microscope, taken during excavation for environmental analysis, has reveled the presence of small granules of metal. These samples came from Julio-Claudian contexts, and their study is unfortunately not yet complete enough to allow a wider correlation. There is no doubt, however, that these small granules of metal are the result of smelting, and their presence can be either corroborative of other types of evidence, or sufficiently diagnostic on their own for the existence of metallurgical activities on these sites in the past to be a distinct possibility. The future application of this method of information retrieval to excavated sites will certainly supply further data towards an identification of metal production centres in this region.

Slags are frequently found in castros but have never been the subject of proper study. Bearing in mind that their macroscopic analysis cannot indicate whether they are the result of bronze or iron working (Bachmann 1982, 4), and considering that the

[64] - In particular by the identification and further excavation of Late Bronze Age settlement sites with evidence of metallurgical activities, or simply bearing the contemporary pottery of "Alpiarça type" (Marques 1972, Marques - Andrade 1974). An updated reference to these sites in northern Portugal can be seen in Martins (1988a).

presence of slag does not necessarily imply that smelting activities were carried out at a site (Cunliffe 1984, 437), little can be added to our knowledge from the source at present. The first few scientific analyses of iron artefacts now available in the region (Cavalheiro 1989)[65] give us some idea of the technological features of Castros iron metallurgy, and indicate how future research may benefit from this type of study. For example, the iron sample from Muro da Pastoria suggests that small furnaces were used. In fact, the characteristics of the metal indicate that comparatively small nodules were produced, owing to the existence of a high percentage of slag inclusions. The overall scarceness of data on this matter makes premature conclusions a great temptation, which one will cautiously avoid. Our knowledge to date raises two surprising conclusions. One is that Castros metallurgy is a late phenomenon (at the turn of the millennium) which had a slow technological development, and the other is that the process of Roman acculturation may have greatly improved techniques of smelting and iron production in the Northwest. However, this model is not isolated, but has been based also on other aspects, such as Castros architecture (Almeida 1983a, 71; Almeida *et al.* 1981, 89-90) and therefore gains support.

The three stages of development in Mediterranean iron technology, as defined by Snodgrass (1980, 336-7) seem applicable to the Northwest, as a whole. In fact, and in spite of the early appearance of this metal in the archaeological record, its rareness and almost exclusive use for ornaments and prestige items (Peña Santos 1988, Ruiz-Gálvez Priego 1980, Silva 1986a, 170) reveal that this was still an unfamiliar commodity in a region with a prosperous bronze industry. For a long time, and until the beginning of the first century BC, iron remained rare and bronze goods, both decorative and utilitarian, are dominant in the Northwest. Only in the last phase, which is coincident with Romanization, did iron become commoner, although bronze artefacts still continued in use.

The jewellery of the Northwest, which had already attained a remarkable quality in the Bronze Age (Ruiz-Gálvez Priego 1989), became an increasing specialized product as the first millennium progressed.

Gold and silver craftsmanship at the castros became highly skilled, with its own set of specialist tools (Taboada Chivite 1977, 71). The Northwest exhibited a wide range of such Iron Age jewellery (Cardozo 1967, 1968a), such as torcs, bracelets and earrings. The industry developed in the later phases, and a renewed vitality resulted from the initial contacts with the Romans. The mastering of filligree and granulation techniques (Teixeira 1946) made possible the production of ornaments such as the torcs from Vilas Boas (Santos Júnior - Freire 1965, 1966) or Lanhoso (Teixeira 1939). Manufacturing techniques varied, from the simple use of a mould and hammering, to the application of filligree, granulation and soldering (Cardozo 1957c) so characteristic of the later phase. Such a degree of craftsmanship again suggests that specialist itinerant goldsmiths did this work, indeed, the distinctive distribution of ornaments with similar typological and decorative features (López Cuevillas 1951c, 40) supports the existence of a group or *school* of goldsmiths (e.g. the Chaves *school*), who would cover one particular area (Cardozo 1942b, 98; 1965c, 168).

7.3.3) Weaving

The activities connected with the production of fabrics in the region (Cardozo 1965a) are not fully documented in the archaeological record. Yet, the production of thread is widely represented by the spindle whorls. These have been found in every castro, and their presence is noticeable from the Late Bronze Age onwards. Spindle whorls are normally made of the same fabric as contemporary pottery and sometimes decorated, but they can also be produced by re-shaping a sherd out of a broken vessel. Evidence for weaving, such as loom weights, does not occur before Romanization. However, the local production of clothing can be inferred from the representation of fabric decoration on some statues of warriors (Figure 56.3) in addition to references in the Classical sources (Strabo III, 3, 7).

7.4) -Raiding

The insistence with which the Classical sources refer to raiding among the Callaicans is enough on its own for us to select this subject as one of the most important in the Castros system. In this section an attempt will be made to define the reasons for this phenomenon by reviewing and re-interpreting the evidence. The extent to which raiding may have affected both the economical and social system will also be analyzed, in order to establish whether it may, to some extent, have been responsible for the self-sufficiency and elaborate fortification of the castros.

65 - Some of the information presented here was disclosed by Eng. Roberto Cavalheiro, Professor at the Department of Metallurgy of Porto University, in the couse of one of his lectures. I am grateful to Eng. Cavalheiro for his important references. However, his reference to an arrowhead from Pastoria (Cavalheiro 1989, 131) does not coincide with the list of finds published by the archaeologist (Soeiro 1985-86b, 25), where only an iron staple is mentioned.

7.4.1) -The evidence

Diodorus Siculus V, 34, 6

"And a peculiar practice obtains among the Iberians and particularly among the Lusitanians; for when their young men come to the bloom of their physical strength, those who are the very poorest among them in worldly goods and yet excel in vigour of body and daring equip themselves with no more than valour and arms and gather in the mountain fastnesses, where they form into bands of considerable size and descend upon Iberia and collect wealth from their pillaging. And this brigandage they continually practise in a spirit of complete disdain; for using as they do light arms and being altogether nimble and swift, they are a most difficult people for other men to subdue."

Julius Caesar "The battle for Gaul" VI, 23 (Wiseman-Cunliffe 1980)

-"... No disgrace attaches to armed robbery, provided it is committed outside the frontiers of the tribe; indeed, the Germans claim that it is good training for the young men and stops them becoming lazy. When one of the chiefs announces at an assembly that he is going to lead a raid, and calls for volunteers to go with him, those who agree with the raid and approve of the man proposing it stand up, and, applauded by the whole gathering, promise him their help. If any of these men then fail to go with him, they are regarded as deserters and traitors and no one ever trusts them again in anything...".

As defined above by two Classical authors, raiding is the association of a variable number of warriors, or armed men, who gather together and form a party, with the main purpose of making a fighting expedition outside the territory of the communnity they belong to. As also referred to by the sources, these raiders are usually young men who have not yet become attached to the social and productive system.

Raiding has been repeatedly referred to by the Classical sources as the men's major and exclusive activity, in which only warriors can take part. However, some aspects of raiding in the Castros Culture are not yet entirely clear, and especially the objectives and social interactions engendered by this activity.

Raiding has been connected with warfare in the course of this study, in the sense that it is one aspect, amongst others, of what are generally considered as warfare activities.

Traditionally, two major motivations for raiding have been put forward: to obtain an economic surplus[66], and to provide a means of escape from internal pressures within the system. However, in communities where the role of warrior is socially dominant, this may be an essential factor for maintaining equilibrium in the system, as an escape from internal pressures. The first argument is the most obvious one, and that usually quoted by the sources. In fact, even in Modern History there is a trend always to bring forward the more obvious arguments for large-scale social conflict, which are explained in economic terms. It is no surprise that the Classical sources justify raiding amongst the *barbarian* peoples as the response to poverty. The approach expressed in this section intends to go beyond this simplistic explanation, of a phenomenon that was so important for the Castros Culture that it developed the characteristics of a warrior society.

Apart from the Classical sources, there is little corroborative archaeological data for raiding amongst the Callaicans.

The extensive and careful fortification of sites throughout the millennium is undoubtedly connected with insecurity amongst peoples. In fact, this does not fit well with the picture provided by the domestic architecture and the subsistence economy.

Such widespread phenomenon as fortification, found throughout the Iberian Northwest, cannot be explained by exogenous factors like invasion. It also has something to do with relationships between neighbours solely and eventually also with competition for resources. Bearing these considerations in mind, a model will be put forward in explanation of raiding in the region.

7.4.2) -Raiding as a system

In the course of this chapter, it was attempted to demonstrate that the explanation of raiding based on the poverty of both the region and the communities can no longer be sustained.

As suggested by the available evidence, raiding can be on either a regional, or a long distance scale. Long distance raiding will be analyzed in Chapter 9. As for regional raiding, it would be directed towards the sites and communities of the region. However, this appears to contradict the idea that cohesion existed in both the social structure (Chapter 8) and material culture. The autonomy of the castros, has been confirmed by an analysis of references to the *castellum*. It will be argued below, that the autonomy of every castro (Pereira Menaut 1983, 206) within the superstructure it belonged to meant that raiding amongst neighbours could take place without disrupting the cohesion of the system, although such pressure may be one of the reasons for fortification of the sites.

Raiding in Castros Culture does not entirely fit with that proposed by Nash (1985, 46-7) for the same phenomenon in Celtic Europe, although important similarities should be pointed out. Mercenarism and raiding as an escape from the

66 - The arguments presented here are intentionally limited. In fact, Anthropology has shown us that there are many other reasons. For example, amongst the *primitives*, raiding can also serve for *interchanging* women between different groups.

internal political and social tensions are, in fact, common features to both Celts and Callaicans.

It has been shown for other areas, that raiding expeditions do not have any particular effect on the overall economic system (Garlan 1972, 12) to which the warriors belong. In fact, as already suggested, agriculture and probably most of the productive economy may have been controlled by female work, to which raiding was alien. Society does not depend economically on raiding for its survival, although social interactions probably do.

There is almost no vidence for a pattern of age, rank and social characteristics amongst the raiders. It is not yet clear if all the armed men would have participated in the expeditions or were in a social position to participate (Fernandez Canosa 1987, 150), or if only a chosen élite of warriors had that privilege. However, the former may be anthropologically correct because in the large indigenous armies referred to by the Classical authors (i.e. long distance raiding) these warriors came from a variety of social backgrounds.

An overview of the Castros productive economy discloses a picture of internal development and self-sufficiency (Vázquez - Pombo 1991b, 199). This contradicts those arguments (Garcia Bellido 1945a, 1946a, 1977 ; Schulten 1927) which present the Callaicans and Lusitanians as victims of a poor and inhospitable region, who had an unshakable devotion to both freedom and their homeland. The recent, and interesting, approach to this subject by Garcia Moreno (1988) is in accordance with my ideas of the detachment of war and brigandage from the economic needs of the Castros Culture.

Certainly pillage was not the most important aspect of the raiding, but rather the achievement of prestige (Fernandez Canosa 1987, 153) and military glory. However, it is known that social and military status can also be achieved through the possession of prestigious items, and this fact is likely to have created different attitudes towards raiding. Mobile goods, such as livestock and jewellery may have been the spoil of short-range action. This might be expected to have an impact in a pastoral society, as the Classical sources imply, in addition to the symbolic value held by ornaments such as torcs for the Celtic warrior in general.

As Diodorus (33, 7, 1) described in the case of Viriato, amongst warriors there was a certain disdain towards those goods that had not been obtained through a prestigious activity, such as raiding, and which the owner would not be able to defend as a warrior. It looks as though the value given to an item mainly depends on the way it was obtained, and how easy it was to keep possession of it. So Viriato despised his father-in-law's wealth as luxury goods that could only be protected by the power of the spear. In this society of warriors, wealth acquired through honest toil was not respected. This implies the existence of some social contradictions: for example, the male status of mobile wealth acquired by raiding versus women's control of land and production (Almeida 1983b), which is so clearly reflected in the Castros Culture.

Raiding could also have been a seasonal event. In fact, seasonal raiding, similarly to the practiced amongst the Celts (Nash 1985, 47), and later in the Middle Ages, may have occurred, thus allowing the warrior some months of involvement with the productive system. As suggested by the Classical sources, pastoralism may have been the productive activity more connected with the warriors.

Being subject to various interpretations, such as an 'ethnic tendency to an adventurous life' (Taboada Chivite 1977, 74), and a side-effect of the matriarchal system (Garcia Bellido 1945a, 177) which was typical of the Northwestern peoples, raiding has never received a proper socioethnic analysis. A number of different factors may all contribute to raiding, e.g. it may be a social activity or an economic supplement to production. If the Classical sources are vague when referring to these practices for the Callaicans, then they are equally generous when it comes to describing the battles of the Lusitanians, especially the so-called *Lusitanian wars*. As discussed above, Strabo's (III, 3, 2-3 and 6) account is rather ambiguous, so that a clear distinction between these peoples and their areas of distribution cannot be made[67], although the following paragraph looks conclusive enough: (Strabo III, 4, 20) "*...to those Lusitanians whose country is situated alongside Baetica and extends as far as the Durius River and its outlets (indeed at the present time they apply the name Lusitania specifically to this country)...beyond the Durius to the north: the inhabitants of this country were spoken of by the people of former times as Lusitanians, but by the people of to-day they are called Callaicans...*"

Should we perhaps consider this warfare and raiding, which the Classical writers attribute to the Lusitanians, as common to all the peoples of the Northwest? Indeed we can, and this point is strongly defended throughout this study and applied to other aspects of the Castros Culture apart from raiding and warfare. Besides, according to Alarcão (1988, 15-6), the term *Lusitanians* might not refer to one people (*populus*) but instead be the generic name for a group of peoples which occupied the same geographical region or shared an ethnic background.

Was this raiding aimed towards the South, or was it merely a random phenomenon? Both the degree of fortification in Northwestern castros and references to an expedition of Celtics and Turdulians

67 - Here, the resemblance between the described Lusitanian warrior and his weaponry, and the "Callaican warrior" statues which have distinct area of distribution between the Douro and Minho rivers, is clear. The same applies to the distribution of public baths structures, as referred to also by Strabo.

in the coastal Northwest (Strabo III, 3, 5) indicate that many regions were indeed affected by external raiding. The direction of these expeditions towards the Iberian South may have occurred for two reasons. Firstly, Baetica was a rich commercial region with a wide range of commodities brought in by Mediterranean trade. This made plundering a rewarding activity. Secondly, the written sources imply that a quest for military glory in the powerful Roman army may have been an attraction for the raiders, as a glimpse at the record of expeditions after 194 BC clearly shows.

7.5) -Long-distance trade and exchange

Trade networks prior to Romanization cannot yet be established with archaeological certainty. Hopefully, scientific methods of analysis may soon change this. For long-distance trade, one can fortunately rely on exogenous finds datable to the second half of the millennium, as shown in Figure 51[68]. Many of the finds included in this figure are recent, and certainly this list will soon be augmented by the results of current research. However, this relative profusion of data should not be allowed to hide the fact that many other types of goods would have been exchanged in the region during the Iron Age. I am, of course, referring to the perishable goods that would leave little trace in the archaeological record.

As an example, let us consider the so-called *sand core glass* of Mediterranean origin. Although attested archaeologically, this attractive object was little more than a container. Its value lies in the fact that it contained perfume or ointment (Feugére 1989), and archaeologists often forget this important aspect.

Thus, there was undoubtedly an intense trade between the region studied and the Mediterranean during the second half of the millennium, in which perishable goods like perfumes may have played a major part, even if metal goods formed the basis of long-distance exchange.

The turning point proposed by Silva (1986a) for some changes in the regional system circa 500 BC can be corroborated by this evidence.

These products were probably transported mainly by sea, as the coastal distribution of the finds (Figure 51) suggests. However, exchange with the hinterland was also necessary, for which the natural routes of communication like rivers were used.

Overland routes, or roads, are an important means of contact betweens populations and cultures, encouraging the spread of goods and ideas. In the Northwest roads also aided Romanization (Almeida 1969, 15).

According to Strabo (III, 3, 7), and supported by Caamaño Gesto (1979b) pre-Roman roads existed in the Northwest, connecting sites and regions. The suggestion that "*the sick were exposed upon the streets so they could get suggestions from those who had experienced the disease*" might imply a reasonable circulation of people within a region.

There is little archaeological evidence for the use of wheeled transport in this region before the Roman Conquest, although carts were apparently known from the early phases of the Castros Culture, but probably not as a means of transport[69].

A bronze votive wagon[70] was found at Costa-Figueira (Penafiel, Portugal) and dated by Cardozo (1946b) to circa 350-250BC. It has recently been referred to as being of an *indigenous tradition* (Piggott 1983, 193).

On the other hand, the use of carts in southern Iberia is well documented, according to various finds datable within the first millennium BC (Serra Ráfols 1948, Blázquez Martínez 1955, Cuadrado 1955, Pijoan 1953, Garcia Bellido 1969, Schüle 1969, Garrido - Orta 1978). Here we are dealing with a world full of Mediterranean parallels (Piggott 1983, 193-4) and close contacts which are supported by eastern colonization[71], of which Tartessos is a good example. As far as Castros Culture of northern Portugal and Galicia is concerned, there is no evidence for the use of wheeled vehicles as there is almost no horse gear, and an absence of wheel-marks[72] in the paved entrances of the excavated castros. Until the end of the millenium therefore, transportation must have been by foot, or using pack animals. The Costa-Figueira wagon is something of a problem. It might be a votive item, used for ceremonial purposes (Piggott 1983, 132), or related to external cultural influences, although native motifs are present on it.

Other evidence for wheeled vehicles includes a wooden wheel, found in a bog at Catoira (Pontevedra) in Galicia. This wheel had a very early radiocarbon date, circa 3670±45 BC (Galhano 1973, fig. 11, Lorenzo Fernandez 1971). If confirmed, this would be a very early date indeed for the use of carts in the Northwest. However, further inspection of this find, (Lorenzo Fernandez 1973) revealed an inscription of the XVIIIth century. A complete revision of the data is thus required.

68 - This list of imported finds is necessarily limited. A more complete list (e.g. including amphorae, glass and beads etc.) can be seen in the extensive study by Naveiro López (1991).

69 - Caamaño Gesto (1979b, 283, foot. 22) accepts the use of wheeled vehicles in Iron Age Callaecia although he does not refer to any relevant evidence.

70 - Referred to by Piggott (1983, 193) as a car, although it is actually a four-wheeled vehicle.

71 - See the recent work of Gamito (1988) on this subject.

72 - In fact, wheel-marks can be seen in a few Galician Castros (Lopez Cuevillas 1955c, 151; Taboada Chivite 1977, 77), but they might fall in the Roman period, considering the evidence for late occupation of these sites.

Trade and cultural contacts with Mediterranean peoples[73] seem to have occurred early in the Northwest, and references to *direct exchange* and the use of silver bars (Strabo III, 3, 7) as currency have been interpreted as an indicator of a pre-monetary exchange economy (Centeno 1987, 187-8, Alonso del Real 1979, 63).

Keeping the above evidence in mind, one is tempted to question whether, in fact, all the exogenous finds in the Castros Culture derive from long-distance trade. It is now clear that raiding expeditions, no matter in what form, reached as far as the Mediterranean shores. Since pillage was one of the motivations for these expeditions, it would not be a surprise if some of the exogenous finds in the Northwest, of first millennium date, had arrived this way. Of course this assumption by no means disminishes the importance of maritime long-distance trade, which the coastal distribution of finds (Figure 51) supports.

8) -Discussion

The evidence now available on the productive system of the Castros Culture is sufficient to refute the traditional picture of under development and poverty, which has been defended by the Classical sources and by archaeologists for so long.

A balanced use of all the available resources allowed the Castros communities to flourish, and a relatively high population density in the region was no obstacle to this. This particular adaptation of the productive economy to the most immediately available local resources meant that the subsistence base varied from castro to castro, i.e. the economic role of cultivation, pastoralism, marine resources or metallurgy, for example, depended entirely on local conditions.

The state of the Castros economy until the turn of the millennium should be understood in terms of a particular social attitude towards the productive activities, rather than as a result of technological or environmental pressure. It appears that the Castros communities were in control of their production but in some cases, the acquisition of an economic surplus for exchange was neglected in favour of warfare, raiding and the building of defensive structures.

Another important point is that a site, or group of sites, appears to be economocally self-sufficient (Vázquez - Pombo 1991b, 199), and thus to some extent independent within the regional system. The available data are still insufficient to take this matter much further, but the existence of an integrated regional system and long-distance trading contacts cannot be denied. Only non-precious metals seem somehow to escape from this general rule, and even these circulated to only a small degree within the region. There is evidence for metallurgical activity at most of the excavated castros, and each site was probably able to supply its own needs. Iron production was, in fact, of little importance in the Iron Age until Romanization. Bronze alloy quality remained high until the end of the Bronze Age, when an unusually high percentage of lead characterized these alloys (Coffyn 1983, 1985). Bronze was in fact preferred to iron for both utilitarian and prestige goods throughout the millennium. The reasons for this be due to the following factors: the superiority of sharpness of bronze blades over non-carbonized iron ones on tools and weapons; the fact that it was easier to decorate bronze than iron; a plentiful supply of raw materials and a highly developed bronze industry to which iron could not easily be accommodated; the relative isolation of this region from outside contacts and ideas concerning ironworking; and the more difficcult procedure involved for the reduction of iron oxides.

During the period of the Castros Culture there was a marked change in pottery manufacture, from the use of sandy clay fabric to one rich in mica, which probably occurred for cultural, rather than technological reasons. This change coincided with the building of the first stone defensive structures[74], and the transformation to an Iron Age cultural setting.

Raiding, even if we view it in strictly economic terms, was not necessitated by any deficiency in the productive system, even though mobile goods may have been transferred through this activity. Raiding was almost independent of the subsistence economy, its value being of a purely social nature. Thus, raiding will be discussed further in Chapter 9, along with other aspects of the role of warfare in society.

The subsistence economy of the Castros Culture has been explored at some length in this chapter, and the castros themselves established as independent and self-sufficient production centres. The next step, which forms the subject of the following chapter, is to examine the social and political structure of the Iron Age Northwest under Roman influence in more detail, and determine the position of the individual castro in this hierarchy.

[73] - Strabo's reference (III, 4, 3) "As for Cantabria, a part of it was held and seized by the Laconians, according to both Asclepiades and others." might imply the Greek trade control of the Atlantic coast.

[74] - The first phase of occupation at Castro de Penices (see Figure 2.2, in Appendix II, for absolute chronology) has an example of an early stone defensive wall, which was built at a time when dwelling structures were still made with perishable materials and the pottery had a sandy fabric. Late pottery was micaceous.

8- THE SOCIAL STRUCTURE

8.1) -Introduction

During the course of the Bronze Age, and in particular during its final phase, Europe witnessed the development of highly structured warrior societies. The rise of these societies and the social dynamics operating within them, was in part responsible for the displacing of human communities throughout Europe. Apart from populations, ideas and artefacts also traveled apace across first millennium Europe, spreading technologies (Coles - Harding 1979, 10), trade and, more importantly, establishing an intricate network of cultural contacts between peoples. This interaction, no doubt, introduced such populations to a new dynamic, where external influences had an increasing role to play in the internal evolution of their communities.

The reasons for this movement of warrior societies have been the subject of many studies. In these studies, models were suggested which might help to explain this phenomenon, whose complexity, together with the patchiness of the archaeological data in Europe, made a full understanding of the situation difficult.

Evidence of a wide range of contacts between the Iberian Northwest and the rest of Europe from the early first millennium onwards has been referred to already in some general studies (Coffyn 1985, Hawkes 1952, MacWhite 1952, Ruiz-Gálvez Priego 1984 and 1986). These focused on the interchange of resources, artefacts, technologies and, more importantly, peoples and ideas. It would not be surprising to learn therefore, that the components of this *prehistoric European community* were subject to varying degrees of external influence, which in turn exerted a considerable effect on its cultural and social development.

It is also generally accepted that, from the Urnfield to the Celts, the overwhelming majority of the social systems in first millennium Europe are highly structured and more or less related to warfare. Competition for resources, and warfare, is reflected in site distribution (Collis 1975, 1984b, 1984c; Härke 1979), in cemeteries (Quesada Sanz 1989), in the economic system and in social structure (Brun 1987, Collis 1986, Cunliffe 1988, Wells 1980).

For Castros Culture, similarly to other aspects focused on above, an image of incipience of the social structure has been brought up by research, thus producing a basic picture of an unsatisfactory duality[1]. The existing model of a society composed of women farmers and men warrior-brigands cannot, in fact, fit with the overall body of evidence available to date, nor be accepted under the theoretical approaches currently used for the study of social systems. Furthermore, there are also many blank spaces which the current picture cannot fill. Aspects such as the high diversity -and technological complexity- of the productive activities, undoubtedly indicate a fuller picture of the Castros social system. But it is not my intention -nor would I have the capabilities- to present a new model. The sections below are concerned with the review of the available data for the social structure of the southwestern area of Castros Culture, expressing, whenever possible, a critical view.

8.2) The Celtic 'family' in Europe: some analogies

Caesar's *De Bello Gallico*

(VI, 11)- "*In Gaul, not only every tribe, every canton, and every subdivision of a canton, but almost every individual household is divided into rival factions. The leaders of these factions are men thought by their followers to have the greatest prestige, and it is to their judgement and assessment that they turn in any question or discussion of policy. The reason for this long-established custom seems to have been to ensure that every man among the common people should have protection against those more powerful than himself.*"

(VI, 15) - "*The second class is that of the knights. Whenever a war breaks out and their services are required -and before my arrival in Gaul, almost every year saw them involved either in an offensive or a defensive war- they are all involved in the campaign, each one attended by as many retainers and dependants as his birth and wealth make possible. The size of a knight's following is the only criterion of influence and power they recognize.*"

(VI, 16) - "*The Gauls believe the power of the immortal gods can be appeased only if one human life is exchanged by another, and they have sacrifices of this kind regularly established by the community.*"

(VI, 17) - "*When they have decided to fight a battle, it is to Mars that they usually dedicate the spoils they hope to win; and if they are successful, they sacrifice the captured animals and collect all the rest of spoils in one place.*"

(VI, 23) - "*No disgrace attaches to armed robbery, provided it is committed outside the frontiers of the tribe; indeed the Germans claim that it is good training for the young men and stops them becoming lazy. When one of the chiefs announces at an assembly that he is going to lead a raid, and calls for*

[1] - However suggestions of higher complexity can be found in recent research (Santos Yanguas 1985, Garcia Fernández 1990).

volunteers to go with him, those who agree with the raid and approve of the man proposing it stand up, and, applauded by the whole gathering, promise him their help. If any of these men fail to go with him, they are regarded as deserters and traitors and no one trusts them again in anything."

For many years in the past, it was believed that Castros society was structurally simple. A matriarchy controlled the social and productive system, while the men's role was to provide an economic surplus through brigandage in neighbouring territories. A substantial body of recent work, mostly composed of monographic studies, has produced quite a different picture of Castros social organization and the interactive effect of Roman contact on it. A study of religious belief and ceremony[2], although not always in accordance with the archaeologically-based interpretation, reveals a structural framework that is remarkably similar to that found in contemporary societies in other regions of Europe. Although the available data are still scarce and fragmentary, it is, perhaps, time to stop considering Castros as an obscure culture confined to the northwestern corner of Iberia by the prevailing geographical conditions.

One of the major problems in modern Portuguese field archaeology is still the coherent interpretation of archaeological data. A clear perception of either the cultural and ethnic movements or of the external influences occurring has seldom been obtained using the archaeological evidence alone. Further, the ideological aspects of the Castros Culture not mentioned in the written sources have been neglected and open to discussion for a long time. Religion, social and political organization and other aspects of life have been subject to various approaches and speculation with little confirmatory support from the data, which all too often have proved unreliable.

Thus, it may be difficult to distinguish any definite evidence for Celtic presence or cultural influence from strictly local innovations in the archaeological record. In the early days of Castros research, the tendency was to emphasize the Celtic contribution[3]. To some extent this is now being redressed by modern research based on the interpretation of data obtained by excavation on a regional scale. Some advance towards a definition of this problem has thus been made. Nowadays, the idea that the Northwest has been an important region since the Neolithic, by virtue of its openness to European and Mediterranean contacts and influences, is beginning to take root in Iberian archaeology.

The above quotes from a written source on the Celts, to which many others could be added, contains an interesting set of observations, which we may compartewith what we know about Callaican society. Of course, it is always tempting to indulge in forced and improper comparisons, or in comparative models to fill the more persistent gaps in one's research. On the other hand, the *barbarian* cultures of the ancient world were often all the same in the eyes of historians and geographers, who were confused by the co-existence of the civilized with the uncivilized. This is an equally difficult problem for modern historians and archaeologists. Bearing this in mind, a multi-disciplinarity approach is vital in any research, in which information available from the classical authors is correlated with the archaeological evidence wherever possible. I shall try to conduct this study along these lines.

To begin with, I shall take a look at one significant aspect of the 'Celtic world' which is common to all European warrior societies. This is the set of rituals necessary for the maintenance of a social hierarchy and its cohesion, namely the custom of feasting and squandering of wealth. If Poseidonius' accounts of the feasts of the Celts in Europe are to be believed, feasting was a ritualized activity, but probably an accurate reflection of the normal social relations and attitudes of the day.

Wine played an important part in these social rituals, as a prerogative of the wealthier warriors. It was consumed immoderately in the eyes of some authors, because the Celts did not have the *civilized habit* of mixing the wine with water, like the Greeks. The social consumption of wine predates the Celtic period. The Celts may have adopted the custom in the 6th century BC, as the presence of drinking sets in graves seems to indicate. This ritual later increased dramatically (Filip 1962, 92) as armed raiding became more common. This is not surprising, as raiding activities would undoubtedly help to strengthen the warfare hierarchy, thus giving a renewed importance to those rituals where these relations were reinforced or defined. In such a process a *catalyst* element, such as wine (Sherratt, 1987)[4] is necessary. Meat was also consumed in copious amounts, either roasted or boiled, during feasting. According to Poseidonius, it served as a regulatory symbol of hierarchy when it came to its sharing out. Pork, mutton and goat's meat were consumed during these feasts. In the Northwest the consumption of goats was common, and sacrifices of goats have been reported by Strabo (III, 3, 7), probably connected with the cult of Ares (Bermejo Barrera 1978, 1979,

---

2 - In general the substantial body of studies produced by Bermejo Barrera and his disciples, referring to the Northwest.

3 - Namely Iberian Castros research up to the decade of the sixties.

4 - One has to bear in mind that not all the European peoples used alcohol for this purpose during the Iron Age. The Germans, according to Caesar (B.G. 4, 2) did not allow wine, and the Scythians preferred hemp in their rituals, although wine was also largely consumed.

1981a) who was the deity connected with warfare[5]; however it could also be explained by the importance of sheep and goat in Castros productive system[6], as argued above.

The widespread distribution of cauldrons, flesh-hooks and roasting-spitts, in Europe, from the Bronze Age onwards (Coffyn 1985; Gerloff 1986; Hawkes 1952; Hawkes-Smith 1957), is undoubtedly connected with communal feasting. Moreover, the frequent occurrence of cauldrons, which often appear to have been deliberately cut into pieces, throughout the Northwest and Meseta strongly suggests a ritual connection, probably funerary in some cases. In support of this idea, there is some evidence for the destruction and intentional deposition of cauldrons[7] in Iberian cremation graves, e.g. in the Spanish Meseta. As far as the Castros Culture is concerned, evidence for the use of cauldrons (Acuña 1976, 330; Hawkes 1952; MacWhite 1931; Ruiz-Gálvez Priego 1979b; Silva 1986a, Schubart 1961) and *situlae* (Carballo Arceo 1983, Hawkes 1952, Monteagudo 1977) in the Northwest is plentiful. Cauldrons and *situlae* may, however, have had very different uses and ritual meaning[8].

When both the classical accounts and the archaeology are taken into consideration, the similarities in certain behavioural characteristics between the Celts from Europe and the Callaicans and Lusitanians are particularly striking and worthy of further comment. It is not the intention here to consider the degree and circumstances of the 'celticization' of the Northwest in any detail, although some considerations will have to be made. When new data comes to light, the situation may have to be re-assessed. In the meanwhile, either the impact of waves of Celtic migrants into the Northwest (*in genere* Bosch Gimpera 1932, 598-602; 1933; 1942, 57-61, 77-83, 91-3; 1945, 133-9; 1950-55, 117-22; 1950), regular visits by small raiding parties, or even some sort of cultural/commercial contact, would probably account for the observed widespread similarities in cultural organization, attitudes towards war, religion and general behaviour. The classical sources provide little relevant detail about the Northwest, and the archaeological data available is still limited. For the time being, therefore, it would not be wise to push the evidence too far by constructing models lacking a coherent basis.

8.3) Castros familial and supra-familial institutions

The peninsular Northwest can be divided into two main areas according to the types of social organization adopted by local peoples during the final stages of the Castros Culture. The two types of social organization are termed: the *gentilitates* and the *castella*. The latter coincides with the central area of the Castros Culture (Lopez Cuevillas 1933) and with what will later be *Gallaecia* (Pereira Menaut 1982, 266). This shows how important the former indigenous organization was for the establishment of the Roman Iberian provinces. The area of the *gentilitas* (Albertos Firmat 1977) is, on the whole, differentiated from the area of the *castella*, as discussed by Caro Baroja (1970, 1976), by its different social organization, from which a more Celtic affiliated ethnic stratum eventually rose.

At the present state of research it is known that the social structure of the *castella* area is organized in families, which are grouped in *castella*. The major ethnic unit is the *populus/civitas*.

8.3.1) Family

The family is the basic unit of Castros society, and its importance and social cohesion are clearly reflected by the existence of the dwelling units we have been calling *familial enclosures*, or simply *house*, thus associating the primary social unit (or a part of it) with the corresponding structure and related space of the castro. A family in Castros society is a group based on links of consanguinity and bound by the authority of the older members. The inscriptions found on two decorated granite door-lintels at Briteiros may add to our understanding of Castros familial structure, although the date of these pieces[9] should be noted. They read, CORONERI / CAMALI / DOMUS (Cardozo 1985, 5-n°3) and CAMALI - DOMI - / CATURO - (Cardozo 1985, 10-n°8), i.e. "*house of Coronerus son of Camalus*" and "*Caturo of the house of Camalus*" respectively. We have then in the first inscription a reference to a *house*, in the sense of *family*[10], whose patron is *Camalus*. The second

5 - Other kind of sacrifice practiced by the Callaicans and connected with the gods of war are the sacrifices of prisoners: the *hecatombs*. The sacrifice of human lives should have been so frequent and widespread in pre-Roman Iberia that, between 96 and 94 BC, the Roman senate has forbidden the inhabitants of Bletissa, in Lusitania, to offer human lives to the gods (Schulten 1937, 153), which has been denounced by Crassus, at the time proconsul of the Ulterior.

6 - As happens with the pig in the 'boars' statues' culture area.

7 - Sabine Gerloff (personal communication), to whom I am most grateful for this information.

8 - Cf. the presence of *situlae* in the diadem of San Martin de Oscos (Figure 60), which appear to be a ritual item within the depicted scene.

9 - The date of the elements associated with these pieces, such as pottery in the Museu Martins Sarmento, Guimarães, suggests that they should be placed in the second half of the first century AD or later.

10 - Regardless of its size and how far back the links of consanguinity can go for the purpose of familial cohesion. The reference to CAMALO found at the Castro de Alvarelhos (Santarém 1977, 164) also draws upon the displacement of people to other sites, probably for business or matrimonial alliances.

inscription refers to a different individual called *Caturo*, who belongs to the 'house of *Camalus*'. One can infer here a familial structure based on blood-ties (Silva 1986a, 268) in descending order, where members of the familial unit refer to themselves as being part of such a *house* (=family)[11], by naming its patriarch[12]. This system may be of limited duration, however. It may have been acceptable for a time, when the familial structure was subject to changes under Roman cultural pressure.

At such times the importance of the father within the family may have increased[13] (Almeida 1983b, 194), diluting the strength of the matrilineal links (Caro Baroja 1970); at least at the juridical level, which has already been referred to as a characteristic of the Northwestern peoples.

The sources are clear when referring to matriarchy. They certainly refer to a familial structure prior to the Conquest, which may have changed substantially at the dawn of the millennium. In fact, the classical sources mention a familial structure based on a matriarchy, where women owned the land (Strabo III, 4, 18; Almeida c.f. above) they worked on (c.f. II, 4, 17; Justinus XLIV, 3, 7, 10), and the transmission of the property was by female succession. This would probably be consistent with a warrior society, in which the male population was unstable, due to either long periods of distant raiding activity or to mortality. Although evidence for such a system in the north of Portugal is lacking, and regional peculiarities make generalization impossible (Silva 1986a, 271), it did survive the first stages of Romanization in Cantabria (Barbero - Vigil 1971, 220-3), where matriarchal links could still be traced in the epigraphy.

The proposed size of a familial group, consisting of the head of the family and his descendants, plus those brought in by matrimonial alliance (Silva 1981-82, 84), would be typical of a kinship group based on blood-ties. It does not agree, however, with the archaeological record, i.e. with the size and function of the 'familial enclosures', but rather with a small castro. As a familial enclosure is viewed as a unit of habitation for an agriculturally-based society, in which a number of different activities were carried out, it cannot be a straightforward indicator of the size of such a group as suggested by Silva (1986a, 268). The group was most probably fragmented for practical reasons[14], and without prejudice to the maintenance of familial cohesion based on blood-ties. The average-sized familial enclosure was probably a dwelling unit of 6-10 people, i.e. a typical rural family consisting of a couple and their children, and eventually their children's spouses.

8.3.2) *Castellum*

The social organization of the peoples that lived to the north of the Douro river may be better expressed by the word *castellum*, whose epigraphic representation is the symbol 'Ɔ '; the "inverted C". A long history of debate precedes the interpretation of this symbol, which is summarized as follows.

The first interpretation of the Ɔ, as *centuria*, was proposed by Hübner (1892, nº 1174) after reading and translating an epigraphic inscription, meaning the basic unit of social organization after the familiar group. This structure is similar to the *gentilitas* referred to in the 'pact of the *Zoelae*' (Schulten 1962, 72-3) and presumably relates to the suggested 'decimal structure' of Castros society (Rodriguez Adrados 1949), in the Indo-european manner.

This model remained for many years, until a synthesis of the known inscriptions bearing Ɔ was presented by Albertos Firmat (1975, 1977), and later supported by the discovery of an inscription in which this symbol was followed by a name ending in *briga*, which is a suffix of clear significance in the Celtic world. This produced a new model for the Ɔ (Albertos Firmat 1981), now meaning *castellum*, in the sense of the primary unit of political organization. In archaeological terms, the primary political unit is represented by the castro. This model was generally accepted and confirmed by subsequent work (Pereira Menaut 1978, 1982, 1983, 1985-86; Pereira - Santos 1980; Santos Yanguas, J., 1985; Silva 1980b, 1981-82, 1986, 272-9). Other interpretations were put forward (Bermejo Barrera 1978-80, 104-6; Rodriguez Colmenero 1979, 1988; Le Roux - Tranoy 1984; Tranoy 1981) but the arguments were not consistent enough to dislodge the Albertos Firmat's model. A last point on the subject was made by Albertos Firmat (1988)[15] in the response to further arguments of Rodriguez Colmenero (1988) on this subject.

[11] - Examples of this social aspect have survived down to the present time in the north of Portugal, in the expression "lar" (home) also used to name the room or part of the house where the hearth is, and therefore where the family gathers to socialize and have its meals.

[12] - This isolated fact could indicate patrilocality, rather than uxorilocality as suggested by Silva (1986a, 270; quoting Bernardi 1978, 276), but the reference to the father's name, which is widespread in northwestern epigraphy, is indicative of an established patrilineac system.

[13] - The particular importance of the crime of parricide can be inferred from Strabo's reference (III, 3, 7) "*...and the parricides they stone to death out beyond their mountains or their rivers.*" and therefore it is plausible that this process evolved prior to the Roman Conquest.

[14] - This fact can be observed on what have been called the "agricultural castros", which are discussed elsewhere. This type of site appears to have started in a time of population growth, and their connection to agricultural practices is undeniable. Moreover, the average size of such a castro suggests that not more that one of this extended families could have lived there.

[15] - In a paper written just before her premature death.

The greater contribution to an understanding of the ethnic and political organization of the peoples of the Northwest, and in particular of the communities connected to the system represented by the symbol Ↄ, has been provided by epigraphic sources, especially tombstones. This is partly the result of acculturation and the adoption of Roman mortuary ritual practices and ways of expressing the identity of the deceased, and dates from Augustan times onwards.

The references to the*castella* found in the Northwest bear, as a rule, the names of the individuals, their basic ethnic origin in the form of the name of the Ↄ they belong to, and normally, though not always, a reference to their major ethnic unit.

One question is raised by the fact that some inscriptions indicate the origin of the individual through a reference to the minor and major ethnic units, whereas others only refer to the minor unit, the *castellum*. A great advance in the understanding of this subject was achieved using a straightforward and simple, yet brilliant, analysis (Pereira - Santos 1980, 122-6) of the evidence. In simple terms, it is based on the assumption that if an individual dies outside his home country, all the references to his origin are mentioned, including his nationality, whereas if he dies inside his country only references to his minor ethnic group are needed. This allows a better definition of ethnic and political boundaries, as well as a neater clarification of the location and territory of the *castella* in the Northwest. As a result, it is now possible to define the area (Figure 52) in which the social organization of the *castellum* system was based. This area mainly coincides with that defined for the Culture, thus reinforcing the point that the Castros Culture has a corresponding and particular social organization.

The *castellum* is then a castro, or group of castros, composed of a variable number of familial units linked by blood-relationship and connected to a territory (Bermejo Barrera 1978-80, 105-6) under the authority of a chieftain, or magistrate.

Archaeological evidence has been demonstrating how the members of a *castellum* are bound to this unit by the way it has been referred to in connection with the origins of its inhabitants.

The social importance of the *castellum* finds its best expression in Pereira Menaut's description (1983, 206): *"...Visto desde dentro, por así decir, cada uno de los castella es el todo: el individuo que pertenece a esta sub-comunidad se considera ajeno a la subcomunidad vecina, sin prejuicio de los lazos que unem a los miembros de cada pueblo."*

The author stresses how the *castellum* works as a whole, to the extent that an individual belonging to this community considers himself disassociated from the surrounding communities. However, this isolation between communities would not interfere with the cohesion of its members towards a super-structure, such as the *populus*.

This social *isolation*, or independence, outlined by Pereira Menaut is well reflected by archaeology, particularly in the way every castro appears to preserve its identity through the self-sufficiency of its manufacturing and the dignity of its defensive structures.

8.3.3) *Populi* and *Civitates*

The pact celebrated between the Zoelae (CIL II 2633) in the *tabula* or *tessera* of Astorga (Garcia Bellido 1966b, nº 9; Ors 1948, 66; 1953, nº 24), now in the Museum of Berlin, can probably give us an idea of the super-structural organization of the Northwestern peoples. This document reveals two pacts, in fact, which were celebrated between the same group of people, in the years AD 27 and AD 152. Differences in terminology are thus important indicators of the social and political changes, which the process of Romanization engendered during the course of the first century.

References to these major units of the region can be found in Pliny (III, 28; IV, 112-3) and Ptolemy (II, 6), where broad information on names and location is provided. Important sources are the numerous epigraphic references[16], amongst which stands out the so-called *'padrão dos povos'*, a milestone-like granite block attached to the Roman bridge of *Aquae Flaviae*. Amongst these *populi*, the *Bracari* are the most related to this study, since they occupied a large area in the core of this region. They were one of the twenty four peoples referred to for the whole *Conventus Bracaraugustanus*.

According to the numerous references made to the *Bracari* by Pliny (III, 28; IV, 112), Appian (73-75), and Ptolemy (II, 6), this people may have been particularly important among the others who occupied the area of the *Conventus*. This impression is reinforced both by the fact that the *Bracari* provided the name for the *Conventus*, and by the related references in the *Itinerary of Antoninus* (420, 8; 442, 1, 2; 423, 6; 427, 4; 429, 5) to roads converging on what has been suggested was their capital,[17] *Bracara Augusta*. The number of roads converging on this city (Figure 50) also draws upon the importance of this particular people in the context of northwestern Iberia.

16 - The most complete synthesis on the subject was produced by Tranoy (1981) and Silva (1986a).

17 - *Bracarum oppidum Augusta*, according to Pliny (IV, 112), and commented on by Tranoy (1981, 64).

Social organization in the area of the *Castella*.

The term *Populus* refers to a social super-structure, probably connected with ethnic links, which will later evolve into a Roman unit called *Civitas*. It was formed by a variable number of *Castella*. One hillfort, or possibly more than one, as a political and social unit, formed one *castellum*. It is not yet certain whether all the hillforts in the Northwest would have had the status of *Castella*. The basic unit is the family. A castro was composed of a number of families that varied according to the size and dwelling structure of the site.

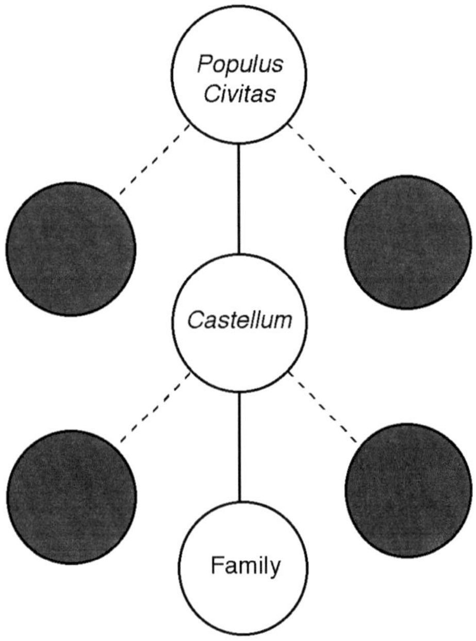

The identification of this people with any sub-culture, or regional cultural aspect, is a problem bound to remain open to discussion. However, considering that the *Bracari* could not occupy such a large area (c.f. Pliny) as the one defined by the statues of warriors and saunas of Type 1b, then perhaps this cultural sub-unit could refer to the Callaicans, as the major social unit.

8.4) Discussion

From what was referred to above, Castros communities were social and politically organized prior to the Roman Conquest. Although most of the evidence dates to the period of Romanization, it contains fossil indicators which may enable fragmentary conclusions about the system to be made. The foundations of this structure and its links have not been properly studied in the past; but recent years have witnessed a substantial advance in this subject, thanks to work done on subjects either directly connected to the political and social institutions, or focusing on subsidiary aspects within the archaeological evidence.

Prior to the contacts with the Romans, the peoples from the Northwest were probably ruled by numerous small chieftains (the *castella*), which would certainly be in accordance with the social organization reflected by the epigraphy. These chieftains would correspond to the basic unit which, as we have seen, is one castro, or a small group of castros, bound by some ethnic cohesion, which may be reinforced by links of dependence derived from political organization. Recent attempts to clarify this point (Silva 1986a, Tranoy 1981) have brought some fresh data. Different levels of chieftain should also be considered during the times prior to the construction of the large castros.

The head of the primary political unit, the chieftain or magistrate, is from early times archaeologically represented by the statues of warriors[18]. The multiplication of these statues in the archaeological record by the turn of the millennium cannot be explained solely by a political reorganization of the chieftains, which was noticeable by this time, but also by renewed emphasis on symbols of power and mastership of masonry. Reference to leaders, *princeps* or magistrates can be found in the epigraphic sources. It may possibly be proposed that the *princeps* referred to in the epigraphy of the Asturian region (Albertos Firmat 1975, 32; Arias *et al.* 1979, 60; Diego Santos 1959, nº 14; Garcia Bellido 1943, 418 foll.) may be an example of this, thus denying the existence of a hierarchical scale also at the political level between the *castella* and the *gentilitas* and *populus*.

It has been repeatedly suggested by evidence that the basic political unit, the *castellum*, was connected with some sort of leadership.

The *tessera* of Carbedo (Caurel, Lugo) (Tranoy 1981, 281-2), dated AD 28, shows the intervention of two indigenous magistrates who act in the celebration of the *hospitium* between one individual and their own community, which is expressed as *castellanis*. If the *castellum* is a political unit, and therefore its representatives have the capacity to celebrate pacts, their status is not entirely clear in all the known cases. The case of the *tesserae* of Monte Murado (Silva 1983b), reflects a more ambiguous status for the interveners of the pacts, who, one assumes, are leaders of the community, since their social status or function is not described. Therefore, it still remains uncertain whether the function of the leader of a castro by the turn of the millennium was exclusively administrative or also included warfare, meaning that warfare and raiding may have been led by distinguished warriors, who may or may not have been chieftains. The answer to this question may

18 - Further information on the statues and their significance will be discussed in the next chapter.

help us to establish who the warrior statues really represent.

Other than the references from Classical sources, the most tangible evidence for the existence of chieftains in castros is provided by iconography and epigraphy, both of which appear to date to the post-Conquest period.

On the other hand, there is no apparent pattern regarding the size and geographical position of the sites in which statues of warriors were found (Figure 58), although the course of rivers could be tentatively suggested.

The presence of magistrates, such as *Abienum*, magistrate of the *Zoelae* in the *tessera hospitales* of Astorga, indicates the existence of a complex political and administrative hierarchy among this people. Whether this early evidence is already a reflection of the imposition of Roman administrative patterns on the indigenous system, which might be supported by the importance of the mineral resources, and by the *hospitium* itself, or whether it refers to a former system, cannot be inferred form the available data. However, the analysis of the various indicators of social and military organization suggest that political, military and juridical authority might have been separate functions prior to the coming of the Romans. The existence of a familial hierarchy, as inferred from Strabo's references (III, 3, 7) may suggest its extension to the site in the form of a council formed by members of the familial units '*according to age and rank*'. This supposition can find corroborative archaeological evidence in the so-called 'house of the council', at Briteiros (c.f. Chapter 4). However, a correlation between the inner space of the *council house* and the number of familial enclosures at Briteiros suggests that this involvement may have been reserved for the patriarchs, instead of the heads of the dwelling units.

The evidence brought up by the several *tesserae hospitales* found in the Northwest[19], or in some way referring to the Northwest, points out the existence of an organized political leadership which can be traced from the Cantabrian wars onwards.

To date, however, evidence has failed to show the relationship between archaeological data and the social system. The existing model is still too ambiguous and fragmentary, not able to show how hierarchy is reflected within the dwelling structures of the site, and particularly where the places for the leaders, the warriors, the farmers and the craftsmen occur.

---

19 - See Pereira Menaut (1985-86) and Silva (1983b; 1986a, *gráfico* 12).

9 - WARFARE AND SOCIETY

9.1.1) Lusitanians and Callaicans: who's who?

In the course of this study it has been stressed how the Classical authors prior to the change of millennium provided very generalized references to the northwestern peoples. As Blázquez Martínez has stated (1975, 17), Strabo (III, 3, 7), when referring to the peoples who lived in the north of Iberia, affirms they all had the same customs, from the Callaicans to the Vasconians. This assertion is applicable to most of the authors.

The Classical sources referring to the warlike attitudes of, and to the robberies conducted by, the peoples from the Northwest mention the *northern peoples*. The Lusitanians are described in more detail. A detailed level of information, such as that about the people 'who dwell next to the Durius river' (Strabo III, 3, 6) is very rare. This ambiguity comprises one of the difficulties in dealing with the ethnic identity of the peoples who inhabited the Northwest according to the Classical writers. The deliberate quotation of long passages in this study is to emphasize that sometimes these generalizations and inaccuracies are interleaved with impressively accurate statements. This does not ease the task of critically accepting or rejecting these sources, particularly in the cases where there is no corroborative archaeological evidence.

Bearing this in mind, to what extent would it be legitimate to ascertain the boundaries between the common practices of *barbarians*, or the ethnic particularities of the Lusitanians and Callaicans regarding warfare and participation in military events? This is a complex subject beyond the range of this study, but it is essential to consider the topic. During the course of this chapter a model will be considered, based on the possibility that, to some extent, the expression 'Lusitanians' may have been misused by some Classical sources. One will therefore assume that, as a general model, the so-called Lusitanian gangs or armies which carried out military expeditions against the Romans, mostly in the south of Iberia, could actually have been composed either of warriors with various Iberian ethnic backgrounds, or solely of Callaicans and Lusitanians; these two groups are generally designated as Lusitanians by the Classical writers before the time of Augustus. The main reasons for this assumption stem from a critical review of some of the Classical sources, particularly those composed closer to the events or those which are based on other relatively contemporary sources.

In fact, Strabo (III, 3, 2-3)[1] suggests that the Lusitanians were also designated as Callaicans, yet he states the very opposite in a subsequent paragraph. He is clearly using the term 'Lusitanians' because he apparently does not feel sure to whom it relates to. Furthermore, we have the unequivocal reference (Strabo III, 3, 3) that the various peoples who inhabited the Northwest, namely the Carpetanians, Vettonians, Vacceans and Callaicans, could be, or are in fact, called Lusitanians. During the expedition of D. Iunius Brutus, as Orosius (V, 5, 12)[2] reports, sixty thousand Callaicans who came to the aid of the Lusitanians were defeated by the Roman army. Brutus, however, was invading the region between the rivers Douro and Minho, the homeland of the Callaicans, and not Lusitania. The other aspect to be borne in mind from Strabo's references (III, 4, 19) is that Lusitania, as a province, was created during the first Augustan provincial reorganization in 27 BC. This was intended to reduce the size of the provinces (Santos Yanguas, N, 1981, 52-3), due to difficulties of political and military control, and especially of the regions which had not yet submitted, in a time of increasing Romanization in a large portion of the Peninsula territory.

Another important aspect is the agreement between the descriptions in the sources of Lusitanian warriors and the so-called Callaican statues, which are found exclusively in a very limited area (Figure 56) to the north of the Douro river. The relation between their distribution and other aspects of material culture and social organization is discussed elsewhere, but they may demonstrate the ethnic limits of a people who may be the Callaicans.

The number of confrontations related to the Conquest of the region north of the Douro river, according to the Classical sources and recent archaeological data, is relatively small in comparison with that during the long opposition carried out by the Lusitanians and the Cantabrians-Asturians. In fact, there were two waves of Conquest. The Callaicans were pacified by two expeditions, one led by D. Iunius Brutus and the other by Caesar, nearly eighty years apart. This contrasts with the nearly one-and-a-half centuries of *Lusitanian wars* in the south, or the hard *Cantabrian wars*, which were led by the Emperor himself and were militarily highly

---

1 -"...the Callaicans themselves have not only furnished the surname for the man who defeated the Lusitanians but they have also brought it about that now, already, the most of the Lusitanians are called Callaicans."..."...Contrary to the men of to-day, however, some call also these peoples Lusitanians..."

2 - *In* Schulten 1937, 140.

effective. This picture does not seem plausible in the light of the evidence, unless one accepts that there were earlier contacts between Romans and Callaicans. The evidence suggests that Roman troops settled in the area of Viseu (Alarcão 1988, 26), on a relatively permanent basis, after the campaigns of Caesar in the region. Considering this location, however, perhaps the Lusitanians were more likely to be affected by the Roman presence than the Callaicans, if one assumes that the currently accepted distribution of these peoples is correct.

The Classical references to the northwestern peoples which date to before the Conquest, should be regarded highly critically when assessing their value for ethnic information. A critical review should be made of the concepts they imply for the time in which they were written, in addition to a permanent calibration with the available archaeological data. A clear ethnic definition of some peoples in relation to the historical events they appear to be connected with is not yet clear. Future monographs on this subject will have the important role of clarifying the situation, so that one can indeed see *who's who* prior to the end of the first millennium BC.

9.1.2) The newcomers: ethnic movements in the Northwest

One of the major arguments for a study of the Castros Culture as a unit has been the presence of external influences in the region, presented in the form of migrations of peoples originating from somewhere in Europe. The definition of *Celtic* has been a major subject for discussion throughout the twentieth century, to the extent that it has blurred many more significant studies and approaches to the subject.

From the very early stages of the study of the Castros Culture, especially after Martins Sarmento had done research on the topic, the similarities between some of its aspects and those of the Celtic Cultures became apparent to archaeologists. Scholars of subsequent generations, in particular Almagro Basch (1982), Bosch Gimpera (1932, 1933, 1942, 1945, 1950, 1950-53), Garcia y Bellido (1950), Lopez Cuevillas (1953d), Martinez Santa-Olalla (1946) and Maluquer de Motes (1975a, 1982), have long championed the idea of migrations of Celtic-affiliated European peoples into Iberia, which provides an explanation for the ethnic and cultural picture presented by the Classical authors.

This line of study has not been followed with any particular enthusiasm during the last twenty years or so. Archaeologists of recent generations appear to be more keen on demonstrating the existence of a stronger indigenous background for the northwestern peoples, as suggested by the apparently uniform cultural development of the Castros Culture. The latest research, now providing more systematic support from radiocarbon dating and scientific methods of data retrieval and study, is once more endorsing the idea of external cultural and ethnic influences.

The Classical references, in fact, do mention that peoples settled in Iberia at various stages, yet they are not particularly accurate concerning the times and circumstances in which these migrations took place. Regarding the Northwest, it is stated that incomers settled there. Earlier studies on this subject have been produced by Bosch Gimpera (1932), López Cuevillas (1953d) and Bouza Brey (López - Bouza 1929), who refer to the *Oestrimnians*, the *Saefes* and the *Cempsi*. Evidence for the origins and locations of these peoples nevertheless still remains dubious, despite the exhaustive discussions presented by these scholars. A more enlightening reference is provided by Strabo (III, 3, 5), regarding the Celts an the Turdulians. Probably based on earlier information from Poseidonius, Strabo states that "*...the country round about the cape itself is inhabited by Celtic people, kinsmen of those on the Anas; for these people and the Turdulians made an expedition thither and then had a quarrel, it is said, after they had crossed the Limaeas River; and when, in addition to the quarrel, the Celtic peoples also suffered the loss of their chieftain, they scattered and stayed there...*".

Some aspects of this narrative, which is also corroborated by Mela (III, 8), are of the utmost importance, since they have been archaeologically substantiated by epigraphic evidence (Silva 1983b, 1986a) for the presence of the *Turduli Veteres*, the 'old Turdulians', on the south bank of the Douro by the turn of millennium. The important find of two epigraphic references (Silva 1983b), dated AD 7 and AD 9 respectively, referring to an equal number of pacts between a Roman citizen and members of the Turdulians, makes it an established fact that the Turdulians had migrated to the lower Douro valley before the turn of the millennium. Further references to the Turdulians can be found in Pliny (IV, 112-3) and Mela (III, 8). The proposed date for the establishment of this group in the region is the V century BC (Silva 1983b, 23). This date sounds acceptable on the grounds of the available evidence, and is now preferred to the date proposed by Garcia Y Bellido (1950, 496), the II century BC, which was based on a careful analysis of the arguments presented in the Classical sources. However, as it is further pointed out (Silva *ibid.*), the expression 'old Turdulians' may well be an indicator of at least two phases of movement, or 'waves', of these settlers: *old* being deliberately contrasted with *new*, or newcomers.

For the purposes of this study, this evidence raises the important question of whether the presence of the *Turduli* and *Celtici* in the Northwest represents a migration of peoples, or a simple military expedition. One has to bear in mind that the Celtic-affiliated finds, which are widespread in the

Northwest[3], are not diagnostic enough for this purpose, since there is a wide range of equally viable reasons for their presence in the region. If, in fact, a migration is the case, then what is the evidence for the extent of the cultural impact on the local communities? Regarding the second hypothesis, how was a group of invaders, formed by either an élite of warriors or a more complex social unit, socially integrated? Most of these aspects cannot be answered at present, but they have to be borne in mind for future research. So far, substantial archaeological evidence has not been forthcoming with which to evaluate the cultural impact of a migration into the Entre-Douro-E-Minho.

The arrival of immigrants in the Northwest from the middle of the millennium onwards may also be related to the long-distance movements of Celtic warriors in Europe, as suggested by Pauli (1985, 26), and substantiated by the sources (e.g. above, Livy 34, 1-4; Pompeius Trogus XX, 5.7-8 and XXIV, 4. 1-3), as a result of demographic growth and internal social pressure.

Apart from the internal significance of this migration for the Northwest, an important aspect to be considered is the cause of the displacement of these communities from their area of origin. An understanding of this is required to corroborate the local evidence.

Both the structure and the size of these groups of migrants remain unknown at present, but the work of Garcia y Bellido (1950), which has without doubt been undervalued in recent decades, may provide a provisional picture. It is based on a relevant text from Caesar's *Bello Civili* (I, 51): *"Nuntiatur Afranio, magnos comitatus qui iter habebant ad Caesarem, ad flumen constitisse. Venerant eo sagittarii ex Rutenis, equites ex Gallia cum multis carris, magnisque impedimentis, ut fert Gallica consuetudo. Erant praeterea cuiusque generis hominum millia circiter VI cum servis liberisque: sed nullus ordo, nullum imperium certum, cum suo quisque consilio uteretur, atque omnes sine timore iter facerent, usu superiorum temporum, atque itinerum licentia. Erant complures honesti adolescentes, senatorum filii et equestris ordinis: erant legationes civitatum: erant legati Caesaris..."*

The information behind this text will now be considered, although no substantial advance on the work of Garcia y Bellido (1950, 490-6) can be made.

The text quoted above refers to a party of warriors, arriving at Ilerda (Lérida) as reinforcements to Caesar's army, during the civil wars. This party probably came from Gaul, and was acompanied by another group, which was not involved in the confrontations, but rather was moving for migratory purposes. This latter group is referred to as numbering six thousand men with their families and slaves. A more surprising detail about this party is the way in which it is stressed that it had no apparent order or leadership, and here one may disagree to some extent with Garcia y Bellido's analysis (*op. cit.* 490). It is not completely clear whether this observation by Caesar meant that the migrants did not have a leader amongst themselves, or whether they were not attached to any leadership in the party of auxiliaries they were travelling with. The other unclear aspect comprises the social attributes of some members of this group, referred to in the last part of the text in terms which Garcia y Bellido interprets as Roman citizens, sons of senators, and of equestrian rank. It literally implies that a group of Roman citizens of prominent rank was migrating from Gaul, this being either their place of origin or an area through which they passed, under the protection of auxiliary troops. This group of migrants, calculated by Garcia y Bellido (*op. cit.*, 491) as twenty thousand individuals, is far too large to be considered as Romans previously settled in south Gaul and being displaced to Hispania in 49 BC. Gaul, however, may have constituted one stage in their migration. The existence of a Roman colony at Narbonne, founded as early as 117 BC, reinforces the importance and effectiveness attributed by the Romans to the control of Mediterranean Gaul, perhaps not exclusively for trade purposes but also for the establishment of a land route to Hispania. It is also known how Caesar's social policy favoured the establishment of a large number of colonists in the recently founded provincial cities (Mossé 1970, 116), such as Hispalis, Urso and Emporium in Hispania.

Another equally plausible hypothesis is that some of the migrants may have been prominent members of a Barbarian people, and referred to by Caesar according to their corresponding Roman social attributes. Whatever conclusion may be drawn, it is undeniable that groups of both warriors and migrants were moving into Iberia. While the warriors were certainly involved in military activities, the movement of the migrants occurred for different reasons. Surprisingly, this does not appear to have been much affected by the instability in many regions of Hispania during the civil wars. Furthermore, most of the war parties would probably have returned to their regions of origin, while the migrants settled in the Peninsula, where they contributed, to some extent, to the development of local culture and language.

The Northwest, just like the other zones of Hispania, also constituted a settlement area for immigrants, who probably arrived both before and after the Conquest.

The internal and external migrations during this period are relatively well known due to the existence of a most important source, the epigraphy, which corroborates other less diagnostic sources. The studies on this subject show widespread internal

3 - See references below in the text and illustrations.

displacements of individuals, which appear to be motivated mostly by economic factors but also by attachment to military units. The internal and external displacements of individuals and groups before the Conquest are not so well known, especially in the Northwest. As has been presented above, there is archaeological and literary evidence for both the Celts and Turdulians, in particular for the latter, but it cannot be assumed that these were the only relevant migratory phenomena operating in the northwest of Iberia during the first millennium BC.

Leaving aside any discussion of the chronology of these migrations until later, the archaeological evidence from the area where the Celts and Turdulians may have settled will now be considered. An overview of the charts, referring to the distribution of features such as Iron Age bath structures of Type 1 (Figure 15) and the statues of warriors (Figures 56 and 57), shows just how similar these distributions are, and how well they compare with the eastern limits of the distribution of the *castella* (Figure 52) in the north of Portugal. It should be stated again that this area is roughly coincident with what will later be the *Conventus Bracaraugustanus*. The point is whether this aspect of the archaeological record will allow any correlation of this very clear cultural unity with an ethnic group to be made; also, whether this cultural unity is the result of an exclusively internal development, or due to the settlement of other peoples at a given period.

Although the available archaeological data are insufficient for tracing evidence of other migrations, linguistic studies, and in particular the onomatic works (Albertos Firmat 1964-65-72, 1976; Hübner 1869-92 and 1899-1903; Tovar 1947, 1950, 1977, 1983, 1985-86; Untermann 1965, 1985), suggest that a generalized Celtic affiliation of many indigenous names was spread evenly throughout the Northwest, which survived until epigraphy came into use in this region.

Since the aim is to build up a picture from the data supplied by both material culture and written souces, the logical starting-point would seem to be a consideration of the concept and image of the warrior.

9.2) The Callaican warrior: a profile

Although accounts from other Classical writers referring to the typical Castros warrior will be noted, the following descriptions from Diodorus[4] deserve a separate transcription for their close agreement with the archaeological evidence.

4 - References to the subject made by Strabo are also quoted in Chapter 2.

Diodorus Siculus

V, 34, 4 - "The most valiant among the Iberians are those who are known as Lusitanians, who carry in war very small shields which are interwoven with cords of sinew and are able to protect the body unusually well, because they are so tough; and shifting this shield easily as they do in their fighting, now here, now there, they cleverly ward off from their person every blow which comes at them."

V, 34, 5 - "They also use barbed javelins made entirely of iron, and wear helmets and swords very much like those of the Celtiberians. They hurl the javelin with good effect, even over a long distance, and, in fine, are doughty in dealing their blows. Since they are nimble and wear light arms, they are swift both in flight and in pursuit, but when it comes to enduring the hardships of a stiff fight they are far inferior to the Celtiberians. In time of peace they practise a kind of elfin dance which requires great nimbleness of limb, and in their wars they march into battle with even step and raise a battle-song as they charge upon the foe."

V, 34, 6 -"And a peculiar practise obtains among the Iberians and particularly among the Lusitanians; for when their young men come to the bloom of their physical strength, those who are the very poorest among them in worldly goods and yet excel in vigour of body and daring equip themselves with no more than valour and arms and gather in the mountain fastnesses, where they form into bands of considerable size and descend upon Iberia and collect wealth from their pillaging. And this brigandage they continually practise in a spirit of complete disdain; for using as they do light arms and being altogether nimble and swift, they are a most difficult people for other men to subdue."

V, 34, 7 - "And, speaking generally, they consider the fastnesses and crags of the mountains to be their native land and to these places, which large and heavily equipped armies find hard to traverse, they flee for refuge. Consequently, although the Romans in their frequent campaigns against the Lusitanians rid them of their great spirit of disdain, they were nevertheless unable, often as they eagerly set about it, to put a complete end to their plundering."

The most accurate picture of the Callaican warrior is certainly the one provided by the various granite statues[5] found in the north of Portugal, although other representations[6] from other areas of the Northwest may also be considered characteristic.

5 - General references to the statues representing Callaican warriors can be found in Almeida, CAB, 1981 and 1983; Almeida 1974b, 28 and 1981; Calo Lourido 1983, 173-85; Cardozo 1973a and 1985; Guerra 1882 and 1982; Martins - Silva 1984; Paço 1966b and 1968c; Peixoto 1899-1903; Pereira 1908 and 1915a; Santos Júnior 1978a; Sarmento 1933a, 36-40, 72-3; Silva 1981-82, 89-90, Figs. 1 and 4, 1983a, Fig. 10; Soeiro 1984, Fig. 127; Sousa, EF, 1954; Taboada Chivite 1965; Tranoy 1981, 351; Vasconcellos 1896 and 1913, 43-62, 615. Furthermore, a complete review can be seen in Silva (1986a, 291-2), with excellent illustrations (figs. CXX to CXXIII).

6 - Such as the warriors from the votive cart from Vilela (Cardozo 1946b), in the Entre-Douro-E-Minho, and also from the gold diadem of San Martin de Oscos (Lopez

The statues of Callaican warriors are thought to be broadly contemporary with a later phase of the Castros Culture (Almeida 1981, 115; Calo Lourido 1991, 672-80; Martins - Silva 1984, 43), but they certainly reflect aspects of a much earlier pattern.

This inference is supported by scarce, but apparently reliable, data from a statue which bears undoubted patterns of archaism: the statue from S. João de Ver (Figure 56.1). The characteristics borne by this make it a predecessor of the statues of Callaican warrior which are dated to the Romanization. No typological connection between the two types has been proposed so far. The statue of S. João de Ver has been included in the group (Figure 57.1 and 57.2) of the *menhir-statues* (Jorge - Jorge 1983, 47), among which the example from Chaves (Jorge - Almeida 1980) is notable. Closer inspection of their characteristics, and further comparison with the one from S. João de Ver will reveal their air of archaism, but also the similarities between some of their symbols.

The statue of S. João de Ver depicts one helmet of *Mediterranean* type which provides a reference for its dating.

In fact, the helmet worn by this statue shows very close typological similarities with the Greek Chalcidian helmets (Snodgrass 1964, 1967), dated around 500 BC, which may lead to interesting conclusions considering the location of this find. However, the complexity of determining the cultural affiliation and date of this statue may not be so straightforward. In fact, one has to bear in mind that in such an accurate iconographic picture there is no evidence for the helmet nose guard, which would then make it comparable to the Samnite-Attic type of helmet (Connolly 1988, 107, 111) of later date. This type of helmet can be placed between the Chalcidian type and the first examples of the Montefortino type datable to the fourth century BC.

The other piece of armour represented on this statue is a type of sword, whose peculiarity may lie in the way some elements are represented rather than in the type they refer to. In fact, the two ends of the weapon are excessively flared, and they may represent both the characteristic wide handle and chape of the early La Tène period; however, this is a matter of interpretation of an apparently imprecise iconographic representation.

In spite of any discrepancies, this sculpture may be considered as belonging to an earlier phase of the *Callaican warrior* group (Silva 1990, 290)[7] which may be dated to the mid-first millennium BC.

Monteagudo, 1977), although this latter may have a different stylistic, ethnic and cultural affiliation.

7 - The dating proposed by this author to the mid first millennium BC (Silva 1986a, 37, 40-3) may be reasonable in the first instance, although the historical boundaries (Silva 1990, 290) may lack coherence. Supported more by possible ethnic movements than by peculiar iconographic elements, of which this statue has many, it is still open to discussion whether this statue can be related to the expedition of the Turdulians or to any other expedition.

These statues are, in general, rather similar in style. They portray a warrior wearing a tunic and kilt (or more probably a tunic held with a belt) in linen or wool[8], occasionally decorated[9] and, in some cases, with the shins protected by greaves. The tunic, which is a garment characteristic of other contemporary European cultures, became well known in the ancient world under the name of *sagus*, to which several Classical authors refer on numerous occasions.

The Callaican war equipment is composed of a helmet[10], a short sword (usually a *falcata* type sword or a dagger), and a small shield, although the association of these elements can vary.

Apart from its obviously utilitarian function, the equipment may also have had some kind of religious or ritual significance beyond prestige display purposes, as has been recorded for the southern peoples (Appian 31, Livy XXXIV, 17). This may be seen in the context of earlier phenomena such as the intentional deposition of weapons in rivers[11], an action which may have been carried out into the Iron Age[12]. Decorative items represented on the statues, considered as prestige ornaments, comprise torcs and bracelets, the latter locally called *viriae* and worn on the upper part of the arms.

It is generally agreed by authors that the Callaicans were foot soldiers. This is corroborated by the statues of the Entre-Douro-E-Minho (Figure 56) and suggested by the significant absence of horse gear in archaeological contexts in this region[13]. The only known representation of mounted warriors within this region may be found on the pottery sherd from Faria (Maluquer de Motes 1948, Martínez Santa-Olalla 1948). Avoiding any discussion of whether this decoration reflects a local or an external *Celtic* cultural feature, I shall contend that the real importance of this element lies in the fact that it was locally, or at least regionally, manufactured (Martinez

8 - Polibius (III, 124) refers to the purple linen shirts of the Iberian troops.

9 - The predominant patterns are 'SS' and lozenges. The other pattern proposed by Silva (1986, 291) for one of the statues of Campos is not, in fact, more than one line of lozenges roughly joined at their sides.

10 - Examples can be seen on the statues found at Capeludos and Sanfins, which can be corroborated by the two bronze helmets found at Castro de Moldes (Castelo de Neiva, Viana do Castelo), studied by Almeida (1980a and 1980b), and by the helmet found at Castro de Lanhoso (Póvoa de Lanhoso, Braga), (Teixeira 1941).

11 - Coffyn 1985 and Meijide Cameselle 1988 for general references to nortwestern river bed deposits, and Wait 1985 for an interpretive approach to this phenomenon.

12 - C.f. the presumed significance of the diadem from San Martín de Oscos (López Monteagudo 1977) and, for the Entre-Douro-E-Minho area, the helmet from Caldelas de Tuy (Santirso *et al.* 1977), found in the bed of the Minho river.

13 - C.f. references in Chapter 7.

Santa-Olalla 1948, 22), as is revealed by its fabric on close inspection. The significance of this decoration is cryptic, and one has to bear in mind that this is merely one small pottery fragment. Yet a further interpretation as a votive item, with similar decoration to that found on the rock carving at Sanfins (Jalhay 1947), which could be part of a hunting scene or even part of a procession such as the one on the diadem of San Martin de Oscos (Figure 60), should be considered as a possibility. However, the fact that this particular piece of pottery was of local manufacture suggests that either the decoration was copied from a similar pattern on another item, or it represented a cultural fact, irrespective of whether it was utilitarian or ritual. Whatever the case, the available evidence is too sparse for any generalization and, considering the actual state of knowledge, one should stick to more reliable data.

With great mobility, and taking advantage of the terrain by frequently laying ambush, the Lusitanian foot soldiers in many ways used the same fighting tactics as the European Celts (Deyber 1987, 170-1). They presented a terrifying aspect with their long hair waving in battle (Appian Iber. 67-9), and in addition they roared and sang.

It is also interesting to consider the accuracy of Strabo's description of the Lusitanian warrior, which is corroborated by Diodorus, as it fits perfectly with the iconographic representations found exclusively in the region of this study (Figure 56), i.e. the statues of Callaican warriors. In my view, this constitutes a further argument in support of the suggestion that the ethnic generalizations of the Classical authors, regarding military events in particular, have become confused, so that Callaicans may have been given the designation of *Lusitanians*. Indeed, this fact might hamper our better understanding of both social and historical development in these communities.

The Callaicans may also have used body-protecting armour, such as a cuirasse, which may be represented by the tunic with a pattern of lozenges on some statues. Appian (Iber, 71) suggests that Viriatus, the Lusitanian leader, was in the habit of sleeping with his weapons and fighting gear on. This is why his murderers had to strike him in the neck. This reference seems to be supported by Diodorus (33, 1, 1-3)[14] who says that Viriatus always wore strong armour. Therefore, it is known that the foot soldiers, and eventually only those with more status, wore some kind of body-protecting gear.

The fact that some statues of warriors have been found without the head is also an interesting subject for discussion. This has been connected by some authors with the Gallo-Celtic cult of the *severed head*[15] (Blanco Freijeiro 1956, 180; Tarracena y Aguirre 1943, 169-71), of which there are various Peninsular examples[16] with a higher concentration in central Iberia. However, the fact that many Callaican statues had their heads removed is probably due to other reasons, e.g. they are more likely to have been accidentally broken across one of the thinnest parts of the piece. As discussed below, this could also have been of symbolic or social, rather than religious, significance, but evidence is far too scarce for a more detailed discussion of this point.

The elements available for the characterization of the weapons of the Callaican warrior in the last few centuries of the millennium are at present fragmentary and the Classical sources are not always consistent in their description. Essentially consisting of light and short-range weapons, as repeatedly referred to by the Classical writers, the equipment of the Callaican warrior reflects the warfare strategies and fighting habits adopted. These included a perfect adaptation to rough terrain, a preference for laying ambush instead of engaging in combat in the open field, the 'false retreat' and, above all, surprisingly fast manoeuvres during confrontation. According to its function, the equipment can be divided into offensive elements such as sword, dagger and spear, and defensive or protective elements such as shield, helmet and cuirass.

More than one type of sword was used by the Callaicans, according to the archaeological evidence. In fact, several variations of the short-sword are known, which would be carried by a foot soldier fighting at close range. Considering the variations of this weapon, it is at present very difficult to determine which type of short sword can be considered as typical of the fighting equipment of the Callaican warrior during the last centuries of the millennium. On the other hand, there is also a striking discrepancy between archaeological data and the iconographic sources in this matter.

One of the types of sword referred to as having been used by the Callaicans is the *falcata* (falchion). Although this is a weapon generally based on Mediterranean prototypes and affiliated with the Greek *kopis*, the Northwestern type of falcata has closer typological parallels with the southern European *mahaire* (Gustin 1974), rather than with the typical falcata of the Iberian South (Figure 63). The Iberian falcata has been widely referred to in the Classical sources (Diodorus V, 33; Livy XXXIV, 14; Polibius II, 33; III, 114; IV, 23; Strabo III, 3, 6; III, 4, 15) as the typical weapon of the foot soldier. It is indeed the commonest and most widespread sword type in

[14] - After Schulten 1937, 328.

[15] - Generally known in Europe under the designation of *têtes coupées*.

[16] - The cases from this region which may have a similar interpretation were published by Almeida, CAB, 1983; Almeida 1971 and Santos Júnior (1978a and 1978b).

Hispania, where it is known from the IVth century AD onwards, and increasingly appears in references to warfare until the end of the millennium, thus demonstrating its military importance. The best examples of these weapons occur in the cremation cemeteries of central and southern Iberia (Figueras Pacheco 1956, 34, fig. VI), in connection with particularly well preserved finds. The military character of the falcata was highly appreciated in the Classical world, to the extent that it served as the model for the sword of the Roman army (Polibius VI, 23) after the second Punic war. This was in recognition of the effectiveness of the weapon, which was re-emphasized by later references[17]. A special steel-hardening process, for which the Iberians were famous (Diodorus V, 33), improved a fighting efficiency already dictated by the characteristics of its design. Analysis done on the distribution of carbon within this implement (Guadán 1979, 38-9; Sandars 1913, 241-2) has shown a higher concentration on the edge of the blade and the external surface, which diminishes towards the core of the sword, thus revealing a specialist skill in hardening iron. However, as demonstrated in Chapter 7, there is no evidence for such a knowledge of ironworking in the Northwest, as no analysis has so far been carried out on Castros falcatas.

The Northwest, by contrast, is relatively poor in such finds (Figure 59). This may be attributed to local funerary practices, in addition to the poor conditions of preservation referred to above.

The falcatas found in archaeological contexts in the Northwest are fairly small and typologically different from the Iberian falcatas[18] (Silva 1986, 177), thus rejecting former suggestions of affiliation (Almeida 1973, 945) between the two. Their small size might sometimes result in their being designated as daggers or small sabres rather than as swords, regardless of how they might have been used in battle. There is no doubt that their closest typological parallels are the *mahaires*, considering the simple construction and angular shape of both types, and not the more elaborate Iberian falcatas. Only a few examples of falcatas are known in the Entre-Douro-E-Minho (Figure 59), of small size and simple workmanship. The advanced state of corrosion of these iron weapons, in addition to poor observation and conservation techniques, precluded the retrieval of any information on decoration of the blade and hilt. The hilt would have been made out of perishable materials, probably wood, and attached to the blade with iron rivets. Known exceptions comprise the two falcatas from Sanfins (Almeida 1973, 295, Fig. I; Jalhay-Paço 1948) which have a socketed hilt, which implies either that further elements were originally present or the weapon had a longer wooden handle. This would augment the range of action of the weapon and facilitate a thrusting action.

Falcatas are represented on the coins of the so-called *caetra mint*, which constitute a special Augustan coinage and which was probably ordered by *Carisius* (Villaronga 1970) during or immediately after the Cantabrian wars. The coins of this group bear representations of Callaican weapons (Caamaño Gesto 1979, 67-76; Cavada Nieto 1972, 211-48; Cortez 1957, 143-6; Guadán 1979, 36-9; Vazquez Seijas 1964, 37-40), namely the *caetra*, falcata and a spear type which, in some cases, may be interpreted as a *soliferreum*. These coins are believed to commemorate the defeat of the Callaicans, and therefore it is plausible that Callaican weapons are represented. However, the *falcata* (Figure 62) and, eventually, the *soliferreum* represented on these coins were not the types used by the Callaicans, as is argued above. This imprecision lacks a suitable explanation, as the argument based on the assumption that the differentiation of the various Northwestern peoples by the Roman writers was not al all clear would not be appreciable here. In fact, it is logical to assume that the military sources would be far more precise about the Callaicans' weapons and characteristics, since they were in frequent contact with them during the campaigns, either as allies or as enemies.

Another important aspect of the diffusion and use of the falcata type dagger in the Northwest is its absence from the local iconographic representations, namely the warrior statues. The number of finds of this item (Figure 59) appears to suggest that it was effectively used by the Callaicans, regardless of whether it should at present be considered a fighting weapon or merely a utilitarian iron tool.

There may be doubts as to whether the falcatas found in the Northwest were, in fact, fighting weapons, or just constituted a particular type of long knife used for special votive or utilitarian purposes. Two examples found which the inner defensive wall at Mozinho[19], Penafiel, and dating to the I[st] century AD well suggest a connection with military activities; it may be assumed that there was a ritual meaning to their deposition in the core of a defensive wall at a time of advanced pacification.

The antennae type sword is relatively rare in the Northwest, and there are no known examples from northern Portugal. Nevertheless, the type is very frequent in some Celtiberian cemeteries (Guadán

17 - See the interesting reference in Seneca (*De Beneficiis* V, 24) as an example of the recognition of the qualities of this weapon by Roman soldiers.

18 - However, rare examples of falcata of Iberian type have been found in the North (v.g. Museum of Lugo), all in ambiguous circumstances. Amongst these, gift, loss during the raiding of southern parties, or loss by Iberian auxiliaries during the Cantabrian Wars, could be considered as plausible reasons for their presence in the region.

19 - Unpublished information kindly provided by Teresa Soeiro, to whom I am most grateful.

1979, 34; Sandars 1913, 219-28) where a multitude of stylistic variations can be recognized. In these context they represent variations on the Hallstatt prototypes (Cabré de Morán 1947, 134-5) created by local craftsmanship. Two small swords were found at the limit of the region under study (Mergelina 1943-44, 37, Lam. 46 a-b) at Santa Tecla, by the northern shore of the Minho river estuary. One of the swords is complete, comprising a bronze hilt and a blade of iron; of the others only an antennae hilt survives, which is very similar to the types referred to above. Although in general the hilt may be composed of several materials in order to achieve more intricate decorative effects, the blade of this type of sword is made predominantly of iron.

Taking into consideration the characteristics of the swords and daggers described above, and bearing in mind their quantitative distribution in this region, it does not seem possible to establish a pattern from such a small sample. One can probably assume, using typology, that the two antennae swords from Santa Tecla are not representative of local production, but the same cannot be stated regarding the sword found at S. Julião (Figure 63). This type is very rare in the Northwest, and its neat parallels in other regions of Iberia and Europe (Cabré de Morán 1947)[20] should be taken as significant. However, a brief review of the iconographic representations of thes daggers on Callaican statues, where a hilt of similar type is shown, should allow some reassessment of the situation; some statues show what seems to be a metal-framed scabbard containing a double-edged dagger with a round pommel (Figure 63).

Apart from the short-sword, the Callaican warrior also used a dagger (Diodorus V, 33), as is indicated on the statue from Santa Comba de Basto (Almeida 1981), where it hangs from a belt on the right-hand side of the body. Several daggers are known from the Entre-Douro-E-Minho region. Not all are easy to identify, as variation in size makes the largest somewhat difficult to distinguish from the short sword. The falcata type of dagger is predominant amongst the examples known from the region (Silva 1986a, 177), with a date range extending from Caesar/Augustus onwards.

Javelins and spears, as mentioned by Strabo (III, 3, 6) were probably used by the warriors of the Northwest. However, the archaeological finds are scarce (Figure 59) and relatively undiagnostic, such is their typological variety and chronological imprecision.

The Callaican spear is characterized by a short head (Figure 64), unlike the one used by the Iberians (Polibius VI-XXV) which is similar to the Greco-Roman type. Its absence from the Callaican statues does not necessarily imply that it had no role as a fighting tool for the particular kind of warfare practised by these peoples. Rather, this may indicate that the spear was not a prestige or display weapon.

Another type of spear, the *soliferreum*, referred to by Diodorus (V, 34, 5) as *barbed javelin*, is a massive spear made entirely of iron. This was well known to the Roman troops and they considered it one of the most terrifying weapons used by the Lusitanians. No archaeological evidence for this weapon has been found in the Entre-Douro-E-Minho region, although it is frequently found in southern Iberia (Menghin 1949), particularly in the graves of warriors (Quesada Sanz 1989).

Some spearheads have been found in this region (Figure 59), in addition to other evidence for the use of the spear, such as the depiction on the decorated pottery sherd from Castelo de Faria (Martinez Santa-Olalla 1948, 27). Further evidence is provided by the rock engraving from Sanfins, where a warrior holding a spear is represented.

During the excavations carried out by Teixeira (1940a and 1940b) at Castro de Lanhoso, an iron trilobate arrowhead was found (Figure 64), but there is no information regarding its stratigraphical position. The trilobate type of arrowhead was well known in the Scythian world, according to Herodotus, and was also used during the Middle Ages. Considering that the find from Lanhoso displays some typological differences from the Scythian type, and bearing in mind the lack of stratigraphical context, it would be wiser to assume that it is medieval.

The *caetra* (Figure 62) was the typical shield of the Callaican warrior. It is generally represented on the Callaican warrior statues (*in genere* Silva 1986a, figs. 120-3), on coins produced by the Publius Carisius mints (Sutherland 1984, 25, fig. 2a), on the Penafiel rock carving (Brandão - Pinto 1966, Brandão - Sousa 1966, Corrêa 1927), on the diadem of San Martin de Oscos (Lopez Cuevillas 1951a and 1951c; Lopez Monteagudo 1977), and also on the cart from Costa-Figueira (Cardozo 1946b).

Archaeological evidence for shields is indeed very scarce in the Northwest (Lopez Cuevillas 1946-47, 554-7). A bronze, butterfly-shaped shield boss was found at Castro de Alvarelhos (Almeida 1973, Fig. 1-3; Soeiro 1980, Fig. 1). This piece (Figure 62) did not belong to a *caetra* type shield but to a La Tène elongated shield, a type common in Gaul during the Gallic wars[21]. This type of shield was also used

[20] - Baring in mind the low level of typological and chronological particularities of this find (Cabré de Morán 1947, 136-7) referred to by this author.

[21] - There are surprisingly similar pieces to this one from Alvarelhos in the museum of St. Germain-en-Laye, France.

in other parts of Iberia (Cabré Aguiló 1940, 78-80) by the indigenous armies.

The iconographic representations of the *caetra* show a concave round shield board with a central *umbo* (shield boss), which may have been reinforced with a metal sheet in order to avoid penetration of the shield by an attacking blade. Made out of hardened leather and wood, the concave shape of the shield board was probably an intentional feature, designed to reduce damage from attack by offering an oblique surface to the striking weapon. An attacker's blade would slide towards the centre of the shield which was reinforced by the *umbo*. Used mostly by men on foot and in hand-to-hand combat, the shield's characteristic deflection of a blade could also be used to momentarily hold it after the striking action i.e. in the angle between the board and the umbo, in order to open up an angle for a return strike. There are fewer examples of *caetrae* which are convex (shown on statues from Capeludos and Meixedo) or planar (shown on two statues from Outeiro de Lezenho).

Strabo (III, 3, 6) refers to a small shield, concave in the front and with grips made of thongs, rather than with solid handles, as would be expected. This is also corroborated by Diodorus (V, 34, 4), who remarks that the grips are interwoven with cords of sinew, thus providing a substantial shock-absorbing ability.

The *caetra* was such a distinctive piece of armour that it defined a special type of warrior (Caesar, b.c. I, 39-1, 48-7, 55; b.c. II 70-4, 75-2, 78-1; Silius Italicus III, 348; Strabo III, 3, 6), namely the *caetrati*. Its particularly skilful use by the Lusitanians/Callaicans (Diodorus V, 34, 4) made it a higly effective defensive weapon, in conjunction with the short-sword or dagger, in hand-to-hand combat where a great deal of mobilily would be necessary. This detail from Diodorus also suggests that the shield was not tied to the arm, but rather held freely by the warrior in order to enable him to change the position of the shield whenever necessary and to protect his body from attacks from different directions.

The helmet is referred to by Diodorus (V, 34, 5) and Strabo (III, 3, 6), and its use among Callaicans is corroborated by some archaeological finds. However, the helmets of the type bearing three crests, as referred to by Diodorus (*ibidem*) and represented in the diadem from San Martin de Oscos[22], are not known in the region and are more probably connected to the Celtiberian world.

The commonest type of helmet among the Callaicans was one made out of sinew (Strabo *ibidem*), a type intended to provide protection from light attack and which had a relatively low resistance to penetration. No archaeological evidence is, however, so far available for these helmets of perishable material. The iconographic representations of helmets worn by the figures of the statues from Sanfins and Capeludos may be of either metal or sinew and leather types.

The metal helmet is surprisingly well represented in the Entre-Douro-E-Minho region by four remarkable and well preserved bronze helmets of the Montefortino type. One (Figure 61c) was found at Castro de Lanhoso (Teixeira 1940c, 1941), two (Figure 61 a-b) at Castelo de Neiva (Almeida 1980a, 1980b), and one was dredged up from the Minho river at Caldelas de Tuy (Santiso *et al.* 1977). In addition, there is one fragment at Citânia de Briteiros (Cardozo 1953b, 715) and another was found at Citânia de Sanfins (Paço 1968c).

The helmets referred to above may be considered masterpieces of bronze craftsmanship. It is known that the two helmets fom Castelo de Neiva were made by hammering (Almeida 1980a, 1980b), apparently from a single sheet of bronze. The skill of the craftsman is also reflected in their profuse and delicate decoration.

The typological and decorative uniformity of these helmets reflects the cultural uniformity of the Entre-Douro-E-Minho region, as is referred to above for other items. Their decorative patterns show a regional stylistic tradition, although some aspects of La Tène art are also present. It has not yet been established whether these helmets were produced within the region. Indeed, their similarity suggest that they were, and there is also evidence that equally skilful craftsmanship in bronze was a product of the region, as shown by the *situlae*, for example (Carballo Arceo 1983).

Among the four helmets from Entre-Douro-E-Minho two variants can be differentiated (Almeida 1980a, 248-9; 1980b, 47-8). The example from Caldelas de Tuy is considered to be earlier within the Montefortino type (Almeida *ibid.*).

The cuirass, to which reference was made above, was also utilized as part of the body-protecting gear of the Callaicans. A particular type of cuirass, probably made of leather or plant fibres, was described by Strabo (III, 3, 6). He also alludes to a widespread use of linen tunics, and to comparatively rare chain mail cuirasses among the Lusitanians.

Greaves are well documented by both literary (Strabo *ibidem*) and archaeological sources. The latter comprise the statues from Cabeceiras and Santa Comba de Basto (Almeida 1981, 114), suggesting the use of this protective element by the foot warriors of this region.

22 - C.f. Lopez Cuevillas 1951a and Lopez Monteagudo 1977, where the particularities of the warriors' gear represented here are analyzed in detail, and parallels are established for the helmets with feathers.

9.3) The nature of the warrior: the mercenary and political cohesion

Thus far, this study has attempted to construct a physical image of what one would assume the typical Callaican warrior to be like, through his aspect, his armour, and the impressions he has left in the Classical sources. Even if the available data were not scarce, the image would be incomplete without corroboration from others areas. In order to determine the nature of the warrior, and ultimately the effects of warfare, it is necessary to examine the relationship between the warrior and the community that supported him, and the type of organization that made this lifestyle possible.

9.3.1) Social hierarchy and institutions

The social and political relationships of the peoples of the Northwest were conducted by rules, or laws, which were sometimes related to, and therefore reinforced by, ritual activities. Our knowledge about this interaction is scarce[23], and the evidence is fragmentary and scattered over the whole region of the Northwest. Considering the number of peoples who are referred to as having inhabited this region, each presumably with its own traits, any general characterization may be elusive whenever the limitations of the available data are set aside. On the other hand, the available data for each ethnic group, or even for the whole Entre-Douro-E-Minho region, are obviously insufficient for any coherent understanding of their hierarchical organization to be achieved. Bearing these limitations in mind, gaps in the data are nevertheless occasionaly filled by information from areas outside the region under study. A tentative picture may be constructed for every particular case discussed below, and the extent to which such an approach is justifiable will be examined.

Hispanic warriors from different ethnic backgrounds are referred to by the Classical sources as having been attached to members of the Roman military hierarchy during the Conquest of Iberia. The institutions which regulated this relationship have been designated by terms such as *fides*, *devotio*, *cliens,* and *hospitium.* The real meaning of these personal links and social attitudes among Hispanic warriors, and between them and the Romans, is of the utmost importance for any understanding both of the warrior hierarchy and of cohesive links among the peoples of the Northwest.

[23] - Asturia is undoubtedly the better known part of the Northwest as far as pre-Roman social organization is concerned. This may be due to the late Romanization and difficculties of Conquest, both of which motivated more detailed references by the Classical authors.

It may be difficult to understand the reasons why these various peoples would participate in the Conquest of the lands of their neighbours by the Roman army, supposedly knowing that they themselves would soon be invaded, without taking into account other more complex aspects of warlike and political behavior. The explanation of this involves two main arguments: the meaning of warlike activities for the Iberian peoples in general; and the extension of their cultural and political identities or, in other words, the degree of cohesion of their political structure and of interaction with their neighbours and those further afield.

Chapter 7, in which the characteristics of the Castros Culture's productive economy is discussed, outlines how a reasonable degree of development had been attained in the Entre-Douro-E-Minho region by the time of the Roman Conquest. Therefore, in plain economic terms, the external accumulation of a surplus was not indispensable for the balance of the system. Accordingly, poverty should no longer be considered as a necessary, or even the major, motivation for the raiding expeditions, although this has been suggested for a rather similar case in Gaul (Deyber 1987, 145-57) where both geographical conditions and economic factors were taken into account. However, the Classical authors (in particular Diodorus, Pliny and Strabo), repeatedly state that the frequent internal wars and raids into the territories outside the castros were motivated by a lack of resources in their homeland. These sources should not be ignored, but re-interpreted in the light of new evidence. Moreover, references of this kind by the Classical writers often reveal deprecatory attitudes towards other cultures which they believed to be in a state of lower civilized development than their own. The 'Lusitanians' were considered to be *disorganized in battle*, without a *clearly established leadership*[24], but these opinions were based on patterns recognizable in contemporary Greco-Roman ideologies. The reasons for the establishment of war parties and expeditions may also have been incomprehensible to contemporary observers in the Classical world. This is probably why there are repeated references to a *certain savageness* (Strabo III, 4, 18) in the northwestern peoples, in contrast to the *togati* or *peaceably inclined* (*ibid.* III, 4, 20) inhabitants of the South.

However, these considerations are not intended to raise doubts about the fact that raiding may have been relatively frequent, and an explanation for this

[24] - The only exception to this suggestion might be the case of Viriato, but it is uncertain whether this is an historical fact, or if Viriato's leadership was emphasized by the sources due to his involvement at a particularly crucial moment. Furthermore, the degree of factual dependence between some Classical authors (c.f. Chapter 2) might also provoke a repeated overstressing of some aspects to the detriment of others.

must lie within the complexity of the social and political structure.

Aspects of the structure and function of the raiding parties, or *communities of warriors* (Garlan 1972, 17), comprised the subject of a recent study by García Fernández, which took as a comparative model the Celtic *fiana* (García Fernández 1990, 207-35), and raised new and interesting assumptions following an analysis based on the religious evidence. García Fernández (*ibid.* 237, 240) concludes that groups of warriors, whose only function was to conduct warlike activities outside their own territory, were bound to a leader by fidelity and obedience and were apart from society to a certain degree. This study raises an important question which has been ignored in previous research, namely the social status of the Castros warrior.

At this stage a model may be constructed to explain the way in which the warrior operated within the Castros society. Evidence has indicated that Castros society is essentially one of farmers, herdsmen and craftsmen. Other than the construction of defensive structures, the production of statues of warriors, and the location of the sites on prominent hills, there are virtually no other known indicators of warlike activity. How, therefore, does the warrior fit into this still uncertain picture of society?

The sources refer to the fact that young men became raiders due to the poverty of their homeland. One cannot necessarily infer that raiding was exclusively connected with youth, although this aspect should be left open to discussion. A man who, at some stage, became involved in warfare, could maintain this status for the rest of his life. Alternatively, a man could hold this status for a limited period during his youth, while he was directly engaged in combat. In the latter case, warfare may have functioned as a regulator of social hierarchy and prestige for the individual.

It is now more difficult to accept that raiding was an activity dictated by the poverty of the region, as it appears to be related to social structure, and organized in a way which has already been established for the Celtic and Germanic tribes. Accordingly, the available data suggest that warfare and raiding were closely connected with social factors, such as the reinforcement of the function of the warrior and the prestige of the warrior status, in which the influence and hierarchical balance of chiefdoms played a predominant and dynamic role. A similar picture has been tentatively put forward for Gaul (Deyber 1987, 159), where the analysis of leadership and social structure is based on a more substantial body of data relating to the Gallic Wars (Wiseman - Cunliffe 1980).

The limits of social identity and obligation did not extend far beyond the group, either in the sense of the single castro (*castellum*), or group of castros as an ethnic superstructure. The pacts referred to above, which established personal links, constituted the regulators of both short and long range political interactions prior to the establisment of Roman law. Moreover, even a brief overview of some details of the Lusitanian wars (Schulten 1937) indicate that leadership was a relatively fluid matter amongst the Lusitanians during the campaigns, apparently depending more on the result of the confrontation than on the institution of leadership. Furthermore, it had little to do with cultural identity either, as the ready acceptance of foreign leaders suggests. This is the case with various groups in the Roman army, such as the Iberian mercenaries, but the acceptance of Sertorius' leadership by the Lusitanians for such a long time and with such strong examples of *devotio* is most significant.

In a wider perspective, was not this scheme applicable to Rome itself? The history of the last centuries of the Roman Republic (Alföldy 1988) reveals a very similar picture of political careers being built and supported by prestige gained through Conquest and army leadership. *Hispania* seems to have been reckoned a privileged place for this purpose (Le Roux 1982, 40), since various Roman politicians (Scipion Emilianus, Tiberius Gracchus, Julius Caesar[25] and Octavius Augustus) gained prestige and political power after their participation in Iberian campaigns.

In general, there is a well established idea of disorganization and a lack of political cohesion among the peoples of the Northwest, which is referred to by Florus (II, 33, 47) and Orosius (VI, 21, 3). On the other hand, an idea of unity is conveyed by the episodic warlike enterprises[26]. This apparent unity was probably connected with political structure, and the real objectives and dynamics of warfare, rather than to any political solidarity.

Aspects of the meaning of warfare can also be understood through the mutual relationships that existed between different communities of warriors, and between the group and the individual, particularly when dependent links had been established.

During the Conquest of *Hispania*, several cases are known of pacts which were celebrated between Romans and indigenous peoples. Some of these pacts were realized in the *tesserae hospitales*. These tables are distributed throughout Iberia in a rather uneven manner[27], but in two main concentrations, namely in the centre/south, and a strip along the northern side of the Douro river.

---

25 - Namely for the questorship of the year 68 BC and the pretorship of 61 BC (Le Roux 1982, 40, foot. 64; quoting Suetonius' *Caesar*, 7, 1 and 18, 1).

26 -*In genere* Shulten's *Fontes Hispaniae Antiquae*, on which comments below in this chapter were based.

27 - Twenty six *tesserae* (if the double pact expressed on the piece from *Asturica Augusta* is taken into account) were compiled by Silva (1986a, graph 12) in the Iberian territory, which reflects how widespread and common these pacts were.

The *tesserae* found at Monte Murado probably constitute the most important document in this region for showing the links of dependence between communities or individuals.

The status and functions of the participants in this pact are not yet completely clear, in spite of the opportune observations on this matter (Silva 1983b). From what the text can tell us, none of them appears to bear official functions. Nevertheless, such an early presence of a private Roman citizen in this region appears enigmatic. If indeed this Roman did not have an official role, then trade appears to be the most plausible hypothesis. It is known that merchants and colonists in Iberia[28] generally followed the legions, either engaged in dealings related to the legion itself, or taking over the economic exploitation of the resources of the conquered regions. The inscription found in *Bracara Augusta* (Hübner 1969-92, CIL II 2423) mentioning the '*Cives Romani qui negotiantur Bracaraugusta*' under Tiberius, suggests that there were Roman merchants settled in the Entre-Douro-E-Minho region at an early stage of Romanization.

Fidelity and devotion, *fides* and *devotio*, are two widely referred to characteristics of the Iberian warriors[29], as may be seen from the first episodes of the Conquest until the complete pacification under Augustus. At the present state of research one cannot assume, however, that this custom "*...to devote their lives to whoever they attach themselves, even to the point of dying for them.*" (Strabo III, 4, 18) applied to all the Iberian peoples. In fact, it is better supported by both Classical references and more recent studies[30] for the Iberian South, but personal links were also established among the Northwestern peoples. The most relevant aspect of this fidelity is the one established between two individuals, i.e. the personal fidelity. It can be most useful as an indicator of hierarchical relationships between the warriors of a group or community and higher status individuals. Some other cases, however, demonstrate a higher degree of complexity, particularly when hierarchical links of dependence (*fides*, *cliens*) are established between individuals of similar rank (Rodriguez Adrados 1946, 166) such as *chieftains* or *princeps*.

The course of the Sertorian wars presents a good example of Iberian military *devotio*,[31] considering the many peoples involved in the events. If the sources are to be believed, the fidelity of indigenous warriors to Sertorius went apparently unchanged throughout the ten years of continuous and violent war, despite both defeats and the unfriendly attitude of Sertorius.

Arguments for the existence of a military-based hierarchy in castros can also be made from some aspects of the warrior statues. At Sanfins, the mortise of one statue was found carved into a rock in front of the western gate of the second defensive wall. The base of the statue found at this site fits this mortise, suggesting that these icons might have been generally displayed in association with symbols of collective power and prestige, such as the defensive walls.

9.3.2) Indigenous *auxilia* during the Conquest of Hispania

The operation of mercenaries, like raiding, is an attempt to export the internal pressures of a social system. However, the rules are different from those of raiding, and so is the final result. Whereas the raider is attached to one particular enterprise, the mercenary is attached to a leader by more permanent and well defined links of obedience, as is discussed above.

The Iberian auxiliaries, due to their outstanding fighting qualities as complementary troops to the regular army, had a very special, yet unstable, status in the Roman legion during the Conquest of *Hispania*. They held a decisive part in the balance of military power during the Iberian episodes of the civil war.

Not much is known about the involvement of the Northwestern peoples as *auxilia* prior to the Cantabrian wars, but inferences may be made from disparate elements in addition to examples from other parts of Iberia.

The auxiliaries from *Hispania* were regarded as well paid (Guadán 1979, 24; Livy XXVI, 45 and XXVII, 17) and indeed pay was an important consideration for them in the early period of the Conquest. Also valued for their particular capabilities in the local terrain and in manoeuvres, the *auxilia* usually retained their traditional armament and combat techniques when fighting alongside the Roman army. Their influence has to some extent affected the development of some aspects of Roman warfare (Roldán Hervás 1972, 117).

The lightness of their armour and their agility enabled them to move quickly through rough terrain. As Caesar (*De bello civ.* I, 70) states, the *cohors* composed of *caetrati* were usually despatched on the most difficult operations.

Their qualities as warriors were widely exploited, and they were often used to save effort and losses by the legion itself. *Perpenna* for example, complained that Sertorius treated his *generals* and *senators* with contempt, making them carry out tasks as unpleasant as the ones reserved for the Iberians and

28 - In particular Roldán Hervás (1972, 81-2) for the Iberian North, with extensive reference to general readings.

29 - General references can be seen in Justinus XLIV, 2; Plutarch (Sert.) 14; Tacitus (Ann.) IV, 45; Valerius Maximus II, 6, 11.

30 - In particular the study by Rodriguez Adrados (1946), with further references.

31 - Auro Gelio XV, 22, 10.

Lusitanians[32]. Furthermore, there are also episodes such as the one from Ategua[33], where the Lusitanian auxiliaries were ordered to execute brutal tasks such as murdering men, women and children by the most barbaric means. Nevertheless, it is also true that *auxilia* were often preferred as the personal guards of generals[34], due to their fighting qualities, their unusual sense of fidelity, and also their lack of concern for political matters. A series of generals, probably beginning with Marius and certainly including Sertorius[35], Caesar[36], Augustus[37], used *auxilia* as personal guard corps. The bravery and perseverance in combat of the indigenous warriors, their devotion to their leaders referred to above and their capacity to establish links of *cliens* with foreign chiefs, made them élite warriors in the context of the legions (Harmand 1967, 178-9; 1970, 188) and fully justified their recruitment in large numbers during the civil wars in *Hispania.*

Apart from general support in battle, the *auxiliary* had the important role of instructing the Roman army in local warfare strategies, some of which, such as ambush and false retreat, proved to be fatal on some Roman campaigns[38]. In addition, they had a special role in fighting indigenous armies (Le Roux 1982, 38)[39] in the rather particular terrain of their homeland.

There were basically three forms of recruitment to the Roman army: as mercenaries[40], as conscripts[41], and a voluntary alliance at the level of chieftains. The first form of recruitment was common, especially in the earlier phases of the Conquest, before it was transcended by the development of other systems such as the application of the right of *imperium* to the conquered populations, and the reinforcement of the links of *fides* and *cliens* with the indigenous leaders. But one has to bear in mind that these latter could only develop from situations of continued contact and effective Conquest, i.e. closer to the end of the Republic.

Another major motivation for the voluntary[42] service of indigenous warriors in the legions, particularly in some Romanized regions of Iberia, was the award of citizenship (Roldán Hervás 1983, 112) with all its inherent rights and privileges, although the motivations for this may have been rather different prior to the Conquest. Unlike raiders, the employment of mercenaries continued through the first stages of Romanization, since it was a useful and effective way of controlling indigenous social pressure by diverting it towards Roman territorial expansion. In its own way, the operation of mercenaries opened a way to the acculturation of local peoples, by diverting their potential for warfare and by affiliating both chieftains and warriors to the Romans. If Romanization and the substantial decrease in the activities of raiders and mercenaries[43] (Nash 1976b, 128) were fundamental motors of change in the Castros system, allowing rapid social and political re-organization, then similar European models may be invoked (Frey 1984). In comparative terms, I would suggest that the model discussed above, although somewhat simplistic, may fit more coherently with the evidence currently available for the Northwest than those proposed by Bintliff (1984, foot. 67), Collis (1984c) and Nash (1976b, 1978) for the Celtic-affiliated peoples.

9.4.1) The military events up to Augustus

The Classical sources all inform us that the peoples inhabiting northwestern Iberia were very attached to warlike activities and were always engaged in expeditions to raid neighbouring sites or regions. There is not much archaeological evidence, however, in support of the involvement of Callaican warriors in specific raiding expeditions, nor a proper reference as to how this operation worked. Literary sources referring to military activities to the north of the Douro river are scarce and ambiguous, and sometimes Callaicans are referred to as Lusitanians when, in fact, the two different ethnic groups occupied separate territories. Only the Cantabrian wars received a more accurate description, although the sources are more concerned with the Asturian region, which no doubt was a reflection of the personal involvement of Augustus in the campaigns. Moreover, the repeated description of strong resistance to Roman Conquest (Schulten 1920, 103-4 ; Appian *Iberica* VI; Orosius

32 - Plutarch, *Sertorius* 25; in Schulten 1937, 234.

33 - Valerius Maximus 9, 2, 4; in Schulten 1940, 153.

34 - This sometimes produced rewarding results. An example is the personal guard of Petreius who, along with the slaves in arms, prevented a mass desertion to Caesar (Harmand 1967, 457) during a crucial moment.

35 - Appian, b.c., 1-112. References to this subject below.

36 - Appian, b.c., 2-109; Caesar, b.h. 2-2; Suetonius, *Caesar*, 86.

37 - Suetonius, *Augustus*, 49.

38 - Schulten 1937, on general references to the Lusitanian wars.

39 - Quoting Roldan 1976, 29-32.

40 - The use of mercenaries was a relatively common practice here, although auxiliaries were of a much lower status than legionaries (Caesar, b.h., 22-7; Harmand 1967, 262-71) as far as pay was concerned.

41 - Military service of this kind was only in use in the pacified areas which were under Roman administrative domination, and indigenous peoples had therefore to live with this important, and rather strict, aspect of the Roman law.

42 - Taking the example of Asturia (Roldán Hervás 1983, 115), however, it appears that compulsory recruitment was the commonest form in this area, and this can be related to the degree of resistance to Conquest presented by each one of the peoples; conscription also operated as a way of a forced acculturation of the most unstable social elements, namely the warriors.

43 - Mercenaries continued to be employed, but their exclusive attachment to the Roman auxiliary brought about substantial changes in the system.

V, Strabo III), often involving a great literary and patriotic devotion (Cardozo 1947, 249-54; Martino 1982), only occasionally refers to the *Callaicans.*

According to the Classical writers, four major military events affected the Castros Culture during the period of this study: the campaigns of Decimo Iunius Brutus, those of Caesar, the Lusitanian wars, and the Augustan campaigns or *Cantabrian wars.* The accounts of these events are the most important source for determining the characteristics of the Castros warriors, as well as the significance and real meaning of their warlike activities.

Iberian warriors were probably involved since the beginning of the Roman campaigns in the Mediterranean (Silius Italicus III, 326-31, V, 195-6; Balil 1956; Garcia Bellido 1963; Roldan 1976, 23-33; Le Roux 1982, 38-41), serving as mercenaries (Spaltestein 1986, 217-8, 352) in which context their combative characteristics were most appreciated. The sources state that Iberian mercenaries were employed in Greece (Garcia y Bellido 1934 and 1974) at *Korintho* in 368 BC, and the three belt-plates of Iberian type found in this area (two in *Korkyra* and one at *Olympia*) lend support to this. Lusitanians served under Hannibal (Livy 21, 43, 8) in the Italian campaign. Amongst those generally designated as *Iberians*, Northwestern warriors should be included as this can be inferred from various references, and particularly from Caesar (B. C. 1, 38, 3[44]; c.f. above Silius Italicus). Furthermore, the long-distance military expeditions of Northwestern warriors, such as the Cantabrians coming to the aid of the Aquitains (*Bello Gallico* III, 26, 6), confirm the wide range of military connections of these peoples, and the large extent of their activities. As discussed above, it is tentatively suggested that the Callaicans could have constituted a substantial part of the *Lusitanians* periodically engaged in raids on the richer areas of the peninsular South (Livy 35, 1). The account of this writer, translated and commented upon by Schulten (1935), reveals just how frequent these raids were. They were most certainly recorded by the sources, because these confrontations with the Roman armies occurred so often, especially during the second century BC when the Romans were enlarging their area of operations in the South.

The first record of these raids dates to the year 194 BC (Livy c.f. above) and were followed by many others. If the first confrontations with the Romans seem occasional, the rich areas of Baetica being a major attraction, it is soon apparent that they were motivated by a different reason. The frequency and extent of the battles, some of them disastrous for the Roman armies, such as the ones of 190 BC at *Castulo* (Livy 37, 46, 7; Orosius 4, 20, 23) and 185 BC at *Toletum* (Livy 39, 30), suggest that from very early on the original objective of the raids was soon superceded by the desire for direct confrontation with the Romans.

The most distinguished figure of the Lusitanian wars is undoubtedly Viriatus. His reputation really derives from the continuous and effective campaigns he led against the Romans, although a certain image was promoted by Shulten (1927), and subsequently reinforced by local nationalism. He is portrayed as a simple and naturalistic king-shepherd-warrior, a liberator of peoples and defender of liberty. This approach has been recently abandoned by Garcia Moreno (1988) and Fernandez Canosa (1987), following a line previously taken by Nash (1976a) which was critical of some pictures drawn by the Classical authors and especially those of Poseidonius, whose opinion was clouded by a particular philosophical stand. It is now necessary to list and analyze these expeditions, after Shulten[45] (1935), in chronological order.

194 - *Cornelius Caepio* was attacked and defeated at *Ilipa* by a group of Lusitanians returning from a raid on *Baetica.*

190 - The Roman army, commanded by *Aemilius Paulus*, was defeated near Lycon in the South,by the Lusitanians.

189 - This same *proconsul* fought another battle against the Lusitanians who were again (or still?) in *Baetica*, possibly raiding the rich cities of the region.

188-87 - The Lusitanians attacked the 'allies of the Romans'.

186 - The Lusitanians were defeated by *Caius Atinius* near *Hasta* in *Baetica*, where they lost six thousand warriors in the battle.

185 - The joint forces of *Lucius Quincius Crispinus* and *Caius Calpurnius Piso* were defeated by Lusitanians and Celtiberians near *Toletum.* The Roman army, however, had possibly recovered from this defeat, as they attained a victory after crossing the Tejo river.

182 - Reinforcement of the Roman military of Citerior, which between 182 and 179 (Knapp 1979, 91) had two legions based here.

181 - Lusitanians and the Roman army under the command of *P. Manlius* clashed again.

44 - Referring to auxiliaries engaged by Afranius and Petreius :"*his rebus constitutis equites auxiliaque toti Lusitaniae a Petreio, Celtiberiae, Cantabris barbarisque omnibus, qui ad Oceanum pertinent, ab Afranio imperantur*", (quoted from Le Roux 1982, 53).

45 - In order to avoid the multiplication of references to Classical authors, I shall refrain from constantly quoting the sources. The major source for the period 237-154 and 154-72 BC is Schulten, respectively 1935 and 1937. The sources from 72 to 19 BC (Schulten 1940), reporting the military events immediately before the campaign of Caesar until the Cantabrian wars, will be dealt with separately, as I consider them to be a different military and social event. The review from Alarcão (1983 and 1988), used more occasionally, will be referred to whenever necessary.

179 - *L. Postumius Albinus* crossed Lusitania an expedition against the territory of the Vaccaeans. It is likely that the Romans (Alarcão 1988, 17) achieved another victory on this expedition occasion.

163 - Further operations between Romans and Lusitanians occurred.

155 - An army composed of Lusitanians and Vettonians, and commanded by *Punicus* defeated the Romans under *Manius Manilius*. The Lusitanians went to the coast after the confrontations, probably as a continuation of the expedition, and lost their chief.

154 - The Lusitanians, led by *Kaisaros*, were defeated by *Lucius Mummius* and his army. However, when chasing the survivors the Romans were defeated and lost nine thousand soldiers. The Conii, who at the time were Roman allies, were attacked and defeated in their capital Conistorgis by a group of Lusitanians from south of the Tejo river, commanded by *Kaukainus*. Appian and Livy[46] both agree in reporting that the party then crossed Gibraltar, but *Mummius* pursued them and engaged them in battle. The Lusitanian army was defeated and lost fifteen thousand warriors.

153 - The Lusitanians seem to have defeated Mummius, thus revenging themselves from the events of the previous year.

152 - The city of Oxthracas, claimed to be the capital of the Lusitanians, was taken by *Marcus Atilius*.

151 - Avoiding direct confrontation, the Lusitanians by-passed the troops commanded by *Servius Sulpicius Galba*. However, while being followed by the Romans, the Lusitanians turned and killed seven thousand of them.

150 - Several Lusitanian armies, apparently not connected in any way, were successively defeated. There were recorded losses of four thousand and fifteen hundred men, and in addition many were imprisioned. *Galba* and *Lucius Licinius Lucus* took advantage of the situation and raided and pillaged Lusitania. Then the Lusitanians proposed peace, but were betrayed by *Galba* who forced them to give up their weapons and killed thousands during the negotiations.

147/146 - Surrounded by the toops of *Vetilius* while raiding Turdetania, ten thousand Lusitanians decided not to surrender and followed a warrior called *Viriatus*, who organized their escape. Sustaining the siege with a reduced number of men, while most of the Lusitanians retreated, Viriatus managed to escape and join the rest of the army at Tribola. *Vetilius* decided to pursue them, but was killed with four thousand men. The Romans sent five thousand Celtiberian auxiliaries against the Lusitanians, but they were massacred.

146 - *Plautius Hypsaeus*, governor of the Ulterior, tried to sustain *Viriatus* and his army who at the time were raiding in Turdetania. The Lusitanians retreated and, being pursued by the Romans, turned unexpectedly to counter-attack and defeat them. The Romans, however, persisted in tailing the Lusitanians, who again inflicted another defeat on them. *Viriatus*, having escaped from the Romans, settled in the region while raiding the nearby cities of Segovia and Segobriga, which were allies of the Romans. It is not clear whether, at this stage the Lusitanians were simply raiding the richer available settlements in the region, or whether these attacks against collaborating cities had a more political or military purpose. Still in this year, or probably in 145, the Romans were defeated again by *Viriatus'* troops.

145 - The consul *Fabius Maximus Emilianus* was designated governor of the Ulterior by the Roman senate, and given an army of seventeen hundred men. It is possible that still during this year (Alarcão 1988, 19) the Lusitanians raided Cordoba.

144 - Appian (Iber. 65) refers to the loss of two cities controlled by *Viriatus*, without providing details about the circumstances of the battles involved. Neither does he refer to the manner of occupation of these two cities by the Lusitanians, who did not thus show any propensity for long-term settlement in conquered areas. A probable hypothesis is that these cities were used as short-term operational bases, and their Conquest by the Roman army is overstressed by Appian.

143 - An uprising of the Arevaci, Belli and Titi against the Romans took place in this year. *Viriatus* seems to have supported it, but it is unclear whether he actually commanded the expedition, or even what sort of insurrection this constituted. *Viriatus* was defeated in a battle against the Romans, but later regained his position and vanquished them; continuing south he took the city of Tucci and raided Bastetania.

142 - The Romans were again defeated at an uncertain location.

141 - *Viriatus*, apparently settled in Tuci, was displaced by the proconsul *Fabius Maximus Servilianus*, who undertook a successful sequence of military actions throughout all southern Iberia at least up to the Tejo river, repossessing various cities and taking ten thousand prisoners. The only defeat was imposed by an unidentified army of ten thousand men headed by *Curius* and *Apuleius*, but it seems that this party ended by being dismantled, due to the dead of *Curius*. Another

46 - Respectively in Schulten 1935, 96-8 and 98-9.

confrontation with a gang commanded by *Connoba* resulted in another victory for *Servilianus*. In a surprising gesture of clemency, *Connoba*'s life was saved but his soldiers' hands were cut off. Was the fighting power of the group more threatening than that of their leader?

It is not clear whether it was possible for *Servilianus* to raid this vast region due to the lack of response from the Lusitanians, who might then have returned to their country; or whether a succession of defeats obliged them to return to the territories to the north of the Tejo. It looks as though *Viriatus* was facing problems of leadership, as the multiplication of indigenous armies and the fast progress of the Roman expeditionary forces may possibly suggest. In fact, having defeated and besieged *Servilianus* and his army, probably in Baeturia (Alarcão 1988, 21), the Lusitanian leader rather surprisingly accepted an armistice, was recognized as '*amicus populi Romani*', and gained the acknowledgment of Lusitanian independence. It seems that the Lusitanians then returned to their homeland.

139 - The consul *Servilius Caepio*, the new governor of the Ulterior, broke the peace and attacked the Lusitanians, Vettonians and Callaicans. Apart from indicating that this expedition reached the region to the north of the Douro river, the reference to the Callaicans raises the question of whether they had so far not participated in the wars; or Appian[47] may have finally decided to provide a more objective definition of the peoples involved in the wars. This reference may also have been used to emphasize the magnitude of this campaign. In addition, this expedition did not necessarily reach the Douro river, but rather the peoples were attacked outside their territories.

The fact that this military enterprise was allowed by the senate suggests that the pacification and Conquest of the whole of Iberia was being taken more seriously by Rome. In fact, if the previous episodes outline a continued effort to sustain the activities of the Lusitanians in the occupied territories, the war now becomes a succession of fast and effective movements and, above all, clearly reflects the use of far greater resources.

Viriatus seems not to have been able to sustain this advance, and negotiations were arranged between *Caepio* and three Lusitanian emissaries who were probably *Viriatus*' *locum tenens*. This resulted in the betrayal and murder of *Viriatus* by the three men. His substitution by *Tautalus* does not seem to have obstructed the establishment of peace.

138-136 - This period is marked by the actions of *Decimus Iunius Brutus*, a proconsul who succeeded *Servilius Caepio*. After having 'pacified Lusitania'[48], his most immediate action was to conduct a military expedition against the Gallaicans, establishing a base in the region of Santarém-Alpiarça (Alarcão 1988, 22) and taking a coastal route towards the North. The importance of this campaign lies in the fact that this is the first event where the military and cultural identity of the Callaicans is recognized by the Romans, to the extent of organizing an expedition against their territory. It also seems surprising that this happens after the last recorded focus of raiding incursions in southern Iberia had been terminated. Why , nevertheless, did *Brutus* choose the Callaican territory instead of, for instance, Asturia? Were the Callaicans causing more military problems to the Romanized areas, or was this just the usual Roman expansion policy towards the mineral-rich region of the Iberian Northwest? The sources are not clear on this point, and they refer mostly to the episodes of battles in and domination of the territory by Brutus' army. According to Appian (73-75), *Brutus* reached the Minho river, after having crossed the Douro and Lima, as it appears that most of the coastal area of present northern Portugal was affected by this expedition. The archaeology does not seem to have provided evidence to support this event, since only the Castro de Terroso (Silva 1984, 128) shows a dense layer of charcoals and ashes, and the Castro de Sabroso (Cardozo 1958c; Hawkes 1984, 193) seems to have reinforced its defensive structures by this time.

114 - *Caius Marius*, praetor of the Ulterior province, fought the Lusitanians. The sources are not descriptive, but it appears to have been a Lusitanian raid on the occupied territories rather than a Roman expedition to the region of the Lusitanians.

112 - The successor of *Marius*, *L. Calpurnius Piso*, died during confrontations with the Lusitanians who were raiding Baetica; *M. Iunius Silanus* is also connected with some similar incursions.

109 - The Lusitanians were subdued by *Q. Servilio Caepione*, presumably in Baetica. His triumph in 107 may suggest the importance of the confrontations.

102 - *Marcus Marius*, the praetor of Ulterior, was apparently aided by Celtiberian auxiliaries[49] against the Lusitanians. Again, this is an

47 - In Schulten 1937, 122.

48 - Appian 73-75.

49 - According to Schulten's (1937, 148) interpretation of Appian (100).

indicator of the lack of political unity among the Iberian peoples, emphasizing their warlike propensity. This episode also reflects the mechanisms of Romanization through the double exploitation of their characteristics by the Romans (recruitment as auxiliaries-acculturation: Mossé 1970, 119), which I believe also occurred in the case of the Callaicans, as is explained elsewhere in this study.

101 (100?) - *L. Cornelius Dolabella*, praetor of Ulterior, apparently defeated the Lusitanians.

99 - Following the indications of a particular mint[50], *C. Coelius Caldus* defeated a culturally varied army, or set of armies, judging by the variety of panoply represented on the coins. In fact, there are some weapons similar to the ones used by Callaicans shown in association with other typically Celtic arms.

The political activities following these events lack significance for the approach of this study. With the onset of the civil war between the followers of *Marius* and *Sulla*, *Sertorius* was appointed praetor of Citerior in 83 BC and a series of events led to his fall in disgrace. After some unpleasant episodes in 80 BC, he was approached by Lusitanian emissaries who offered him the leadership of the Lusitanian groups against the Romans.

It appears at first to be rather strange that a '*dissident*' Roman general had been chosen to lead the Lusitanians against his own former comrades. Probably the explanation lies in several facts. First of all, *Sertorius* had gained the friendship of the Iberian chieftains by his pleasant and considerate attitude towards them in 83 BC, as is recalled by Plutarch[51]. In addition, his fame as a warrior and a general raised him above the military vicissitudes he had recently been through. It is clear, however, that for the Lusitanians warlike activities were becoming less defined by any national or cultural identity. This suggests that these groups did not constitute one single army, but a gathering of several armies from different ethnic backgrounds, who had a common objective and were led by a chief chosen according to the circumstances. On the other hand, *Sertorius* himself adopted the famous Lusitanian fighting technique of guerrilla warfare in rough terrain, and was aware of its effectiveness in that particular context. This supplied another reason for both his popularity and military success from 80 to 72 BC. Sertorius surrounded himself with other dissident Romans and Iberian *princeps* with whom he established a small *imperium*, organizing a senate and appointing senators, praetors and questors.

Relations between *Sertorius* and his companions and allies, however, became more tense as a result of both a sequence of defeats, which occurred from 74 BC onwards with the loss of Celtiberia, and also of a dramatic change in *Sertorius'* attitudes towards his followers[52] which caused deep discontent among them. According to Plutarch (Sertorius 25) and Appian (1, 13), *Sertorius* had lost his moral balance after a series of defeats, and the desertions which followed of both Romans and Iberians were dealt with cruelly.

In the year 72 BC, the two *locum tenens* of *Sertorius*, *Perpena* and *Tarquicius*, agreed to assassinate him during a banquet (Schulten 1949, 135). This was followed by the disintegration of his army, whose elements were dispersed after their last defeat by *Pompeius*. Here, according to Appian (1, 114), there is an example of the fidelity of the Lusitanians, who stayed with *Sertorius* until his end.

The active participation of Lusitanians in the Sertorian wars should by no means mask the predominantly exogenous character of this conflict. In fact, although Plutarch[53] clearly says that the Lusitanians had asked *Sertorius* to lead them against the Romans, the subsequent campaigns appear to be more of a fight between two wings of the Roman army in the context of the resolution of internal Roman political quarrels; the Lusitanians merely had the role of auxiliaries. This suggestion may be illustrated by a text of Plutarch[54] (Sertorius 25) reporting the complaints of one of *Sertorius'* staff, who said that their tasks and treatment were as unpleasant as the ones reserved for the Iberians and the Lusitanians. Moreover, the cruelty and contempt shown by *Sertorius* towards the indigenous auxiliaries in 73 BC (Diodorus 37, 22) is a good indicator of the role of low-ranking subordinates in the conflict.

In any case, the greatest significance of these military events for the Iberians was the close contact with Roman warfare and culture, which opened the way for the tribes of *Hispania* to an easier acceptance of *romanitas*.

A Roman permanent military camp, funded by *Q.C. Metelus Pius* in the 70s (Alarcão 1988, 35), was established in the vicinity of *Egitania* (Idanha-A-Velha) in the heart of Lusitania. This camp, together with the ones settled further north in the region of Viseu, suggest that military control had been effected over the region to the south of the Douro river by the time of *Caesar's* campaigns.

In the year 61 BC *Julius Caesar* was appointed *propraetor* of Ulterior, while facing fiscal problems due to his heavy debts. According to Suetonius

50 - See Schulten (1937, 149), for a more detailed description.

51 - Sertorius 6 (after Schulten 1937, 160).

52 - Appian 1, 112-113; Diodorus 37, 22; Plutarch, Sertorius 25.

53 - In Schulten 1937, 168-70.

54 - In Schulten 1937, 234-40

(Caesar 18)[55], he was invoked by the inhabitants of this province, who were being raided by Lusitanians. Plutarch[56] states that *Caesar* had twenty *cohorts* and, adding ten more to this number, ended up with fifteen thousand men. *Caesar* faced and defeated the Lusitanians in their own country, that is, the region between the Tejo and Douro rivers. When they retreated across the Douro, *Caesar* followed and vanquished them, after which he returned back south where he obtained ships and followed the Atlantic coast[57] up to *Brigantium* (La Coruña), in Galicia.

This fast but effective campaign stresses the fact that on several occasions the Lusitanians seemed to be unable to defend their homeland with the same vigour with which they raided the rich regions of the South or had faced the Roman armies in *Baetica*. The military control of Lusitania seems to be assured from now on by the establishment of a military settlement in *Scalabis* (Alarcão 1988, 26), present Santarém, which undoubtedly assisted the process of cultural assimilation.

After the division of the empire in 56 BC by the Conference of Luca, *Pompeius* ruled *Hispania* and appointed *Lucius Afranius* and *Marcus Varro* as governors of the Citerior and Ulterior provinces respectively[58]. *Petreius* controlled Lusitania and Vettonia with the command of two legions. It has been suggested that one of these two legions would have been settled in Lusitania in the region of Viseu, and more particularly at 'Cava de Viriato' (Alarcão 1988, 26). There is no doubt that such a large establishment of troops in this area produced enormous pressure for the submission of the peoples between the Tejo and Douro rivers. This may also be connected with the fact that Lusitania and Vettonia formed a separate district for the first time, although only for military purposes. Furthermore, the fact that *Petreius* demanded that Lusitania provide a large number of auxiliary troops[59] to help the Pompeian devotees in the confrontations with *Caesar*'s armies, indicates a relative control of the population. It seems that this recruitment was extended to Cantabria, which was not yet subdued, suggesting that most of this recruitment was voluntary, taking advantage of the fondness of these peoples for warfare.

The defeat of *Petreius'* troops in Illerda[60] in 49 BC by rather unusual means does not seem to have much changed the strategy of the military or political occupation of Lusitania in the short term. The fact, however, that *Cassius Longino*, *Caesar*'s governor for Ulterior, attacked *Mons Herminius* in pursuit of the *Medubrigenses*[61], suggests renewed problems in Lusitania. The recruitment of new auxiliaries in Lusitania is once more reported[62], in spite of this apparent instability. It is curious, however, that although the Lusitanians were traditionally considered as allies of *Pompeius*, as is referred to by the Classical sources, they were recruited under *Caesar's* army, unless they were compelled. The relevance of this lies in the fact that it is either an indicator of the strengthening of *romanitas* and its civic obligations, such as serving in the army, through warfare links such as the recognition and observance of former relations of *fides* and *cliens* with the Romans, or merely an acceptance of punctual recruitment, involving no political preferences and simply an occupation of war.

Other details of episodes of the civil war in *Hispania* show a growing tendency of the attachment of the indigenous peoples to the Romans through the establishment of links of dependence. Their characteristics and function will be analyzed in a different section.

By 47 BC the civil war had broken out again in *Hispania*, and it did not end with the victory of *Caesar* at Munda two years later. The episode of *Philonem*, a Pompeian who regained *Hispalis* (Seville) by hiring a Lusitanian gang[63], eventually an army, shows the remaining independence and availability of the Lusitanian warriors for warfare activities of any kind or objective, although one should also restress their preference for the Pompeian side. The end of the civil war in *Hispania* by the end of 44 BC, brought relative stability to the South. The establishment of legions attached to the governor of Ulterior in Lusitanian territory[64] may possibly be an indicator of instability in the region, but no further reference is given to this fact by any source.

Between 29 and 19 BC there was a renewal of the war in *Hispania*. These events have been designated as the '*Augustan campaigns*' or, more frequently, as the '*Cantabrian wars*'. One of their peculiarities, as the first designation suggests, is that the campaigns were commanded by Augustus himself. This draws upon the importance of the definitive Conquest of *Hispania* by the submission of

55 - In Schulten 1940, 11.

56 - Plutarch *Caesar* 12, in Schulten 1940, 11.

57 - The suggestion that Caesar's campaign was conducted entirely by sea (Rodriguez Colmenero 1977, 64-5) does not seem to have much supporting evidence either from the sources or from Caesar's political position in the Senate.

58 - Caesar, B.C. I, 38; in Schulten 1940, 33-4, 230.

59 - C.f. above footnote.

60 - The versatility in combat, courage and loyalty of the *auxilia* is illustrated here with new episodes (Harmand 1970, 192).

61 - *De Bello Alexandrino* 48, 2; in Schulten 1940, 81.

62 - *De Bello Alexandrino* 51, 3; in Schulten 1940, 82.

63 - *Bellum Hispaniense* 35, 3-4; in Schulten 1940, 120.

64 - Cicero *ad fam.* 10, 33, 3; in Schulten 1940, 174. Asinius reports to Cicero that he was so removed from any suspicion of civil war that he settled his legions in 'the end' of Lusitania. This laconic reference does not reveal whether this was the northern or the southern end. In any case, there is no doubt that the legions were controlling almost permanently the region between the Douro and Tejo at this time.

the last bastions of resistance against Roman occupation. Of course, one should also bear in mind how important it was for the young emperor to increase his influence in the senate by achieving some victories, and in particular to retain large armies and win rich imperial provinces.

In 29 BC the Vaccaeans joined the Cantabrians and Asturians and started a war that lasted for ten years. This was the last conflict involving the Romans against the tribes of *Hispania*.

The division of the provinces of *Hispania* between Augustus and the Senate was made in 27 BC. Tarraconensis and Lusitania became imperial provinces, whereas the Senate kept Baetica. Lusitania was split into two parts[65]. The region to the north of the Douro river, Callaecia and Asturia, joined Citerior, whereas the region to the south of the Douro river remained a province in its own right.

Should this fact be interpreted as an indicator of a special situation or status of the region to the south of the Douro in the process of Conquest and Romanization? Indeed, as has already been pointed out, the Roman military camps established in the region of *Beiras* during the Lusitanian wars had undoubtedly contributed to the pacification of the region and parallel control of the warrior élites from the castros to the south of the Douro river.

The importance of the Cantabrian wars and the definitive pacification of the Northwest may be evaluated by the presence of the emperor between the years 27 BC and 25 BC, although this may also be due to other reasons stemming from the particular development of Roman internal political affairs. The young *Tiberius* himself had participated in the Cantabrian wars[66] from 26 to 25 BC.

The Cantabrian wars have been the subject of detailed studies (Martino 1982, Rodriguez Colmenero 1979, Schulten 1962) which focus on the location and development of its episodes, most of which are not relevant to this study. The expedition against the Cantabrians early in 25 BC was slow and ineffective, as the Cantabrian tactics of guerrilla warfare (Dio 53, 25, 2) did not allow direct and decisive confrontations.

Augustus had to withdraw to *Tarraco* due to a sudden illness. His generals continued operations and managed to obtain a victory. The campaign was organized in three columns, all marching to Cantabria. The army from Lusitania, commanded by *Publius Carisius*, may have established a base in *Bracara Augusta* (Balil 1976, 47; Schulten 1962, 174)[67] or, more probably, in its surroundings, and from there invaded and subdued the western side of the Asturian region. The degree to which the Callaicans, settled between the Douro and Minho rivers, resisted the invasion is not known because the sources are few. The victory of *Carisius* over the Callaicans is a well known fact, which may have been overstressed by the Roman sources, and is commemorated by the so-called '*Carisius mint*' (Sutherland 1984, 25; Villaronga 1970) from Emerita. It is worth noting the vitality of the Castros Culture in this region, its degree of Romanization, and in particular the increasing dynamics reflected in many aspects of the archaeological record for the last decades of the millennium. This shows a degree of pacification which undoubtedly reflects the result of *Caesar's* campaigns in the region (Almeida 1983a, 1983b) and a more permanent contact with Romans. Moreover, the assumption that *Bracara Augusta* was used as a base in these campaigns, with no record of local incidents, indicates of a certain effectiveness in *Caesar*'s incursion of 61 BC, which eventually led to greater collaboration with the Callaicans.

The peace in *Hispania* was not completely achieved by the time of the departure of *Augustus*. In 24 BC there was another Cantabrian insurrection, apparently with some initial military success, but it was promptly suppressed by the *Legatus Augusti Lucium Aemilium* in a most severe manner[68].

In 22 BC the war against the Cantabrians and Asturians recommenced, but they were respectively defeated and punished[69] by *C. Furnius* and *P. Carisius* who re-established peace, albeit temporarily. The last, but dramatic, Cantabrian rebellion was fought by *Agrippa* in 19 BC and, according to Dio Cassius (54, 11, 1), it was originated by the escape of Cantabrians who had been made prisioners during the course of previous wars and sold as slaves. Having murdered their owners, they returned to Cantabria and recommenced the war, causing some psychological distress in the Roman army. With this confrontation ends a process that began as far back as 218 BC; i.e. the Conquest of *Hispania*.

9.4.2) Castros under Augustus: permanence or change?

In previous chapters an attempt was made to show how important the period of Augustus was in the Northwest, as defining a distinct turning-point in the development of the Castros Culture. It was shown that the changes were so remarkable and sudden that they marked a revolution within the Castros system. Previous chapters have also shown that these changes were referred to in virtually every aspect of the archaeological record. The understanding of this phase is fundamental in

65 - Dio 53, 12, 4 (*in* Schulten 1940, 184), and Strabo III, 4, 20.

66 - Suetonius, *Tib.*, 9; in Schulten 1940, 189.

67 - Alternatives to the hypothesis of these authors are presented by Torres Rodríguez (1976 and 1980, 111-2), and are the subject of revision by Martino (1982).

68 - Dio 53, 29 (in Schulten 1940, 190).

69 - Dio 54, 5, 1 (in Sculten 1940, 191).

determining the mechanisms of development in the Castros Culture by the turn of the millennium.

The analysis of the strategies of Romanization is a most definitive method of establishing the previous state of the Castros Culture and its interactions with resistance to, or acquiescence with, the dominant culture. An overall view of Roman attitudes towards the indigenous peoples indicates a process which is characterized by a relative freedom allowed to, and a respect of, the various cultural identities; but there is also a strong and inexorable progression towards the Roman concept of ways of living. This is why the Romanization of the Northwest is so uniform, effective, and long-lasting.

The successful spread of the Imperial cult in the Northwest had, obviously, the purpose of reinforcing obedience to Rome through the deification of its ruler, thus strengthening his authority. Furthermore, this strategy had the parallel result of appealing to the indigenous upper levels of the social hierarchy (Alarcão - Etienne 1976, 185), as the spread of the concept of deification of the chief would no doubt confirm the status of local leaders.

Although the available evidence indicates that Romanization was the motor of this sudden and intense change in the Castros Culture, other factors must not be neglected. The Romanization of the region to the north of the Douro is characterized by the use of distinct strategies, in the context of Iberia, due to the very particular social organization and culture of the peoples who inhabited this region. Contrary to what has been suggested (Alarcão 1988, 54), one could accept that some *oppida* of the type at Briteiros, Mozinho or Sanfins had already achieved the embryonic characteristics of an urban central settlement on indigenous lines, prior to the Conquest. The deep reforms perceptible in these sites by the turn of the millennium imply the existence of some previous social, economic and political status which made them eligible to become large cities among the castros. It should be clarified by future research whether this urban centralization was, in fact, exclusively promoted by the Romans, or if there was a previous trend towards the development of such urban central places by the indigenous system (Frey 1984, for Gaul) prior to the Conquest. This indigenous trend towards urbanization and nucleated settlement has been observed in Britain (Cunliffe 1976, 151-3) for the period preceding the Roman Conquest. The immediate development of these centres under Roman occupation was, in this particular case, the result of previous dynamics, and therefore the Roman effect in this process should not be over-evaluated. A similar picture has been proposed for Gaul (Nash 1976b, 129), where 'urbanization had only recently begun at the time of the Conquest...'

The dating for the beginning of the process of acculturation in the region is not clear. The state of the Castros system at the start of the Conquest is still ambiguous. Obvious similarities, however, between the Castros Culture and contemporary European societies can no longer be rejected by the argument that the Northwest had always been the *finisterra* where systems developed differently. Recent discussion has insistently pointed out the many relations between the regional elements of what may be called the 'European World System' (Kristiansen 1992). Moreover, the traditional arguments are less consistent in defending this isolation of the Castros Culture from the European cultures of the Iron Age.

At the present state of research the reign of Augustus or, more precisely, the end of the Cantabrian wars, may be considered as the turning point for change in the Castros system, as many authors have already suggested (c.f. above). However, this situation must not obstruct future research from attempting to ascertain earlier evidence for the change in the system, and from determining within the mechanisms of this change between what was promoted by the Romans from what was the result of the internal dynamics of the Castros communities.

9.5) War, economy, prestige and social order

Although several theories were built to explain the causes of war in primitive societies, it became clear that war could not be solely connected to biological and economical factors, as indeed war is a '*social invention*' (Mead 1971, 38-9), and accordingly relates to the way the social systems are organized.

War can have different motivations, and its impact in the involved communities cannot be analysed in a schematic way. In the course of this study, the analysis of war in Castros Culture has been deliberately supported by two major references, which are considered here as being complementary: the economic factors, which were formerly believed to be the major cause of war and the social factors, in particular those related to social pressure and hierarchy. The overview presented in Chapter 7, in which recent data is evaluated, reflects a pattern where subsistence economy is ecologically balanced and is capable of supporting productive independent communities and their population growth. It is, therefore, necessary to properly evaluate the social and political factors as major motivations for warlike activities.

The literary sources, which were, as a rule, under the dependence of particular philosophical views and also tied to political propaganda needs (Rodrigues 1962, 320-1), as discussed above, tend to provide a derogatory image of the warfare systems of the indigenous warriors. The overwhelming majority of the studies done on this subject have followed this line, thus spreading the idea that the *Lusitanians* were robbers and shepherds because their homeland could not fulfil their basic economic needs, and that they raided Baetica for pillage and to procure new lands. This picture can no longer be accepted in the light of new evidence. Moreover, the various Conquests of territories and cities during the *Lusitanian wars* were rather ephemeral and usually ended by voluntary withdrawal; a recurrence that certainly would not happen if the search for new land was one of the major motivations.

The available data indicates that warfare may have been common and permanent throughout the first millennium, to the extent that it has survived, probably in a symbolic manner, in the massive defensive walls and statues of warriors of the post-pacification period. So strongly has warfare marked Castros Culture that the symbols of warfare prestige will be maintained and valorized in a period when the goals are completely different from previous times.

War in Castros Culture, then, can be related to various aspects, all seeking the same objectives: competition between communities and balance of the social system. War, as an economic activity, could have produced spoil, slaves, tributes and links of dependence. The latter were capable of being transformed into goods or services at some stage. The possession of part or all of the above would enhance hierarchy at local and regional levels. Archaeology can reflect these aspects of social complexity through the increase of the communal and personal symbols of power and prestige, such as those summarized earlier: unequal defensive structures and patterns of territorial occupation, public architecture, statues and war gear.

In spite of the comparative model of typological and stylistic characteristics in the statues of warriors (Calo Lourido 1991), the available data suggests that they are indigenous in all their typology and significance. Evidence from the earlier phases demonstrates that the statues of Callaican warriors, even if they were made during, or after, a period of reorganization of the social system, contain fossilized elements which reflect an earlier system (Figures 56.1, 57.1 and 57.2).

The system, based on links of clientage between warriors, chieftains and communities, might have continued during the decades following the Conquest, and was therefore based on reminiscent patterns of prestige, which were then changing their social layout and becoming more extensive amongst the group. It is hard to believe that all the massive stone structures could have been built exclusively using the manpower and economic surplus of the community that lived inside them[70], but rather in cooperation with neighbouring villages, where the participation was certainly ruled by hierarchical relations. This might have been the major line for the evolution and establishment of an urbanism, such as Sanfins and Briteiros, with an increasingly neat establishment of site hierarchy, noticeable in the major archaeological and cultural aspects. The same pattern might be suggested for earlier phases; however, sufficient data is as yet unavailable.

The relative uniformity of the sites before the Conquest suggests the existence of various chieftains, eventually one per site, although there is little indication of their hierarchy, if any. In fact, the fast and sudden urban development of some sites, during or immediately after the Conquest, cannot be exclusively explained by Roman protectionism, dominant geographical location or even special availability of important resources. The existence of some kind of previous site hierarchy should also be borne in mind as a motor for this trend.

Other communities had reached similar states of centralization by the end of La Tène (Collis 1982, Frey 1984, Cunliffe 1976 and 1986), and yet all with different timing towards Romanization. This internal process of change was probably what Almeida (1983b, 189-91) has suspected to be the *early Romanization*, which may have started with the expedition of Caesar to the Northwest.

Furthermore, what of the suggested ßwidespread appearance of new castros (Almeida 1986), amongst which the so-called *'agricultural Castros* stand out (Almeida, CAB, 1987, 333-4), known for their location in agricultural river valleys? Can Almeida's model (1983a and 1983b), being based on an economic development, fully explain the whole process of both population, economy and urban development? I believe it can, although the motor for the whole process is not yet clear. Events such as the Lusitanian wars, where long and distant campaigns were carried out and reasonably large armies involved, suggest that large numbers of lives were lost during combat, which should have affected the populational stability of many castros. This may have sustained the creation of new sites prior to the change of millennium.

It has been suggested (Almeida 1984, 38; reinterpreting Dio and Florus) that the remarkable growth of the number of sites during the Augustan period, and the foundation of new castros closer to the valleys[71], were a result of the emperor's policy and pressure, being motivated by the distribution of the *ager*. There is no doubt, however, that the reduced size and the agriculturally-bound location of these valley hillforts seems to give them an appearance of

70 - Recalling the '*tribal communitarism*' referred by Taboada Chivite (1977, 74).

71 - However Almeida's model can be coherent with one aspect of the available data, one must not forget that the exploitation of the valleys by the small defended settlements has started early in the Ist millennium BC (Soeiro 1985-86a).

units of production, or *defended hamlets*, rather than genuine castros, in the classic meaning of the expression. Almeida's remarks (c.f. above) on the *political disruptions* of the traditional castros raises an interesting set of questions. Was this disruption caused by a direct political attitude towards the sites which had a certain level of Romanization at the time? Or was it a side-effect of other Roman pressures which are more subtle in the historical record? Was it a rule dictated by a military conqueror, or was it a result of the natural development of the system?

One must not forget that, before the award of of the *ius latii*, the patterns of acculturation should have been very much along the same lines as those of the aspirations of the Castros ruling classes, rather than generalized to all the population, meaning that a change in form would not necessarily have brought an immediate change in the basic structure.

How then does a warrior fit into a society now without war and the resource of neighbouring brigandage? What are the new patterns of social prestige? The answers to these problems necessitate a deep and wide analysis of Castros society, in particular of the social organization, which is a complex subject that extends outside the area of this study. There is no doubt, however, that after the Conquest there is a renewal of the prestige of the oppida, as the epigraphic evidence for the Ɔ, or *castella* (Albertos 1975, Beltrán Lloris 1988, Pereira Menaut 1978 and 1982, Rodriguez Colmenero 1988, Tranoy 1981, Le Roux - Tranoy 1984), suggests. The apparent self-confinement of the castros throughout the millennium may suggest that the reality expressed by the Ɔ may not be exclusive to the later phase. Furthermore, the multiplication of pacts (or alliances) between Romans and Callaicans, as seen in the *tesserae hospitales*, show that the names of the *castella* and their peoples are mentioned along with the names of their leaders, who often used a Roman name from a very early stage (Pereira Menaut 1985-86) in the period of Conquest.

This situation might imply a reorganization of the warrior status, now more settled in a stable political, and probably economic, basis. Within this model, prestige could be gained by different methods:

1-By participation in the warfare activities as auxiliaries in the Roman army, which had a double function as a prestigious activity, both for Roman and indigenous patterns, given the army's political importance and influence on the organization of the first provincial urban settlements. This Roman-regulated warfare would present a new opportunity in the political careers of the local hierarchies. This participation in the conqueror's army would lead directly to a wider social and political participation in Roman provincial citizenship, which could to some extent explain the emergence of an indigenous provincial aristocracy. In fact, it has been suggested by Le Roux (1980; 1982, 41; 1985) that local *auxiliaries* helped the Romanization of Iberia by supporting the stabilization of the Conquest, by working as an interface between local hierarchies and the agents of Rome.

2-By maintenance of the prestige through the political influence which could have often been gained or reinforced by collaboration with Romans, as some *tesserae hospitales*[72] seem to reflect.

3-By control of the means of production, and reinforcement of status through possession of wealth. This process can be complementary to the others.

The whole picture is still unclear. However, this process is undoubtedly reflected by the sudden multiplication, by this time, of the Callaican statues which are known to be related to the display of the prestige of a chieftain in association with the prestige of the hillfort. The warrior's statue from Citânia de Sanfins, which was standing by the site entrance, clearly suggests a convergence of significance of both the statue and the defensive wall. The same model would explain why so many defensive walls were built after the pacification, when supposedly there would be no need for them for exclusively warfare purposes. Moreover, some defensive walls are so extensive that the site could not possibly have held the population necessary to build them.

The answer seems to lie in the rebuilding of the model, by starting with a point that seems fundamental to me. The end, or reduction, of raiding and mercenarism (Nash 1976b, 128) ended an important outlet for social tensions and conflicts. These were now confined to the system. This situation had various effects. The means of production, first belonging to women[73], began to be controlled by men, and this is why the father now gains increasing power within the familial structure (Almeida 1983b, 195). With the reduction of warfare there is an obvious decrease in mortality. Pottery production, masonry and metallurgy show a remarkable evolution in techniques, as well as a trend towards mass production. In fact iron metallurgy becomes more common, jewelry becomes more refined through the mastership of metal alloys and soldering, as well as of filigree and granulation. There are innumerable examples to show a

[72] - Silva 1983c for the important case of Monte Murado, Vila Nova de Gaia. In fact, the two reported pacts made between the *Turduli Veteres* and the Roman consul evolves from the simple hospitality in the first *tessera*, dating from 7 AD, to the *fides* and *cliens* two years after. It may also be of relevance the fact that the indigenous intervenients, presumably the political representatives of the social group, have changed within these two years.

[73] - This idea repeatedly spread through the Classical sources is schematic and may be true only to a limited extent. As Almeida has once said (1983b, 195): in Castros society "house, tenancy of the fields -as agriculture was women's task- belonged to women's world", whereas men's were the mobile goods, the ones achievable through brigandage, the external territories and the expeditions.

development of productive activities. This, however, is the economic side of the whole process. As far as the political side is concerned, the foundation of new sites by this time may be of relevance, although still somewhat enigmatic, as one can judge by the proud designation of *castellum* given to these new sites. Apart from a search for new exploitable land, these new castros may also correspond to a populational disunion of some sites, but they will take an important part in the new (or at least now neater) process of hierarchical distribution within a region, and in the establishment of a clear and organized urban development. The proximity of the sites, all of which mantained visual contact with surrounding neighbours, as all were located on hilltops, suggests that defence could have been communitarian within the same region and/or ethnic group, and was therefore one factor of cohesion.

If one forgets for a moment that the walls of castros may have been defensive structures, then the ambience of the sites reflects a structure related to communitarism and production, just as many European hillforts were (Cunliffe 1978, 278) at this time.

9.6) Discussion

The first defensive structures in this region seem to be closely connected to a large spread of metal extraction and metallurgical activities at a local scale, and probably each defended site would have been a centre of production. It would not be surprising if the need for the first palisades corresponded to the establishment of a new social order, where a structured warlike society was taking its first steps towards centralization through activities such as raiding (Champion 1982), both for controlling territory and retrieving goods. These two factors would prove essential for the maintenance of the hierarchical balance of the system.

The fortification of sites has to be seen as a side effect of various aspects, not all of which are completely understood and evaluated at present. A brief overview of the material culture of the early first millennium shows that specialized productive activities, such as agriculture, metallurgy and pottery, and long-distance trade and contacts, were widespread and developed in the Northwest. We have then a social system on a course of fast development towards the establishment and reinforcement of political and territorial hierarchy.

Of course economic aspects cannot be neglected. Internal competition for power and control of the economic sources is the more coherent approach, considering the relatively even evolution of material culture from the previous cultural background; but the arrival of new ethnic elements should also be borne in mind. This competition for power and the development of social complexity were necessarily connected to the control of the metallurgical sources and the production of metal. Evidence is categorical on this point. Still, the model is not completely convincing, and exogenous factors can be taken into account for this discussion, although their importance should not be overstressed. Unfortunately, here one can play with data on a theoretical basis. In fact, if a group of people can settle in a different region leaving little or no evidence of their presence in the archaeological record, it is also true that an enormous amount of exogenous archaeological evidence in a region does not necessarily signify a migration or invasion. Other explanations, such as an intense trade network, can also be considered. Even trade could leave no archaeological traces as a two-way process, if perishable goods, or raw materials, were predominant. Therefore, some empty spaces and possible ways out have to be left when it comes to analysing the social factors that lead the evolution of the Northwestern populations in the early first millennium.

With the first military expeditions to the Northwest, and in particular after Caesar's campaign, there is a growing participation of indigenous auxiliaries in the legions, and their identity by the Classical writers became neater to, at least judging by the increasing quantity and accuracy of the references.

The warrior statues, with their well-defined distribution and their accurate iconographic representation, are an invaluable help for the characterization of the Callaican warrior. But can we assume that the fighting gear from these statues indeed reflects the entire, or even the more common, paraphernalia of weapons of the given region? Archaeological evidence suggests it does not, pointing, on the contrary, to a rather uneven relationship between these two sources. Bearing in mind, however, that these statues certainly portray chieftains or high-status warriors, to which are usually attached prestige items definining their status, it might be wiser to consider that they do not necessarily represent the common, or low-status, warrior's gear. As a matter of fact, spears and falcatas are not represented in the statues, yet this does not allow any reliable assumption concerning the presumably low status of these weapons (Quesada Sanz 1990).

Evidence for external influences can also be seen in some of the war gear. The helmets of *Montefortino* type (Figure 61) are probably the most diagnostic. Also, the shield boss from Castro de Alvarelhos (Figure 62) raises interesting questions about its origins. Its near photographic similarity to the Gaulish and Iberian types would not suggest it may be a local copy, but rather some kind of import; either it has arrived with a warrior or with a merchant. On the other hand, its chronology would range between Caesar's campaigns in Iberia and the end of the century. The more feasible explanations for the presence of such a type of shield in the north of Portugal would then be trade, brought by a Gaulish or Iberian auxiliary in the Roman legions, or brought by raiding parties of the time.

The range of action of the groups of Iberian warriors that sought battle in foreign territories, as mercenaries, extended as far as Aquitania (Caesar B.G. III, 26)[74] in aid of the Gaulish tribes against the Romans. But they were also recorded in Greece (Garcia y Bellido 1934 and 1974) and in Italy, under Hannibal (Livy 21, 43, 8).

The wars in the Iberian South may have involved several peoples, amongst which were the Callaicans. Even accepting the normal Roman habit of exaggerating the facts when it comes to revealing the triumphs of their generals against the barbarians, it seems that there is an enormous number of participants -and losses- as well as a remarkable continuity in the wars in the South. It looks, in fact, as if there was too much warfare to be attributable to one single people.

Raiding parties may have continuously crossed Iberia during the Castros period. Raiding and mercenarism, however, certainly increased with the growth of internal pressure in the system by the last few centuries of the millennium.

War, in its different manifestations and side aspects, is strongly reflected in Castros society and material culture, as suggested by the currently available data. The major difficulties are, on the one hand, determining social mechanisms and their interactions with warfare and, on the other, evaluating how.

Warfare was exclusive to men, and the participation of women in defence of their homeland with the greatest *barbaric* courage (Appian 71-73; Strabo III, 4, 17) may have occurred, but no more than as an episodic and should not imply the existence of a class of female warriors[75].

The assumption that cattle-herding communities were militarily stronger than the communities of farmers (Bouzek 1982, 187) can barely be defended here, given the relative balance of these two productive strategies in the Northwest, as shown in Chapter 7. Cattle-breeding cannot, in fact, be considered a major activity of Castros folk, on the light of recent evidence.

There are some indicators that, although the warriors' class might have been highly hierarchical, a relatively vertical permeability may have allowed the eventual rise in rank of some lower status warriors. It is not yet completely clear whether the military status amongst the Callaican warriors was determined either by familial or social inheritance, or attained in warfare expeditions. However, the relative frequency in which we see different leaders conducting war parties seems to indicate that leadership prior to the Cantabrian wars could have been relatively circumstantial.

The claimed detachment of the Lusitanian soldier from political power (Fernandez Canosa 1987, 154) does not seem to be expressed in either the Classical sources or by the models of Celtic society. Treason and multiplication of indigenous armies during the wars analyzed above show that there was a strong and permanent political tension towards leadership. Moreover, the characteristics of the hierarchical order inside the castro is not yet completely understood. Of course this would have to be analysed in a very strict diachronic perspective, as this factors are likely to have changed very much with time. But perhaps one should consider two spheres in Castros social organization: the more permanent hierarchy in the site and in the homeland, in opposition to the more fluid and ephemeral hierarchy related to the *communities of warriors* and to raiding.

In the Northwestern world of the second half of the millennium, warfare and raiding are often equal in their military and social significance.

As has already been outlined, after the first few years of confrontations in the Lusitanian wars, the fight against the Romans seems to become the major motivation for these expeditions, replacing that of prestige/pillage. On the other hand, it is also clear how the leadership of the *Lusitanians* becomes more important and worthy of reference with the evolution of the situation. This might reflect a development of leadership or hierarchical organization among the *Lusitanian* parties as well as the military inconvenience of their actions. In fact, from 155 BC onwards, the leader of the indigenous armies begins to be mentioned by the Classical sources. They also show how short the career of these leaders usually are as many of them die during combat. That, by no means, seems to endanger the continuation of the campaigns, and the easy acceptance of the various leaderships by such heterogeneous armies during the second century BC is surprising. On the other hand, another important novelty is the increase in the number of warriors involved in these expeditions. This can be evaluated by the recorded number of losses amongst the Lusitanians, as well as by the incremental participation of other peoples, such as the Vaccaeans and Celtiberians.

The Conquest and the virtually immediate abandonment of the cities and surrounding territory in Baetica, as recorded by the sources, not only indicate that the search for new territories was not the leading motivation of the raids, but also that their degree of political organization was unable to cope with a continued control of the conquered areas. A scenario of exclusively warrior parties, therefore, has to be clearly differentiated from a warlike community in search of new territories, as the latter would involve a

---

74 - Referring to Cantabrian cavalry.

75 - Alonso del Real 1979 draws upon the excessive importance paid to this subject by Strabo, which he connects to mythological interpretations.

different and more systematic displacement of populations and would not solely be restricted to warriors.

The greater significance of the history of these wars is, however, the long and apparently uneven sequence of contacts between Romans and *Lusitanians*, sometimes as enemies, at other times as allies, and a process of acceptance and acculturation developed after this nearly permanent contact (Mossé 1970, 119)[76] prior to and during the Conquest. Several Iberian natives aspired to (and obtained) Roman citizenship prior to Romanization, and many Roman generals adopted their particular fighting strategies and weapons which had so many times proved to be effective. This 'brotherhood in arms' has led to a very important way of control of the populations through their warrior élites, regardless of whether they were the rulers of the sites, through the establishment and reinforcement of the links of dependence with members of the community.

After the Cantabrian Wars, although Caesar's campaign could be considered a turning-point (Almeida 1983b, 188-9; Silva 1986a) in this region, there is a general 'pacification' which brings a turnover in the whole raiding system, as referred to by Strabo. Its disappearance no doubt carried disturbance to the system.

The role of the warrior should have survived by attaining a more prestigious status, now with less emphasis on external raiding, but with increased interest in possession and control of prestigious items. This is perhaps why most of the defensive walls seem to be built during and after this period, as well as the warrior statues representing heroes or chiefdoms (Almeida 1986, 166-7; Martins-Silva 1984; Silva 1986a). The prestige of the warriors could now be based more on the display of the prestige items connected with their rank, which is no longer connected with warfare, but rather with the control of the means of production in order to produce surplus, as well as political power. Some others, more attached to the warlike life-style, or with less chances of social integration, are welcome to join the Roman army where corps of Callaican auxiliary (Le Roux 1980 and 1985) are known.

In fact, the Conquest of Hispania, and the western Mediterranean, has stopped the substantial recruitment of auxiliaries (Nash 1976b, 128), either Celts or Iberian; however, the disruptions caused in the respective systems may have been different, as seen above.

The Conquest of the Northwest has originated such a dramatic change in Castros Culture that it would be fair to call it an 'Augustan revolution'. In fact, it was a revolution in techniques and craftsmanship, in production, demography, mentalities and social organization, and certainly in politics. It can no longer be sustained that the Northwest remained an isolated region throughout the millennium, although in a very irregular way. Castros peoples certainly have retained most of their regional cultural identities, but they were not isolated from the rest of Iberia, or the Mediterranean, or Atlantic Europe. Goods and ideas have reached this *finisterra* at various stages, by travelling both with merchants and immigrants, and probably the whole of Europe was then a closer community than we now dare to accept.

The reinterpretation of the available data suggests that Castros Culture can no longer be depicted as a society based on the bipolarity of women-producers and men-warriors or raiders, but rather as a more complex and organized society, highly structured at both local and territorial level.

76 - This author also proposes that economic mechanisms have conducted this system.

> Entraînés à exploiter les traces des actions quotidiennes, nous sommes trés à l'aise dans un domaine familier et cette aisance nous engage à projeter notre propre vision, puis à la justifier. Le risque est de ne trouver que ce que nous connaissons. L'homme de la préhistoire n'est-il pas notre ombre? (Taborin 1989, 80)

10- CONCLUSIONS

During the course of this study, it has been shown that the wide range of natural resources available in the Northwest of Portugal made the region particularly desirable for settlement in the first millennium BC. Fertile soils, abundant supplies of water and a mild climate enabled a broad spectrum of subsistence activities to be practised in a number of different environments, including those on the coast, river valleys, hill slopes and plateaux. The topographical variety of the area produced several distinctive micro-climates, which were exploited to the full. In some parts of the region, conditions were probably ideal for the cultivation of species such as vine. Indeed, Mediterranean species like olive began to be grown soon after the arrival of the Romans.

Of the other resources of the region, metals were probably the most important. Copper, tin, lead and gold were extracted and a thriving metallurgical industry developed. In some circumstances, metals may even have been obtainable without the need for extensive mining.

Easy access by sea and river attracted much external interest in the region, and an extensive trade in metal products resulted. In fact, the Northwest became something of a meeting point between the various foreign influences, particularly those of the Mediterranean and Atlantic Europe.

From at least the beginning of the first millennium BC (supported by radiocarbon dates), defended sites were built on the top of hills. This may have been in direct response to intense interest in the resources of the region, and especially metal ores and products. A link between defended sites and metallurgical activities can be demonstrated. Thus, protection of these resources may have been a major factor, though not the sole reason for the construction of these hillforts. The hillforts were also strategically located, able to command large areas of the surrounding landscape, and to exert control over the routeways. The many bronze hoards and finds characteristic of the Atlantic Bronze Age in this region further substantiate the economic dependence of the hillforts on the metal trade.

The unique geographical identity of the region is corroborated by additional aspects of the material culture. Amongst these, the occurrence of stone statues of the Callaican warrior (Figure 58), and bath structures of type 1b (Figure 18), are worthy of note because their distributions coincide almost exactly with the boundaries of the area studied in nortwestern Portugal.

The principal aim of this study has been to understand the social role of warfare and its effects on a population which settled in the northwest of Portugal during the Iron Age. It has been assumed that a detailed study of the social and cultural interactions of the Castros Culture in its wider context is essential to this.

The effects of warfare can be so important for a community that they pervade the entire system. Such significant changes might be expected to leave their mark in the archaeological record, and to be remaked on by the Classical writers. This is why such a broad overview has been attempted, and so many apparently irreconcilable subjects analyzed. It has been the intention of this study to collate and evaluate the many scattered and diverse fragments of data available, in order to build up a picture of Castros society during the second half of the millennium. Whenever possible, these data have been supplemented by the results of my own excavations, fieldwork and environmental studies.

The second objective was to investigate the long-accepted idea that warfare and raiding on the castros was basically motivated by the poverty of their inhabitants, whose means of production was so inefficient that their needs could only be supplied through brigandage.

Further topics of study included: an evaluation of the characteristics of a progressive society in development; whether the establishment of hillforts was an indicator of warfare as an endemic phenomenon; how many of the earlier patterns were retained at the end of the Iron Age; and what the real effects of Romanization might have been.

It should perhaps be made clear at this stage that the following conclusions are the result of my own re-interpretation of the data in the light of the evidence currently available to me.

Some very interesting similarities and contrasts between the castros have come to light during the course of this research. The more significant of these, at the risk of making sweeping generalizations, are as follows:

Iron Age social structure should be considered as arising from, and further developing, that of the

Bronze Age. The system may, of course, have been modified by external influences through time.

Radiocarbon dates suggest that a major expansion of fortification in the northwest of Portugal occurred during the last few centuries of the second millennium BC.

The force motivating the intensification of fortification was probably connected with the production and distribution of metal goods.

During the Late Bronze Age the defended site is not the only type of settlement. Open sites may have specialized in agricultural and mining activities, whereas smelting and the production of metal goods were carried out solely at the hillforts.

Increasing centralization is a feature of this period, and central places may have been responsible for any redistribution of the surplus, particularly of metals. These fortified central places were not, however, exclusively concerned with the control of resources and the redistribution of metals. A rise in bellicosity caused by the arrival of immigrants may have played a role. Alternatively, centralization may have been a suitable response to competition for long-distance trade, which may have included perishable goods in addition to prestige items. Amongst the latter, it should be noted that amber reached the area studied in the form of a bead found at Castelo de Matos.

This trend towards centralization favoured the spread of the defended settlements. In the oldest phase, which one can relate to the Late Bronze Age, the defences are relatively simple. They are timber-built, and consist of a palisade and external ditch. The sites themselves are small, suggesting a low population density.

By the VIIth-VIth centuries BC all sites appear to be hillforts and no open sites are known. At the same time there is a gradual introduction of stone into the architecture. At Castro de Penices, for example, communal structures such as the defences were built in stone (Figures 32 and 33), whereas the domestic dwellings were still made out of wood and clay.

As most of these sites were situated on granite hilltops where suitable building stone was plentiful, it is perhaps surprising that so much construction was in perishable materials. In fact, the intensive use of stone did not begin until the early Iron Age. The chronology of this phenomenon is still imprecise, however, but when the use of stone becomes commoner and the pottery intentionally micaceous, then we can refer to the hillfort as a castro.

The multiplication of hillforts within the region and a general economic self-sufficiency are the main differences between this period and the Bronze Age. It should be noted at this stage that the high level of production of metal goods in the Atlantic Bronze Age and the appearance of early castros with stone construction and micaceous pottery, may indeed be related.

This early, and apparently generalized, adoption of stone in construction may be due to environmental, rather than cultural, factors. High rainfall, mild temperatures and acidic soils probably all contributed to the relatively short life of wooden structures in this region. The use of stone also becomes common long before iron, which is necessary for so much masonry work to become widespread.

The timber necessary for the construction of palisades and huts was probably gathered over an area quite close to the site. Increasing deforestation in the site catchment area would inevitably have lead to a considerable degree of erosion. Exposure of the surface mineral resources may also have occurred. However, such denudation of the woodland would have created better conditions for stock raising, and the pastoral component of the economy may have gained in importance. This point cannot, however, be proved with any certainty.

For the the earlier period of the Iron Age, the site distribution suggests that good accessibility to resources, rather than interests related to hierarchy or a uniform distribution within the region, determined location. This is probably why the higher hilltops were neglected at this time. The coming of the Romans broke this pattern by introducing various other factors, amongst which trade stands out. The former self-sufficiency of the sites also changed, and a new hierarchical order can be traced in the archaeological record from the Augustan period onwards.

The earlier castros were relatively small; which may have been the result of a stable population, which did not change substantially until Romanization. This is not true in all cases, however. For example, recent research at Castro de Penices has shown that the size of the site remained the same from the VIth century BC until the mid Ist century AD, when this castro appears to have been abandoned.

A decrease in the production of metal goods might have been responsible for the substantial decrease in both external contacts[1] and the self-sufficiency of the majority of sites in this region between the emergence of the Castros Culture and the dynamic period of Romanization.

One might suggest, therefore, that the slow and uneven spread of iron metallurgy through the Northwest has not been of much relevance to social and economic development in the first few centuries of the Iron Age. On the other hand, bronze metallurgy seems to have played a most significant role during this period, since it is the metal most often encountered in the archaeological record until Romanization.

---

[1] - As has been suggested by Almeida, CAB, (1990a, 270) for the Lima river basin.

There is general agreement that iron metallurgy came late to the Castros Culture. However, iron was known from a very early stage, as the finds from Castro de Torroso show. Ideas, techniques and goods travelled quickly in the first millennium BC, and this region was not isolated. Iron ores were also available. Probably the late development of iron metallurgy is due more to the internal limitations of the system than to any scarcity of craftsmanship, metallurgical knowledge or raw material. The continued use of bronze, on which the system had been based for so long, is also another factor to be considered.

Evidence has shown that agriculture was able to provide all subsistence needs before the Conquest.

By the end of the millennium, many sites had enlarged their areas and reorganized their inner space. This is a two-way process undoubtedly connected with the noticeable population growth. Agricultural production, the development of iron metallurgy, the mass production of pottery, the elaboration of craftsmanship, and masonry techniques are but a few examples of the many changes that had occurred by the end of the millennium.

The changes to the system were so substantial with Romanization that one could refer to an *Augustan revolution* in the Castros Culture. The emergence of large urban castros can be read as an indication of a reorganization of the hierarchy, both in sites and society. The former hierarchy which was based on the prestige acquire through warfare and communal acts, now gave way to political prestige and wealth. This political prestige may still owe something to the symbolic image of the warrior, however. This is indicated by the various statues of warriors, and the inscriptions of the *house* of *CAMALUS*.

The idea 'women work in the fields-men are warriors' cannot be accepted any more, even when describing social structure and the division of labour. The available data demand a more coherent model, since a place has to be found for some specialized craftsmanship, e.g. to carry out metallurgical activities, and to incorporate leadership, warriors and farmers. The role of ritual and religion has yet to be accounted for, even though many studies and Classical references indicate the existence of both gods and religious rituals.

The subsistence production system was apparently self-contained. The importance given to local resources and their well-balanced use is another relevant characteristic of the pre-Roman system. As an example of this balance, some sites had a great economic emphasis on cultivation, others on pastoralism.

Trading contacts did of course exist, with both the Mediterranean and Atlantic Europe. Much of the volume of trade may have been in perishable goods. For example, what would survive from the wine trade if wooden barrels were used instead of amphorae?

The social organization of the peoples who lived in this region, as based on the *castella*, is another indication of the fact that the hillforts may have been self-sufficient, from their period of formation until the coming of the Romans. The importance paid to the *castella* in epigraphy as a mark of personal origin well reflects its significance within the political system, i.e. as distinct from that major political unit, the *populus*.

Some vague ethnic references in the Classical sources prior to Augustus leave some room for the belief that warriors from the Northwest were involved in warfare in the penninsular South during the second century BC. As presented in the previous chapter, there is evidence to show that Iberian warriors travelled far during the course of their expeditions as mercenaries, during the second half of the millennium, e.g. trough Iberia, and even to Aquitania, Italy and Greece. During the second half of the millennium at least, Iberia was being continually crossed by parties of warriors, who occasionally settled down. The Turdulians and the Celts are examples of this phenomenon in the Northwest. There may have been others before them, such as the Oestrimnians and the Saefes, as suggested by López - Bouza (1929), but archaeological evidence is not explicit on this point.

Although the direction of these expeditions may have been predominantly directed to the richer lands of the Iberian South, the episode of the Turdulians shows that movements towards the Northwest[2] also took place.

Probably much of the evidence which has been interpreted in the past as the result of Celtic migrations can, in fact, be explained by the presence of parties of foreign warriors in the Northwest throughout the millennium. Among these, there may have been Celtic-affiliated people.

Always located inside the defensive structures, the dwelling architecture exhibits aspects which apparently contradict our previous images of this society. It suggests patterns of life more connected with a society based on agriculture and craftsmanship, rather than with a warrior society, as we now understand it. This contradiction may lead to a different understanding of Castros society. It would seem that a significant part of the system was turned towards production, with the remainder dealing with warfare.

Previous suggestions, that warfare may have been conducted by *communities of warriors* which were separated from the main community whilst engaged in warfare, is now more acceptable. But

---

2 - Directions of these movements cannot be traced at present. Let us take, however, the example presented by the distribution map of La-Tène type helmets, produced by Silva (1986a, Est. VI), where the northern distribution of this find goes from Ampurias to the Entre-Douro-E-Minho, following the course of the Douro.

these two groups within the system did not, apparently, interfer with one another, as the dwelling units within defensive walls are, above all, consistent with a farmstead function. Similar models have been put forward for Wales (Figure 20) in the Iron Age.

The existing cohesion between the inhabitants of a castro may be further illustrated by its communal features. Yet, the available evidence refers to a later period. For the early stages, archaeologists still cannot see the organization of the site clearly.

The overall change in the Castros system by the end of the millennium cannot be explained by the dynamic impact of Roman policy and acculturation alone. It also has to do with an internal change of attitude towards warfare. The social role previously developed through effective warring now acquires a more ritual pattern. The prestige gained by fighting can instead be obtained by a display of attributes, such as those shown on warrior statues and represented by defensive walls. Rituals may be reinforced, and saunas may be related to this. Changes in leadership may be indicated by the presence of four statues at the site Outeiro do Lezenho.

This replacement of war by display was induced by Romanization, but it was not a compulsory event. Previous contacts with the indigenous warriors may have led the Romans not to force the issue. Their strategy, as shown by the *tesserae*, was to control a society through its leaders. This may have been achieved by reinforcing the leader of the group, particularly when this leader was recognized by the conquering side. In addition to its leaders, the *castellum* also obtained prestige through the construction of communal symbols, such as magnificent defensive walls or elaborate saunas. By such means, the chieftain gradually becomes less of a warrior and more of a political figure, until he becomes a *togati*.

This reduction in warfare allowed economic production and craftsmanship to flourish. Both population and surpluses increased as a result, and new sites were founded lower down the valleys. Prestige still mattered however, and these new sites were built with defences, imitating Roman forts.

The continuity, or otherwise, of occupation in the defended settlements is another important aspect which recently has had to be re-evaluated, after the assumption that some hilltops may have had a continuous occupation throughout the first millennium BC (see Chapter 6).

It has now become a widely accepted fact that the process of Romanization of the Castros peoples was not particularly violent[3]. Rather, it was conducted towards establishing the cultural patterns of one dominant culture, and this is why the expression 'Romanization' is such a proper term. The process which produced the new society was two-way (Almeida 1983a and 1983b), i.e. its goals may have been Roman but its relationship with the environment was clearly indigenous and local cultural identity was attained. This is probably why the acculturated features (Calo Lourido 1991), such as saunas, warrior statues, art and architecture, still reflected an indigenous cultural identity. Only the award of the *Ius Latii* could weaken, but not sever, this unique cultural identity.

3 - Perhaps with the exception of the Cantabrians

There are many limitations to the data currently available. For example, the scarcity of suitable data prevents us from drawing any conclusions about patterns of site distribution prior to Romanization. There are still many excavations to be done on sites occupied during this epoch. This is an important task for future research.

The present state of research does not allow us to establish with certainty whether a site was continuously occupied or was abandoned at some stage, even for relatively long periods, say one or two centuries prior to the IInd century BC, unless imported datable finds occur.

In fact, stone structures have left their imprint in the archaeological record, whereas the remains of structures built out of perishable materials are almost non-existent[4], just to provide but one example. Archaeological research in the Northwest has not reached a level capable of solving any of the above problems, and this will probably remain the case for many years to come. The domestic settlements of the first millennium, up to the IInd-I^st^ centuries BC, are little known, in spite of many recent contributions on the subject, which include regional monographs and data on site continuity. Even these, it should be understood, are insufficient for any establishment of patterns for this period.

From what we have seen of this broad analysis, probably the less satisfactory conclusions are the ones relating to the social function of the warrior, and his position within the system. What kind of warrior are we dealing with?

I trust that future progress in the subject might alter or develop what I have tentatively proposed here.

The techniques used to determine whether a site was abandoned or continuously occupied are predominantly, if not exclusively, archaeological. To resolve this, it will be necessary to start by building a chronological and typological framework, i.e. by relative and absolute dating of the material culture. Aspects of the material culture, such as ceramics, metallurgy or architecture, will have to be well-

4 - Not only on the obvious difference of evidence left by each of these, but more importantly on what concerns durability, effects on sedimentation, flexibility to re-use and socio-technical implications.

known on points of internal and regional variability, in order to establish continuity or discontinuity. Without the help of well cross-dated elements, the process of construction, destruction, and reconstruction[5], and even the sedimentation mechanisms themselves, are virtually impossible to establish at present. There exists a very complex dynamic of understanding, and the technical and cultural mechanisms which are inherent to a given period are particularly difficult to interpret.

The last ten years or so have seen a major development in the establishment of the ethnicity of the Castros Culture, by redefining what is meant by a cultural identity[6], and especially in the construction of models of cultural and ethnic typologies for the Northwestern Iron Age. Still, in spite of various contributions, the problem remains open to discussion. If it is a fact that the Northwest may be considered as one coherent unit during the Iron Age, it is no less true that recent research has demonstrated the existence of various regional sub-cultures, as the peculiarities of the archaeological plans indicate. The concept that cultural units, which can be large in spatial and ethnic terms, can be adapted to regional conditions and produce their own particular cultural answers is not new in anthropology. A greater challenge is the one of determining the guidelines, or boundaries, for dealing with problems of archaeological association in a large group.

The trend over the last decade has been a division according to geomorphological regional units, such as river valleys. Having been organized mostly on a basis of the characterization of the material culture, and groupings by archaeological features commonly found in a geographic framework, research has barely managed to establish even a tenuous definition of Castros social systems. This concentration of effort on field research has furthermore neglected the important fact that archaeological evidence provides us with fossilized elements of social systems. Thus, their study and interpretation should also take the human factor into account, rather than being concerned exclusively with a schematic study of material culture,[7] as is the usual method. Such an approach becomes even more striking when recalling Trigger's statement (1978, 117) that "*...a uniform material culture does not constitute proof that the people associated with it necessarily had a strong sense of common identity...*". Could this occur the other way around? Can relatively different communities, as shown by material culture, have a strong sense of common identity? They can, in fact, if one accepts expressions such as 'sub-culture' or 'regional culture'.

Recent advances on the thorny topic of defining the identity of the ***Castros Culture*** have made the use of this expression at the present time a tenuous act. It is particularly frustrating to review that, over one century of research on a single culture, it may turn out not to be acceptable in real terms. Considering the insistence with which this subject is brought up in discussion, certainly by the end of this decade the expression ***Castros Culture*** will either have survived as an archaeological term, or then have been substituted by a more *innocuous* one.

5 - The expression *destruction* used here does not necessarily mean intentional destruction, but rather any factor that promotes the sudden or progressive decay of an archaeological structure.

6 - Mostly following the theoretical framework proposed by the European and American *schools* of archaeological ethno-anthropology, and in particular, the *New Archaeology*.

7 - Exceptions are the contributions made in Ancient History research at Porto and Santiago de Compostela Universities, in subjects such as Epigraphy and Religion.

BIBLIOGRAPHY

Abasolo - Pérez 1980: Abásolo, J.A.; Pérez Rodriguez, F. El casco celtico de Botorrita (Valladolid). *BSAA* 46, 93-114

Abrantes 1945: Abrantes, A.M. Ara votiva aos Lares Marinhos. *O Tripeiro*, 5ª série 1 (6), 125

1966: Abrantes, A.M. Cale. Sua origem e localização. *Lucerna* 5, 96-104

Acuña Castroviejo 1976: Acuña Castroviejo, F. Excavaciones en el castro de O Neixón Campaña de 1973. *NAH* (Prehistoria 5), 25-30

1977: Acuña Castroviejo, F. Panorama de la cultura castrexa en el N. O. de la Península Ibérica. *BA* 31, 235-253

Aira - Guitian 1985-86: Aira Rodriguez, M.J.; Guitian Ojea, F. Contribuicion al estudio de la Cultura Castreña Gallega: analisis palinologico de los castros de Vixil y Penarrubia (Lugo). *Pontevedra Arqueologica* 2, Pontevedra, 191-200

Aira Rodriguez *et al.* 1989: Aira Rodriguez, M.J.; Saa Otero, P.; Taboada Castro, T. *Estudios paleobotanicos y edafologicos en yacimientos arqueologicos de Galicia.* Arqueoloxía/Investigación, Xunta de Galicia - *DXPHD*, A Coruña

1990: Aira Rodriguez, M.J.; Ramil Rego, P.; Alvarez Nuñez, A. Estudio paleocarpologico realizado en el castro de Penalba (Campo Lameiro, Pontevedra). *Botánica Complutensis* 16, Madrid, 81-9

Aira - Vazquez 1985: Aira Rodriguez, M.J.; Vazquez Varela, J.M. Nuevos Datos Palinológicos sobre la Agricultura Prehistorica en Galicia (España). *TAE* 15 (2-4), 241-52

Aitken 1990: Aitken, M.J. *Science-based Dating in Archaeology.* Longman, London

Alarcão 1983: Alarcão, J. *Portugal Romano.* Lisboa (3ª ed.;1ª ed.1973)

1988: Alarcão, J. *O Domínio Romano em Portugal.* Publicações Europa-América, Mem Martins

Alarcão - Alarcão 1963: Alarcão, J.; Alarcão, A.M. Vidros romanos do Museu Martins Sarmento. *RG* 73 (1-2), 175-209

Alarcão - Étienne 1976: Alarcão, J.; Étienne, R. Le Portugal a l'epoque augusteene. *Ciudades Augusteas de Hispania I*, Zaragoza, 171-87

Albertos Firmat 1964-65-72: Albertos Firmat, M.L. Nuevos antroponómios hispánicos. *Emerita* 32-33-40, (32) 229-352, (33) 109-43, (40) 287-318

1974: Albertos Firmat, M.L. El culto a los montes entre los Galaicos, Astures y Berones y algunas de las deidades más significativas. *EAA* 6, 147-57

1975: Albertos Firmat, M.L. Organizaciones suprafamiliares en la Hispana antigua. *Studia Archaeologica* 37, Valladolid (=*BSAA* 40-41, 1975, 5-66)

1976: Albertos Firmat, M.L. La Antroponimia prerromana en la Península Ibérica. I *CLCPPI*, 57-86

1977: Albertos Firmat, M.L. Perduraciones indígenas en la Galicia romana: los castros, las divindades y las organizaciones gentilícias en la epigrafia. *Bimilenario de Lugo*, 17-27

1981: Albertos Firmat, M.L. Organizaciones suprafamiliares en la Hispana antigua II. *BSAA* 47, 208-14

1988: Albertos Firmat, M.L. Sobre los 'castella' del Noroeste Peninsular. Pereira Menaut, G. (ed.) 1º *Congreso Peninsular de Historia Antigua* 2, Santiago de Compostela, 191-5

Albuquerque, M 1980: Albuquerque, M. A cerâmica da Citânia de Briteiros. *RG* 80 (1-2), 105-122

Alföldy 1988: Alföldy, G. *The social History of Rome.* London (1st ed. 1975)

Allan 1965: Allan, J.C. A Mineração em Portugal na Antiguidade. *BM* 2 (3), 139-75

1970: Allan, J.C. Considerations on the antiquity of mining in the Iberian Peninsula. *Journal of the Royal Anthropological Institute of Great Britain and Ireland*, London, (=Allan 1965)

Almagro Basch 1940: Almagro Basch, M. El hallazgo de la Ria de Huelva y Final de la Edad del Bronce en el occidente de Europa. *Ampurias* 2, 85-143

1958: Almagro Basch, M. *Inventaria Archaeologica, (E.1), Depósito de la Ría Huelva.* Madrid

1982: Almagro Basch, M. La invasíon céltica en España. Menéndez Pidal (Ed.) *Historia de España. La historia prerromana* 1 (3) 1-278

Almagro Gorbea 1972: Almagro Gorbea, M. Nuevas fechas para la Prehistoria y la Antropología Peninsular. *TP* 29, 228-42

1973: Almagro Gorbea, M. Cascos del Bronce final en la Península Ibérica. *TP* 30, 349-62

1974a: Almagro Gorbea, M. Los asadores de bronce del Suroeste peninsular. *RABM* 77, 351-95

1974b: Almagro Gorbea, M. Orfebrería del Bronce final en la Península Ibérica. El tesoro de Obispalía, la orfebrería tipo Villena y los cuencos de Axtroki. *TP* 31, 39-100

1977: Almagro Gorbea, M. El Bronce final y el perído orientalizante en Extremadura. *BPH* 14, Madrid

Almagro *et al.* 1978: Almagro Gorbea, M.; Bernaldo Quirós, F.; Fernandez Miranda, M.; Lopez, P. *Catálogo de yacimentos arqueologicos con datacion mediante C14 de la Peninsula Ibérica e Islas Baleares e Canarias.* Instituto Español de Prehistoria, Barcelona

Almeida, CAB 1979: Almeida, C.A.B. A Rede Viária do Conventus Bracaraugustanus. Via Bracara Asturicam Quarta. *Minia* 3 (2ªs), 61-163

1981: Almeida, C.A.B. Uma escultura antropomórfica da Citânia de Roriz (Barcelos). *CEG* 32 (96-97), 467-74

1982a: Almeida, C.A.B. Castelo de Faria. Campanha de escavações de 1981. *Barcellos Revista* 1 (1), 79-88

1982b: Almeida, C.A.B. Castelo de Neiva. *BCE* 1 13-33

1983: Almeida, C.A.B. A cabeça de guerreiro galaico da Citânia de Roriz-Oliveira. *Barcellos Revista* 1 (2) 81-90

1985a: Almeida, C.A.B. Barcelos. Castelo de Faria 1982-1983. *IA* 5, 50-52

1985b: Almeida, C.A.B. Carta arqueológica do Concelho de Esposende. *BCE* 7-8, 27-51

1985c: Almeida, C.A.B. Ponte de Lima. Vitorino das Donas 1983. *IA* 5, 136-137

1986: Almeida, C.A.B. Carta Arqueológica do concelho de Esposende. *BCE* 9-10, 39-59

1987: Almeida, C.A.B. Carta Arqueológica do concelho de Esposende. *BCE* 11-12, 93-108

1990a: Almeida, C.A.B. *Proto-História e Romanização da Bacia Inferior do Lima.* Centro de Estudos Regionais, Estudos Regionais 7-8 (nº esp.), Viana do Castelo, (= 1987 typescript)

1990b: Almeida, C.A.B. Inventário arqueológico: uma viagem ao passado arqueológico de Balugães. *Barcellos Revista* 1 (2ª s), Barcelos, 131-46

Almeida - Abreu 1987: Almeida, C.A.B.; Abreu, A.A. A necrópole de incineração de Gondomil, Valença. *Lucerna* 2 (2ª s)

Almeida *et al.* 1985: Almeida, C.A.B.; Abreu, A.A.; Baptista, A.J. Relatório das escavações do Castro do Peso. *CV* 9, 263-281

Almeida - Baptista 1981: Almeida, C.A.B.; Baptista, A.J. Castros e castelos de Ponte de lima. *I Colóquio Galaico-Minhoto* 2, 287-309

1984: Almeida, C.A.B.; Baptista, A.J. Inventário Arqueológico - O Castro de S. Simão de Cossourado - Barcelos. *Barcellos Revista* 2 (1), 99-110

Almeida - Neiva 1982: Almeida, C.A.B.; Neiva, M.A.P. O Castro de S. Lourenço, Vila Chã - Esposende. *BCE* 2, 5-17

Almeida - Soeiro 1980: Almeida, C.A.B.; Soeiro, M.T.C.M. Sondagens nos Castros de Abade do Neiva e Roriz (Barcelos, 1978). *SANP* 2, 29-35

Almeida 1959: Almeida, C.A.F. Um aspecto do culto dos rios na Lusitânia. *DL* 8(9), 873-887

1965: Almeida, C.A.F. O problema das casas redondas castrejas. *Lucerna* 4, 196-204

1969: Almeida, C.A.F. *Romanização das terras da Maia.* Maia

1971: Almeida, C.A.F. Uma escultura castreja de Calheiros, Ponte de Lima. Duas considerações. II *CNA* 1, 293-6

1972: Almeida, C.A.F. A Póvoa de Varzim e o seu aro na Antiguidade. *BC-PV* 11(1), 5-34

1973a: Almeida, C.A.F. Influências meridionais na cultura castreja. *RFLUP* (Série História) 4, 197-207, (= XIII *CAN*, 1975, 491-8)

1973b: Almeida, C.A.F. Aspectos da mineração romana de ouro em Jales e Trêsminas (Trás-os-Montes). XII *CAN*, Zaragoza, 553-62

1974a: Almeida, C.A.F. Cerâmica castreja. *RG* 84 (1-4), 171-97

1974b: Almeida, C.A.F. *Escavações no Monte Mozinho* (1974). Penafiel

1974c: Almeida, C.A.F. O monumento com forno de Sanfins e as escavações de 1973. III *CNA*, 149-72

1977: Almeida, C.A.F. *Escavações no Monte Mozinho*, 2 (1975-1976). Penafiel

1978a: Almeida, C.A.F. *Arquitectura românica de Entre- Douro-e-Minho.* (Doctoral thesis, typescript), Porto

1978b: Almeida, C.A.F. *Castelologia medieval de Entre-Douro-e-Minho.* (typescript), Porto

1980a: Almeida, C.A.F. Dois capacetes e três copos em bronze, de Castelo de Neiva. *Gallaecia* 6, 245-55

1980b: Almeida, C.A.F. Importantes objectos em bronze de Castelo de Neiva. *Arqueologia* 2, Porto, 45-9

1980c: Almeida, C.A.F. O templo do Mozinho e o seu conjunto. *Portugália* (Nova Série) 1, 51-6

1981: Almeida, C.A.F. Nova estátua de guerreiro galaico-minhoto (Refojos de Basto). *Arqueologia* 3, Porto, 111-16

1983a: Almeida, C.A.F. Cultura Castreja. Evolução e problemática. *Arqueologia* 8, Porto, 70-4

1983b: Almeida, C.A.F. O castrejo sob o domínio romano: a sua transformação. in Pereira Menaut, G. (ed.) *Estudos de Cultura Castrexa e de Historia Antiga de Galicia.* Santiago de Compostela, 187-98

1984: Almeida, C.A.F. A casa castreja. *MHA* 6, 35-42

1986: Almeida, C.A.F. Arte castreja. *Arqueologia* 13, Porto161-172

1987: Almeida, C.A.F. *Alto Minho.* Editorial Presença, Porto

Almeida - Jorge 1979: Almeida, C.A.F.; Jorge, V.O. *A estátua-menir de Faiões (Chaves).* Offp. G.E.A.P. Porto

Almeida - Lopes 1981-82: Almeida, C.A.F.; Lopes, F.G.A. *Eja (Entre-os-Rios). A civitas e a igreja de S. Miguel. Portugália* 2-3 (N/S), 131-6

Almeida *et al.* 1981: Almeida, C.A.F.; Soeiro, T.; Almeida, C.A.B.; Baptista, A.J. *Escavações arqueológicas em Santo Estevão da Facha.* Arquivo de Ponte de Lima, 3 (sep.), Ponte de Lima

1982: Almeida, C.A.F.; Soeiro, T.; Almeida, C.A.B.; Baptista, A.J. Duas datações de C14 para o castro de Santo Estevão da Facha. *Arqueologia* 6, Porto, 79

Almeida, F 1970: Almeida, F. Minas de ouro na *Gallaecia* portuguesa. *Legio VII Gemina*, León, 289-99

Alonso del Real 1979: Alonso del Real, C. Estrabon Revisitado. *Gallaecia* 3/4, 53-69

Alonso Nuñes 1975: Alonso Nuñes, J.M. El Noroeste de la Peninsula Ibérica en Estrabón. *BAur* 5, 343-7

Alonso Romero 1975: Alonso Romero, F. Estudio sobre las antiguas embarcaciones de cuero del Atlántico. *Gallaecia* 1

1976a: Alonso Romero, F. Antecedentes, realización y finalidad de la expedicion Breogan. *Gallaecia* 2, 173-91

1976b: Alonso Romero, F. La navegación durante la prehistoria en el Atlántico. *BAur* 6, 73-81

1976c: Alonso Romero, F. *Relaciones atlánticas prehistóricas entre Galicia y las Islas Británicas y medios de navegación.* Gráficas Numen, Vigo

Altuna 1980: Altuna, J. Historia de la domesticacion animal, en el Pais Vasco, desde sus origenes hasta la romanizacion. *Munibe* 32 (1-2), San Sebastian, 9-107

Alvar Ezquerra 1980: Alvar Ezquerra, J. *La navegacion prerromana en la Peninsula Iberica: colonizadores e indigenas.* (doctoral thesis, typescript), Universidad Complutense, Madrid

Alvarez Nuñez 1986: Alvarez Nuñez, A. *Castro de Penalba.* Campaña 1983. Arqueoloxía/Memorias 4, Xunta de Galicia - *DXPHD*, Santiago de Compostela

Alves, FM 1931: Alves, F.M. *Chaves. Apontamentos arqueológicos.* Gaia

Alves, L 1980: Alves, L. Aspectos da cultura castreja no Alto Minho. *Caminiana* 3, 105-41

Andrade 1918: Andrade, V. Castro de Entre-os-Rios. *AP* 23, 74-6

1920: Andrade, V. Monte Mozinho. *BRAACAP* 13, 275-7

Anderson 1984: Anderson, A. *Interpreting Pottery.* Batsford. London

Annonimous 1971: Annonimous *Carta dos Solos de Portugal.* Serviço de Reconhecimento e de Ordenamento Agrário

1985: Annonimous *Identification of Textile Materials.* The Textile Institute, Manchester (7ª ed., 1ª ed. 1975)

Araújo 1980: Araújo, I. Castros, outeiros e crastos na paisagem de Entre-Douro-e-Minho. *Minia* (2ª s.) 3 (4), 101-17

Araújo, JR(ibeiro) 1920: Araújo, J.R(ibeiro) *Perosinho - Apontamentos para a sua monographia.* Porto

Araújo, JR 1948: Araújo, J.R. Estudos de cerâmica castreja. *AAM* 2, 42-51

1960-61: Araújo, J.R. Sobre algumas 'pedras' de Vila Mou. *BA* 11-12, 85-90

Argote 1732-34: Argote, J.C. *Memorias para a história eclesiastica de Braga, primaz das Hespanhas.* 4 vol. Lisboa

Arias Vilas 1979: Arias Vilas, F. El castro de Penarrubia (Lugo) y la novedad de su datacion por C14. XV *CAN*, Zaragoza, 613-22

Arias *et al.* 1979: Arias Vilas, F.; Le Roux, P.; Tranoy, A. *Inscriptions romaines de la province de Lugo.* Paris, (=IRPL)

Aubet Semmler *et al.* 1983: Aubet Semmler, M.E.; Remedios Serna, M.; Escacena, J.L.; Ruiz Delgado, M.M. *La mesa de Setefilla, Lora del río (Sevilla). Campanã de 1979.* EAE 122

Audouze 1989: Audouze, F. Foyers et structures de combustion domestiques aux âges des métaux. Olive, M.; Taborin, Y. (eds.), *Nature et fonction des foyers préhistoriques.* Mémoires du Musée de Préhistoire d'ile de France 2, Nemours, 327-34

Azevedo 1939: Azevedo, A. *A terra da Maia*, 1. Porto

1946: Azevedo, A. O Monumento funerário da Citânia (nova interpretação). *RG* 56 (1-2), 150-64

Azevedo, AP 1896-1902: Azevedo, A.P. Extractos archeologicos das "memorias parochiaes de 1758". AP 2 (1896, 62-4, 89-92, 136-41, 177-92, 252-64, 305-18), AP 3 (1897, 101-6, 149-54, 193-208, 225-44), AP 4 (1898a, 135-53, 245-53, 315-29), AP 5 (1900, 26-31, 49-52, 90-3, 153-60, 187-92, 254-6, 297-304, 343-52), AP 7 (1902, 74-9, 123-8, 190-2, 237-40)

1898b: Azevedo, A.P. O territorio de Anegia. *AP* 4, 193-221

Azevedo, M 1895: Azevedo, M. Noticias archeologicas de Trás-os-Montes. *AP* 1, 93-6, 130-6

Azevedo, R 1960: Azevedo, R. O Porto desde a Proto-História à época do infante D. Henrique. *SG* 7, 104-46

1965: Azevedo, R. O termo Cale. *Lucerna* 4, 101-6

Bachiler Gil 1987: Bachiler Gil, J.A. Los castros del alto Duero. *Anales de Prehistoria y Arqueologia* 3, Murcia, 77-84

Bachmann 1982: Bachmann, H.-G. *The identification of slags from archaeological sites.* Institute of Archaeology Occasional Publication N° 6, London

Balil 1956: Balil, A. Un factor difusor de la romanizacion: las tropas hispanicas al servicio de Roma (siglos III - I a de J. C.). *Emerita* 24 (1), 108-34

1976: Balil, A. Bracara Augusta y el Conventus Bracarus. *CSECS* 16, 44-57

Baptista 1986: Baptista, A.M. *Adenda à notícia explicativa da Carta Geológica de Portugal, folha 1-D, (Arcos de Valdevez) - Arqueologia.* Terra de Val de Vez 9 (sep.)

Barber 1990: Barber, J. Scottish burnt mounds: variations on a theme. Buckley, V. (Ed.) *Burnt Offerings.* Wordwell Ltd. Dublin, 98-104

Barbero - Vigil 1971: Barbero, A.; Vigil, M. La organización social de los Cántabros y sus transformaciones en relación con los orígenes de la Reconquista. *HA* 1, 197-232

Barbi Alonso 1991: Barbi Alonso, A. Estudo dos materiais do castro de Fazouro (Lugo). Xunta de Galicia, *Arqueoloxia/Informes* 2, A Coruña, 319-21

Barfield - Hodder 1981: Barfield, L.H.; Hodder, M. Birmingham's Bronze Age. *Current Archaeology* 78, London, 198-200

1987: Barfield, L.H.; Hodder, M.A. Burnt mounds as saunas, and the prehistory of bathing. *Antiquity* 61, 370-9

1989: Barfield, L.H.; Hodder, M.A. 1: Burnt mounds in the West Midlands: Surveys and excavations. Gibson, A. (ed.) *Midlands Prehistory*, BAR (BS) 204, Oxford, 5-13

Barreiros 1915: Barreiros, F.B. Ensaio de inventário dos castros do concelho de Montalegre. *AP* 20, 211-3

1920: Barreiros, F.B. Materiais para a arqueologia do Concelho de Montalegre. *AP* 24, 58-87

Barroca 1988: Barroca, M.J. A ocupação Medieval em Castelo de Matos. Primeira abordagem. *Arqueologia* 17, Porto, 59-71

Barros 1919: Barros, J. *Geografia de Entre Douro e Minho e Trás-os-Montes.* Porto

Basto 1963a: Basto, A.M. Ainda o lugar da cividade. *Sumário de Antiguidades* (2ª ed.), 143-7

1963b: Basto, A.M. O lugar da cividade do Porto. *Sumário de Antiguidades* (2ª ed.), 137-41

Bell 1982: Bell, M. The effects of land-use and climate on valley sedimentation. Harding, A. (ed.) *Climatic Change in Later Pre-History. Edinburgh*, 127-42

Bellino 1902: Bellino, A. Cidade Velha de Santa Luzia. *A Aurora do Lima* (12 Set.)

1909: Bellino, A. Cidades mortas. *AP* 14, 1-28

Beltran Lloris 1988: Beltran Lloris, F. Un espejismo historiográfico. Las organizaciones gentilicias hispanas. Pereira Menaut, G. (ed.) 1° *Congreso Peninsular de Historia Antigua* 2, Santiago de Compostela, 197-37

Bermejo Barrera 1978-80: Bermejo Barrera, J.C. Variaciones sobre el tema de la centuria. *CEG* 31 (93-95), 95-116

1978: Bermejo Barrera, J.C. *La sociedad en la Galicia castreña.* Santiago de Compostela

1979: Bermejo Barrera, J.C. Três notas sobre Estrabón. *Gallaecia* 3-4, 71-90

1981a: Bermejo Barrera, J.C. La Función Guerrera en la Mitología de la *Gallaecia* Antigua. Contribuición a la sociologia de la Cultura Castreña. *Zephyrus* 32-33, 262-75

1981b: Bermejo Barrera, J.C. Los excrementos y la política. Una nota a Estrabón: III, 4, 16. *Caesaraugusta* 53-54, 77-90

1983: Bermejo Barrera, J.C. Etnografía castreña e historiografía clásica. Pereira Menaut, G. (ed.) *Estudos de Cultura Castrexa e de Historia Antiga de Galicia.* Santiago de Compostela, 129-46

Beug 1982: Beug, H.J. Vegetation history and climatic changes in central and southern Europe. Harding, A. (ed.) *Climatic Change in Later Pre-History.* Edinburgh, 85-102

Biek 1963: Biek, L. *Archaeology and the microscope.* Lutterworth Press, London

Bintliff 1981: Bintliff, J.L. Theory and reality in Palaeoeconomy: some words of encouragment to the Archaeologist. Sheridan, A.; Bailey, G. (eds.), *Economic Archaeology. Towards an integration of Ecological and Social approaches.* BAR (15) 96, Oxford, 35-50

1982: Bintliff, J.L. Climatic change, archaeology and Quaternary science in the eastern Mediterranean region. Harding, A.F. (ed.) *Climatic Change in Later Pre-History.* Edinburgh, 143-61

1984: Bintliff, J.L. Iron Age Europe in the context of social evolution from the bronze Age through to historic times. Bintliff, J.L. (ed.) *European Social Evolution: archaeological perspectives. University of Bradford.* Bradford

Blanco Freijeiro 1956: Blanco Freijeiro, A. Cabeza de un castro del Narla. Notas sobre el tema de la cabeza humana en el arte céltico. *CEG* 11 (34), 159-80

1957a: Blanco Freijeiro, A. Exvoto con scena de sacrificio. *RG* 67, 499-516

1957b: Blanco Freijeiro, A. Orígen y relaciones de la orfebrería castreña. *CEG* 12 (36, 5-28), (37, 137-57), (38, 267-301)

1960: Blanco Freijeiro, A. La cultura castreña. *ISPPI*, Pamplona, 179-94

Blázquez Martinez 1955: Blázquez Martinez, J.M. Los carros votivos de Mérida y Almorchón. *Zephyrus* 6, 41-60

1962a: Blázquez Martinez, J.M. Estado de la Romanizacion de hispania bajo César y Augusto. *Emerita* 30, 71-129

1962b: Blázquez Martinez, J.M. *Religiones primitivas de Hispania. 1. Fuentes literarias y epigráficas.* Roma

1968: Blázquez Martinez, J.M. Economia de los pueblos prerromanos del área no ibérica hasta la época de Augusto. Tarradell, M. (ed.) *Estudios de economia antigua de la Península Iberica.* Barcelona, 191-269

1971: Blázquez Martinez, J.M. La Iberia de Estrabón. *HA* 1, 11-94

1972: Blázquez Martinez, J.M. Economía de Hispania al final de la República Romana y a comienzos del Imperio según Estrabón y Plinio. *RUM* 78 (20), 57-143

1974: Blázquez Martinez, J.M. *Ciclos y temas dela historia de España: La romanización.* (2 vols.), Madrid

1975: Blázquez Martinez, J.M. *Diccionario de las religiones prerromanas de Hispania.* Madrid

1977: Blázquez Martinez, J.M. La romanización del Noroeste de la Península Ibérica. *Bimilenario de Lugo*, 67-81

Blázquez - Aguilera 1915: Blázquez, A.; Aguilera, D. Las Casitérides y el comercio del estaño en la Antigüedad. *BRAH* 67, 165-87

Boaventura 1954: Boaventura, M. Uma ara votiva no castro de S. Lourenço. *Correio do Minho* 21 Out. Braga

Bosch Gimpera 1932: Bosch Gimpera, P. *Etnología de la Península Ibérica.* Barcelona

1933: Bosch Gimpera, P. Los celtas en Portugal y sus caminos. *HMS*, 54-72

1942: Bosch Gimpera, P. Two Celtic Waves in Spain. *Proceedings of the British Academy* 26, London

1945: Bosch Gimpera, P. *El poblamiento antiguo y la formación de los pueblos de España.* México

1950-55: Bosch Gimpera, P. Les mouvements celtiques. *Études Celtiques* {5 (2), 1950-51, 352-400; 6 (1), 1952, 71-126; 6 (2), 1953-54, 328-55; 7 (1), 1955, 146-84}

1950: Bosch Gimpera, P. Infiltrações germânicas entre os Celtas penínsulares. *RG* 60 (3-4), 339-49

1961: Bosch Gimpera, P. *Les Indo-Européens. Problèmes archéologiques.* Paris

1974: Bosch Gimpera, P. *Paletnología de la Península Ibérica.* Graz

Botelho 1897: Botelho, H. Antiguidades de Trás-os-Montes. 1-Castros. *AP* 3, 69-72

1902-05: Botelho, H. Archeologia de Trás-os Montes. *AP* 7 (1902, 149-55), 9 (1904, 49-59, 166-70), 10 (1905, 237-8)

Bouhier 1979: *La Galice. Essai géographyque d'analyse et d'interprétation d'un vieux complexe agraire.* La Roche-sur-Yon (Vendée)

Bouza Brey 1942: Bouza Brey, F. *La mitología del agua en el Noroeste hispánico.* La Coruña

Bouzek 1982: Bouzek, J. Climatic changes and central European prehistory. Harding, A.F. (ed.) *Climatic Change in Later Pre-History.* Edinburgh, 179-91.

Bradley 1978: Bradley, R. *The Prehistoric Settlement of Britain.* Routledge & Kegan Paul, London

1980: Bradley, R. Subsistence, exchange and technology. A social framework for the Bronze Age in southern England c. 1400-700 BC. Barrett, J.; Bradley, R. (eds.), *Settlement and Society in the British Later Bronze Age.* BAR (British Series) 83 (1), Oxford, 57-75

	1984:	Bradley, R. *The Social Foundations of Prehistoric Britain: themes and variations in the archaeology of power.* Longman Archaeology Series, London
	1985:	Bradley, R. *Consumption, Change and the Archaeological Record.* University of Edimburgh Occasional Paper 13, Edimburgh
	1988:	Bradley, R. Hoarding, recycling and the consumption of prehistoric metalwork: technological change in Western Europe. *WA* 20 (2), 249-60
	1990:	Bradley, R. *The passage of arms: an archaeological analysis of prehistoric hoards and votive deposits.* Cambridge University Press, Cambridge
Brandão, AM	1891:	Brandão, A.M. Apontamentos Folkloricos Famalicenses. *RG* 6
Brandão	1962:	Brandão, D.P. O Monte do Castelo de Penha Longa (Marco de Canaveses) (Resumo). XXVI *CLEPC* 2 (5), 461
	1963a:	Brandão, D.P. Ara dedicada a Júpiter e pedra sepulcral insculturada de Sanfins, Paços de Ferreira. *Lucerna* 3, 232-5
	1963b:	Brandão, D.P. Notas de Arqueologia portuense. Novos achados arqueológicos no Morro da Sé do Porto. *Lucerna* 3, 236-45
	1965:	Brandão, D.P. Bibliografia sobre Portucale. *Lucerna* 4, 110-24
	1973:	Brandão, D.P. O Alto ou Monte do Crasto Penha Longa (Marco de Canaveses). *TAE* 22 (3), 289-94
	1985:	Brandão, D.P. *A Citânia de Safins na história da arqueologia portuense.* Paços de Ferreira
Brandão -Pinto	1966:	Brandão, D.P.; Pinto, S.S. Nova iconografia do petróglifo do Monte de Penha fidelis e suas implicações interpretativas. *RFLL* 10, 261-7
Brandão - Sousa	1966:	Brandão, D.P.; Sousa, E.F. A Penha-Fidelis, o seu petróglifo e os seus problemas. *Lucerna* 5 434-41
Brindley - Lanting	1990:	Brindley, A.L.; Lanting, J.N. The dating of fulachta fiadh. Buckley, V. (ed.) *Burnt Offerings.* Wordwell Ltd. Dublin, 55-6
Brito	1908:	Brito, M.J.C. Achado de moedas romanas. *AP* 8, 94-8
Brothwell - Higgs	1969:	Brothwell, D.; Higgs, E. *Science in Archaeology.* Thames & Hudson, London
Brun	1987:	Brun, P. *Princes et Princesses de la Celtique: le premier Âge du Fer (850-450 av J.-C.).* Les Hespérides, Paris
	1988:	Brun, P. Les 'Résidences princières' comme centres territoriaux: éléments de verification. *Les princes Celtes et la Mediterranée.* Rencontres de l'École du Louvre, Paris, 129-43
Butzer	1971:	Butzer, K.W. *Environment and Archaeology: An Ecological Approach to Prehistory.* Aldine, Chicago
	1975:	Butzer, K.W. The Ecological Approach to Prehistory: Are we Really Trying?. *American Antiquity* 40, 106-11
	1982:	Butzer, K.W. *Archaeology as human ecology.* Cambridge University Press, Cambridge
Buxó i Capdevilla	1988:	Buxó i Capdevila, R. La reconstrucció de la vegetació prehistòrica: el cas de les llavors i els fruits a Catalunya. *Cota Zero* 4, 39-45
Caamaño Gesto	1979a:	Caamaño Gesto, J.M. Aportaciones al estudio de las monedas de la caetra: Las monedas de la Collección Blanco Cicerón (La Coruña). *BAur* 9, 67-76
	1979b:	Caamaño Gesto, J.M. Posible reutilizacion de caminos prerromanos en epoca romana. *Gallaecia* 3/4, 281-5
Cabral - Soares	1984:	Cabral, J.M.P.; Soares, A.M. Datação pelo radiocarbono. II - Sobre a estimação do verdadeiro valor das datas convencionais de radiocarbono e a comparação de duas datas. *Arqueologia* 10, Porto, 89-99
Cabré de Morán	1931:	Cabré de Morán, A.P. Excavaciones en los Dólmenes de Salamanca. Memoria. *MJSEA* 113
Caesar (b.c.)	1959:	Caesar *Comentarii de Bello Civili.* Kraner, F.; Hofmann, F. (eds.), Berlin
Caldas	1975:	Caldas, J.A. *Monografia de S. João de Longos Vales - Monção.* Braga
Calheiros	1910:	Calheiros, A. *Refoios através dos tempos.* Almanaque de Ponte de Lima, Ponte de Lima
Calo Lourido	1983:	Calo Lourido, F. Arte, decoración, simbolismo e outros elementos da cultura material castrexa. Ensaio de sintese. Pereira Menaut, G. (ed.) *Estudos de Cultura Castrexa e de Historia Antiga de Galicia.* Santiago de Compostela, 159-85
	1991:	Calo Lourido, F. *A Plástica da Cultura Castrexa Galego-Portuguesa.* Universidade de Santiago de Compostela, Pontevedra, (Typescript)
Calo - Sierra	1983:	Calo Lourido, F.; Sierra Rodriguez, X.C. As orixenes do castrexo no Bronce final. Pereira Menaut, G. (ed.) *Estudos de Cultura Castrexa e de Historia Antiga de Galicia.* Santiago de Compostela, 19-86
Cameron	1988:	Cameron, E. Acerca da preservação de resíduos orgânicos nas superfícies corroídas dos metais. Queiroga, F.; Sousa, I.M.A.R.; Oliveira, C.M. (eds.) *Paleoecologia e Arqueologia.* Vila Nova de Famalicão, 81-8

Cano - Vázquez 1991: Cano Pan, J.; Vázquez Varela, J.M. El aprovechamiento del mar en los castros costeros de Lugo (Galicia). Queiroga, F.; Dinis, A.P. (eds.), *Paleoecologia e Arqueologia* 2, Vila Nova de Famalicão, 71-5

Carballo Arceo 1983: Carballo Arceo, L.X. Aportación al estudio de las sítulas en el Occidente de la Península Ibérica. *CEG* 34 (99), 7-32

1986: Carballo Arceo, L.X. *Povoamento castrexo e romano da Terra de Trasdeza.* DXPAM, Xunta de Galicia, Santiago de Compostela

1987: Carballo Arceo, L.X. Avance ao estudo dos materiais arqueoloxicos do castro de Santa Trega. *Lucerna* 2 (2ª série), 95-117

1989a: Carballo Arceo, L.X. *A Cultura Castrexa na bacia media do rio Ulla.* (Doctoral thesis presented to the University of Santiago de Compostela, typescript)

1989b: Carballo Arceo, L.X. Coelho Ferreira da Silva, Armando: A Cultura Castreja no Noroeste de Portugal. Museu Arqueológico da Citânia de Sanfins. Paços de Ferreira, 1986; 367 páxinas. *Gallaecia* 11, 363-4

Cardoso, AP 1966: Cardoso, A.P. Pesquizas Arqueológicas na Freguesia da Ribeira. *O Lima,* 92

Cardozo, L 1747-51: Cardozo, L. *Diccionario geographico de Portugal.* 2 vols.

1758: Cardozo, L. *Memorias paroquiaes.*

Cardozo 1926: Cardozo, M. Consagrado às ninfas. *RG* 36 (1-2), 64-9

1927-32: Cardozo, M. Bibliografia Sarmentina. *RG* {nº37, 1927, (3), 115-9; (4), 185-201}, {nº38, 1928, (1-2), 85; (3-4), 241-2}, (39, 1929, (1-2), 119}, {41, 1931, (1-2), 101}, {42, 1932, (1-2), 91}

1930: Cardozo, M. Jóias arcaicas encontradas em Portugal. *Nós* 11 (72), 207-18

1931; 1932; 1942: Cardozo, M. A última descoberta arqueológica na Citânia de Briteiros e a interpretação da Pedra Formosa. *RG* 41 (1-2); 41 (3); 41 (4); 42 (1-2); 42 (3-4), 55-60; 201-9; 250-60; 7-25; 127-39

1935a: Cardozo, M. *Notícias resumidas das ruínas arqueológicas de Briteiros e Sabroso e descrição dos objectos provenientes destes castros, expostos no Museu da Sociedade Martins Sarmento, em Guimarães.* Sociedade Martins Sarmento, 4ª ed., Guimarães

1935b: Cardozo, M. Possível identificação do primitivo local de Pedra Formosa na Citânia de Briteiros. *RG* 45 (3-4), 150-3

1936a: Cardozo, M. Novas descobertas arqueológicas na Citânia de Briteiros. *RG* 46 (3-4), 159-63

1938: Cardozo, M. Jóias áureas proto-histórico da Citânia de Briteiros. *RG* 48 (1-2), 5-40, (= Petrus Nonius 1 (3), 254-60)

1942a: Cardozo, M. Uma notável peça de joalharia primitiva. *AFCP* 27 (2), Porto, 89-100

1942b: Cardozo, M. Una pieza notable de la orfebreria primitiva. *AEA* 15 (47), Madrid, 93-103

1943a: Cardozo, M. A origem da palavra Citânia comentada por Martins Sarmento, Leite de Vasconcelos e Adolfo Coelho. *PN* 5 (1-2), Lisboa, 10-21

1943b: Cardozo, M. *Algumas inscrições lusitano-romanas da região de Chaves. I- As inscrições do extinto Museu. II- Inscrições de Chaves e vizinhanças.* Câmara Municipal, Chaves

1943c: Cardozo, M. Antiguidades transmontanas. (I- Fragmento de um torques. II- Contas policromas de pasta vítrea). *RG* 53 (1-2), 109-16

1943d: Cardozo, M. Escavações na Citânea de Briteiros. *RG* 53 (3-4), 247-56

1944: Cardozo, M. Novo achado de jóias pré-romanas. *RG* 54 (1-2), 19-28

1946a: Cardozo, M. Arquitectura citaniense. (O problema das casas com tecto de abóbada). *Minia* 1 (3-4), 243-57

1946b: Cardozo, M. Carrito votivo de bronce del Museo de Guimarães (Portugal). *AEA* 19 (62), Madrid, 1-28

1946c: Cardozo, M. O Monumento Funerário da Citânia. *RG* 56 (3-4), 289-308

1948: Cardozo, M. Escavações arqueológicas na Citânia de Briteiros. Campanha de 1948. *RG* 58 (3-4), 343-8

1949a: Cardozo, M. Algunas observaciones sobre el arte ornamental de los castros del Noroeste de la Península. IV CASE, 345-67

1949b: Cardozo, M. Escavações na Citânia de Briteiros. Relatório da campanha de 1949. *RG* 59 (3-4), 406-14

1949c: Cardozo, M. Nova estela funerária do tipo da Pedra Formosa. *RG* 59 (3-4), 487-516

1950a: Cardozo, M. Escavações na Citânia de Briteiros. Breve notícia da campanha de 1950. *RG* 60 (3-4), 518-26

1950b: Cardozo, M. Mais uma achega para o estudo da joalharia pré-histórica portuguesa. *BGAF* 2 (2), Barcelos, 17-23

1951: Cardozo, M. Escavações na Citânia de Briteiros. Campanha de 1951. *RG* 61 (3-4), 455-72
1952a: Cardozo, M. A organização defensiva dos castros do Norte de Portugal e a sua romanização. *AHM* 22, 121-52
1952b: Cardozo, M. A romanização dos castros do Norte. *Lisbon Courier* 74, Lisboa, 15-6; 39-41
1954a: Cardozo, M. Escavações na Citânia de Briteiros. Campanha de 1954. *RG* 64 (3-4), 412-20
1954b: Cardozo, M. O IV Congresso Internacional de Ciências Pré e Proto-históricas, realizado em Madrid, desde 21 a 27 de Abril de 1954. *RG* 49, 99-12
1955a: Cardozo, M. Breves notas acerca de um tipo conhecido de antigas contas polícromas de vidro. *RG* 65 (1-2), 133-6
1955b: Cardozo, M. Exploração arqueológica na Citânia de Briteiros. Relatório da 23ª. Campanha (Out. 1955). *RG* 65 (3-4), 431-8
1956a: Cardozo, M. Exploração arqueológica na Citânea de Briteiros. Relatório da 24ª Campanha (1956). *RG* 66 (3-4), 508-12
1956b: Cardozo, M. Martins Sarmento em terras da Galiza. *RG* 66, 109-24
1956c: Cardozo, M. Notícia de duas novas arrecadas de ouro antigas. *RG* 66 (3-4), 449-62, (= XXIII CLEPC 7ª Sec., 55-60)
1957a: Cardozo, M. Breve análise do espólio de uma sepultura lusitano-romana. *Homaxe a Florentino Lopez Cuevillas*, Vigo, 41-52
1957b: Cardozo, M. Citânia de Briteiros. 25ª Campanha arqueológica (Set.-Out de 1957). *RG* 67 (3-4), 551-3
1957c: Cardozo, M. Das origens e técnica do trabalho do ouro e a sua relação com a joalharia arcaica peninsular. *RG* 67 (1-2), 5-46
1957d: Cardozo, M. Notícia de uma jóia antiga adquirida pelo Museu de Martins Sarmento. *RG* 67 (1-2), 179-84
1957e: Cardozo, M. Uma aquisição preciosa do Museu de Martins Sarmento. (Tesouro de Lebução). *RG* 67 (3-4), 417-42
1958a: Cardozo, M. Citânia de Briteiros. 26ª Campanha arqueológica. (Set.-Out. de 1958). *RG* 68 (3-4), 455-9
1958b: Cardozo, M. Les ruines archéologiques de la Citânia de Briteiros. Description d'un oppidum lusitanien du NW de la Péninsule Ibérique. *CISPP* 5
1958c: Cardozo, M. Missão inglesa de escavações num castro do Norte de Portugal (Sabroso) (3 a11 de Abril de 1958). *RG* 68 (3-4), 439-54
1959a: Cardozo, M. 2ª Campanha de escavações arqueológicas em castros do Norte de Portugal (Cividade de Âncora e Monte do Cútero) dirigida pelo prof. Dr. Christopher Hawkes da universidade de Oxford (7 a 22 de Setembro de 1959). *RG* 69 (3-4), 522-46
1959b: Cardozo, M. A mó e a farinha, o forno e o pão. *TAE* 17 (1-4), Porto, 235-48
1959c: Cardozo, M. Citânia de Briteiros. 27ª Campanha Arqueológica. (Set.-Out. de 1959). *RG* 69 (3-4), 517-20
1959d: Cardozo, M. Joalharia lusitana. *Conimbriga* 1, 13-27
1959e: Cardozo, M. Um novo achado em Portugal de jóias de ouro proto-históricas. *RG* 69, 127-38
1960a: Cardozo, M. A tradição náutica na mais antiga história da Península Hispânica. *RG* 70 (3-4), 491-552
1960b: Cardozo, M. Breves observações a propósito das análises espectrográficas de alguns instrumentos metálicos da idade do bronze pertencentes ao Museu de Martins Sarmento. *RG* 70 (1-2), 169-84
1960c: Cardozo, M. Cartas de Ricardo Severo para Martins Sarmento. *RG* 70 (1-2), 5-20
1960d: Cardozo, M. Citânia de Briteiros. 28ª Campanha arqueológica (Set.-Out.1970). *RG* 70 (3-4), 551-4
1961a: Cardozo, M. Citânia de Briteiros. 29ª Campanha arqueológica.(Out.- Nov. 1961). *RG* 71 (3-4), 451-4
1962a: Cardozo, M. Alguns problemas da cultura dos castros do Norte de Portugal. *XXVI CLEPC* 2 (5), 391-423
1962b: Cardozo, M. *Sociedade de Martins Sarmento, Citânia de Briteiros e Castro de Sabroso*. Guimarães
1965a: Cardozo, M. A fiação e a tecelagem na antiguidade peninsular. *Cong. Internacional de Etnografia* 2, 29-36
1965b: Cardozo, M. A metalurgia na Proto-história da Península Ibérica. *Dédalo* 1 (2), 29-49
1965c: Cardozo, M. A perda frequente de especímenes preciosos da nossa joalharia arcaica. *RG* 75 (1-4), 153-68
1965d: Cardozo, M. A romanização do Noroeste da Península Ibérica. *Ethnos* 4, 75-82

1965e: Cardozo, M. *Citânia e Sabroso. Notícia descritiva para servir de guia ao visitante.* Sociedade Martins Sarmento, 5ª ed. Guimarães

1965f: Cardozo, M. Curtimento de peles e aplicações de couro na antiguidade peninsular. *Lucerna* 4, 155-9

1966a: Cardozo, M. As indústrias de cerâmica e vidreira na antiguidade peninsular particularmente na cultura dos castros no Noroeste (Considerações Gerais). *Lucerna* 5, 469-75

1966b: Cardozo, M. Torques encontrado em Vila Flor. *Lucerna* 5, 459-60

1967: Cardozo, M. Elementos bibliográficos para o estudo da joalharia arcaica luso-espanhola. *RG* 77 (3-4), 329-76

1968a: Cardozo, M. Bibliografia das jóias arcaicas da Península Ibérica. *RG* 78 (1-2), 85-8

1968b: Cardozo, M. Escavações arqueológicas na Citânia de Briteiros (32ª. campanha). *RG* 78 (3-4), 291-6

1968c: Cardozo, M. Sabroso, Castro de. Serrão, J. (ed.) *Dicionário de História de Portugal* 3, Lisboa, 705-7

1970: Cardozo, M. Die Vorgeschichtliche Höfensiedlung von Penha bei Guimarães (Portugal). *MM* 11, 90-5

1971: Cardozo, M. A Estação préhistórica da Serra da Penha (Guimarães). II *CNA* 1, 239-59

1972: Cardozo, M. *Bio-bibliografia de Mário Cardozo.* Guimarães

1973a: Cardozo, M. Breves considerações sobre escultura primitiva da Lusitânia pré e protohistórica. *CEG* 28 (85), 153-60

1973b: Cardozo, M. La culture des castros du Nord du Portugal. *TAE* 22 (3), 261-82

1980: Cardozo, M. *Citânia de Briteiros e Castro de Sabroso.* Guimarães, (8ª ed.)

1985: Cardozo, M. *Catálogo do Museu de Martins Sarmento. Secção de Epigrafia Latina e de Escultura Antiga.* Sociedade Martins Sarmento, Guimarães (3ª ed.)

Cardozo - Santos 1953: Cardozo, M.; Santos Junior, J. R. Ex-votos às ninfas em Portugal. *Zephyrus* 4, Salamanca 53-68

Caro Baroja 1944: Caro Baroja, J. *Regímenes sociales y económicos de la España prerromana.* Madrid

1970: Caro Baroja, J. Organización social de los pueblos del Norte de la Península Ibérica en la Antiguedad. *Legio VII Gemina*, León, 9-62

1974: Caro Baroja, J. *Los pueblos del Norte de la Península Ibérica.* Madrid (1ª ed. 1943)

1976: Caro Baroja, J. *Los pueblos de España. Ensayo de etnología.* 2 vols. (2ª ed., 1ª ed. 1946), Ediciones Istmo, Madrid

Cartaillac 1886: Cartaillac, E. *Les âges préhistoriques de l'Espagne et du Portugal.* Paris

Castelo-Branco 1963b: Castelo-Branco, F. Subsídios para a história das escavações de Briteiros. *Lucerna* 2, 304-17

Castillo 1908: Castillo, A. *Los castros gallegos.* La Coruña

Castro *et al.* 1980: Castro, J.S.; Correia, S.H.; Oliveira, E.P. O Castrum Maximum (Monte Castro), Braga - Arqueologia e história. *SANP* (vol. 2), Guimarães, 37-53

Castro, LFM 1968: Castro, L.F.M. Carta Litológica de Portugal. Sup. *Agronomia Lusitana*, vol. 29-4, 289-374

Cavada Nieto 1972: Cavada Nieto, M. Hallazgos monetarios en castros de Galicia. *BSAA* 38, 211-48

Cavalheiro 1989: Cavalheiro, J. O Ferro na Antiguidade. *Arqueologia* 19, 124-32

Centeno 1987: Centeno, R.M.S. *A circulação monetária no Noroeste de Hispânia até 192.* Porto

Centeno - Silva 1978: Centeno, R.M.S.; Silva, A.C.F. Corte estratigráfico na Citânia de Briteiros (Guimarães) 1977-78. *RG* 88, 421-30

Cepa 1939: Cepa, M.M. *Monografia de S. Miguel de Alvarães.* Braga

1944: Cepa, M.M. *Monografia de S. Bartolomeu do Mar.* Braga

Chamoso Lamas 1955: Chamoso Lamas, M. Santa Mariña de Aguas Santas (Orense). *CEG* 10 (30), 41-88

Champion 1982: Champion, T.C. Fortification, ranking and subsistence. Renfrew, C.; Shennan, S.J. (eds.). Ranking, Resource and Exchange. Cambridge University Press, Cambridge, 61-6

Champion *et al.* 1984: Champion, T.C.; Gamble, C.; Shennan, S.; Whittle, A. *Prehistoric Europe.* Academic Press, London

Champion - Megaw 1985: Champion, T.C.; Megaw, J.V.S. (eds.). *Settlement and Society. Aspects of West European prehistory in the first millennium B.C.* Leicester University Press, Cambridge

Charlesworth 1924: Charlesworth, M.P. *Trade-routes and Commerce of the Roman Empire.* Cambridge University Press, Cambridge

Cherry 1990: Cherry, S. The finds from fulachta fiadh. Buckley, V. (ed.) *Burnt Offerings.* Wordwell Ltd., Dublin, 49-54

Clark	1986:	Clark, G.A. El nicho alimenticio humano en el Norte de España desde el Paleolítico hasta la romanización. *TP* 43, 159-84
Clark, RM	1975:	Clark, R.M. A calibration curve for radiocarbon dates. *Antiquity* 49 (196), 251-66
Coelho, JF	1962:	Coelho, J.F. Breves considerações sobre achados e lendas de Codeços, Paços de Ferreira. *SG* 9 (1), 137
	1963:	Coelho, J.F. Será obra celta a casa circular do castro da freguesia de Codeços, Paços de Ferreira?. *Lucerna* 3, 179-80
	1965:	Coelho, J.F. Considerações sobre achados arqueológicos em Codeços, Paços de Ferreira. *Lucerna* 4, 226-8
Coelho *et al.*	1959:	Coelho, M.E.R.O.; Martins, J.C.M.; Simões, M.S.G.; Farinha, A.M.D.; Magalhães, E.C. Nota descritiva da Citânia de Sanfins e do seu espólio arqueológico. I *CNA* 1, 271-79
Coffyn	1983:	Coffyn, A. La fin de l'âge du Bronze dans le centre - Portugal. *AP* 1 (4ª Série), 169-96
	1985:	Coffyn, A. *Le Broze final atlantique dans la Péninsule Ibérique.* Publ. Centre Pierre Paris 11, Paris
Coles - Harding	1979:	Coles, J.M.; Harding, A.F. *The Bronze Age in Europe.* London
Collis	1975:	Collis, J.R. *Defended sites of the Late la Tène in Central and Western Europe.* BAR 15 (SS) 2
	1981:	Collis, J.R. A theoretical study of hill-forts. Guilbert, G. (ed.) *Hill-Fort Studies*, Leicester University Press, 66-76
	1982:	Collis, J.R. Gradual growth and sudden change: urbanisation in temperate Europe. Renfrew, A.C.; Shennan, S. (eds.). *Ranking, resources and exchange: aspects of the archaeology of early European society.* Cambridge University Press, Cambridge
	1984a:	Collis, J.R. Aulnat (Puy-de-Dôme) and urbanisation - The theoretical problems. *Études Celtiques* 21, 111-17
	1984b:	Collis, J.R. *The European Iron Age.* Batsford, London
	1984c:	Collis, J.R. *Oppida. Earliest Towns North of the Alps.* University of Sheffield, Sheffield
Conde Valvís	1952:	Conde Valvís, F. *La Cibdá de Armea, en Santa Marina de Aguas Santas.* Orense
	1955:	Conde Valvís, F. Las termas romanas de la Cibdá de Armea en Santa Marina de Aguas Santas. III *CAN*, 432-46
Connolly	1988:	Connolly, P. *Grece and Rome at War.* Macdonald & Co. London
Corrêa	1924:	Corrêa, A.A.M. *Os povos primitivos da Lusitânia.* Porto
	1927:	Corrêa, A.A.M. O petroglifo do guerreiro Lusitano no Monte do Castelo de Penafiel. *Broteria* 4, 16-29
	1928:	Corrêa, A.A.M. Lusitânia pré-romana. Damião P. (ed.) *História de Portugal* 1. Barcelos, 77-214
	1933-34:	Corrêa, A.A.M. Cale, Portucale e Porto. *ASEG* 6, 159-215
	1935a:	Corrêa, A.A.M. Cale in *Gallaecia. Memorias da Academia de Ciências de Lisboa* 1, 137-42
	1935b:	Corrêa, A.A.M. *As origens da cidade do Porto.* Porto, (2ª ed.)
	1936:	Corrêa, A.A.M. A antiguidade do Porto. *TAAP* 2, 21-57
Corrêa - Teixeira	1946:	Corrêa, A.A.M.; Teixeira, C. A lenda e as ruínas de Calcedónia, na Serra do Gerez. *Minia* 1 (3-4), 212-22
Correia, JM	1942	Correia, J.M. Povos primitivos de Riba-Coa. *Altitude* 2 (12), 336-9
Cortez	1945:	Cortez, F.R. A ara votiva romana de Reiriz-Monção. *BCMO* 15 (1), 12-15
	1946a:	Cortez, F.R. El Castro de Argifonso (Vila do Conde). *AEA* 19, 149-55
	1946b:	Cortez, F.R. *Machados e outros objectos de bronze.* Museu Nacional de Soares dos Reis, Porto
	1948a:	Cortez, F.R. Actividad arqueológica de Portugal durante 1947. *AEA* 21 (72), 269-81
	1948b:	Cortez, F.R. O Castro de Esturãos. *AAM* 2, 15-21
	1949a:	Cortez, F.R. Actividades arqueológicas de Portugal em 1948. *AEA* 22 (77), 400-13
	1949b:	Cortez, F.R. *Novos Materiais para o estudo da Idade do Bronze do Trás-os-Montes e Alto-Douro.* Museu Etnológico do Douro, Régua
	1949d:	Cortez, F.R. Restos arqueológicos de Vila do Conde. I - O Castro de Argifonso. *TAE* 12 (3-4), 270-8
	1951b:	Cortez, F.R. Das populações pré-celtas do Norte de Portugal. *BAFN* 2, 161-84
	1957:	Cortez, F.R. Moedas Ibericas nos Castros Bragançanos. *Nummus* 4 (3), 143-6
Costa	1868-69:	Costa, A.C. *Corografia portuguesa e descriçam topografica do Reyno de Portugal.* Braga (2ª ed.)

Costa *et al.* 1980: Costa, A.G.; Faria, A.M.; Carvalho, J.S. Sondagens arqueológicas no concelho de Barcelos: Abade de Neiva, Faria, Roriz. *SANP* 2, 5-28

Costa, AR 1945: Costa, A.R. *Descrição topográfica e histórica da cidade do Porto.* Porto

Costa, HM 1965-66: Costa, H.M. Notas sobre as escórias encontradas no Castro de Carvalhelhos. *TAE* 20 (1-2), 173-80

Costa, JG 1963: Costa, J.G. Achado arqueológico encontrado em Solveira concelho de Montalegre em Abril de 1961. *Lucerna* 3, 119-25

1968: Costa, J.G. Montalegre e Terras de Barroso. Montalegre

Costa, MCG 1969: Costa, M.C.G. *Achados arqueológicos de Monte Mózinho* (Penafiel). Coimbra

Cuadrado 1955: Cuadrado, E. El carro ibérico. III *CAN*, 116-34

Cubero Corpas 1991: Cubero Corpas, C. Panicum miliaceum en la Edad de Hierro en el Noreste de la Península Ibérica. Queiroga, F.; Dinis, A.P. (eds.). *Paleoecologia e Arqueologia* 2, Vila Nova de Famalicão, 269-80

Cunha 1946a: Cunha, A.R. Um velho 'castro'. *Diário do Minho* 2 Jun. 1946, 3

1947: Cunha, A.R. Cividade do Monte das Caldas. *Diário do Minho* 3 Nov. 1947, 3, 6

1948b: Cunha, A.R. O Monte da Santinha e o de S. Miguel. *Diário do Minho* 23 Ago. 1948, 3-4

1949: Cunha, A.R. Santa Marta de Cortiças. *Diário do Minho* 21 Ago. 1949, 2

1950a: Cunha, A.R. Cividade de Chacim. *Diário do Minho* 19 Mar. 1950, 1-4

1950b: Cunha, A.R. O Monte do Castelo em Roças. *Diário do Minho* 28 Abr. 1950, 2, 4

1950c: Cunha, A.R. A Calcedónia do Gerês. *Diário do Minho* (5 Mai. 1950), 2, 4

1952: Cunha, A.R. A pré-história bracarense. O castro de Santa Marta. *Correio do Minho* (20 Ago. 1952), 3

1953-54: Cunha, A.R. Escavações no oppidum de Santa Marta. *BA* 5 (1-3), 241-43

1953a: Cunha, A.R. A Santa Marta do Alto. *Correio do Minho* (12 Mar. 1953), 4

1953b: Cunha, A.R. Novíssimas inscrições romanas de Braga. *BA* 5, 249-52

1954-55: Cunha, A.R. No oppidum de Santa Marta. *Correio do Minho* (1954, 14 Out., 11 Nov., 25 Nov.), (1955, 20 Jan., 24 Fev.), Braga

1955a: Cunha, A.R. A cidade velha de Santa Marta. *Diário do Minho* 14 Jan. 1955, 1, 3

1955b: Cunha, A.R. No Castelo de Barbudo. *Diário do Minho* 7 Fev. 1955, 1, 3

1956: Cunha, A.R. A cidade de Cossourado. *Diário do Minho* 6 Abr. 1956, 1

1957: Cunha, A.R. A arqueologia de Paredes Secas. *Diário do Minho* (23 Dez. 1957), 1, 4

1961: Cunha, A.R. Um miliário inédito. *O Distrito de Braga* 1 (3-4), Braga, 319-44

1975: Cunha, A.R. Trepando aos montes. *O Distrito de Braga* 1 (1-4), Braga, 485-535

Cunha, NCA 1979: Cunha, N.C.A. *Paredes de Coura.* Braga, 2ª ed. (1ª ed. 1909)

Cunliffe 1971: Cunliffe, B. Some aspects of hillforts and their cultural environments. Hill, D.; Jesson, M. (eds.) *The Iron Age and its Hill-Forts.* 53-70

1973: Cunliffe, B. Chalton Hants: The evolution of a Landscape. *AJ* 53 (2)

1976: Cunliffe, B. The origins of urbanisation in Britain. Cunliffe, B.; Rowley, T. (eds.) *Oppida: the beginnings of urbanisation in barbarian Europe.* BAR (SSII), Oxford, 135-61

1978: Cunliffe, B. *Iron Age Communities in Britain.* Routledge (2º ed. - 1º ed. 1974), London

1982: Cunliffe, B. Settlement hierarchy and social change in southern Britain in the Iron Age. *Analecta Praehistorica Leidensia* 15, 161-81

1984: Cunliffe, B. *Danebury. An Iron Age Hillfort in Hampshire. The excavations.* 1969-1978 Vol. 1: the site. Vol. 2: the finds. The Council for British Archaeology, Research Report nº 52

1986: Cunliffe, B. *Danebury: anatomy of an Iron Age hillfort.* Batsford, London

1988: Cunliffe, B. *Greeks, Romans and Barbarians: Spheres of interaction.* Batsford, London

1990: Cunliffe, B. Before hillforts. *OJA* 9 (3), 323--36

Damon *et al.* 1972: Damon, P.E.; Long, A.; Wallick, E.I. Dendrochronology calibration of the carbon-14 time scale. Rafter, A.; Grant-Taylor, T. (eds) *Proceedings of the 8th International Conference on Radiocarbon Dating.* Royal Society of New Zealand, Wellington, 44-59

Daveau 1977: Daveau, S. L´Évolution géomorphologique Quaternaire au Portugal. Principaux aspects et problémes posés par son étude. Recherches Françaises sur le Quaternaire INQUA 1977, Spp. au *Bulletin AFEQ*, 1977-1 nº 50, 11-21

1980: Daveau, S. Espaço e tempo. Evolução do ambiente geográfico de Portugal ao longo dos tempos pré-históricos. *Clio* 2, Lisboa, 13-37

1988: Daveau, S. Aspectos do ambiente quaternário no Noroeste de Portugal. *TAE* 28 (1-2), Porto, 193-200

Delarue 1979: Delarue, J. *Le métier de bourreau.* Librairie Athéme Fayard. Paris

Deyber 1987: Deyber, A. La Guérilla Gauloise Pendant la Guerre des Gaules (58-50 avant J.-C.). *Études Celtiques* 14, 145-83

Dias 1946: Dias, A.J. Las construcciones circulares del Noroeste de la Península Ibérica y las Citanias. *CEG* 2 (6), 173-94

1947: Dias, A.J. Construções circulares no litoral português. Contribuição para o estudo das construções circulares no Noroeste da Península Ibérica. *TAE* 11 (1-2), 192-95

1948: Dias, A.J. Las Chozas de los Cabeçudos e las construcciones circulares de las citanias españolas y potuguesas. Contribución etnográfica para la reconstrucción de la vida en las citanias. *AEA* 21 (71), 164-72

1949: Dias, A.J. O problema da reconstituição das casas redondas castrejas. *TAE* 12 (1-2), 126-68

1982: Dias, A.J. *Os arados portugueses e as suas prováveis origens.* Imprensa Nacional - Casa da Moeda. Maia (1ª ed. 1948)

Dias, L.A.T. 1984: Dias, L.A.T. Estação Romana do Freixo - Marco de Canavezes. *Arqueologia* 9, Porto, 86-90

Diego Santos 1959: Diego Santos, F. *Epigrafía romana de Asturias.* (=ERA), Oviedo

Dinis, AP 1993: Dinis, A.P. (Report on the excavations at Castro do Crastoeiro, forthcoming).

Dinis 1954: Dinis, M.V. Lendas da citânia de Sanfins (Paços de Ferreira). *DL* 5 (9), 63-73

1962: Dinis, M.V. Achados arqueológicos em Paços de Ferreira. *SG* 9 (1), 223-9

1963: Dinis, M.V. Manifestações neoliticas na região de Paços de Ferreira. *Lucerna* 3, 88-95

1965: Dinis, M.V. O Castro de Busto (Penamaior - Paços de Ferreira). *Lucerna* 4, 189-95

1966: Dinis, M.V. O Castro da Vila (Penamaior - Paços de Ferreira). *Lucerna* 5, 512-8

1971: Dinis, M.V. O Castro do Capelo Vermelho (em Codessos - Paços de Ferreira). II *CNA* 2, 287-91

1976: Dinis, M.V. O Castro de S. Brás (Frazão, Paços de Ferreira). *RG* 86, 217-21

1978: Dinis, M.V. As ruínas luso-romanas dos Castêlos (na Vila de Paços de Ferreira). *RG* 88, 187-90

1980: Dinis, M.V. Achados de numismas romanos (no termo de Paços de Ferreira). *RG* 90, 389-93

Diodorus 1961: Diodorus Siculus *The Library of History.* Vols. I-VI, Oldfather, C.H. (ed.) The Loeb Classical Library, London

Dion 1952: Dion, R. Le problème des Cassitérides. *Latomus* 12, 306-14

1953: Dion, R. Le problème des Cassitérides. *Offerts à Ph. Arbos Mélanges Geographiques* IV, Pub. de la Faculté des Lettres de Clermont-Ferrand, Fasc. 8, Clermont-Ferrand, 23-9

Domergue 1970a: Domergue, C. Introduction à l`étude des mines d`or du Nord-Ouest de la Péninsule Ibérique dans l`Antiquité. *Legio VII Gemina*, Leon, 225-86

1970b: Domergue, C. Les explotations aurifères du Nord-Ouest de la Péninsule Ibérique sous l`occupation romaine. *La minería hispana e iberoameriacana 1*, 151-93

1972-74: Domergue, C. À propos de Pline, "Naturalis Historia" 33, 70-78 et pour illustrer sa descripcion des mines d`or romaines d`Espagne. *AEA* 45-47 (125-130), 499-528

1987: Domergue, C. *Catalogue des mines et des fonderies antiques de la Peninsule Iberique.* 2 vols. Diffusion E. de Boccard, Madrid

1990: Domergue, C. *Les mines de la Peninsule Iberique dans l'antiquité Romaine.* École Française de Rome, Rome

Dueñas *et al.* 1988: Dueñas, J.A.; Freixa, A.; Moreno, V; Pedro, P.; Rodriguez, P. Estimulos endogenos y exogenos en el desarrollo de la jerarquizacion de la Laietania oriental en epoca iberica. Pereira Menaut, G. (ed.) 1º *Congreso Peninsular de Historia Antigua* 2. Santiago de Compostela, 39-56

Duval 1971: Duval, P.M. *La Gaule jusqu'au V siècle.* Picard, Paris

Ede 1988: Ede, J.E. Técnicas de amostragem para o estudo de sementes. Queiroga, F.; Sousa, I.M.A.R.; Oliveira, C.M. (eds.) *Paleoecologia e Arqueologia.* Vila Nova de Famalicão, 71-80

Eiroa 1975: Eiroa, J.J. Aportación a una ordenación de las plantas castreñas del Noroeste peninsular. *Miscelanea Arqueológica*, Zaragoza, 159-63

1980: Eiroa, J.J. Notas sobre la cronología de los castros del Noroeste de la Península Ibérica. *SANP* 1, 71-83

1988: Eiroa, J.J. La Cultura Castreña del Noroeste de la Península Ibérica 15 años después. *Anales de Prehistoria y Arqueología* 4, Murcia, 111-23

Esparza Arroyo 1980: Esparza Arroyo, A. Nuevos castros con piedras hincadas en el borde occidental de la Meseta. *SANP* 2, 71-83

1983: Esparza Arroyo, A. Sobre el limite oriental de la cultura castreña. II *SANP*, Madrid, 103-19

	1986:	Esparza Arroyo, A. *Los castros de la Eda del Hierro del Noroeste de Zamora.* Zamora
Fariña Busto	1979:	Fariña Busto, F. As fíbulas de longo travessão sem espira nos castros do NW peninsular. *BAur* 9, 27-49
Fariña *et al.*	1983:	Fariña Busto, F.; Arias Vilas, F.; Romero Masiá, A.M. Panorámica geral sobre la cultura castrexa. Pereira Menaut, G. (ed.) *Estudos de Cultura castreja e de história antiga de Galicia*, 87-127
Fernandes	1989:	Fernandes, M.A. Felgueiras de ontem e de hoje. Câmara Municipal de Felgueiras, Felgueiras
Fernandez Canosa	1987:	Fernandez Canosa, J.A. As maneiras de combate na Lusitânia. *TAE* 27 (1-4), 149-58
	1988:	Fernandez Canosa, J.A. Rutura das hostilidades, alianza e paz e fides na Hispania pre-romana: instituizons de guerra. Pereira Menaut, G. (ed.) 1° *Congreso Peninsular de Historia Antigua* 2, Santiago de Compostela, 239-48
Fernández Fuster	1953:	Fernández Fuster, L. Sobre la interpretación de los monumentos con 'pedras formosas'. *AEA* 26(88), 379-84
	1954:	Fernández Fuster, L. Interpretación de la Piedra Formosa de Briteiros. *CEG* 9 (27), 15-33
	1955:	Fernández Fuster, L. La Pedra Formosa de Briteiros. III *CAN*, 372-74
Fernández Gomez	1989:	Fernández Gómez, F. Orfebrería indígena en época prerromana. García Castro, J.A. (ed.) *El oro en la España Prerromana.* Revista de Arqueologia, Madrid, 82-9
Ferreira, JAP	1959:	Ferreira, J.A.P. Arqueologia portuense. Novos achados arqueológicos na Avenida de D. Afonso Henriques, quando do seu rompimento em 1950-54. *BC-CMP* 22 (1-2), 367-68
Ferreira	1966:	Ferreira, O.V. Os artefactos pre-historicos de âmbar e a sua distribuição em Portugal. *RG* 76, 61-6
	1968:	Ferreira, O.V. Algumas notas acerca da pesca na Antiguidade. *AP* 2 (3ª Série), 113-33
Ferreira - Viana	1956:	Ferreira, O.V.; Viana, A. L'importance du Cuivre Peninsulaire dans les Ages du Bronze. *CISPP* 4 (Madrid 1954), Zaragoza, 521-33
Ferreiro Lopez	1988:	Ferreiro Lopez, M. La campaña militar de Cesar en el año 61. Pereira Menaut, G. (ed.) 1° *Congreso Peninsular de Historia Antigua* 2. Santiago de Compostela, 363-72
Feugére	1989	Feugére, M. Les vases en verre sur noyan d'argile en Méditerranée nord-occidentale. Feugére, M (ed.) Le verre préromain en europe occidentale. Montagnac, 29-62
Figueiral	1990:	Figueiral, I.M.A.C.R. *Le Nord-Ouest du Portugal et les modifications de l'ecosysteme, du Bronze Final à l'époque Romaine, d'aprés l'anthracoanalyse de sites archéologiques.* 2 vols. Université Montpellier II, (typescript)
	1991:	Figueiral, I.M.A.C.R. Buraco da Pala: um meio-ambiente vegetal explorado pelo homem. Resultados da análise antracológica. Queiroga, F.; Dinis, A.P. (eds.). *Paleoecologia e Arqueologia* 2, Vila Nova de Famalicão, 13-29
Figueiral - Queiroga	1988:	Figueiral, I.M.A.C.R.; Queiroga, F. Castelo de Matos, 1982-86. *Arqueologia* 17, Porto, 137-50
Figueiredo	1895-98:	Figueiredo, A.M. Informações arqueológicas recolhidas no Diccionario Geographico de Cardozo. 1895, *AP* 1, 142-4, 241-3, 316-20; 1896, *AP* 2, 162-5; 1897, *AP* 3, 218-23, 281-6; 1898, *AP* 4, 53-8
Figueras Pacheco	1956:	Figueras Pacheco, F. *La Necropolis Ibero-Púnica de la Albufera de Alicante.* Estudios Ibericos 4, Valencia
Filgueiras	1974:	Filgueiras, O.L. Remanescentes de formas de navegar pré-romanas em uso no Noroeste peninsular. III *CNA*, 223-48
Filip	1962:	Filip, J. *Celtic Civilization and its Heritage.* New Horizons. Prague (orig. ed. 1960)
Flam-Zuckermann	1970:	Flam-Zuckermann, F. À propos d'une inscription de Suisse (CIL, XIII, 5010): étude du phénomène du brigandage dans l'Empire romain. *Latomus* 29, 451-73
Fonseca	1936:	Fonseca, T. *Esposende e o seu concelho.* Esposende
	1948:	Fonseca, T. *O concelho de Barcelos aquém e além Cávado.* Barcelos
Fontes	1980:	Fontes, A.L. Culto ao deus Larouco, Júpiter e Atégina. *SANP* 3, 5-20
Fontes, LFO	1990:	Fontes, L.F.O. Carta Arqueológica do Concelho de Braga. *Forum* 8, Braga, 107-32
Fortes	1899:	Fortes, J. *A estação archeologica d'Alvarelhos.* Porto
	1902:	Fortes, J. Instrumentos de bronze. *AP* 7, 102-6
	1905-08a:	Fortes, J. As fíbulas no Noroeste da Península. *Portugália* 2, 15-33
	1905-08b:	Fortes, J. Thesouro de Viatodos da Idade do Bronze. *Portugália* 2, 110-11
	1909:	Fortes, J. Gaya no passado. *Mea Villa de Gaya.* Porto, 9-28
Frankowski	1918:	Frankowski, E. *Hórreos y palafitos de la Península Ibérica.* Madrid

Freire 1965: Freire, O. Algumas fivelas de bronze do Museu Antropológico da Universidade do Porto. *Lucerna* 4, 205-15

1967-68: Freire, O. Análise química de quatro fivelas de bronze. *TAE* 20 (3-4), 381-82

Freitas - Santos 1980: Freitas, A. M.; Santos Júnior , J.R. O Castro da Curalha - 5ª Campanha de escavações, 1979. *TAE* 23 (4), 293-405

1981: Freitas, A. M.; Santos Júnior , J.R. O Castro da Curalha - 6ª Campanha de escavações,1980. *TAE* 24 (1), 59-86

1982: Freitas, A. M.; Santos Júnior , J.R. O Castro da Curalha - 7ª Campanha de escavações, 1981. *TAE* 24 (2), 265-91

1983: Freitas, A. M.; Santos Júnior , J.R. O Castro da Curalha - 8ª Campanha de escavações, 1982. *TAE* 24 (3), 453-74

1984: Freitas, A. M.; Santos Júnior , J.R. O Castro da Curalha - 9ª Campanha de escavações, 1983. *TAE* 24 (4), 589-617

Freitas 1923: Freitas, E. *Estudo sobre dois bronzes arqueológicos.* Porto

1960: Freitas, E. *Felgerias Rubias. Subsídios para a história do conselho de Felgueiras.* Porto

Freitas, EAC 1949: Freitas, E.A.C. Santa Marinha de Retorta. *DL* 3 (5), 20-27

Freitas, J 1971: Freitas, J. Citânia de S. Julião de Caldelas. *AP* 5 (3ª Série), 133-8

Frenzel 1966: Frenzel, F. Climatic change in the Atlantic/sub-Boreal transition on the Northern Hemisphere: botanical evidence. Sawyer, J.S. (ed.) *World climate 8000-0 BC.* London, 99-123

Frey 1984: Frey, O.-H. Die Bedeutung der Gallia Cisalpina für die Enstehung der Oppida-Kultur. Dobiat, C. (ed.) *Studien zu Siedlungsfragen der La Ténezeit.* Marburg

Galhano 1973: Galhano, F. *O Carro de bois em Portugal.* Lisboa

Gallay 1989: Gallay, A. Vivre autour du feu. Recherche d'une problématique d'analyse archéologique. Olive, M.; Taborin, Y. *Nature et fonction des foyers préhistoriques.* Mémoires du Musée de Préhistoire d'ile de France 2. Nemours, 101-11

Gamito 1988 Gamito, T.J. *Social Complexity in Southwest Iberia 800-300 B.C. The case of Tartessos.* BAR (IS) 439

Ganderton 1981: Ganderton, P.S. *Environmental Archaeology. Site methods and interpretation.* Vorda Research Series 2, Highworth

Garcia Alonso 1981-82: Garcia Alonso, F. Ordenacion tipologica del instrumental de pesca en bronze ibero-romano. *Pyrenae* 17-18 315-28

Garcia Fernández 1986: Garcia Fernández-Albalat, B. Las llamadas divinidades de las aguas. Bermejo Barrera (ed.) *Mitología y Mitos en la Hispania prerromana.* Madrid, 141-92

1990: Garcia Fernández-Albalat, B. *Guerra y religión en la Gallaecia y la Lusitania antiguas.* Ediciós do Castro, A Coruña

Garcia Moreno 1988: Garcia Moreno, L.A. Infancia, juventud y primeras aventuras de Viriato, caudillo Lusitano. Pereira Menaut, G. (ed.) 1º *Congreso Peninsular de Historia Antigua* 2, Santiago de Compostela, 373-82

Garcia y Bellido 1934: Garcia y Bellido, A. Los Iberos en la Grecia propria y en el oriente Helenistico a traves de los escritores antiguos. *IP* 8, 100-2

1940-41b: Garcia y Bellido, A. El caldero de Carbaceno y la diadema de Ribadeo. *AEA* 14 (45), 560-63

1943: Garcia y Bellido, A. Los Albiones del Noroeste de España y una estela hallada en el occidente de Asturias. *Emerita* 11, 418-30

1944: Garcia y Bellido, A. Una nueva citania portuguesa en curso de excavación. *CEG* 1 (2), 297-8

1945a: Garcia y Bellido, A. España y los españoles hace dos mil años según la ' Geografía' de Estrabón. Madrid, 4ª ed. 1968

1945b: Garcia y Bellido, A. La navegabilidad de los ríos de la Península Ibérica en la Antigüedad . *IP* 16 (35), 115-22

1946a: Garcia y Bellido, A. Bandas y guerrillas en las luchas con Roma. Madrid

1946b: Garcia y Bellido, A. El casco de Lanhoso. *AEA* 19 (65), 356-58

1950: Garcia y Bellido, A. Algunos problemas relativos a las invasiones indoeuropeas en España. *AEA* 23, 487-96

1963: Garcia y Bellido, A. Los auxiliares hispanicos en los ejercitos romanos de ocupacion (200 al 30 antes de J. C.). *Emerita* 31 (2), 213-26

1966a: Garcia y Bellido, A. O problema dos enterramentos na cultura castreja. *RG* 76 (1-2), 5-24

1966b: Garcia y Bellido, A. Tessera hospitalis del año 14 de la era hallada en Herrera de Pisuerga. *BRAH* 99, 149-66

1968a: Garcia y Bellido, A. Cámara funeraria de la cultura castreña . *AEA* 41 (117-18), 16-44

1968b: Garcia y Bellido, A. *El urbanismo en España. La edad antigua.* Madrid, 3-64

	1969:	Garcia y Bellido, A. Los bronces Tartessicos. Maluquer de Motes, J. (ed.) *Tartessos y sus problemas.* Barcelona, 163-71
	1971:	Garcia y Bellido, A. Orígens de la casa redonda de la cultura castreña del N.O. de la Península. *RG* 81 (1-2), 25-35
	1974:	Garcia y Bellido, A. Otros testimonios más de la presencia de mercenarios españoles en el Mediterráneo. *Simposio Internacional de Colonizaciones (1971)*, Barcelona, 201-3
	1977:	Garcia y Bellido, A. La España del siglo primero de nuestra era (según P. Mela y C. Plinio). Madrid
Garlan	1972:	Garlan, Y. *La guerre dans l'Antiquité.* Paris
Garrido - Orta	1978:	Garrido Roiz, J.P.; Orta Garcia, E.M. Excavaciones en la necropolis de La Joya Huelva, II. *EAE* 96
Gerloff	1986:	Gerloff, S. Bronze Age Class A Cauldrons. Typology, Origins and Chronology. *JRSAI* 116, 84-115
Girão	1958:	Girão, A.A. *Atlas de Portugal.* Coimbra
Gomes, CA	1979:	Gomes, C.A. O Castro de Álvora (subsídios para o seu estudo). *CV* 3, 161-75
Gomez Moreno	1941:	Gomez Moreno, M. Oro en España. *AEA* 14(45), 461-70
Gomez Tabanera	1975:	Gomez Tabanera, J.M. La cultura castreña astur y los modos de producción en las sociedades castreñas del N.W. hispano. XIII *CAN*, 557-77
	1980:	Gomez Tabanera, J.M. Aspectos de la cultura castreña en sus manifestaciones en Asturias y de los modos de producción en las sociedades protohistóricas del NW de la Península Ibérica. *SANP* 2, 87-110
	1964:	Gonçalves, F. A cividade de Terroso. *BC-PV*, 307-22
Gonzalez Fernandez	1983:	González Fernandez, X.M. O comércio romano na ría de Vigo. *Brigantium* 4, A Coruña, 73-98
González - Pina	1983:	González Prats, A.; Pina Gosalbez, J.A. Analisis de las pastas ceramicas de vasos hechos a torno de la fase orientalizante de Peña Negra. *Lvcentvm* 2, 115-45
Gossé	1942:	Gossé, G. Las minas y el arte minero de España en la Antigüedad. *Ampurias* 4 43-68
Goudineau	1968:	Goudineau, C. *La ceramique aretine lisse.* Boccard, Paris
Grosse	1959:	Grosse, R. Las fuentes desde Cesar hasta el siglo V de J.C. *FHA* 8, Barcelona
Guadán	1979:	Guadán, A.M. *Las armas en la moneda ibérica.* Cuadernos de Numismatica, Madrid
Guerra	1882:	Guerra, L.F. A estátua callaica de Vianna. *Pero Gallego* 1 (15), 3-4, (= *RSNS* 4, 1896, 291-4)
	1900a:	Guerra, L.F. Limia e Brutobriga. *AP* 5, 2-7
	1900b:	Guerra, L.F. Os castellos de Fraião e de Pena da Rainha. *AP* 5, 134-6
	1900c:	Guerra, L.F. Vestígios romanos no concelho de Vianna do Castello. *AP* 5, 175-7
	1909:	Guerra, L.F. O Crasto de Areosa. *AP* 14, 59-60
	1982:	Guerra, L.F. *Notícias históricas do concelho e vila de Boticas.* Boticas, (1° ed., Viana do Castelo, 1911)
Guilaine	1976:	Guilaine, J. Systèmes chronologiques et séquences culturelles des civilisations néolitiques et protohistoriques de la France. Guilaine, J. (ed.) *La Préhistoire française* 2, 17-23
Guillén Oterino	1982:	Guillén Oterino, A. Sobre la introducción del Castaño, Castanea sativa, en el Mediterráneo Occidental. *Zephyrus* 34-35, 99-102
Guimarães, JS	1970:	Guimarães, J.S. O Picoto de Santo Amaro (Guimarães). *RG* 80 (1-2-3-4), 141-50, 381-86
	1971:	Guimarães, J.S. O Picoto de Santo Amaro (Guimarães). II *CAN* (2), 436-70
Guitian - Vazquez	1975:	Guitián Rivera, F.; Vázquez Varela, J.M. Contribuición al estudio de la metalurgia del bronce en la cultura castreña. *Gallaecia* 1
	1979:	Guitián Rivera, F.; Vázquez Varela, J.M. Sobre la tecnologia de la cerâmica castreña (ceramica de las islas Cies. Pontevedra). *Gallaecia* 3/4, 275-9
Gündel	1968:	Gündel, H.G. Viriato - Lusitano, caudillo en las luchas contra los Romanos. *Caesaraugusta* 31-32, 175-98
Gustin	1974:	Gustin, M. Mahaire: Doprino k povezom Picena, Slovenijein Srednejga Podonavja v 7 stol pr. n. st. (Mahaire Krummschwerter: urgeschichtliche Verbindungen Picenum-Slowenien-Basarabi), *Situla* 14-15, Ljubljana, 77-94
Harding	1982:	Harding, A.F. Introduction: climatic change and archaeology. Harding, A.F. (ed.) *Climatic Change in Later Pre-History.* Edinburgh, 1-10
Härke	1979:	Härke, H.G.H. *Settlement types and settlement paterns in the West Hallstatt Province.* BAR (IS) 57, Oxford
	1989:	Härke, H.G.H. Transformation or collapse? Bronze Age to Iron Age settlement *in west central Europe. Sørensen, M.L.; Thomas, R. (eds.) The Bronze Age-Iron Age Transition in Europe: aspects of continuity and change in European societies c. 1200 to 500 BC.* BAR (IS) 483, Oxford
Harmand	1967:	Harmand, J. *L'armée et le soldat à Rome de 107 à 50 avant notre ère.* Paris

	1970:	Harmand, J. César et l'Espagne durant le second *Bellum Civile. Legio VII Gemina*, Leon, 181-203
Hartmann	1970:	Hartmann, A. *Prähistoriche Goldfunde aus Europa.* Studien zu den Anfängen der Metallurgie 3, Berlin
	1971:	Hartmann, A. Análises de alguns objectos pré-históricos de ouro, procedentes do Norte de Portugal. *RG* 81 (1-2), 129-38
Hartmann - Kalb	1969:	Hartmann, A.; Kalb, Ph. Investigaciones espectro-analíticas sobre hallazgos hispánicos de oro. *APL* 12, 93-8
Haselgrove	1986:	Haselgrove, C. Central Places in British Iron Age studies: a review of some problems. Grant, E. (ed.) *Central Places, Archaeology and History.* Department of Archaeology and Prehistory, Sheffield, 3-12
	1987:	Haselgrove, C. Culture process on the periphery: Belgic Gaul and Rome during the late Republic and early Empire. Rowlands, M.; Larsen, M.; Kristiansen, K. (eds.) *Centre and Periphery in the Ancient World.* Cambridge University Press, Cambridge, 104-24
Hastorf - Popper	1988:	Hastorf, C.A.; Popper, V.S. (eds.) *Current Paleoethnobotany. Analytical Methods and Cultural Interpretations of Archaeological Plant Remains.* The University of Chicago Press, Chicago
Hawkes	1952:	Hawkes, C.F.C. Las relaciones atlánticas en el Bronze final, entre la Península Ibérica y las Islas Británicas con respecto a Francia y la Europa Central y Mediterránea. *Ampurias* 14, 81-119
	1954:	Hawkes, C.F.C. Archaeological theory and method: Some suggestions from the Old World. *AA* 56, 155-68
	1958:	Hawkes, C.F.C. Escavações no Castro de Sabroso (Abril de 1958). *RG* 68 (3-4), 445-53
	1971:	Hawkes, C.F.C. North-western castros: excavation, archaelogy and history. II *CNA* 1, 183-86
	1984:	Hawkes, C.F.C. The castro culture of the peninsular Nort-West: fact and inference. Blagg, T.F.C.; Jones, R.F.J.; Keay, S.J. (eds.) *Papers in Iberian Archaeology.* BAR (IS) 193 (1), 187-203
Hawkes - Smith	1957:	Hawkes, C.F.C.; Smith, M.A. On some Buckets and cauldrons of the Bronze and Early Iron Ages. *AJ* 37, 131-98
Hedges *et al.*	1990:	Hedges, R.E.M.; Housley, C.R.; Bronk, C.R.; Van Klinken, G.J. Radiocarbon dates from the Oxford AMS system: Archaeometry datelist 11. *Archaeometry* 32 (2),211-37
Hencken	1956:	Hencken, H. Carp's Tongue Swords in Spain, France and Italy. *Zephyrus* 7,125-78
	1971:	Hencken, H. *The earliest european helmets. Bronze age and early Iron age.* Cambridge Massachussets
Hernandez Hernandez	1982:	Hernandez Hernandez, F. La escultura zoomorfa del Occidente peninsular. *TP* 39, 211-39
Hidalgo Cuñarro	1980:	Hidalgo Cuñarro, J.M. La ceramica con decoracion bruñida en el noroeste peninsular. *Gallaecia* 6, 81-100
	1983:	Hidalgo Cuñarro, J.M. *Excavaciones arqueológicas en el Castro de Vigo.* Vigo
	1984:	Hidalgo Cuñarro, J.M. El Castro de Vigo y el comercio atlántico en Noroeste peninsular. *RG* 94, 371-87
	1985a:	Hidalgo Cuñarro, J.M. *Castro de Troña, Ponteareas - Pontevedra. Campaña 1983.* Arqueoloxía/Memorias, Xunta de Galicia - *DXPHD*, Arqueoloxía / Memorias 3, Santiago de Compostela
	1985b:	Hidalgo Cuñarro, J.M. *Castro de Vigo. Campaña 1983.* Arqueoloxía / Memorias 1, Santiago Compostela
	1987:	Hidalgo Cuñarro, J.M. Una fecha de C-14 del Castro de Troña (Ponteareas, Pontevedra). *CEG* 37, 31-9
	1988-89	Hidalgo Cuñarro, J.M. Excavaciones arqueológicas en castro de Troña (Ponteareas-Pontevedra). Campañas: de 1984-1986. *Castrelos* 1-2, Vigo, 81-108
Hidalgo - Rodríguez	1988:	Hidalgo Cuñarro, J.M.; Rodríguez Puentes, E. Dos modelos de habitat castreño: Castro de Troña y Castro de Fozara. *TAE* 28 (3-4), 133-44
Hipólito	1960-61:	Hipólito, M.C. Dos tesouros de moedas romanas encontradas em Portugal. *Conimbriga* 2-3, 1-165
	1983:	Hipólito, M.C. As moedas gregas da Serra do Pilar (Vila Nova de Gaia). *Arqueologia* 8, 75-82
Höck	1980:	Höck, M. Corte estratigráfico no Castro de S. Juzenda (concelho de Mirandela). *SANP* 2, 55-70
Hodder - Orton	1976:	Hodder, I.; Orton, C. *Spatial Analysis in Archaeology.* University Press, Cambridge

Hodder, MA	1990:	Hodder, M.A. Burnt mounds in the English West Midlands. Buckley, V. (ed.) *Burnt Offerings*. Wordwell Ltd. Dublin, 106-11
Hogg	1966	Hogg, A.H.A. Native Settlement in Wales. Thomas, C.(ed.) *Rural Settlement in Roman Britain*. CBA research report 7, London, 28-38
Hübner	1869-92:	Hübner, E. *Corpus Inscriptionum Latinarum* II. Berlin, 1869; Supplementum. Berlin, 1892, (=CIL)
	1871:	Hübner, E. *Notícias archeologicas de Portugal*. Lisboa
	1897:	Hübner, E. Bracari. *RE* 3
	1897:	Hübner, E. Callaeci. *RE* 3
	1899-1903:	Hübner, E. *Addimenta nova ad corporis vol. II*. (EE 8, 1899; EE 9, 1903)
Jäger - Lozek	1982:	Jäger, K.D.; Lozek, V. Environmental conditions and land cultivation during the Urnfield Bronze Age in central Europe. Harding, A.F. (ed.) *Climatic Change in Later Pre-History*. Edinburgh, 162-78
Jalhay	1944:	Jalhay, E. Citânia de Sanfins (Paços de Ferreira). Escavações... 1ª Campanha - 1944. *Brotéria* 39 (5), 414-35
	1947:	Jalhay, E. Uma notável gravura rupestre da Citânia de Sanfins. *Brotéria* 44 (5), 554-63
	1949a:	Jalhay, E. Un umbral de puerta zoomórfica de la Citania de Sanfins. IV *CASE*, 300-3
	1949b:	Jalhay, E. Um vaso de olaria rutena na Citânia de Sanfins. *Brotéria* 48 (3), 317-25
	1950a:	Jalhay, E. Nova marca de oleiro na Citânia de Sanfins. *Zephyrus* 1, 29-31
	1950b:	Jalhay, E. Uma jóia de oiro na Citânia de Sanfins. *Brotéria* 50 (1), 38-49
	1951:	Jalhay, E. Um tipo curioso de escada na Citânia de Sanfins. *BAP* 1, 5-9
Jalhay - Paço	1948:	Jalhay, E.; Paço, A. A Citânia de Sanfins (Paços de Ferreira), 2ª, 3ª e 4ª campanhas. *Brotéria* 46 (6), 699-717
Janssen - Woldringh	1981:	Janssen, C.R.; Woldringh, R.E. A preliminary radiocarbon dated pollen sequence from the Serra da Estrela, Portugal. *Finisterra* 16 (32), Lisboa, 299-309
Jarman *et al.*	1972:	Jarman, N.H.; Legge, A.J.; Charles, J.A. Retrieval of plant remains from archaeological sites by froth flotation. Higgs, E.S. (ed.) *Papers in Economic Prehistory*. Cambridge University Press, Cambridge, 39-48
Jato - Vazquez	1972:	Jato Rodriguez, M.V.; Vázquez Varela, J.M. Correlación entre los datos palinológicos y hallazgos prehistóricos en Galicia. *Compostellanum* 17, 14-5
Joleaud	1929:	Joleaud, M. *L'anciannité de l'exploitation de l'étain dans le NW de l'Espagne*. L'Anthropologie, Paris
Jones	1983:	Jones, M. (ed.) *Integrating the Subsistence Economy*. BAR (IS) 181
Jordá Cerdá	1969:	Jordá Cerdá, F. *Guia del Castrillón de Coaña* . Salamanca
	1985-86:	Jordá Cerdá, F. Sobre la celtización tardía de Asturias. *Veleia* 2-3 (IV CLCPPI), 261-4
Jorge	1983:	Jorge, S.O. *Duas datas de C14 para a sepultura 1 da estação do Tapado da Caldeira (Baião)*. *Arqueologia* 8, Porto, 55-6
	1985:	Jorge, S.O. Datas de carbono 14 para a Pré-história recente do Norte de Portugal: os dados e os problemas. *Arqueologia* 12, 154-183
	1988a:	Jorge, S.O. *O povoado da Bouça do Frade (Baião) no quadro do Bronze Final do Norte de Portugal*. Monografias Arqueológicas 2, G.E.A.P. Porto
	1988b:	Jorge, S.O. Reflexões sobre a Pré-História Recente do Norte de Portugal. *TAE* 28 (1-2), 85-112
	1988c:	Jorge, S.O. O povoado da Bouça do Frade (Baião) - breve apontamento. *Arqueologia* 17, Porto, 134-7
Jorge, VO	1982:	*Megalitismo do Norte de Portugal: o distrito do Porto - os monumentos e a sua problemática no contexto europeu*. Porto (typescript)
Jorge - Sousa	1980:	Jorge, S.O.; Sousa, J.J.R. Resultados preliminares de uma sondagem arqueológica da Chã do Castro (Amares, Braga). *SANP* 1, 121-31
Jorge - Almeida	1980:	Jorge, V.O.; Almeida, C.A.F. *A estátua-menir fálica de Chaves*. Offp. G.E.A.P, Porto
Jorge - Jorge	1983:	Jorge, V.O.; Jorge, S.O. Nótula preliminar sobre uma estátua-menir do Norte de Portugal. *Arqueologia* 7, Porto, 44-7
Kalb	1974-77:	Kalb, Ph. Uma data C-14 para o Bronze Atlântico. *AP* 7-9 (3ª Série), 141-4
	1980a:	Kalb, Ph. O 'Bronze atlântico' em Portugal. *SANP* 1 113-20
	1980b:	Kalb, Ph. Zur Atlantischen Bronzezeit in Portugal. *Germania* 58, 25-115
Kalb - Höck	1980:	Kalb, Ph.; Höck, M. Ausgrabungen in der Grabbügelnekropole Fonte de Malga (Viseu, Portugal). *MM* 20, 43-55
Kelly	1990:	Kelly, R.S. Recent work in north-west Wales: the excavation of a burnt mound at Graeanog, Clynnog, Gwynedd, in 1983. Buckley, V. (ed.) *Burnt Offerings*. Wordwell Ltd. Dublin, 117-28

Klein *et al.* 1982: Klein, J.; Lerman, J.C.; Damon, P.E.; Ralph, E.K. Calibration of radiocarbon dates: tables based on the consensus data of the workshop on calibrating the radiocarbon time-scale. *Radiocarbon* 24, 103-50

Knapp 1979: Knapp, R.C. The Legions in the Spain. *RUC* 18 (118), 87-95

Kristiansen 1992: Kristiansen, K. (forthcoming) The emergence of the european world system in the bronze age: divergence, convergence and social evolution during the first and second millennia B.C. in Europe. Kristiansen, K., Jansen, J. (eds.) *Europe in the first millennium.* Sheffield University Press, Sheffield

Lamb 1982: Lamb, H.H. Reconstruction of the course of postglacial climate over the world. Harding, A.F. (ed.) *Climatic Change in Later Pre-History.* Edinburgh, 11-32

Lanhas - Brandão 1967: Lanhas, F.; Brandão, D.P. Inventário de objectos e lugares com interesse arqueológico. *RE* 8 (1), 5-74

1969: Lanhas, F.; Brandão, D.P. Inventário de objectos e lugares com interesse arqueológico. *RE* 12 (2), 295-344

Larsson 1990: Larsson, T.B. Skärvstenshögar - the burnt mounds of Sweden. Buckley, V. (Ed.) *Burnt Offerings.* Wordwell Ltd., Dublin, 142-53

Lasserre 1966: Lasserre, F. *Strabon, Géographie.* (vol. 2), Paris

Le Roux, F 1965: Le Roux, F. Aspects de la fonction guerrière chez les Celtes. *Ogam* 17 (1-2), 175-92

Le Roux 1975: Le Roux, P. Aux origines de Braga (Bracara Augusta). *BA* 29 (67-68), 155-9

1980: Le Roux, P. Les *auxilia* romains recrutés chez les *Bracari* et l'organisation politique du Nord-ouest Hispanique. *SANP* 3, 43-66

1982: Le Roux, P. *L'armée romaine et l'organisation des provinces ibériques d'Auguste à l'invasion de 409.* Publications du Centre Pierre Paris 8, Paris

1985: Le Roux, P. Provincialisation et recrutement militaire dans le N.O hispanique au Haut-Empire romain. *Gerión* 3, 283-308

Le Roux - Tranoy 1984: Le Roux, P.; Tranoy, A. Ɔ le mot et la chose. Contribution au débat historiographique. *Lucerna* (s/n), 239-55, (= Tranoy, A.; Le Roux, P. 1983: ibidem. *AEA* 56)

Leal - Ferreira 1873-90: Leal, A.P.; Ferreira, A. *Portugal antigo e moderno.* Lisboa

Lema 1978: Lema, P.B. *Tourém. Uma aldeia raiana do Barroso.* Lisboa

Lemos *et al.* 1981: Lemos, F.S.; Martins, M.; Delgado, M. *Actividade arqueológica 1976-1980.* Braga

Lima 1926-28: Lima, A.P. Uma visita a S. Miguel-o-Anjo. *TAE* 3, 289-98

1956: Lima, A.P. O lugar da Torre. *BC-CST* 4 (3), 222-3

Little 1990: Little, G.M. *The technology of pottery production in Northwestern Portugal during the Iron Age.* Cadernos de Arqueologia (Monografias), Braga

López Cuevillas 1926: López Cuevillas, F. *La Edade do Ferro en Galiza.* Nós 25, La Coruña

1932a: López Cuevillas, F. Los brazaletes posthallstátticos del Noroeste hispánico. *AEAA* 22, 225-36

1932b: López Cuevillas, F. Os torques do Noroeste hispánico. *ASEG* 4, 37-132

1933: López Cuevillas, F. A área xeografica da cultura norte dos castros. *HMS*, 99-107

1935: López Cuevillas, F. O culto das fontes no Noroeste hispánico. *TAE* 7, 73-104

1946-47: López Cuevillas, F. Armería posthallstáttica del Noroeste hispánico. *CEG* 2 (8), 543-89

1947a: López Cuevillas, F. Las cabanas de los castros. Homenagem a Julio Martinez Santa Olalla, *AMSEAEP* 22 (1-4), 141-6

1950b: López Cuevillas, F. Un barrio inédito del castro de Troña. *CEG* 5 (16), 258-60

1951a: López Cuevillas, F. La diadema áurea de Ribadeo. *CEG* 6 (18), 23-31

1951b: López Cuevillas, F. Esculturas zoomorfas y antropomorfas de la cultura de los castros. *CEG* 6 (19), 177-203

1951c: López Cuevillas, F. *Las joias castreñas.* Madrid

1952: López Cuevillas, F. La etnología de la cultura castreña. *Zephyrus* 3, 5-13

1953a: López Cuevillas, F. Cerámicas castrexas de la ciudad y provincia de Lugo. *BCML* 5, 253-60

1953b: López Cuevillas, F. *La civilización céltica en Galicia. Santiago de Compostela* (2ª ed. 1988)

1954: López Cuevillas, F. La edad del Hierro en el Noroeste. *La Prehistoria de España* II (sep. IV Congreso Internacional de Ciencias Prehistoricas y Protohistoricas), Madrid, 1-34

1955a: López Cuevillas, F. Armas de bronce ofrendadas al río Sil. *Zephyrus* 6, Salamanca, 233-40

1955b: López Cuevillas, F. La 'Cidá do Castro' de San Millán. *CEG* 10 (30), 138-40

1955c López Cuevillas, F. El comercio y los medios de transporte de los pueblos castreños. CEG 10 (31), 145-58

1956: López Cuevillas, F. Os intrumentos de traballo dos castrexos. *BRAG* 27, 270-6

1958: López Cuevillas, F. La hoz de hierro de la 'Cidá do Castro' (Cualedro, Orense). *CEG* 13 (41), 329-39

1966: López Cuevillas, F. A economia nos tempos da cultura castrexa. *Grial* 13, 216

Lopez - Bouza 1929: López Cuevillas, F.; Bouza Brey, F. Os Oestrimnios, os Saefes e a ofiolatría en Galicia. *ASEG* 2, 29-193

Lopez - Lorenzo 1946: López Cuevillas, F.; Lorenzo Fernández, J. Las habitaciones de los Castros. *CEG* 2 (5), 7-74

Lopez - Taboada 1958: López Cuevillas, F.; Taboada Chivite, J. Nuevos Hallazgos en el Outeiro de Baltar. *CEG* 13 (41), 314-20

Lopez Monteagudo 1973: Lopez Monteagudo, G. Cabeza zoomorfa inédita del castro de Santa Tecla. XII *CAN*, 341-3

1977: Lopez Monteagudo, G. La Diadema de San Martin De Oscos. *RUC* 26 (109), Madrid, 99-108

1979: Lopez Monteagudo, G. Particularidades escultóricas de los verracos. XV *CAN*, 721-6

1986: Lopez Monteagudo, G. Relaciones entre la Cultura de los Verracos y la Cultura Castreña. *TP* 43, 211-26

Lorenzo Fernandez 1948: Lorenzo Fernandez, J. El monumento proto-historico de Aguas Santas y los ritos funerarios en los castros. *CEG* 2 (10), 157-211

1971: Lorenzo Fernandez, J. Una rueda fósil en Catoira (Galicia). *BAur* 1

1973: Lorenzo Fernandez, J. Metamorfosis de unha casa castrexa. *TAE* 22 (3), 255-9

1973: Lorenzo Fernandez, J. Sobre la rueda de Catoira. *BAur* 3, 173-5

Luengo y Martínez 1950: Luengo y Martínez, J.M. Excavaciones arqueológicas en el castro y su necrópolis de Meirás (La Coruña). *IMCGEA* 23

Lumley *et al.* 1976: Lumley, H; Renault-Miskovsky, J.; Miskovsky, J.-C.; Guilaine, J. Le cadre chronologique e paléoclimatique du Postglaciaire. Guilaine, J. (ed.) *La Préhistoire française* 2, 3-16

Macedo 1896: Macedo, M.D. Alguns apontamentos archeológicos relativos às duas freguesias de Sobreposta e Pedralva. *RG* 13 (3), 121-3

Machado 1955: Machado, A.S. Novas versões sobre o problema de Cale e Portucale. *BC-CMP* 18 (3-4), 344-79

1956: Machado, A.S. *As origens da cidade do Porto*. O problema de Portucale. Porto

1962: Machado, A.S. A arqueologia e as origens do Porto. *SG* 9 (1), 213-6

Machado, F 1951: Machado, F. *Monografia do concelho de Esposende*. Esposende

Machado, FSL 1920: Machado, F.S.L. *Uma cidade morta no Monte Mozinho*. Coimbra

MacWhite 1951: MacWhite, E. *Estudios sobre las relaciones atlánticas de la Península Hispánica en la edad del Bronce*. Publicaciones del Seminario de Historia Primitiva del Hombre, Madrid

Magalhães *et al.* 1977: Magalhães, A.; Freitas, A.M.; Santos Júnior, J.R. O Castro da Curalha - 2ª e 3ª campanhas de escavações, 1975 e 1976. *TAE* 23 (1), 19-40

1978: Magalhães, A.; Freitas, A.M.; Santos Júnior, J.R. O Castro da Curalha - 4ª campanha de escavações, 1977. *TAE* 23 (2-3), 267-77

Magny 1982: Magny, M. Atlantic and Sub-boreal: dampness and dryness? Harding, A.F. (ed.) *Climatic Change in Later Pre-History*. Edinburgh, 33-42

Maluquer de Motes 1948: Maluquer de Motes, J. A proposito de umas cerámicas procedentes del Castelo de Faria. *BGAF* 1 (1), 33-8

1950: Maluquer de Motes, J. Hallazgos monetarios de la Citanea de Sanfins. *Zephyrus* 1, 60-1

1955a: Maluquer de Motes, J. Contribución al estudio de la primitiva casa indoeuropea en la Península. III *CAN*, 109-14

1955b: Maluquer de Motes, J. El processo histórico de las primitivas poblaciones peninsulares, *Zephyrus* 6, 145-69, 241-55

1968: Maluquer de Motes, J. Panorama económico de la primera edad de Hierro. Tarradell, M. (ed.) *Estudios de economía antigua de la Península Ibérica*, Barcelona, 61-79

1973: Maluquer de Motes, J. La originalidad de la cultura castreña. *TAE* 22 (3), 335-44

1975a: Maluquer de Motes, J. Formación y desarrollo de la cultura castreña. I *JMACH* 1, 269-84

1975b: Maluquer de Motes, J. La edad del Bronce en el Occidente Atlántico. I *JMACH* 1, 129-45

1977: Maluquer de Motes, J. El mundo indígena del Noroeste hispánico antes de la llegada romana. *Bimilenario de Lugo*, 7-15

1982: Maluquer de Motes, J. Los Pueblos de la España céltica. Pueblos celtas. in Menéndez Pidal (ed.) *Historia de España. La Protohistória*, 1 (2). (La cultura material de los pueblos del Noroeste Peninsular), Madrid, 1-90

Mañanes 1976: Mañanes, T. Asturica Augusta y su convento juridico. *CSECS* 16, 39-43

Margalef 1956: Margalef, R. Oscilaciones del clima Portglaciar del Noroeste de España, registradas en los sedimentos de la ría de Vigo. *Zephyrus* 7, 5-9

Marinval 1988a: Marinval, P. *Cueillette, agriculture et alimentation vegetale de L'épipaleolithique jusqu'au 2º age du Fer en France meridionale. Apports palethnographiques de la carpologie.* (These Doctorat, École de Hautes Études en Sciences Sociales), Paris

1988b: Marinval, P. *L'Alimentacion Végétale en France du Mésolitique jusqu'à L'Âge du Fer.* Éditions du CNRS, Toulouse

Marques 1972: Marques, G. Arqueologia de Alpiarça. A estações representadas no Instituto de Antropologia do Porto. *TIADMC* 13

Marques-Andrade 1974: Marques, G.; Andrade, M. Aspectos da proto-história do terrtório português. 1 - Definição e distribuição geográfica da cultura de ALPIARÇA (Idade do Ferro). III *CNA,* 125-48

Marques, JATM 1984a: Marques, J.A.T.M. Elementos para o estudo da idade do Bronze no Alto Minho (Ponta de lança do Castro da Senhora da Graça, Monção). *Humanidades* 4, 40-1

1984b: Marques, J.A.T.M. Inventário arqueológico do concelho de Monção. Estado da questão. *Revista de História - U.L.* 1, 73-110

1984c: Marques, J.A.T.M. Monção. Monte da Senhora da Assunção. *IA* 4, 118-21

1985a: Marques, J.A.T.M. Monção. Sª da Assunção - 1982 e 1983. *IA* 5, 134-6

1985b: Marques, J.A.T.M. Monção. Sª da Graça - 1983. *IA* 5, 136

1987a: Marques, J.A.T.M. Assentamentos castrejos do concelho de Monção. *RCHUP* 2, 77-120

Marques - Silva 1985: Marques, J.A.T.M.; Silva, E.J.L. Viana do Castelo. Castro de Moldes (Guilheta) - 1982. *IA* 5 138

Martínez Fernández 1971: Martínez Fernández, J. Castro de Mohias. Resultado de una investigación geo-cronológica. *BIEA* 25, 351-6

Martínez Santa-Olalla 1933: Martínez Santa-Olalla, J. Monumentos funerários célticos. As pedras formosas e as estrelas em forma de casa. *HMS*, 226-35

1946: Martínez Santa-Olalla, J. *Esquema paletnológico de la Península Hispánica.* Madrid

1948: Martínez Santa-Olalla, J. El fragmento cerámico céltico. *BGAF* 1 (1), 21-8

Martino 1982: Martino, E. *Roma contra Cantabros y Astures. Nueva lectura de las fuentes.* Santander

Martins, JB 1984: Martins, J.B. *Levantamento aqueológico de Chaves, Boticas, Valpaços e Vila Pouca de Aguiar.* Chaves

Martins 1981: Martins, M. O povoado fortificado de Santo Ovídio. *Arqueologia* 3, 103-10

1983a: Martins, M. Distrito de Braga. Amares, a) Lago. *IA* 3, 22-4

1983b: Martins, M. Distrito de Braga. Fafe, a) Monte de Santo Ovídio. *IA* 3, 25-7

1984a: Martins, M. A Citânia de S. Julião, Vila Verde. Primeiras sondagens. *CA* 1 (S 2), 11-27

1984b: Martins, M. Distrito de Braga. Fafe, Castro de Santo Ovídio. *IA* 4, 53-4

1985a: Martins, M. A Ocupação do Bronze Final da Citânia de S. Julião, em Vila Verde. Caracterização e cronologia. *TAE* 25 (2-4), 197-240

1985b: Martins, M. Amares. Povoado do Lago - 1982. *IA* 5, 49-50

1985c: Martins, M. Fafe. Povoado de Santo Ovídio - 1982 e 1983. *IA* 5, 54-6

1985d: Martins, M. Vila Verde. Castro do Barbudo - 1983. *IA* 5, 57-8

1985e: Martins, M. Vila Verde. Citânia de S. Julião - 1982 e 1983. *IA* 5, 58-60

1986a: Martins, M. Duas datas de C14 para a ocupação do Bronze Final do povoado de S. Julião, Vila Verde. *Arqueologia* 13, Porto, 159-60

1986b: Martins, M. O povoado proto-histórico do Lago (Amares). Sistemas de defesa e fases de ocupação. *AP* 4 (S 4), 149-84

1987a: Martins, M. *O povoamento proto-histórico e a romanização da bacia do curso médio do rio Cávado.* Braga, Typescript, (= 1990)

1987b: Martins, M. *A cerâmica proto-histórica do vale do Cávado. Tentativa de sistematização.* *CA* 4 (S 2), 9-62

1988a: Martins, M. A Arqueologia dos Castros no Norte de Portugal: balanço e perspectivas. *TAE* 28, 11-35

1988b: Martins, M. *O povoado fortificado do Lago, em Amares.* CA (Monografias 1), Braga

1988c: Martins, M. *A Citânia de S. Julião, Vila Verde. Memória dos trabalhos realizados entre 1981 e 1985.* CA (Monografias 2), Braga

1989: Martins, M. *O Castro do Barbudo, em Vila Verde. Resultados das campanhas de 1983, 1984 e 1985.* CA (Monografias 3), Braga

1990: Martins, M. *O povoamento proto-histórico e a romanização da bacia do curso médio do rio Cávado.* (doctoral thesis presented to Universidade do Minho. CA (Monografias), Braga, (= 1987a, typescript)

Martins - Silva	1984:	Martins, M.; Silva, A.C.F. A estátua de guerreiro galaico de S. Julião (Vila Verde). *CA* 1 (S 2), 29-47
Mead	1971:	Mead, M. *L' Anthropologie comme Science Humaine.* Payot, Paris
Medeiros *et al.*	1980:	Medeiros, A.C. (†); Pereira, E.; Moreira, A. *Carta geológica de Portugal (1/50.000). Notícia explicativa da folha 9-D, Penafiel.* Lisboa
	1964:	Medeiros, A.C.; Pilar, L., Fernandes, A.P. *Carta geológica de Portugal (1/50.000). Notícia explicativa da folha 13-B, Castelo de Paiva.* Lisboa
	1975:	Medeiros, A.C.; Teixeira, C.; Lopes, J.T; Perez, T.P. *Carta geológica de Portugal (1/50.000). Notícia explicativa da folha 5-B, Ponte da Barca.* Lisboa
Meijide Cameselle	1988:	Meijide Cameselle, G. *Las Espadas del Bronce Final en la Península Ibérica.* Universidade de Santiago, Santiago de Compostela
Meira	1945:	Meira, A.R. *Afife.* Porto
Menendez Amor	1969:	Menendez Amor, J. Análisis esporo-polínicos de tres perfiles situados en la cuenca hidrográfica del río Deo (La Coruña). *Boletin de la Real Sociedad Española de Historia Natural (Geologia)* 67, 61-7
Menendez Pidal	1982:	Menéndez Pidal, R. *Historia de España.* (España prehistórica 1 (1), La protohistoria 1 (2), La historia prerromana 1 (3)). Madrid
Meneses	1938:	Meneses, A. Reflexões sobre a Citânia (de Briteiros). *Correio do Minho* 1-8 Julho
Menghin	1949:	Menghin, O.F.A. *El soliferreun de los Hispanos.* BSAA 15, 19-23
Meneses, M	1929:	Meneses, M. Notícias arqueológicas do concelho de Ribeira de Pena. *AP* 27, 29-48
Mergelina	1943-44:	Mergelina, C. La Citania de Santa Tecla, La Guardia (Pontevedra). *BSAA* 11 (37-39), 13-54
Miranda Junior *et al.*	1983:	Miranda Júnior, A.; Santos, J.N.; Santos Júnior, J.R. Castros do Concelho de Boticas. *TAE* 24 (3), 401-51
Miranda	1954:	Miranda, A. *Mozinho.* Penafiel
Moita	1962:	Moita, I. Notícia de um catálgo e carta de distribuição dos castros no território português. XXVI *CLEPC* 2 (5), 453-60
	1966:	Moita, I. A toponímia castreja. Distribuição e significado. *Lucerna* 5, 526-49
	1971:	Moita, I. Povoados, redutos e santuários castenses (Ensaio de classificação). II *CNA* 1 271-81
Montalvão	1971:	Montalvão, A. *Visita a castros nos arredores de Chaves.* Chaves
Monteagudo	1945:	Monteagudo, L. Nuevas joyas prerromanas del Norte de Portugal. *AEA* 18 (58), 81-89
	1952:	Monteagudo, L. Monumentos propriedad de la Sociedad Martins Sarmento. *AEA* 25 (85), 112-116
	1954:	Monteagudo, L. Metalurgía hispánica de la edad del Bronce, con especial estudio de Galicia y Norte de Portugal . *Caesaraugusta* 4, 55-95
	1955:	Monteagudo, L. Cale, Callaici y Callaecia. III *CAN*, 413-4
	1977:	Monteagudo, L. *Die Beile auf der Iberischen Halbinsel.* Prähistoriche Bronzefunde IX (6), München
	1981:	Monteagudo, L. Koiné del bronce atlántico. I *Colóquio Galaico-Minhoto* 2, 366-395
Monteiro	1985:	Monteiro, J.N. *Apontamentos para uma monografia de Sequeirô.* Boletim Paroquial de Sequeirô, Santo Tirso
Moreira, DA	1969:	Moreira, D.A. *Paisagem toponímica da Maia.* Maia
Moreira, MAF	1981a:	Moreira, M.A.F. *A cristianização do conventus bracaraugustanus. Caminiana* 5, 31-89
	1981b:	Moreira, M.A.F. As arrugiae da época romana no vale do rio Tinto. *Actas do colóquio Galaico-Minhoto.* Ponte de Lima, 395-423
	1982:	Moreira, M.A.F. A romanização do litoral do Alto Minho. *Caminiana* 6, 31-96
Morzadec-Kerfourn	1974:	Morzadec-Kerfourn M.T. Variations de la ligne de rivage armoricaine au Quaternaire; Analyse pollinique de dépots organiques littoraux. Rennes, 208
	1983:	Morzadec-Kerfourn M.T. A análise polínica - sua aplicação ao estudo do meio-ambiente do homem pré-histórico na Bretanha litoral. *Arqueologia* 8, Porto, 44-54
Mossé	1970:	Mossé, C. *La colonisation dans l'Antiquité.* Fernand Nathan, Paris
Müller Karpe	1959:	Müller-Karpe, H. *Beiträge zur Chronologie der Urnenfelderzeit Nördlich und Südlich der Alpen.* Römisch-Germanische Forschungen 22, Berlin
Nash	1976a:	Nash, D. Reconstructing Poseidonius' Celtic Ethnography: some Considerations. *Britannia* 7, London, 111-26
	1976b:	Nash, D. The growth of urban society in France. Cunliffe, B.; Rowley, T. (eds.) *Oppida: the beginnings of urbanisation in barbarian Europe.* BAR (SSII), Oxford, 95-133

	1978:	Nash, D. Territory and State Formation in Central Gaul. Green, D.R.; Haselgrove, C.; Spriggs, C.C. (eds.) *Social Organization and Settlement.* BAR (IS) 47, Oxford, 445-82
	1985:	Nash, D. Celtic territorial expansion and the Mediterranean World. Champion, T.C.; Megaw, J.V.S. (eds.) *Settlement and Society: aspects of West European prehistory in the first millennium BC.* Leicester University Press, 45-67
Naveiro López	1991:	Naveiro López, J.L. *El comercio antiguo en el N.W. Peninsular. Lectura Histórica del Registro Arqueológico.* Monografias Urxentes do Museu, Nº5, Museu Arqueolóxico, A Coruña
Neiva	1944:	Neiva, J.M.C. Jazigos Portugueses de Cassiterite e de Volframite. *CSGP* 25, 1-251
Neiva, MAP	1982:	Neiva, M.A.P. *O Castro de S. Lourenço - Vila Chã.* Esposende
	1987:	Neiva, M.A.P. *Esposende. Breve roteiro histórico.* Esposende
Neves	1942:	Neves, L.Q. Para a carta arqueológica do Minho. Depondo num inquérito. O castro do monte de Santinho ou Roques (Viana do Castelo), *Diário do Minho* (27 Fev., 12 Mar.)
	1948:	Neves, L.Q. Na Citânia de Roques. *AAM* 2, 56-61
	1959:	Neves, L.Q. O Castro do 'Santinho' ou Roques, no termo de Viana do Castelo. I *CAN* 1, 381-87
	1962:	Neves, L.Q. Breve notícia, até ao presente inédita, do achado de instrumentos de bronze no concelho de Melgaço. *SG* 9 (1), 94-99
	1965:	Neves, L.Q. Os castros do Norte de Portugal. *Lucerna* 4, 172-180
Noronha-Ribeiro	1983:	Noronha, F.; Ribeiro, M.L. *Carta geológica de Portugal (1/50.000). Notícia explicativa da folha 6-A, Montalegre.* Lisboa
Nunes	1973:	Nunes, H.B. O castro de Nª Senhora da Assunção (Barbeita). *A Terra Minhota* (30 Abr.), Monção
	1979:	Nunes, H.B. Uma visita ao Castro de Santinho (Roques). Festas das Neves, *Viana do Castelo* 6, 18-24
	1984:	Nunes, H.B. *Félix Alves Pereira e a arqueologia do Alto Minho. Uma breve introdução à sua obra.* Terras de Val de Vez 4 (1), Braga
Oliveira	1971:	Oliveira, A.L. *Terras de Bouro e o seu concelho.* Braga
	1976:	Oliveira, A.L. *Terras de Coura.* Póvoa de Varzim
	1978:	Oliveira, A.L. *Valença do Minho.* Póvoa de Varzim
	1980:	Oliveira, A.L. *Terras de Bouro.* Braga
	s.d.:	Oliveira, A.L. *Fafe e o seu concelho.* Braga
Oliveira, C	1934:	Oliveira, C. *O concelho de Gondomar.* Porto, 2 vols.
Oliveira, EP	1982:	Oliveira, E.P. Estudos Bracarenses, 4 -Para uma bibliografia do cónego Arlindo Ribeiro da Cunha. *BBUC* 37, 161-88
	1984:	Oliveira, E.P. *Bibliografia arqueológica portuguesa (1935-69).* Lisboa
	1985:	Oliveira, E.P. *Bibliografia arqueológica portuguesa (1970-1979).* Lisboa
Oliveira *et al.*	1969:	Oliveira, E.V.; Galhano, F.; Pereira, B.E. *Construções primitivas em Portugal.* Lisboa
	1983:	Oliveira, E.V.; Galhano, F.; Pereira, B.E. *Alfaia agrícola portuguesa.* Lisboa
	1991:	Oliveira, F.; Queiroga, F.; Dinis, A.P. O pão de bolota na Cultura Castreja. Queiroga, F.; Dinis, A.P.(eds.). *Paleoecologia e Arqueologia* 2, Vila Nova de Famalicão, 251-68
Oliveira, M	1908:	Oliveira, M. Thesouros encontrados em alguns castros do Norte de Portugal. *Portugália* 2, Porto, 667-8
Ors	1948:	Ors, A. d' Una nueva tabla emeritense de hospitium publicum. *Emerita* 16, 46-74
	1953:	ORS, A. d' *Epigrafía jurídica de la España romana.* Madrid
Osborne	1982:	Osborne, P.J. Some British later prehistoric insect faunas and their climatic implications. Harding, A.F. (ed.) *Climatic Change in Later Pre-History.* Edinburgh, 68-74
Paço	1952:	Paço, A. Citânia de Sanfins. I - Notícia histórica. *BC-CST* 1 (3), 357-384
	1953a:	Paço, A. Citânia de Sanfins . II - Muralhas. *Zephyrus* 4, 489-94
	1953b:	Paço, A. Citânia de Sanfins. III - Breve notícia de um tesouro monetário. *Brotéria* 56 (6), 673-89
	1953c:	Paço, A. Do Castro de Vila Nova de S.Pedro à Citânia de Sanfins. *RG* 63 (3-4), 683-697
	1954a:	Paço, A. Citânia de Sanfins. *Numisma* 2, 139-53
	1954b:	Paço, A. Citânia de Sanfins. IV - Alguns documentos históricos. *BC-CST* 3 (2), 197-212
	1954c:	Paço, A. Citânia de Sanfins. V - Nota sobre algumas moedas. *Nummus* 2 (7), 139-53
	1955:	Paço, A. Nota sobre sementes proto-históricas e outras encontradas em Portugal. III *CAN*, Zaragoza, 510-15

1956: Paço, A. Citânia de Sanfins. Intensité de sa romanisation. IV *CISPP*, 943-6
1957: Paço, A. Citânia de Sanfins. Alguns dos seus problemas. XXIII CLEPC (7ª Secção), 353
1959: Paço, A. Escavações e problemas do Castro de Vila Nova de S. Pedro e da Citânia de Sanfins. I *CNA* 1, 261-9
1960: Paço, A. Citânia de Sanfins (Paços de Ferreira). Sua importância turística e cultural. Revista Turismo 24 (6) (3ª Série), Lisboa, 19-20
1961a: Paço, A. Citânia de Sanfins. VI - Fundo pré-histórico. *Lucerna* 1 (3-4), 106-19
1963: Paço, A. O padre Eugénio Jalhay e a Citânia de Sanfins. *Lucerna* 3, 71-7
1965b: Paço, A. Citânia de Sanfins. VII - Escultura de cabeça de ofídeo. *Lucerna* 4, 167-71
1968a: Paço, A. A Citânia de Sanfins. *RE* 10, 329-54
1968b: Paço, A. Citânia de Sanfins (Paços de Ferreira, Portugal). IX - Una rectificación histórica. *AEA* 41 (117-118), 45-59
1968c: Paço, A. Citânia de Sanfins. VIII - Fragmentos de estátuas de guerreiros calaicos. *Brotéria* 76, 710-25
1974: Paço, A. La 'Citânia' de Sanfins e le lotte tra 'castrejos' e romani. Estudos Italianos em Portugal 37, 3-9

Paço - Jalhay
1955: Paço, A.; Jalhay, E. Tesouro monetário da Citânia de Sanfins. Anais da Academia Portuguesa da História 6 (2ª Série), 189-275

Paço - Quesado
1956: Paço, A.; Quesado, A.P. Digressões arqueológicas pelo Alto-Minho. *AAM* 6 (1-2), 80-90, 168-79
1957: Paço, A.; Quesado, A.P. Considerações acerca de algumas cerâmias dos castros de Perre e Nogueira (Viana do Castelo). XXIII *CLEPC* (7º sec.), 387-93

Paris
1903: Paris, P. Statues lusitaniennes de style primitif. *AP* 8, 1-8

Pauli
1980: Pauli, L. (ed.) Die Kelten in Mitteleuropa: Kultur, Kunst, Wirtschaft. Keltenmuseum Hallein, Österreich, Salzburg
1985: Pauli, L. Early Celtic society: two centuries of wealth and turmoil in central Europe. Champion, T.C.; Megaw, J.V.S. (eds.) *Settlement and Society: aspects of West European prehistory in the first millennium BC.* Leicester University Press, 23-43

Pautreau
1989: Pautreau, J.P. The Transition from Bronze Age to Iron Age in France: Economic, Cultural a*nd Spiritual Change. Sørensen, M.L.; Thomas, R. (eds.) The Bronze Age-Iron Age Transition in Europe: aspects of continuity and change in European societies c. 1200 to 500 BC.* BAR (IS) 483, Oxford, 204-62

Pautreau - Queiroga
1989: Pautreau, J.P.; Queiroga, F. Le Castro das Ermidas. Village fortifié du Portugal, *Archeologia* 253, Paris, 44-9

Peixoto
1899-1903: Peixoto, R. O Basto. *Portugália* 1, 832-3
1905-08a: Peixoto, R. 'Prisões de gado'. *Portugália* 2, 78-9
1905-08b: Peixoto, R. A Cividade de Riodouro. *Portugália* 2, 284
1905-08c: Peixoto, R. Beneméritos de archeologia - as explorações da Cividade de Terroso e do Castro de Laundos no concelho da Póvoa de Varzim. *Portugália* 2, 677-80

Peña Santos
1986: Peña Santos, A. *Yacimiento Galaico-Romano de Santa Trega. Campaña 1983.* Arqueoloxía/Memorias 5, Xunta de Galicia - DXPHD, A Coruña
1987: Peña Santos, A. El castro de Torroso (Mos, Pontevedra). *TAE* 27 (1-4), 113-36
1988: Peña Santos, A. Los objectos metálicos del castro de Torroso (Mos, Pontevedra). *TAE* 28, 113-32

Penas Truque
1986: Penas Truque, M.A. Los dioses de la montaña. Bermejo Barrera, J.C. (ed.) *Mitología y mitos en la Hispania prerromana.* Madrid, 117-40

Penedo Romero
1988: Penedo Romero, R. Datos paleontológicos sobre la ganadería de la cultura castreña de Galicia. *TAE* 28, 325-40

Pereira
1895: Pereira, F.A. Castelo de S. Miguel-oAnjo (Notas de um reconhecimento). *AP* 1, 161-75
1898b: Pereira, F.A. Insculturas em rocha em castros de Valdevez, ou vários penedos com pias. *AP* 4, 289-303
1902: Pereira, F.A. Um passeio arqueológico no concelho de Arcos de Valdevez. *AP* 7, 193-209
1904: Pereira, F.A. Um castro com muralhas. *AP* 9, 214-19
1907a: Pereira, F.A. Ara Celtibérica da época romana. Um novo Génio. *AP* 12, 36-52
1907b: Pereira, F.A. Geografia proto-histórica da Lusitânia. Situação conjectural de Talabriga. *AP* 12, 129-58
1908: Pereira, F.A. Novo material para o estudo da estatuária e arquitectura do Alto Minho. *AP* 13, 202-44
1909: Pereira, F.A. As lagaretas do Castro de S. Miguel-o-Anjo em Ázere. *AP* 14, 310-4
1910: Pereira, F.A. Habitações castrejas do N. Portugal - Uma hipótese de reconstituição. *Límia* 1 (1ª Série), 11-4

1911: Pereira, F.A. Sobrevivência pitorescas de uma arquitectura arcaica. *Límia* 1 (7-8) (Série 2ª), 101-5
1914: Pereira, F.A. Estudos do Alto Minho. Habitações castrejas do Norte de Portugal. *Límia* 14, 17
1915a: Pereira, F.A. Novas figuras de guerreiros lusitanos descobertas pelo Dr. L. de Figueiredo da Guerra. *AP* 20, 1-16
1915b: Pereira, F.A. Cinegética e Arqueologia. *AP* 20, 224-58
1929: Pereira, F.A. Jornadas de um curioso pelas margens do Lima. *AP* 28, 1-51
1930: Pereira, F.A. Padre João Luís Lourenço Loução. *Portucale* 3, 372-3
1933: Pereira, F.A. Os vestíbulos das habitações citanienses. *HMS*, 27-30
1938-41: Pereira, F.A. Páginas inéditas de Félix Alves Pereira . *RG* {48 (4), 1938, 277-82; 48 (4), 277-82}, {49 (1-2), 1930, 35-41; 49 (3-4), 103-11), (51 (1-2), 1941, 27-40; 51 (4) 312-30}

Pereira - González 1988: Pereira, A.S., González, C.M. Castro do Cruito. *Arqueologia* 17, Porto, 151-8

Pereira Menaut 1978: Pereira Menaut, G. Caeleo Cadraiolonis f. Cilenus Ɔ Berisamo *et al.*: centuria or castellum? A discussion. *HA* 8, 271-80
1982: Pereira Menaut, G. Los castella y las comunidades de *Gallaecia*. *Zephyrus* 34-35, 249-67, (=*SANP* 2, Madrid, 167-92)
1983: Pereira Menaut, G. Las comunidades galaico-romanas. Habitat y sociedad en transformacion. Pereira Menaut, G. (ed.) *Estudos de Cultura Castrexa e de Historia Antiga de Galicia*. Santiago de Compostela, 199-212
1985-86: Pereira Menaut, G. Nueva Tabula Patronatus del Noroeste de Hispania. *Veleia* 2-3 (IV CLCPPI), 299-302

Pereira - Santos 1980: Pereira Menaut, G.; Santos Yanguas, J. Sobre la romanización del Noroeste de la Península Ibérica: las inscripciones con mención del origo personal. *SANP* 3, 117-30

Pérez Outeiriño 1986: Pérez Outeiriño, B. Torques. *Gran Enciclopedia gallega* vol 29, 107-10
1987: Pérez Outeiriño, B. A Cidade de San Cibran de Las. Objectivos e resultados das ultimas intervencions arqueologicas (1982-1983). *Lucerna* 2 (2ª Serie), 5-39
1989: Pérez Outeiriño, B. Orfebrería Castreña. García Castro, J.A. (Ed.) *El oro en la España Prerromana*. Revista de Arqueologia, Madrid, 90-107

Pericot - Lopez 1931: Pericot Garcia, L.; Lopez Cuevillas, F. Excavaciones en la Citania de Troña. *MJSEA* 115

Piggott 1983: Piggott, S. *The Earliest Wheeled Transport*. Thames & Hudson, London

Pijoan 1953: Pijoan, J. *Summa Artis VI. El arte prehistorico Europeu*. Madrid

Pilcher - Hughes 1982: Pilcher, J.R.; Hughes, M. The potencial of dendrochronology for the study of climate change. Harding, A.F. (ed.) *Climatic Change in Later Pre-History*. Edinburgh, 75-84

Pimentel 1902: Pimentel, A. *Santo Thyrso de Riba d'Ave*. Santo Tirso

Pina 1928: Pina, L. A Penha eneolítica. *RG* 38 (3-4), 135-8
1936: Pina, L. Novos elementos para o estudo da Penha eneolítica (Guimarães). *TAAP* 2, 81-4
1940: Pina, L. Tentame topográfico de arqueologia pré-romana e luso-romana do concelho de Guimarães. *CMP*, 503-8
1965: Pina, L. A arqueologia portuense na 'Anacrisis' de Pereira de Novais. *Lucerna* 4, 65-74

Pinho 1905-08: Pinho, J. Castros do concelho de Amarante. *Portugália* 2, 476, 673-5
1928: Pinho, J. A ara de Marecos. Penha-fidelis. *Penha-fidelis* 1, 95-7,116-20,182-7
1929a: Pinho, J. A cividade de Eja. *Penha-fidelis* 1 (11), 205-12
1929b: Pinho, J. O tesouro de Gondeiro. *Penha-fidelis* 1 (10), 182-7

Pinto 1927: Pinto, R.S. Introdução à arqueologia portuense. *O Tripeiro* 26, 27, 24-5, 46
1928-29: Pinto, R.S. Museu de Martins Sarmento. *RG* (38 (3-4), 1928, 192-6), (39, 1929, (1-2), 27-43; (3-4), 169-81)
1929a: Pinto, R.S. Machados de bronze do Museu Municipal do Porto. *Portucale* 2, 421
1929b: Pinto, R.S. Petróglifos de Sabroso e a arte rupestre em Portugal . *Nós* 62, 19-26
1930a: Pinto, R.S. Machados de bronze das margens do Ave. *TAE* 4 (3), 306
1930b: Pinto, R.S. Museu Municipal do Porto - achegas para um catálgo. *O Tripeiro* 4ª série 1 (171), 15
1932: Pinto, R.S. A Cividade de Terroso e os Castros do Norte de Portugal. *RG* 42 (1-2), 81-91
1933a: Pinto, R.S. Activité minière et métallurgique pendant l'âge du Bronze au Portugal. *AFCP* 18, 6-8
1933b: Pinto, R.S. Explotaciones mineras de la edad del bronce en Portugal . *IP* 6, 177-81
1933c: Pinto, R.S. O Castro de Sendim, Felgueiras (Nota Preliminar). *HMS*, 376-80

Pintor 1975: Pintor, M.A.B. Melgaço Medieval. Braga

Pleiner 1980: Pleiner, R. Early Iron Metallurgy in Europe. Wertime, T.A.; Muhly, J.D. (eds.) *The coming of the age of iron.* Yale University, New Haven, 375-415

Pliny 1958: Pliny (The Elder) *Natural History.* Rackham, H. (ed.) The Loeb Classical Library, Cambridge (Mass.), London (4° ed.)

Ponte 1980: Ponte, S. A génese das fíbulas do Noroeste Peninsular. *SANP* 2, 111-9

1981: Ponte, S. Quatro fíbulas da Citânia de Santa Luzia. *Minia* 4 (5) (2ª Série) , 49-52

1984: Ponte, S. Fíbulas de sítios a Norte do rio Douro. *Lucerna* (s/n), 111-44

Prieto Arciniega 1978: Prieto Arciniega, A. La devotio Ibérica como forma de dependencia en la Hiapania prerromana. *MHA* 2, 131-5

Proença Júnior 1908: Proença Júnior, F.T. *Ensaio de classificação dos castros potugueses.* Leiria

Queiroga 1983: Queiroga, F.M.V.R. Património Histórico de Vila Nova de Famalicão. *BC-CMVNF* 4, 47-51

1984: Queiroga, F.M.V.R. Escavações arqueológicas em Castelo de Matos. Notícia preliminar . *Arqueologia* 9, 105-16

1985a: Queiroga, F.M.V.R. Escavações arqueológicas no castro das Ermidas. A campanha de 1983. *BC-CMVNF* (sep.), 1-42

1985b: Queiroga, F.M.V.R. Vila Nova de Famalicão. Castro das Ermidas - 1983. *IA* 5, 56-7

1985c: Queiroga, F.M.V.R. Vila Nova de Famalicão. Castro de Vermoim - 1982. *IA* 5, 56

1987: Queiroga, F.M.V.R. Corte estratigráfico no castro de Penices. *BC-CMVNF* 7, 3-22

1992: Queiroga, F.M.V.R. (forthcoming) The Late Iron Age 'Monumentos com Forno' in Northwest Iberia. *Proceedings of the Symposium Europe in the Ist Millennium B.C.*, Oxford, 3-4 April 1986 (forthcoming).

Queiroga *et al.* 1992: Queiroga, F.M.V.R.; Dinis, A.P.; Oliveira, F. (forthcoming) A estrutura de banhos castreja do castro das Eiras. *RG*

Queiroga - Figueiral 1992: Queiroga, F.M.V.R.; Figueiral, I.M.A.C.R. (forthcoming) Datações de Carbono 14 para Castelo de Matos. *BC-CMVNF* 9

Queiroga - Silva 1991: Queiroga, F.M.V.R.; Silva, L.J.M.P. As corrosões metálicas. Seu potencial no estudo dos metais da cultura castreja. Waldren, W.H.; Ensenyat, J.A.; Kennard, R.C. (eds.) *IInd Deya International Conference of Prehistory,* I. BAR (IS) 573, Oxford, 165-77

Quesada Sanz 1989: Quesada Sanz, F. *Armamento, Guerra y Sociedad en la Necrópolis Ibérica de El Cabecico del Tesoro (Murcia, España).* (2 vols.), BAR (IS) 502, Oxford

1990: Quesada Sanz, F. La falcata Ibérica: ¿un arma de origen Ilírio y procedencia Itálica? *AEA* 63, 65-93

Quesado 1955: Quesado, A.P. Quatro castros descobertos no concelho de Viana do Castelo. *Notícias de Viana* (6 Nov.), 4

Raddatz 1969: Raddatz, K. *Die Schatzfunde der Iberischen Halbinsel.* Madrider Forschungen 5, Berlin, 2 vol.

Ralph *et al.* 1973: Ralph, E.K.; Michael, H.N.; Han, M.C. Radiocarbon dates and reality. *MASCA Newsletter* 9, 1-20

Ramil González 1990: Ramil González, E. Castro de Prado-Espasante. Excavación de su Pedra Formosa. *Revista de Arqueologia* 108, Madrid, 38-9

Ramos 1943: Ramos, A.F. Lavra. *Apontamentos para a sua monografia.* Porto

Real 1984: Real, M.L. Arqueologia portuense em 1984. Notícias sobre as operações no Morro da Sé, em áreas de intervenção do CRUARB. *Arqueologia* 10, 75-9

Rees 1979: Rees, S.E. *Agricultural Implements in Prehistoric and Roman Britain.* BAR (BS) 69, Oxford

Regalo 1986: Regalo, H.J.L. Levantamento Arqueológico do Concelho de Vila Verde. *Minia* 8, 70-109

Reis 1978: Reis, A.A.M. *A romanização no concelho de Ponte de Lima.* Ponte de Lima

Reis, JAL 1904: Reis, J.A.L. *A villa de Valongo.* Valongo

Renfrew 1979: Renfrew, C. Before civilization: *The radiocarbon revolution and prehistoric Europe.* Harmondsworth

Renfrew, J 1973: Renfrew, J. *Palaeoethnobotany. The prehistoric food plants of the Near East and Europe.* London

Ribeiro, F 1930-34: Ribeiro, F. Novas descobertas arqueológicas na Citânia de Briteiros. *RG* (40 (3-4), 1930, 171-5), (44 (3-4), 1934, 205-8)

Ribeiro, J 1949: Ribeiro, J. Miniatura monográfica do concelho de Chaves. *Almanaque de Chaves* 2° ano, Chaves, 31-90

Ribeiro - Moreira 1986: Ribeiro, M.L.; Moreira, A. *Carta geológica de Portugal (1/50.000). Notícia explicativa da folha 1-B, Monção.* Lisboa

Ribeiro 1945: Ribeiro, O. *Portugal. O Mediterrâneo e o Atlântico.* Coimbra

1952-58: Ribeiro, O. Portugal. Terán, M. (ed.) *Geografía de España y Portugal* 5, Barcelona

	1961:	Ribeiro, O. *Geografía e civilização* - Temas portugueses. Lisboa
Rives	1975:	Rives, M. Les origines de la vigne. *La Recherche* 53 (6), 120-9
Rocha	1895:	Rocha, A.S. Elementos para o estudo comparativo de alguns objectos recolhidos no castro de S. Miguel. *AP* 1, 262-4
Rodrigues	1958:	Rodrigues, A.V. Cabeça de guerreiro lusitano da Guarda (Portugal). *Zephyrus* 9, 227-8
	1962:	Rodrigues, A.V. O problema lusitano. *SG* 9 (1), 319-29
Rodriguez Adrados	1946:	Rodriguez Adrados, A. La 'fides' Ibérica. *Emerita* 14, 128-209
	1949:	Rodriguez Adrados, F. *El sistema gentilicio decimal de los indoeuropeos occidentales y las origines de Roma.* Madrid
Rodriguez Colmenero	1972:	Rodriguez Colmenero, A. Aspectos geográfico-histórico en torno al convento jurídico bracaraugustano. *HA* 2, 135-63
	1977:	Rodriguez Colmenero, A. *Galicia meridional romana.* Bilbao
	1979:	Rodriguez Colmenero, A. *Augusto e Hispania. Conquista y organización del Norte Peninsular.* Cuadernos de Arqueologia de Deusto, Seminario de Arqueología 7, Bilbao
	1988:	Rodriguez Colmenero, A. Nuevos 'termini' territoriales entre unidades gentilicias galaico-romanas. Pereira Menaut, G. (ed.) 1º *Congreso Peninsular de Historia Antigua* 2, Santiago de Compostela, 271-89
Rodriguez - Fontes	1980:	Rodriguez Colmenero, A.; Fontes, A.L. El culto a los montes entre los galaico-romanos. *SANP* 3, 21-35
Roldan Hervas	1972:	Roldan Hervas, J.M. El elemento indígena en las guerras civiles en Hispania: aspectos sociales. *HA* 2, 77-124
	1974:	Roldan Hervas, J.M. *Hispania y el exercito romano.* Salamanca
	1983:	Roldan Hervas, J.M. El ejercito como factor de la romanizacion de Asturia. *Indigenismo y romanización en el conventus Asturum*, 109-22
	1984:	Roldan Hervas, J.M. Ejercito y poblamiento en el norte de la peninsula iberica bajo dominio romano. *MHA* 6, 67-84
	1986:	Roldan Hervas, J.M. La incidencia del ejercito romano en el poblamiento del N.O. de hispania. Castillo, A. (ed.) *Ejército y Sociedad. Cinco estudios sobre el mundo antiguo.* Universidad de León, León, 13-49
Romaní Martinez	1974-75:	Romaní Martinez, M. Las anforas romanas de Isorna (Rianxo, La Coruña). *CEG* 39 (87-89), 318-27
Romero Masiá	1976:	Romero Masiá, A. *El habitat castreño.* Santiago de Compostela
	1987:	Romero Masiá, A. Dous instrumentos de traballo no castro de Borneiro. *Gallaecia* 9/10, Santiago de Compostela, 229-33
Rosário	1973:	Rosário, A. *Falam documentos.* 2-3 (31-66), Braga
Ruiz-Gálvez Priego	1979a:	Ruiz-Gálvez Priego, M. El Bronce antiguo en la fachada atántica peninsular: un ensayo de perodización. *TP* 36, 151-72
	1979b:	Ruiz-Gálvez Priego, M. El depósito de Hio (Pontevedra) y el final de la edad del Bronce en la fachada atlántica peninsular. *MP* 33, 3-22
	1980:	Ruiz-Gálvez Priego, M. Consideraciones sobre el origen de los puñales de antenas gallego-asturianos. *SANP* 1, 85-111
	1984:	Ruiz-Gálvez Priego, M. *La Peninsula Iberica y sus relaciones con el circulo cultural atlántico.* (Doctoral thesis, typescript), Universidade Complutense, Madrid
	1986:	Ruiz-Gálvez Priego, M. Navegacion y comercio entre el Atlántico y el Mediterráneo a fines de la Edad del Bronce. *TP* 43, 9-42
	1989:	Ruiz-Gálvez Priego, M. La orfebrería del Bronce Final. El poder y su ostentación. García Castro, J.A. (Ed.) *El oro en la España Prerromana.* Revista de Arqueologia, Madrid, 46-57
Ryder -Gabra-Sanders	1987:	Ryder, M.L.; Gabra-Sanders, T. A microscopic study of remains of textiles made from plant fibres. *OJA* 6, Oxford, 91-108
Rye	1981:	Rye, O.S. *Pottery Technology.* Manuals on Archaeology 4, Washington
Salinas de Frias	1983:	Salinas de Frias, M. La funcion del hospitium y la clientela en la conquista y romanizacion de celtiberia. *Stvdia Historica* 1 (1), Salamanca, 21-41
Sampaio	1893-97:	Sampaio, A. *Estudos históricos e económicos, 1, As 'villas' do Norte de Portugal. RG* (10 (3-4), 1893, 161-76, 209-21), (11 (3), 1894, 139-47), (12 (1-2-4), 1895, 5-10, 65-90, 155-73), (14 (4), 1897, 161-86)
San Valero-Aparisi	1950:	San Valero-Aparisi, J. Restos del castro de Faria. *BGAF* 2 (2), 11-15
Sanahuja	1971:	Sanahuja Yll, M.E. Instrumental de hierro agrícola e industrial de la época ibero-romana en Cataluña. *Pyrenae* 7, 61-110
Sanches	1988:	Sanches, M.J. O povoado da Lavra (Marco de Canaveses). *Arqueologia* 17, Porto, 125-34
Sánchez Palencia	1983a:	Sánchez-Palencia Ramos, F.J. Explotaciones auriferas en el Conventus Asturum. *Indigenismo y romanización en el Conventus Asturum.* Oviedo, 67-87

	1983b:	Sánchez-Palencia Ramos, F.J. La explotacion prerromana del Oro del Noroeste de la Peninsula Iberica. *BAur* 13, 31-67
Sánchez - Fernández	1985:	Sánchez-Palencia Ramos, F.J.; Fernández Posse, M.D. La Corona y el castro de Corporales I. Truchas (León). Campañas de 1978 a 1981. *EAE* 141, Madrid
Sánchez - Pérez	1983:	Sánchez-Palencia Ramos, F.J.; Pérez García, L.C. Las explotaciones auriferas y la ocupacion romana del noroeste de la Peninsula Ibérica. II *SANP*, Madrid, 225-46
	1989:	Sánchez-Palencia Ramos, F.J.; Pérez García, L.C. Los yacimientos auríferos de la Península Ibérica. García Castro, J.A. (ed.) *El oro en la España Prerromana.* Revista de Arqueologia, Madrid, 16-23
Sandars	1913:	Sandars, H. The weapons of the iberians. *Archeologia* 64, Oxford, 205-94
Santarém	1951:	Santarém, C.M.F. O castro do Monte do Padrão. *BC-CST* 1 (1), 49-66
	1952:	Santarém, C.M.F. Agumas peças inéditas do Museu Abade Pedrosa. *BC-CST* 2 (1), 105-11
	1955a:	Santarém, C.M.F. O castro do Monte do Padrão. Campanhas de 1952-53-54. *BC-CST* 3 (4), 397-429
	1977:	Santarém, C.M.F. Uma inscrição romana de Alvarelhos. Santo Tirso. *BC-CST* 1 (1) (2ª Série), 161-70
Santiso et al.	1977:	Santiso, A.G.; Sobriño, G.; Tamuxe, X. Casco celta de Caldelas de Tuy. *Museu y Archivo Histórico Diocesano* 2, Tuy, 39-48
Santos Júnior	1940:	Santos Júnior, J.R. Sobrevivência folclórica dos berrões da Vilariça. *CMP* 2, 368-71
	1957:	Santos Júnior, J.R. O Castro de Carvalhelhos. *TAE* 16 (1-2), 25-62
	1959-60:	Santos Júnior, J.R. As rampas de acesso às muralhas do Castro de Carvalhelhos. *BCMO* 20, 361-68
	1963a:	Santos Júnior, J.R. Escavações no Castro de Carvalhelhos - Campanha de 1963. *TAE* 19 (2), 187-93
	1963b:	Santos Júnior, J.R. Escavações no Castro de S. Vicente da Chã (Barroso). *TAE* 19 (1), 71-80
	1964:	Santos Júnior, J.R. Escavações no Castro de Carvalhelhos - Campanha de 1964. *TAE* 19 (3-4), 360-5
	1966a:	Santos Júnior, J.R. Dois fornos do Povo em Trás-os-Montes. *TAE* 20 (1-2), 119-46
	1966b:	Santos Júnior, J.R. Duas campanhas de escavações no Castro de Carvalhelhos (1965-1966). *TAE* 20 (1-2), 181-90
	1971:	Santos Júnior, J.R. Escavações no Castro de Carvalhelhos - Campanha de 1970. *TAE* 22 (1), 72-5
	1973a:	Santos Júnior, J.R. As notáveis condições de defesa do Castro de Carvalhelhos, *TAE* 22 (3), 207-219
	1973b:	Santos Júnior, J.R. Alguns problemas castrejos: cobertura das casas. *TAE* 22 (3), 231-5
	1975a:	Santos Júnior, J.R. A cultura dos berrões no Norte de Portugal. *TAE* 22 (4), 353-516
	1975b:	Santos Júnior, J.R. Castro de Carvalhelhos - Campanha de escavações em Agosto de 1975. *TAE* 22 (4), 559-66
	1975c:	Santos Júnior, J.R. A cultura dos berrões proto-históricos do Nordeste de Portugal (Conferência). *RG* 85, 127-32
	1977a:	Santos Júnior, J.R. Castro de Carvalhelhos - Campanha de 1976. *TAE* 23 (1), 161-5
	1977b:	Santos Júnior, J.R. Novos elementos da remota zoolatria em Trás-os Montes. *TAE* 23 (1), 5-18
	1978a:	Santos Júnior, J.R. A estação arqueológica de Vilarelhos e a cabeça de guerreiro lusitano. *TAE* 23 (2-3), 345-51
	1978b:	Santos Júnior, J.R. Cabeça-troféu do Museu Municipal de Chaves. *TAE* 23 (2-3), 352-56
	1978c:	Santos Júnior, J.R. 27ª campanha de escavações no Castro de Carvalhelhos (1977). *TAE* 23 (4), 323-33
	1980a:	Santos Júnior, J.R. As fíbulas de prata do Museu de Chaves. *TAE* 23 (4), 599-607
	1980b:	Santos Júnior, J.R. 28ª campanha de escavações no Castro de Carvalhelhos (1979). *TAE* 23 (4), 607-19
	1980c:	Santos Júnior, J.R. O Castro de Fontes (Santa Marta de Penaguião). *TAE* 23 (4), 621-27
	1981a:	Santos Júnior, J.R. Um perdido berrão transmontano, assinalado por um grande arqueólogo. *TAE* 24 (1), 156-9
	1981b:	Santos Júnior, J.R. 29ª campanha de escavações no Castro de Carvalhelhos (1980), *TAE* 24 (1), 140-7

1982: Santos Júnior, J.R. 30ª Campanha de escavações no Castro de Carvalhelhos (1981). *TAE* 24 (2), 249-63

1983a: Santos Júnior, J.R. 31ª campanha de escavações no Castro de Carvalhelhos (1982). *TAE* 24 (3), 511-9

1983b: Santos Júnior, J.R. The 'berrões' proto-historic stone statues of the Boars, worshipped as idols, object of zoolatry. *TAE* 24 (3), 533-7

1984a: Santos Júnior, J.R. 32ª campanha de escavações no Castro de Carvalhelhos (Julho-Agosto 1983). *TAE* 24 (4) 673-82

1984b: Santos Júnior, J.R. O Castro da Curalha - Chaves vivente na época suévica?, *TAE* 24 (4), 683-4

1984c: Santos Júnior, J.R. Trinta anos de escavações no Castro de Carvalhelhos (Boticas - Vila Real). *RG* 94, 411-24

1985: Santos Júnior, J.R. A cultura dos berrões proto-históricos fundamente radicada em Trás-os-Montes. *TAE* 15 (1), 31-40

Santos - Freire 1964: Santos Júnior, J.R.; Freire, O. O castro de S. Vicente da Chã (Barroso). Campanha de escavações. 1964. *TAE* 19 (3-4), 336-71

1965: Santos Júnior, J.R.; Freire, O. O Torques de ouro de Vilas Voas (Vila Flor). *RG* 75 (1-4), 137-52

1966: Santos Júnior, J.R.; Freire, O. O Torques de Vilas Voas (Vila Flor, Trás-os-Montes). *Lucerna* 5 (1-4), 443-58

Santos - Isidoro 1963: Santos Júnior, J.R.; Isidoro, A.F. Escavações no Castro de S. Vicente da Chã (Barroso). *TAE* 19 (2), 178-86

Santos Yanguas 1978: Santos Yanguas, J. Contribuicion al estudio de los restos de formas de dependencia en el area celtica peninsular en epoca romana. *MHA* 2, 137-45

1985: Santos Yanguas, J. *Comunidades indígenas y administración romana en el Noroeste hispánico*. Universidad del País Vasco, Lejona/Leioa

Santos Yanguas, N 1981: Santos Yanguas, N. La administracion romana del N.O. de la Peninsula Iberica hasta finales del siglo I d.C. *Brigantium* 2, A Coruña, 49-71

1984: Santos Yanguas, N. La economia agropecuaria en la Asturias castreña prerromana. *BIEA* 112, Oviedo, 485-533

Santos, JN 1955: Santos, J.N. *Notas arqueológicas, históricas e etnográficas. I - Castrum Quiffionis.* Matosinhos

1962a: Santos, J.N. Altar com covinhas no Castro de Guifões. *SG* 9, 111-16

1962b: Santos, J.N. Marcas dedadas nas telhas romanas do Castro de Guifões. *SG* 9 (1), 230-82

Santos, MC 1969: Santos, M.C. Subsídios para o estudo arqueológico de Montalegre, Mealhada e Viseu. *Ethnos* 6, 201-18

Sarmento 1880: Sarmento, F.M. Observações acerca do Vale de Âncora. *O Pantheon* 1, 63-67

1881: Sarmento, F.M. A estátua do Pátio da Morte. *O Pantheon* 1, 72-73

1884-1909: Sarmento, F.M. Materiaes para a arqueologia do concelho de Guimarães. *RG* {1 (4), 1884, 161-89}, {2 (4), 1885, 189-202}, {5 (3), 1888, 109-21}, {13 (1, 4), 1896, 5-18, 149-68}, {15 (3), 1898, 152-67}, {16 (1), 1889, 5-22}, {18 (1-4), 1901, 8-29, 117-35}, {19 (1, 3), 1902, 19-33, 109-19}, {22 (1-4), 1905, 5-32, 97-123}, {23 (2), 1906, 41-51}, {24 (2-4), 1907, 53-66, 113-22}, {26 (1-2, 4), 1909, 5-19, 129-39}

1894-95: Sarmento, F.M. Materiaes para a archeologia da comarca de Barcelos. *RSNS* 3, 186-98

1895: Sarmento, F.M. 'Cidade Velha' de Monte-Córdova *AP* 1 (6), 145-151

1896: Sarmento, F.M. A propósito de Valábriga. *Nova Alvorada* 5 (11), 428-30

1903-05: Sarmento, F.M. Materiaes para a arqueologia do concelho de Guimarães, Citânia. *RG* {20 (1-4), 1903, 5-16, 57-71, 112-25}, {21 (1-4), 1904, 5-9, 49-63, 97-120}, {22 (1-4), 1905, 5-32, 97-123}

1906-09: Sarmento, F.M. Materiaes para a arqueologia do concelho de Guimarães, Sabroso. *RG* {23 (1), 1906, 45-51}, {24 (2-4), 1907, 53-66, 113-22}, {26 (1-2, 4), 1909, 5-19, 129-39}

1922-27: Sarmento, F.M. Cartas ao Professor Pereira Caldas. *RG* {32 (1-4), 1922, 5-9, 109-13, 241-6, 365-9}, {33 (4), 1923, 199, 202}, {34 (1), 1924, 5-9}, {35 (2-3), 1925, 77-9, 137}, {37 (1-3), 1927, 5-10, 97-100}

1930-38: Sarmento, F.M. Cartas ao padre Martins Capela. *RG* {40 (3-4), 1930, 81-7}, {44 (3-4), 1934, 161-2}, {48 (1-3), 1938, 6}, 81-87

1933: Sarmento, F.M. Dispersos. Coimbra

1939-41: Sarmento, F.M. Correspondência entre Martins Sarmento e o Marquês de Sousa Holstein. *RG* (49 (1-4), 1939, 12-4, 95-101}, {51 (1-2), 1941, 15-7}

1940: Sarmento, F.M. A correspondência do padre Joaquim Pedrosa a Martins Sarmento. *RG* 50 (3-4), 206-7

1970: Sarmento, F.M. Os inéditos de Martins Sarmento. *RG* 80 (1-2), 5-72

Saunier 1910-11: Saunier, L. Investigacion historica acerca de las antiguas explotaciones de oro en España. *BCMO* 4-5 (55-70)

Savory 1949: Savory, H.N. The atlantic Bronze age in South Western Europe. *PPS* 12, 128-55 (=1951: A idade do Bronze atlântico no Sudoeste da Europa. *RG* 61 (3-4), 323-77)

1966: SAVORY, H.N. A idade do Ferro - B do Noroeste da Península Ibérica. Novas luzes acerca de um antigo problema. *RG* 76 (1-2), 117-46

1968: SAVORY, H.N. *Spain and Portugal.* Thames & Hudson, London

Schubart 1961: Schubart, H. Atlantische Nietenkessel von der Pyrenaehalbinsel. *MM* 2, 35-54

Schüle 1969: Schüle, W. *Die Meseta-Kulturen der Iberischen Halbinsel.* 2 vols. Madrider Forschungen 3, Berlin

Schulten 1920: Schulten, A. *Hispania.* Barcelona

1922: Schulten, A. Avieno, Ora Maritima. *Fontes Hispaniae Antiquae* 1, Barcelona

1925: Schulten, A. 500 a.C. hasta César. *Fontes Hispaniae Antiquae* 2, Barcelona, 162-85

1927: Schulten, A. *Viriato.* Porto

1935a: Schulten, A. Las guerras de 237-154 a.C. *Fontes Hispaniae Antiquae* 3, Barcelona

1937: Schulten, A. Las guerras de 154-72 a.C. *Fontes Hispaniae Antiquae* 4, Barcelona

1940: Schulten, A. Las guerras de 72-19 a.C. *Fontes Hispaniae Antiquae* 5, Barcelona

1949: Schulten, A. *Sertorio.* Barcelona

1952: Schulten, A. Estrabón, Geografia de Iberia. *Fontes Hispaniae Antiquae* 6, Barcelona

1959-63: Schulten, A. *Geografía y etnografía antiguas de la Península Iberica .* Madrid, 2 vol.

1962: Schulten, A. *Los Cántabros y Astures y su guerra con Roma.* Madrid (2ª ed.)

Serra Ráfols 1924: Serra Ráfols, J. Els començos de la mineria i la metallúrgia del coure à la península ibérica. *BACAEP* 2, 146-86

1948: Serra Ráfols, J. Carrito Iberico de Bronce, del Museo de granollers. *AEA* 21

Serrano 1957: Serrano, A. Observaciones sobre la distribución geográfica de la escultura zoomorfa prerromana. *Zephyrus* 8, 103-10

Severo 1899-1903: Severo, R. Exvoto de bronze da Colecção Manuel Negrão. *Portugália* 1, 325-31

1905-08e: Severo, R. As arrecadas d'ouro do Castro de Laundos. *Portugália* 2, 403-12

Severo - Cardozo 1889-1903: Severo, R.; Cardoso, F. O ossuário da freguesia de Ferreiró. *Portugália* 1, Porto, 177-200

Shackley 1981: Shackley, M.L. *Environmental Archaeology.* Allen & Unwin. London

1985: Shackley, M.L. Using Environmental Archaeology. Batsford

Sheridan - Bailey 1981: Sheridan, A.; Bailey, G. (eds.). *Economical Archaeology. Towards an Integration of Ecological and Social Approaches.* BAR (IS) 96, Oxford

Sherratt 1981: Sherratt, A. Plough and pastoralism: aspects of the secondary products revolution. Hodder, I; Isaac, G.; Hammond, N. (eds.) *Patterns of the Past: Studies in Honour of David Clarke.* Cambridge University Press, Cambridge, 261-305

1987: Sherratt, A. Cups that cheered. Waldren, W.H.; Kennard, R.C. (eds.). *Bell Beakers of the Western Mediterranean.* BAR (IS) 331, Oxford 81-114

1991: Sherratt, A. Sacred and Profane Substances: the Ritual Use of Narcotics in Later Neolithic Europe. Garwood, P.; Jennings, D.; Skeates, R.; Toms, J. (eds.) *Sacred and Profane.* OUCA, Monograph Nº 32, Oxford, 50-64

1992: Sherratt, A. What can archaeologists learn from Annalistes?. Knapp, A.B. (ed.) *Archaeology, Annales, and ethnohistory.* Cambridge University Press, Cambridge, 135-42

Silius Italicus 1987: *Silivs Italicvs Pvnica.* Iosephus Delz (ed.) Stutgard

Silva 1980a: Silva, A.C.F. Campanha de trabalhos arqueológicos na Cividade de Terroso (Póvoa de Varzim). *BC-PV* 20 (2), 305-15

1980b: Silva, A.C.F. Organizações gentilícias entre Leça e Ave. *Portugália* 1 (N/S), 79-90

1981-82: Silva, A.C.F. Novos dados sobre a organização social castreja. *Portugália* 2-3 (N/S), 83-94

1983-84: Silva, A.C.F. A cultura castreja no Noroeste de Portugal: habitat e cronologias. *CIUAN*, 121-29

1983a: Silva, A.C.F. A idade dos metais em Portugal. *História de Portugal* 1, Alfa, 100-47

1983b: Silva, A.C.F. As tesserae hospitales do Castro da Senhora da Saúde ou Monte Murado (Pedroso, V. N. Gaia). Contributo para o estudo das instituições e povoamento da Hispânia antiga. *Gaya* 1, 9-26

1984: Silva, A.C.F. Aspectos da proto-história e romanização no concelho de Vila Nova de Gaia e problemática do seu povoamento. *Gaya* 2, 39-58

1986a: Silva, A.C.F. *A Cultura Castreja no Noroeste de Portugal.* Paços de Ferreira

	1986b:	Silva, A.C.F. *Paços de Ferreira: As origens do povoamento: do megalitismo à romanização.* Estudos Monográficos. Paços de Ferreira
	1990:	Silva, A.C.F. A Idade do Ferro em Portugal. Serrão, J.; Marques, A.H.O.; Alarcão, J. (Eds.). *Nova História de Portugal* 1. Presença, Lisboa, 259-341
Silva - Centeno	1977:	Silva, A.C.F.; Centeno, R.M.S. Sondagem arqueológica na Citânia de Briteiros (Guimarães). Notícia sumária. *RG* 87, 277-80
	1980:	Silva, A.C.F.; Centeno, R.M.S. Escavações arqueológicas na Citânia de Sanfins (Paços de Ferreira) 1977-78. *Portugalia* 1 (N/S), 57-78
Silva *et al.*	1986c:	Silva, A.C.F. *et al.* Arqueologia do Vale do Douro. Exposição e bibliografia. I *Congresso Internacional sobre o rio Douro.* Vila Nova de Gaia, 97-120
	1984:	Silva, A.C.F.; Silva, C. T.; Lopes, A.B. Depósito de fundidor do final da idade do Bronze do Castro da Senhora da Guia (Baiões, S. Pedro do Sul, Viseu). *Lucerna* (s/n), 73-95
Silva, ARP	1976:	Silva, A.R.P. Carbonized grains and plant imprints in ceramics from the castrum of Baiões (Beira Alta, Portugal). *Folia Quartenaria* 47, Krakow
	1988:	Silva, A.R.P. A etnobotânica na arqueologia portuguesa. Resultados desde 1931 a 1987. Queiroga, F.; Sousa, I.M.A.R.; Oliveira, C.M. (eds.). *Paleoecologia e Arqueologia*, Vila Nova de Famalicão, 5-36
Silva, C.T	1979:	Silva, C.T. O Castro de Baiões (S.Pedro do Sul). *Beira Alta* 38 (3), 509-531
	1980:	Silva, C.T. Contribuição para o estudo da cultura castreja na Beira Alta. *SANP* 2, 171-81
Silva, DM	1985:	Silva, D.M. *Entre Homem e Cávado: 1 - 2. Monografia do concelho de Amares; 3. Monografia do concelho de Terras do Bouro.* Amares
	1984:	Silva, E. J. L.; Marques, José Augusto T. M. Escavação de uma Cista em Lordelo (Anha - Viana do Castelo). *Revista de História* - U.L.P. 1, 51-72
Silva, EJL	1982:	Silva, E.J.L. *Escavações arqueológicas no castro do Monte do Castelo.* Monte do Castelo, 95
Silva - Marques	1984:	Silva, E.J.L.; Marques, J.A.T.M. Escavações no Monte do Castelo de Moldes - Castelo de Neiva, Viana do Castelo (notícia preliminar). *CV* 8, 33-43
Silva, J	1963:	Silva, J. Os recintos fortificados de Baltar e Montemuro. *Lucerna* 3, 126-35
Simpson *et al.*	1979:	Simpson, G.; Hawkes, C.F.C.; Hull, M.R. Some British and Iberian Pennanular Brooches and Other Early Types in the Rhineland and the 'Decumates Agri'. *AJ* 59 (2), 319-42
Snodgrass	1964:	Snodgrass, A.M. *Early Greek Armour and Weapons.* Edimburgh University Press
	1967:	Snodgrass, A.M. *Arms and Armour of the Greeks.* Thames and Hudson. London
	1980:	Snodgrass, A.M. Iron and Early Metallurgy in the Mediterranean. Wertime, T.A.; Muhly, J.D. (eds.). *The coming of the age of iron.* Yale University. New Haven, 335-74
Soares - Cabral	1984:	Soares, A.M.; Cabral, J.M.P. Datas convencionais de radiocarbono para estações arqueológicas e a sua calibração: revisão crítica. *AP* 2 (4ª Série), 167-213
Soares - Dias	1985:	Soares, R.M.; Dias, L.A.T. *Área arqueológica do Freixo, Marco de Canaveses. Contributo documental 1758 a 1984.*
Soeiro	1980:	Soeiro, T. Objectos em bronze do Castro de Alvarelhos. *Gallaecia* 6, 237-43
	1981-82:	Soeiro, T. Monte Mozinho: cerâmica cinzenta fina. *Portugália* 2-3 (N/S), 97-120
	1981:	Soeiro, T. Castro de Peso em Santa Leocádia de Geraz do Lima. *Arqueologia* 3, 99-102
	1984:	Soeiro, T. Monte Mozinho. *Apontamentos sobre a ocupação entre Sousa e Tâmega.* Penafiel - Boletim Municipal de Cultura 3ª Série 1, Penafiel
	1985-86a:	Soeiro, T. A Suvidade de S. Mamede de Recezinhos. Campanha de escavações 1985. *Penafiel - Boletim Municipal de Cultura* 3ª Série 2/3, Penafiel, 15-32
	1985-86b:	Soeiro, T. Chaves. Muro da Pastoria, Chaves - Campanha de escavações de 1982-83. *Portugália* (NS) 6-7, 21-8
Soeiro *et al.*	1981:	Soeiro, T.; Centeno, R.M.S.; Silva, A.C.F. Sondagem arqueológica no Castro de Sabroso (Guimarães) 1981. *RG* 91, 341-50
Sousa	1957:	Sousa, A. *Estudos de arqueologia, etnologia e história. Antiguidades do município de Gaia: civilizações pré-romanas, romana e romano-portuguesa.* Rio de Janeiro
Sousa, EF	1954:	Sousa, E.F. Relatório das escavações levadas a efeito no Monte Mózinho. *DL* 6 (5-6), 136-49
	1962:	Sousa, E.F. A citânia (Cidade Nova) situada no sul do concelho de Penafiel e seus castros. *SG* 9 (1), 126-32
	1965:	Sousa, E.F. As moedas encontradas na Citânia de Mósinho (Cidade Morta) e as suas possíveis conclusões. *Lucerna* 4, 249-69
Sousa, JJR	1968-70:	Sousa, J.J.R. A estação arqueológica da Falperra. Notas para a sua história. *AB* (25-27), 57-64
Strabo	1960:	Strabo *Geography (book III).* Jones, H.L. (ed.) The Loeb Classical Library 2. London

	1966:	*Strabon Géographie.* (vol. 2). Lasserre, F. (ed.) Les Belles Lettres. Paris
Suess	1970:	Suess, H.E. Bristlecone-pine calibration time 5200 BC to present. Berger, R.; Suess, H.E. (eds.). *Radiocarbon Dating.* University of California Press, Berkeley, 777-84
Sutherland	1984:	Sutherland, C.H.V. *The Roman Imperial Coinage.* (vol. I). Spink, London
Taboada Chivite	1955:	Taboada Chivite, J. Carta arqueológica de la comarca de Verín. III *CAN*, 476-80
	1965:	Taboada Chivite, J. *Escultura celto-romana.* Vigo
	1977:	Taboada Chivite, J. El aspecto agropecuario de la economia castreña. *RUC* 26 (109), 71-83
Taborin	1989:	Taborin, Y. Le foyer: document et concept. Olive, M.; Taborin, Y. (eds.). *Nature et fonction des foyers préhistoriques.* Mémoires du Musée de Préhistoire d'ile de France 2, Nemours, 77-80
Tarracena y Aguirre	1943:	Tarracena y Aguirre, B. Cabezas-trofeo en la españa céltica. *AEA* 16, 157-71
Tarradell	1965:	Tarradell, M. Enterramientos infantiles en el interior de habitaciones ibéricas. *Pyrenae* 1, 174-5
Tavares	1970:	Tavares, A.A. Citânia de Sanfins. Habitação e modo de vida. *JA* 2, 347-57
Teixeira	1939:	Teixeira, C. Os torques do Castro de Lanhoso (Póvoa de Lanhoso). *AFCP* 24 (4), 245-52
	1940a:	Teixeira, C. Molde de fundição para machados de bronze de duplo anel. *TAE* 9 (1-2), 126-30
	1940b:	Teixeira, C. Notas arqueológicas sobre o Castro de Lanhoso. *TAE* 9 (1-2), 117-25
	1940c:	Teixeira, C. O Castro de Lanhoso e seu espólio. *CMP* 1 515-29
	1941:	Teixeira, C. Hallazgo arqueológico notable. Un casco céltico en Portugal. *Ampurias* 3, 138-9
	1943-44:	Teixeira, C. Cerâmica lusitano-romana de Rossas (Vieira do Minho). *TAE* 10 (2), 165-7
	1946:	Teixeira, C. A arte das filigranas nos castro minhotos. *Minia* 1(3), 169-74
	1956:	Teixeira, C. *Carta geológica de Portugal (1/50.000). Notícia explicativa da folha 1-A, Valença.* Lisboa
	1969:	Teixeira, C. *Carta geológica de Portugal (1/50.000). Notícia explicativa da folha 2-A, Portos.* Lisboa
Teixeira-Assunção	1961:	Teixeira, C.; Assunção, C.T. *Carta geológica de Portugal (1/50.000). Notícia explicativa da folha 1-C, Caminha.* Lisboa
Teixeira *et al.*	1974:	Teixeira, C.; Assunção, C.T.; Coelho, A.V.P. *Carta geológica de Portugal (1/50.000). Notícia explicativa da folha 6-B, Chaves.* Lisboa
Teixeira *et al.*	1967:	Teixeira, C.; Fernandes, A.P., Peres, A.; Assunção, C.T.; Pilar, L. *Carta geológica de Portugal (1/50.000). Notícia explicativa da folha 10-C, Peso da Régua.* Lisboa
	1969:	Teixeira, C.; Medeiros, A.C.; Alves, C.A.M.; Moreira, M.M. *Carta geológica de Portugal (1/50.000). Notícia explicativa da folha 5-C, Barcelos.* Lisboa
	1965:	Teixeira, C.; Medeiros, A.C.; Assunção, C.T. *Carta geológica de Portugal (1/50.000). Notícia explicativa da folha 9-A, Póvoa de Varzim.* Lisboa
	1972:	Teixeira, C.; Medeiros, A.C.; Coelho, A.P. *Carta geológica de Portugal (1/50.000). Notícia explicativa da folha 5-A, Viana do Castelo.* Lisboa
	1973:	Teixeira, C.; Medeiros, A.C.; Macedo, J.R. *Carta geológica de Portugal (1/50.000). Notícia explicativa da folha 5-D, Braga.* Lisboa
Terán	1952-58:	Terán, M. (ed.) *Geografía de España y Portugal.* Barcelona
Terán - Otero	1952-58:	Terán, M.; Otero Pedrayo, R. Galicia, región astur-cántabra, la Meseta septentrional. Terán , M. (ed.) *Geografía de España y Portugal.* Barcelona
Thadeu	1965:	Thadeu, D. *Carta Mineira de Portugal, na escala 1/500 000. Notícia Explicativa.* Serviços Geológicos de Portugal. Lisboa
Torras Troncoso et al.	1980:	Torras Troncoso, M.L.; Diaz-Fieros Viqueira, F.; Vazquez Varela, J.M. Sobre el comienzo de la agricultura en Galicia. *Gallaecia* 6, 51--9
Torres Rodriguez	1976:	Torres Rodriguez, C. La conquista romana de Galicia. La Romanización de Galicia *CSECS* 16, 11-30
	1980:	Torres Rodriguez, C. La tragedia del Monte Medulio y su ubicacion. *Gallaecia* 6, 111-20
Torres	1978-79:	Torres, J. Tesouro monetário do Castro de Alvarelhos. *BC-CST* 1 (1-3), 15-250
Tovar	1947:	Tovar, A. Notas sobre la fijacíon de las invasiones indoeuropeas en España. *BSAA* 13, 21-35
	1950:	Tovar, A. Sobra la compleijidad de las invasiones indoeuropeas en nuestra Península. *Zephyrus* 1, 33-37
	1977:	Tovar, A. El nombre de celtas en Hispania, *RUC* 26 (109), 163-78
	1983:	Tovar, A. Etnia y lengua en la Galicia antigua: El problema del celtismo. Pereira Menaut, G., (ed.) *Estudos de Cultura Castrexa e de História Antiga de Galicia.* Santiago de Compostela, 247-82

	1985-86:	Tovar, A. Lenguas y pueblos de la antigua Hispania: lo que sabemos de nuestros antepassados protohistóricos. *Veleia* 2-3, (IV CLCPPI), 15-34
Tranoy	1977:	Tranoy, A. A propos des 'Callaeci' de Pline. Épigraphie et peuplement. *BA* 31 (71-72), 225-33
	1981:	Tranoy, A. *La Galice Romaine. Recherches sur le Nord-Ouest de la Péninsule Ibérique dans l'Antiquité.* Publications du Centre Pierre Paris 7, Paris
	1974:	Tranoy, A. (ed.) *HYDACE, Chronique.* (Sources Chrétiennes 218-219), Paris
	1988:	Tranoy, A. Du Heros au Chef. L'image du guerrier dans les sociétés indigènes du nord-ouest de la péninsule ibérique (IIe siècle avant J.-C. — Ier siècle après J.-C.). *Caesarodunum* 23, Paris, 219-27
Trigger	1978:	Trigger, B.G. *Time and traditions.* Columbia University Press, New York
Untermann	1965:	Untermann, J. *Elementos de um atlas antroponímico de la Hispania antigua.* Madrid
	1985:	Untermann, J. Los teónimos de la región lusitano-gallega. III *CLCPPI*, 343-63
Uria Riu	1941a:	Uria Riu, J. Excavaciones en el Castellón de Coaña. *RUO* 2
	1941b:	Uria Riu, J. Ritos funerários en las cámaras de Briteiros y Coaña. *RUO* 3
	1956:	Uria Riu, J. Nuevos hallazgos esqueléticos de la Edad del Bronce en la mina del Aramo (Asturias). *CISPP* 4 (Madrid 1954), Zaragoza, 631-5
Utrilla Miranda	1987:	Utrilla Miranda, P. G.A. Clark (T.P. 1986) El nicho alimenticio humano en el Norte de España desde el Paleolítico a la Romanización. *TP* 44, 337-40
Vasconcellos, JC	1950:	Vasconcellos, J.C. Cereais e Leguminosas do Castelo de Faria. *BGAF* 2, 37-8
Vasconcellos	1895:	Vasconcellos, J.L. Notícias arqueológicas em Penafiel . *AP* 1, 15-6
	1896:	Vasconcellos, J.L. Estátuas de Guerreiros Lusitanos. *AP* 2, 29-32
	1897, 1905, 1913:	Vasconcellos, J.L. *Religiões da Lusitania.* 1897 vol. 1, 1905 vol. 2, 1913 vol. 3, Lisboa
	1900:	Vasconcellos, J.L. Archeologia do Alto Minho. *AP* 5, 33-9
	1901:	Vasconcellos, J.L. De 28 de Agosto de 1880 (Briteiros): Ainda os congressistas. O Castro de Guifões. *AP* 6, 330-48
	1902a:	Vasconcellos, J.L. Cidade Velha de Santa Luzia. *A Aurora do Lima* (23 Agosto) 1
	1902b:	Vasconcellos, J.L. Estátua de um guerreiro lusitano. *AP* 7, 23-6
	1902c:	Vasconcellos, J.L. Notícias várias. *AP* 7, 283-8
	1905:	Vasconcellos, J.L. Aquisições do Museu Ethnológico Português. *AP* 10, 381-2
	1933:	Vasconcellos, J.L. Castros lusitanos I. Cividade de Paderne, II. Castro de Belinho. *AP* 29, 31-49
	1938a:	Vasconcellos, J.L. Cale e Portugale. Opúsculos 5, Lisboa, 28-38
	1938b:	Vasconcellos, J.L. Primórdios do Porto. Opúsculos 7, Lisboa, 1394-1400
Vasconcelos, J	1905-08:	Vasconcelos, J. Materiaes para o inventário arqueologico do concelho de Baião. *Portugália* 2, 669-72
Vasconcelos	1914-16:	Vasconcelos, M. Apontamentos arqueológicos do concelho do Marco de Canaveses. *AP* 19 (1914, 12-31), 21 (1916, 319-31)
Vázquez Seijas	1964:	Vázquez Seijas, M. Posibles emisiones de Augusto en Lugo. Numisma 14 (70), 37-40
Vázquez Varela	1973:	Vázquez Varela, J.M. Bases paleontológicas para el estudio de la ganadería de la cultura castreña de Galicia. Compostellanum 18 (1-4), 309-16
	1974-75:	Vázquez Varela, J.M. Hallazgos de bellotas en el Castro de Vixil: reflexiones sobre la agricultura castreña. *BCML* 9 (81-84), 195-8
	1975a:	Vázquez Varela, J.M. El conchero del castro de Queiruga (La Coruña). *BSAA* 40-41, 500-3
	1975b:	Vázquez Varela, J.M. Estudio de la fauna marina de los concheros de los castros de A Peneda y Montealegre (Pontevedra). *Gallaecia* 1, Santiago de Compostela, 141-46
	1975c:	Vázquez Varela, J.M. Estudio del conchero protohistorico de las Cíes. I JMACH 1, Vigo, 291-5
	1976:	Vaáquez Varela, J.M. Bases paleontológicas para el estudio de la pesca en la cultura castreña: una investigación preliminar. *BAur* 6, 83-6
	1977:	Vázquez Varela, J.M. La ganadería de la cultura castreña en Galicia. XIV *CAN*, 641-4
	1986:	Vázquez Varela, J.M. Dieta real y dieta imaginaria. Bermejo Barrera, J.C. (ed.) *Mitología y mitos en la Hispania prerromana* 2. Madrid, 231-39
	1988:	Vázquez Varela, J.M. Aplicación de un modelo etnoarqueológico a la agricultura castreña. *TAE* 28 (3-4), 99-105
Vázquez - Aira	1988:	Vázquez Varela, J.M.; Aira Rodriguez, M.J. La economía de los pueblos prerromanos del Norte de Hispania según los testimonios palinologicos. Pereira Menaut, G. (ed.) 1º *Congreso Peninsular de Historia Antigua* 2, Santiago de Compostela, 291-8

Vázquez - Pombo	1991a:	Vázquez Varela, J.M.; Pombo Mosquera, X.A. A recolleita de landras na Protohistoria de Galicia. Queiroga, F.; Dinis, A.P. (eds.). *Paleoecologia e Arqueologia* 2, Vila Nova de Famalicão, 121-7
	1991b:	Vázquez Varela, J.M.; Pombo Mosquera, X.A. La economia castrexa no N.O. de A Terra Cha (Lugo). Queiroga, F.; Dinis, A.P. (eds.). *Paleoecologia e Arqueologia* 2, Vila Nova de Famalicão, 191-200
Veiga	1941:	Veiga, A.B.C. Ourique - Vale de Vez. *Anais* 1, Lisboa, 9-186
Veloso - Cardoso	1965:	Veloso, F.J.; Cardoso, J. Estrabão: Livro III da 'geografia'. *Amphitheatrum* (supp. *SG*), Porto
Vernet	1988:	Vernet, J.-L. A antracologia préhistórica. Queiroga, F.; Sousa, I.M.A.R.; Oliveira, C.M. (eds.). *Paleoecologia e Arqueologia*, Vila Nova de Famalicão, 45-50
Viana	1926:	Viana, A. Através do Minho II - A exploração metodica dos nossos castros. *Gente Minhota* (6, 88-90), (7, 111-13)
	1932:	Viana, A. Justificação de um cadastro de monumentos arqueológicos para o estudo de Arqueologia do Alto Minho. *ADVC* 1, Viana do Castelo, 1-24
	1937:	Viana, A. Achados arqueológicosno Monte de Santo António- Afife. *TAAP* 3, Viana do Castelo, 75-83
	1955a:	Viana, A. Citânia de Santa Luzia. *Zephyrus* 6, Salamanca, 61-88
	1955b:	Viana, A. A cova da Moura. III *CAN*, 481-97
	1955c:	Viana, A. Um 'lapidarius' de Afife (Viana do Castelo). III *CAN*, 525-8
	1955d:	Viana, A. Notas sobre a cerâmica do Castro de Vilar de Mouros (Caminha, Portugal). III *CAN*, 529-34
	1959-60:	Viana, A. Insculturas rupestres do Alto Minho. *BCMO* 20 (1-4), 209-31
	1960-61:	Viana, A. Cividade de Ancora. Notícia sobre a actividade de II Campo Internacional de Trabalho Arqueológico Promovido em 1960 pela Associação Académica de Coimbra. *Conimbriga* 2-3, 247-70
	1962:	Viana, A. *Algumas noções elementares de arqueologia prática.* Beja
	1963:	Viana, A. Cividade de Âncora: sua importância. Explorações de 1960 e 1961. *Lucerna* 3, 167-78
Viana - Neves	1959:	Viana, A.; Neves, L.Q. Nota sobre o Castro de S. Caetano (Longosvales - Monção). I *CAN* 1, 389-96
Viana - Oliveira	1954:	Viana, A.; Oliveira, M.S. 'Cividade Velha' de Santa Luzia (Viana do Castelo). *RG* 64 (1-2), 40-72
	1955:	Viana, A.; Oliveira, M.S. Sobre a citânia de Santa Luzia (Viana do Castelo, Portugal). III *CAN*, 541-51
Vieira	1886-87:	Vieira, J.A. *O Minho pitoresco.* (2 vols) Lisboa
Vigarello	1988:	Vigarello, G. *Concepts of cleanliness. Changing attitudes in France since the Middle Ages.* Cambridge University Press, Cambridge
Villaronga	1970:	Villaronga, L. Emisión monetaria augustea con escudo atribuible a P. Carisio y a la zona Norte de Hispania. XI *CAN*, 591-600
Villas Bôas	1948a:	Villas Bôas, J.S.P. Um machado de bronze. *BGAF* 1(1), 13-20
	1948b:	Villas Bôas, J.S.P. Uma nova machadinha 'Castreja'. *BGAF* 1 (1), 29-32
Viterbo	1983-84:	Viterbo, J.S.R. *Elucidário de palavras, termos e frases que em Portugal antigamente se usaram e hoje se ignoram.* Porto-Lisboa
Vitorino	1917-18:	Vitorino, P. Notas artísticas e arqueológicas - exploração arqueológica. *Lusa* 1, 156
	1929:	Vitorino, P. 'Cidade Velha' de Santa Luzia (Viana do Castelo). *Portucale* 2, 448
Wait	1985:	Wait, G.A. *Ritual and Religion in Iron Age Britain.* (2 vols.), BAR BS 149
Walker	1985:	Walker, M. 5.000 anos de viticultura en España. Revista de *Arqueologia* 53, Madrid, 44-7
Wells	1980:	Wells, P. *Culture Contact and Culture Change: Early Iron Age central Europe and the Mediterranean world.* New Studies In Archaeology, Cambridge University Press, Cambridge
	1984:	Wells, P. *Farms, Villages and Cities: commerce and urban origins in late prehistoric Europe.* Cornell University Press, Ithaca
Williams	1990:	Williams, G. Burnt mounds in south-west Wales. Buckley, V. (ed.) *Burnt Offerings.* Wordwell Ltd., Dublin, 129-40
Wiseman - Cunliffe	1980:	Wiseman, A. & P.; Cunliffe, B. *Julius Caesar. The Battle for Gaul.* London

PERIODICALS

AA	American Anthropologist
AAM	Arquivo do Alto Minho (Viana do Castelo)
AB	Arquivo de Beja (Beja)
ADVC	Arquivo do Distrito de Viana do Castelo (Viana do Castelo)
AEA	Archivo Español de Arqueología (Madrid)
AFCP	Anais da Faculdade de Ciências do Porto (Porto)
AHM	Arquivo Histórico Militar (Lisboa)
AJ	The Antiquaries Journal (London)
AMSEAP	Actas y Memorias de la Sociedad Española de Antropología y Prehistoria (Madrid)
AP	O Arqueólogo Português (Lisboa)
ASEG	Arquivos do Seminario de Estudos Galegos (Santiago)
BA	Bracara Augusta (Braga)
BACAEP	Buttleti de Associacó Catalana d'Antropologia, Etnología i Prehistoria (Barcelona)
BAFN	Boletim da Associação de Filosofia Natural (Porto)
BAP	Boletim dos Amigos do Porto (Porto)
BAR	British Archaeological Reports (Oxford)
BAur	Boletín Auriense (Orense)
BBUC	Boletim da Biblioteca da Universidade de Coimbra (Coimbra)
BC-CMVNF	Boletim Cultural da Câmara Municipal de Vila Nova de Famalicão (Vila Nova de Famalicão)
BC-CST	Boletim Cultural do Concelho de Santo Tirso (Santo Tirso)
BCE	Boletim Cultural de Esposende (Esposende)
BC-CMP	Boletim Cultural da Câmara Municipal do Porto (Porto)
BCML	Boletín de la Comisión Provincial de Monumentos Historicos y Artísticos de Lugo (Lugo)
BCMO	Boletín de la Comisión Provincial de Monumentos Historicos y Artísticos de Orense (Orense)
BC-PV	Boletim Cultural da Póvoa do Varzim (Póvoa de Varzim)
BGAF	Boletim do Grupo dos Alcaides de Faria (Barcelos)
BIEA	Boletín del Instituto de Estudios Asturianos (Oviedo)
BM	Boletim de Minas (Lisboa)
BPH	Bibliotheca Praehistorica Hispaña
BR	Barcellos Revista (Barcelos)
BRAACAP	Boletim da Real Associação dos Architectos Civis e Arqueólogos Portugueses (Lisboa)
BRAH	Boletín de la Real Academia de Historia (Madrid)
BSAA	Boletín del Seminario de Estudios de Arte y Arqueología (Valladolid)
CA	Cadernos de Arqueologia (Braga)
-	Caesaraugusta (Zaragoza)
CAN	Congreso Nacional de Arqueología (Espanha) III -(Galicia 1953) Zaragoza 1955 XI -(Mérida 1968) Zaragoza, 1970 XII -(Jaén 1971) Zaragoza, 1973 XIII -(Huelva 1973) Zaragoza, 1975 XIV -(Vitoria 1975) Zaragoza 1977 XV -(Lugo 1977) Zaragoza, 1979
CEG	Cuadernos de Estudios Gallegos (Santiago)
CISPP	Congrès International des Sciences Prehistoriques et Protohistoriques
CIUAN	Colóquio Inter-Universitário de Arqueologia do Noroeste -Porto, 1984 (Portugália, Nova Série, 3/4, Porto 1983/84)
CMP	Congresso do Mundo Português -Lisboa 1940 (Lisboa 1940)
CNA	Congresso Nacional de Arqueologia (Portugal)
CSECS	Cuadernos del Seminario de Estudios Cerámicos de Sargadelos (La Coruña)
CV	Cadernos Vianenses (Viana do Castelo)
DL	Douro Litoral (Porto)
DXPHD	Dirección Xeral do Patrimonio Artístico e Monumental, Xunta de Galicia (La Coruña)
EAA	Estudios de Arqueología Alavesa (Alava)
EAE	Excavaciones arqueológicas en España (Madrid)
FHA	Fontes Hispaniae Antiquae
HA	Hispania Antiqua (Valladolid)
HMS	Homenagem a Martins Sarmento (Guimarães)
IMCGEA	Informes y Memorias de la Comisión general de Excavationes Arqueológicas (Madri)
JA	Jornadas Arqueológicas (Lisboa)
JRSAI	Journal of the Royal Society of Antiquaries of Ireland (Dublin)
MHA	Memorias de Historia Antigua (Oviedo)
MJSEA	Memoria de la Junta Superior de Excavaciones y Antigüedades (Madrid)
MM	Madrider Mitteilungen (Madrid, Heidelberg)

MP	El Museo de Pontevedra (Pontevedra)
NAH	Noticiario Arqueológico Hispánico (Madrid)
OJA	Oxford Journal of Archaeology (Oxford)
PN	Petrus Nonius (Lisboa)
PPS	Proceedings of the Prehistoric Society (London)
RABM	Revista de Arquivos, Bibliotecas y Museos (Madrid)
RA	Revista de Arqueologia (Lisboa)
RCHUP	Revista de Ciências Históricas, Universidade Portucalense (Porto)
RE	Revista de Etnografia (Porto)
RFLL	Revista da Faculdade de Letras de Lisboa (Lisboa)
RFLUP	Revista da Faculdade de Letras do Porto (Porto)
RG	Revista de Guimarães (Guimarães)
RSNS	Revista de Sciencias Naturaes e Sociaes (Porto)
RUC	Revista de la Universidad Complutense (Madrid)
RUO	Revista de la Universidad de Oviedo (Oviedo)
SANP	Seminário de Arqueologia do Noroeste Peninsular -Guimarães, 1978 (Guimarães, 1980)
SG	Stvdivm Generale (Porto)
TAAP	Trabalhos da Associação dos Arqueólogos Portugueses (Lisboa)
TAE	Trabalhos da Sociedade Portuguesa de Antropologia e Etnologia (Porto)
TIADMC	Trabalhos do Instituto de Antropologia Dr. Mendes Corrêa (Porto)
TP	Trabajos de Prehistoria (Madrid)
WA	World Archaeology (London)

APPENDIX I

1) Introduction

The main purpose of this appendix is to provide a fairly complete and accurate list of sites showing evidence of Iron Age occupation. Although this may not have been entirely achieved, the present list is one step forward in the unrewarding task of making any list of sites for such a vast region. Some of the studies of the kind produced to date have shown errors and inaccuracies which tend to eternise the vagueness of our knowledge of some areas.

These inaccuracies may be cultural and chronological, often being referred to as castros sites showing occupation datable to the Bronze Age, Middle Age, or even exclusively to the Roman period. However, other than this, the toponymical ambiguity and the inaccuracy of location has greatly contributed to some of the confusion. Cases were recorded when sites bore more than one name and were located in different administrative areas. These cases often resulted in multiplication of sites in the published listings of a larger scope, where field confirmation is rare. This situation has decreased in recent years due to the publication of studies based on accurate field work and which are confined to smaller geographical areas.

The major contribution of this appendix is to present an updated list where particular attention has been paid to two major aspects: the geographical/administrative location of each site, and the cultural/chronological phase.

A major list of the kind produced in recent years (Silva 1986a, 69-94) was extensively used here as a work basis, mainly for the list of sites as well as for directional bibliographical references. Visits to many of the listed sites have shown, however, inaccuracies in both their location and period of occupation which would not be within the time-range of this study. It is also fair to point out that these misleading data do start in earlier studies, as one can see by the bibliographical references.

The lists recently produced, as a result of the study of smaller geographical or administrative areas (Almeida, CAB, 1985b, 1986, 1987, 1990a and 1990b; Almeida - Baptista 1981; Marques, JATM, 1984a; Martins 1990; Martins, JB, 1984; Montalvão 1971a; Regalo 1986; Soeiro 1984) have proved to be far more accurate. They cover mostly the coastal area of the Entre-Douro-E-Minho, since only Montalvão (1971a) and Martins, JB, (1984) refer to the region around Chaves.

The list of sites presented below is the result of a study based on bibliographical references which was corroborated by visits to many of the sites. Unfortunately, many of them also could not be visited.

The sites are listed in alphabetical order of the major administrative references: the districts[1]. Further divisions are the Councils and the Parishes.

The name by which the site is commonly known is mentioned, bearing in mind the eventual discrepancies between the local denomination and the name referred to in studies, and therefore sometimes more than one name is referred to for the same site. There are other cases where a site has not yet been named, and there is no nearby name of place which could be applied. In such cases a question mark between brackets will be used.

Further data on the location of the sites is provided by the geographical coordinates. For the Longitude the international meridian was used, and the altitude is above sea level. The sites in which the coordinates are not referred to were not visited or located, and in these cases reference is made to the available bibliography.

The bibliography listed may not be exhaustive or updated, since an enormous number and variety of references can be found in local and regional periodicals. However, it may be sufficient for the present evaluation of the number and characteristics of the castros in the Northwest of Portugal.

1 - They are organized according to the order defined by the "National Geographic Code"

2) LIST OF SITES

2.1) District of Braga

Nº: 1
Name: Monte da Santinha
Parish: Amares
Council: Amares
Latitude: 413816 N
Longitude: 082041 W
Altitude: 195
Bibliography: Azevedo, PA, 1896, 253-4. Cunha 1948b, 3, 4; 1975, 525-6. Martins 1987, 105. Silva 1986a, nº 226.

Nº: 2
Name: Monte de S. Miguel
Parish: Bouro Sta. Maria
Council: Amares
Latitude: 414013 N
Longitude: 081512 W
Altitude: 566
Bibliography: Cunha 1975, 415, 516.

Nº: 3
Name: Monte do Outeiro
Parish: Bouro Sta. Maria
Council: Amares
Latitude: 413945 N
Longitude: 081711 W
Altitude: 175
Bibliography: Martins 1987, 119

Nº: 4
Name: Grovos
Parish: Caires
Council: Amares
Latitude: 413853 N
Longitude: 082111 W
Altitude: 320
Bibliography: Cardozo 1980, 174, nº 105. Martins 1987, 106. Silva 1986a, nº 224. Silva, DM, 1985, 222-3.

Nº: 5
Name: Castro
Parish: Carrazedo
Council: Amares
Latitude: 413736 N
Longitude: 082302 W
Altitude: 96
Bibliography: Martins 1987, 110. Silva 1986a, nº 221. Silva, DM, 1985, 135-48.

Nº: 6
Name: Requiem
Parish: Ferreiros
Council: Amares
Latitude: (?)
Longitude: (?)
Altitude: (?)
Bibliography: Silva 1986a, nº 222. Silva, DM, 1985, 135-48.

Nº: 7
Name: Castro da Ponte
Parish: Lago
Council: Amares
Latitude: 413636 N
Longitude: 082438 W
Altitude: 65
Bibliography: Silva, DM, 1985, 139. Sousa 1971-72, 180-181. Martins 1983a, 22-4; 1985b; 1987, 113. Silva 1986a, nº 220.

Nº: 8
Name: Monte das Caldas (Castro de Caldelas)
Parish: Outeiro (S. Sebastião)
Council: Amares
Latitude: 414010 N
Longitude: 082217 W
Altitude: 220
Bibliography: Silva, DM, 1985, 229. Jorge - Sousa 1980, 120. Martins 1987, 108

Nº: 9
Name: Castro do Seramil
Parish: Outeiro da Vila/ Seramil
Council: Amares
Latitude: 414023 N
Longitude: 081842 W
Altitude: 458
Bibliography: Martins 1987, 121. Cunha 1961, 321.

Nº: 10
Name: Castelo de Espinho
Parish: Portela

Council: Amares
Latitude: 413918 N
Longitude: 082139 W
Altitude: 480
Bibliography: Cunha 1957, 1, 4. Martins 1987, 105, 117. Silva 1986a, nº 227. Silva, DM, 1985, 135-48. Leal 1973, 93. Belino 1909, 6.

Nº: **11**
Name: Chã do Castro (Joubreia)
Parish: Portela
Council: Amares
Latitude: 413850 N
Longitude: 082203 W
Altitude: 311
Bibliography: Alarcão-Alarcão 1963, 199-200. Jorge-Sousa 1980. Martins 1987, 116-7. Silva 1986a, nº 225, pp 202, est. 90 (5a and 5b).

Nº: **12**
Name: Monte de Sequeiros
Parish: Sequeiros
Council: Amares
Latitude: 414105 N
Longitude: 082143 W
Altitude: 228
Bibliography: Silva, DM, 1958-59, 353-55. Martins 1987, 121

Nº: **13**
Name: Sequeade
Parish:
Council: Barcelos
Latitude: 413010 N
Longitude: 083210 W
Altitude: 200
Bibliography: Martins 1987, 138. Soeiro 1983, 63.

Nº: **14**
Name: Monte do Facho - Alto da Torre
Parish: Abade de Neiva
Council: Barcelos
Latitude: 413247 N
Longitude: 083836 W
Altitude: 170
Bibliography: Almeida-Soeiro 1980, 29-36. Costa et al. 1980, 10-4. Fonseca 1948, 55. Martins 1987, 123. Silva 1986a, nº 201.

Nº: **15**
Name: Queijeiros
Parish: Abade do Neiva
Council: Barcelos
Latitude: 413336 N
Longitude: 083917 W
Altitude: 150
Bibliography: Martins 1987, 124. Costa et al. 1980, 11.

Nº: **16**
Name: Alto dos Castelos
Parish: Aguiar/Durrães
Council: Barcelos
Latitude: 413745 N
Longitude: 083921 W
Altitude: 160
Bibliography: Fonseca 1948, 69. Silva 1986a, nº 190.

Nº: **17**
Name: Outeiro do Castro (Castro de Airó; Castelo de Bastuço)
Parish: Airó (S. Jorge de)
Council: Barcelos
Latitude: 413120 N
Longitude: 083230 W
Altitude: 222
Bibliography: Calo Lourido 1991, 53-5. Figueiredo 1896, 165. Fonseca 1948, 13. Martins 1987, 125. Silva 1986a, nº 209. Soeiro 1983.

Nº: **18**
Name: Monte de Lousado - Cidade Grande
Parish: Alheira
Council: Barcelos
Latitude: 413724 N
Longitude: 083454 W
Altitude: 300
Bibliography: Martins 1987, 126. Silva 1986a, nº 195, 193. Figueiredo 1895, 242. Fonseca 1948, 93.

Nº: **19**
Name: Senhora da Aparecida (Senhora do Castro)
Parish: Balugães
Council: Barcelos
Latitude: 413831 N
Longitude: 083850 W
Altitude: 155
Bibliography: Almeida, CAB, 1990b, 134-5. Fonseca 1948, 138. Silva 1986a, nº 192.

Nº: **20**
Name: Castro
Parish: Carapeços
Council: Barcelos
Latitude: 413525 N
Longitude: 083745 W
Altitude: 200

Bibliography: Martins 1987, 127. Silva 1986a, nº 196. Villas Boas 1948b, 43.

Nº: **21**
Name: Castro de S. Simão
Parish: Cossourado
Council: Barcelos
Latitude: 413749 N
Longitude: 083657 W
Altitude: 236
Bibliography: Almeida-Baptista 1984. Azevedo, AP, 1897, 234-5. Silva 1986a, nº 194.

Nº: **22**
Name: Alto do Senhor do Lírio
Parish: Durrães
Council: Barcelos
Latitude: (?)
Longitude: (?)
Altitude: (?)
Bibliography: Silva 1986a, nº 188.

Nº: **23**
Name: Giesta
Parish: Durrães
Council: Barcelos
Latitude: 413700 N
Longitude: 083801 W
Altitude: 81
Bibliography: Silva 1986a, nº 189.

Nº: **24**
Name: Picoto dos Mouros
Parish: Durrães
Council: Barcelos
Latitude: (?)
Longitude: (?)
Altitude: (?)
Bibliography: Silva 1986a, nº 187.

Nº: **25**
Name: S. Mamede
Parish: Feitos
Council: Barcelos
Latitude: 413345 N
Longitude: 084051 W
Altitude: 409
Bibliography: Fonseca 1948, 232. Martins 1987, 130. Silva 1986a, nº 199.

Nº: **26**
Name: Pena Grande
Parish: Galegos (Santa Maria)
Council: Barcelos
Latitude: 413355 N
Longitude: 083411 W
Altitude: 140
Bibliography: Martins 1987, 136. Silva 1986a, nº 202.

Nº: **27**
Name: Castelo de Faria
Parish: Gilmonde
Council: Barcelos
Latitude: 412947 N
Longitude: 083845 W
Altitude: 298
Bibliography: Almeida, CAB, 1982a, 79-88; 1985b, 50-2. Costa et al. 1980, 15-22. Fonseca 1948, 183-4. Maluquer de Motes 1948, 33-38. Martinez Santa Olalla 1948, 21-28. Martins 1987, 131. San Valero Aparisi 1950. Silva 1986a, nº 206. Vasconcellos, JC, 1950.

Nº: **28**
Name: Monte da Saia (Fralães)
Parish: Grimancelos
Council: Barcelos
Latitude: 412715 N
Longitude: 083511 W
Altitude: 3
Bibliography: Cardozo 1957d, 179-84. Cunha, AR, 1946b, 3. Fonseca 1948, 121. Kalb 1980b, nº 17, Abb. 3. Sarmento 1933a, 160. Martins 1987, 128. Silva 1986a, nº 211. Vasconcellos 1897 (1), 375, 379. Garcia e Bellido 1968, 33-34. Torres 1876; 1877.

Nº: **29**
Name: Monte de Santa Marinha
Parish: Quintiães
Council: Barcelos
Latitude: (?)
Longitude: (?)
Altitude: (?)
Bibliography: Silva 1986a, nº 191.

Nº: **30**
Name: Santa Eulália
Parish: Rio Côvo
Council: Barcelos
Latitude: (?)
Longitude: (?)
Altitude: (?)
Bibliography: Fonseca 1948, 335. Silva 1986a, nº 207.

Nº: **31**
Name: Facho (Citânia de Roriz)
Parish: Roriz/Oliveira
Council: Barcelos

Latitude: 413439 N
Longitude: 083332 W
Altitude: 324
Bibliography: Almeida, CAB, 1981; 1983. Almeida-Soeiro 1980. Costa et al. 1980, 23-8. Fonseca 1948, 69. Kalb 1980b, nº 12, Abb. 2. Maluquer de Motes 1948. Martins 1987, 134. Monteagudo 1977, nº 1702. Sarmento 1970. Silva 1986a, nº 203. Villas Boas 1948b, fig. 3.

Nº: 32
Name: Monte do Castro
Parish: Ucha
Council: Barcelos
Latitude: 413423 N
Longitude: 083111 W
Altitude: 50
Bibliography: Fonseca 1948, 397. Martins 1987, 139. Silva 1986a, nº 205.

Nº: 33
Name: Cabeço da Ribeira
Parish: Adaúfe
Council: Braga
Latitude: 413606 N
Longitude: 082416 W
Altitude: 73
Bibliography: Martins 1987, 142.

Nº: 34
Name: Cividade
Parish: Cividade
Council: Braga
Latitude: 413242 N
Longitude: 082535 W
Altitude: 183
Bibliography: Bellino 1909. Castro et al. 1980, 39. Martins 1987, 144. Silva 1986a, nº 234.

Nº: 35
Name: Monte da Falperra (Sta. Marta)
Parish: Esporões
Council: Braga
Latitude: 413050 N
Longitude: 082338 W
Altitude: 563
Bibliography: Bellino 1909, 18-19. Castro et al. 1980, 39. Cunha 1948a, 3, 6; 1948b, 3-4; 1949, 2; 1952, 3; 1953a, 4; 1953-54, 241-3; 1954-55, 1-4; 1955a, 1, 3; 1975, 487-96. Ponte 1984, nº 63. Silva 1986a, nº 240. Sousa, JJR, 1968-70.

Nº: 36
Name: Castro
Parish: Ferreiros
Council: Braga
Latitude: 413209 N
Longitude: 082626 W
Altitude: 155
Bibliography: Teixeira 1955-56, 7. Castro et al. 1980, 39. Martins 1987, 149. Silva 1986a, nº 233.

Nº: 37
Name: Monte Redondo
Parish: Guizande
Council: Braga
Latitude: 412823 N
Longitude: 082648 W
Altitude: 427
Bibliography: Bellino 1909, 7-15, 19-28. Castro et al. 1980, 39. Cunha 1975, 497-8. Silva 1986a, nº 239. Vasconcellos 1905 (2), 334; 1913 (3), 175.

Nº: 38
Name: Castro Mau
Parish: Merelim - S. Paio
Council: Braga
Latitude: 413519 N
Longitude: 082649 W
Altitude: 50
Bibliography: Martins 1987, 160.

Nº: 39
Name: Castro da Consolação
Parish: Nogueiró
Council: Braga
Latitude: 413301 N
Longitude: 082340 W
Altitude: 175
Bibliography: Bellino 1909. Castro et al. 1980, 39. Cunha 1975, 501-2. Rosário 1973, 23-4, 49, 62. Silva 1986a, nº 237.

Nº: 40
Name: Castro da Sola
Parish: Palmeira
Council: Braga
Latitude: 413517 N
Longitude: 082428 W
Altitude: 99
Bibliography: Martins 1987, 152.

Nº: 41
Name: Eiras Velhas
Parish: S. Mamede
Council: Braga
Latitude: 413506 N

Longitude: 082032 W
Altitude: 493
Bibliography: Martins 1987, 159. Moita 1966, 544. Teixeira et al. 1973, 47. Silva 1986a, nº 241.

Nº: **42**
Name: Castro Máximo (Monte Castro)
Parish: S. Vicente/Real
Council: Braga
Latitude: 413335 N
Longitude: 082535 W
Altitude: 198
Bibliography: Almeida 1974a, 187. Castro et al. 1980, 37-52. Cortez 1954. Lemos et al. s/d, 41-3. Martins 1987, 156. Silva 1986a, nº 235, pp 204, fig. 91/12. Sousa 1973, 10. Teixeira 1936 (=1955-56). Freitas 1890, 320. Belino 1909, 5.

Nº: **43**
Name: Castro de Pedroso
Parish: S. Victor (Braga)
Council: Braga
Latitude: 413422 N
Longitude: 082344 W
Altitude: (?)
Bibliography: Fontes, LFO, 1990, 131.

Nº: **44**
Name: Monte das Caldas
Parish: Sequeira/Gondizalves
Council: Braga
Latitude: 413213 N
Longitude: 082804 W
Altitude: 304
Bibliography: Bellino 1909, 18-9. Castro et al. 1980, 38-9. Cunha 1947, 3, 6; 1975, 499-500. Martins 1987, 150. Silva 1986a, nº 232.

Nº: **45**
Name: Monte do Castro
Parish: Sobreposta
Council: Braga
Latitude: 413352 N
Longitude: 082031 W
Altitude: 567
Bibliography: Bellino 1909, 1-28. Castro et al. 1980, 39. Macedo 1896,121-3. Silva 1986a, nº 238. Teixeira 1936.

Nº: **46**
Name: Castelhão (Castro de Sta. Lucrécia de Algeriz)
Parish: Sta. Lucrécia de Algeriz
Council: Braga
Latitude: 413528 N
Longitude: 082232 W
Altitude: 200
Bibliography: Martins 1987, 158.

Nº: **47**
Name: Cividade de Chacim
Parish: Cabeceiras de Basto
Council: Cabeceiras de Basto
Latitude: 413101 N
Longitude: 075726 W
Altitude: 414
Bibliography: Araújo 1980, 105. Cunha 1950a, 1, 4. Silva 1986a, nº 309.

Nº: **48**
Name: Moimenta
Parish: Cavez
Council: Cabeceiras de Basto
Latitude: 413121 N
Longitude: 085410 W
Altitude: 468
Bibliography: Meneses, M, 1929, Fig. 1. 33. Silva 1986a, nº 313.

Nº: **49**
Name: Outeiro dos Mouros
Parish: Pedraça
Council: Cabeceiras de Basto
Latitude: 412958 N
Longitude: 075712 W
Altitude: 385
Bibliography: Sarmento 1933a, 166. Silva 1986a, nº 312.

Nº: **50**
Name: Castro de Chamoim
Parish: Refojos de Basto
Council: Cabeceiras de Basto
Latitude: (?)
Longitude: (?)
Altitude: (?)
Bibliography: Cunha 1975, 531-2. Silva 1986a, nº 311.

Nº: **51**
Name: Santa Comba
Parish: Refojos de Basto
Council: Cabeceiras de Basto
Latitude: 413021 N
Longitude: 075828 W
Altitude: 310
Bibliography: Almeida 1981, 111-6. Martins-Silva 1984, 29-40. Silva 1981-82; 1986a, nº 310.

Nº: **52**
Name: Cividade (Cambeses)
Parish: Rio Douro

Council: Cabeceiras de Basto
Latitude: 413219 N
Longitude: 085720 W
Altitude: 695
Bibliography: Peixoto 1899-1903, 832-3. Silva 1986a, nº 308.

Nº: **53**
Name: Alto do Crasto (Murgido)
Parish: Borba da Montanha
Council: Celorico de Basto
Latitude: 412342 N
Longitude: 080603 W
Altitude: 718
Bibliography: Silva 1986a, nº 316.

Nº: **54**
Name: Alto do Crasto (Quintela)
Parish: Borba da Montanha
Council: Celorico de Basto
Latitude: 412339 N
Longitude: 080501 W
Altitude: 675
Bibliography: Silva 1986a, nº 317.

Nº: **55**
Name: Roda da Santinha (Borrega)
Parish: Borba da Montanha
Council: Celorico de Basto
Latitude: 412327 N
Longitude: 080603 W
Altitude: 740
Bibliography: Silva 1986a, nº 318.

Nº: **56**
Name: Crasto
Parish: Canedo de Basto
Council: Celorico de Basto
Latitude: 412603 N
Longitude: 075637 W
Altitude: 336
Bibliography: Silva 1986a, nº 315.

Nº: **57**
Name: Alto do Ladário
Parish: Ribas
Council: Celorico de Basto
Latitude: 412743 N
Longitude: 075925 W
Altitude: 639
Bibliography: Silva 1986a, nº 314.

Nº: **58**
Name: Castro de Belinho
Parish: Antas (S. Paio)
Council: Esposende
Latitude: 413553 N
Longitude: 084624 W
Altitude: 137
Bibliography: Almeida, CAB, 1985b, nº 32; 1986, 53-5. Araújo 1980, 105. Fonseca 1936, 201-2. López Cuevillas 1946, 21. Neiva, MAP, 1987, 16, 40. Pinto 1932, 7, 11-12. Silva 1986a, nº 181. Vitorino 1917-18, 156. Vasconcellos 1934, 45-9.

Nº: **59**
Name: Monte
Parish: Antas (S. Paio)
Council: Esposende
Latitude: 413610 N
Longitude: 084541 W
Altitude: 90
Bibliography: Almeida, CAB, 1985b, nº 2; 1986, 47-8. Silva 1986a, nº 179. Pinto 1932, 7, 11-12.

Nº: **60**
Name: Outeiro dos Picotos
Parish: Fonte Boa
Council: Esposende
Latitude: 413047 N
Longitude: 084425 W
Altitude: 22
Bibliography: Almeida, CAB, 1985b, nº 58. Fonseca 1936, 49. Neiva, MAP, 1987, 75. Silva 1986a, nº 186.

Nº: **61**
Name: Castro da Cova da Bouça
Parish: Mar (S. Bartolomeu)
Council: Esposende
Latitude: 413454 N
Longitude: 084650 W
Altitude: 120
Bibliography: Almeida, CAB, 1985b, nº 3; 1987, 94-5. Cepa 1944, 20. Silva 1986a, nº 182.

Nº: **62**
Name: Castro de N. S. da Paz
Parish: Marinhas
Council: Esposende
Latitude: 413416 N
Longitude: 084624 W
Altitude: 161
Bibliography: Almeida, CAB, 1985b, nº 41; 1987, 98. Boaventura 1965. Cepa 1944, 24. Silva 1986a, nº 183.

Nº: **63**
Name: Castro do Senhor dos Desamparados
Parish: Palmeira de Faro
Council: Esposende
Latitude: 413220 N

Longitude: 084402 W
Altitude: 126
Bibliography: Almeida, CAB, 1985b, nº 48; 1987, 104-5. Neiva 1987, 107. Silva 1986a, nº 184.

Nº: **64**
Name: Castro de S. Lourenço
Parish: Vila Chã
Council: Esposende
Latitude: 413320 N
Longitude: 084533 W
Altitude: 200
Bibliography: Almeida, CAB, 1985b, nº 20; 1986, 41-3. Almeida - Neiva 1982, 5-18. Araújo 1980, 103. Boaventura 1954; 1965, 270-3. Encarnação 1975, 172-3. Fonseca 1936, 143. Leal 1886, 681. Machado,F, 1951. Neiva, MAP, 1982; 1987, 16-8, 119. Neves 1965, 172-80. Rosário 1973, 62-3. Santos et al. 1983, 12-3. Sarmento 1933, 159-60. Silva 1986a, nº 185. Vieira 1886, 199.

Nº: **65**
Name: Monte de Santo Ovídio
Parish: Fafe
Council: Fafe
Latitude: 412727 N
Longitude: 081005 W
Altitude: 332
Bibliography: Cardozo 1972b, 153, nº 93. Lemos et al. s.d., 41-3. Martins 1981; 1983b, 25-7; 1984b, 53-4; 1985c, 54-6. Oliveira s.d., 65-6, 233-30. Pereira 1915a. Ponte 1984, nº 27. Sarmento 1933a, 36-40, 72-3, 165-72. Silva 1986a, nº 305. Vasconcellos 1896, 29-32; 1905 (2), 40; 1913 (3), 53-4.

Nº: **66**
Name: Castelo
Parish: Moreira de Rei
Council: Fafe
Latitude: (?)
Longitude: (?)
Altitude: (?)
Bibliography: Oliveira s.d., 81-4. Silva 1986a, nº 304.

Nº: **67**
Name: Castro Formigoso
Parish: Revelhe
Council: Fafe
Latitude: 412834 N
Longitude: 080915 W
Altitude: 424
Bibliography: Oliveira, s.d., 94-5. Silva 1986a, nº 302.

Nº: **68**
Name: Crasto
Parish: Ribeiros
Council: Fafe
Latitude: 412905 N
Longitude: 080750 W
Altitude: 469
Bibliography: Oliveira s.d., 95-7. Silva 1986a, nº 303.

Nº: **69**
Name: Cividade (Paredes)
Parish: S. Gens
Council: Fafe
Latitude: 412702 N
Longitude: 080622 W
Altitude: 560
Bibliography: Oliveira s.d., 98-100. Silva 1986a, nº 306.

Nº: **70**
Name: Deveza Escura
Parish: Abação (S. Tomé)
Council: Guimarães
Latitude: 412427 N
Longitude: 081548 W
Altitude: 434
Bibliography: Azevedo, PA, 1896, 63. Pina 1940, 96-107. Silva 1986a, nº 293.

Nº: **71**
Name: Lapa da Mulher
Parish: Azurém
Council: Guimarães
Latitude: (?)
Longitude: (?)
Altitude: (?)
Bibliography: Sarmento 1933a, 234-5. Silva 1986a, nº 282.

Nº: **72**
Name: Citânia
Parish: Briteiros (Salvador)
Council: Guimarães
Latitude: 413136 N
Longitude: 081914 W
Altitude: 337
Bibliography: Alarcão-Alarcão 1963, 183-97. Albuquerque, M, 1980, 105-22. Almeida 1974a. Blanco Freijeiro 1957a, 292. Cardozo 1934-36; 1936a, 159-63; 1938, 35-40; 1943c, 247-56; 1948, 343-8; 1949b, 406-14; 1950a, 518-26;

1951, 455-72; 1952a, 348-58; 1952b, 711-17; 1954a, 412-20; 1955b, 431-8; 1956a, 508-12; 1957b, 551-3; 1958a, 455-9; 1959d, 517-20; 1960d, 551-4; 1961a, 451-4; 1968b, 291-6; 1980. Castelo Branco 1963b, 304-17. Fariña 1979, nº 3. Kalb 1980b, nº 15, Abb. 2. Monteagudo 1977, 268, nº 1797. Perez Outeiriño 1982, 48-52. Pina 1940, 506. Ponte 1984. Ribeiro, F, 1930-34. Sarmento 1903-05. Silva 1986a, nº 262. Silva - Centeno 1977. Taboada Chivite 1977, fig. 5.

Nº: **73**
Name: Montezelo
Parish: Briteiros (Santa Leocádia)
Council: Guimarães
Latitude: 413101 N
Longitude: 082028 W
Altitude: 273
Bibliography: Pina 1940, 506. Silva 1986a, nº 261.

Nº: **74**
Name: Penedo da Mata do Vale
Parish: Candoso (S. Martinho)
Council: Guimarães
Latitude: (?)
Longitude: (?)
Altitude: (?)
Bibliography: Sarmento 1933a, 234-6. Silva 1986a, nº 286.

Nº: **75**
Name: Matamá / Monte de Sto. Antoninho
Parish: Costa
Council: Guimarães
Latitude: 412630 N
Longitude: 081412 W
Altitude: 526
Bibliography: Coffyn 1985, 223. Kalb 1980b, nº 21, Abb. 4. Sarmento 1933a, 223. Silva 1986a, nº 284.

Nº: **76**
Name: Monte de Santa Eulália
Parish: Fermentões
Council: Guimarães
Latitude: 412710 N
Longitude: 082305 W
Altitude: 285
Bibliography: Silva 1986a, nº 281. Sarmento 1933a, 262-64.

Nº: **77**
Name: Castro
Parish: Gandarela
Council: Guimarães
Latitude: (?)
Longitude: (?)
Altitude: (?)
Bibliography: Sarmento 1933a, 250-1. Silva 1986a, nº 296.

Nº: **78**
Name: Castelo
Parish: Gominhães
Council: Guimarães
Latitude: 412909 N
Longitude: 081624 W
Altitude: 399
Bibliography: Silva 1986a, nº 271.

Nº: **79**
Name: Castro
Parish: Gonça
Council: Guimarães
Latitude: 413024 N
Longitude: 081522 W
Altitude: 485
Bibliography: Silva 1986a, nº 267.

Nº: **80**
Name: Castro
Parish: Guardizela
Council: Guimarães
Latitude: (?)
Longitude: (?)
Altitude: (?)
Bibliography: Sarmento 1933a, 256. Silva 1986a, nº 295.

Nº: **81**
Name: Castro de Sabroso
Parish: Longos
Council: Guimarães
Latitude: 413037 N
Longitude: 082024 W
Altitude: 278
Bibliography: Cardozo 1934-36; 1980. Hawkes 1958, 446-53. Kalb 1980b, nº 16, Abb. 2. Lopez Cuevillas 1955b, 149-50. Monteagudo 1977, 278, nº 1798. Pina 1940, 505, 513. Pinto 1928-29. Ponte 1984. Sarmento 1906, 41-51; 1907, 53-66, 113-22; 1909, 5-19, 129-39; 1933a, 22-35. Silva 1986a, nº 265. Soeiro et al. 1981, 341-50. Taboada Chivite 1977, Fig. 3.

Nº: **82**
Name: Pedrais

Parish: Longos
Council: Guimarães
Latitude: 413104 N
Longitude: 082101 W
Altitude: 268
Bibliography: Silva 1986a, nº 264, 198, 211-28. Cardozo 1980, 1935 p. 294; Cuevillas 1951c, p. 61; Monteagudo 1977 p. 268; 1798; Vasconcelos 1930a, p. 19; Ponte 1984; Fortes 1905, 1904 p. 5; 1908a p. 17-18

Nº: **83**
Name: Picoto de Santo Amaro
Parish: Mascotelos
Council: Guimarães
Latitude: 412521 N
Longitude: 081904 W
Altitude: 309
Bibliography: Guimarães, JS, 1970, 141-50, 381-6; 1971, 436-70. Pina 1940, 513. Ponte 1984, nº 31. Sarmento 1933a, 237-9. Silva 1986a, nº 287.

Nº: **84**
Name: Cemitério dos Mouros
Parish: Moreira de Cónegos
Council: Guimarães
Latitude: 412234 N
Longitude: 082104 W
Altitude: 150
Bibliography: Sarmento 1933a, 217, 252, 256. Silva 1986a, nº 297.

Nº: **85**
Name: Castro
Parish: Nespereira
Council: Guimarães
Latitude: (?)
Longitude: (?)
Altitude: (?)
Bibliography: Sarmento 1933a, 251. Silva 1986a, nº 290.

Nº: **86**
Name: Monte de S. Miguel
Parish: Oleiros
Council: Guimarães
Latitude: 412729 N
Longitude: 082304 W
Altitude: 385
Bibliography: Kalb 1980b, nº 18, Abb. 4. Silva 1986a, nº 277.

Nº: **87**
Name: Monte de S. Tiago
Parish: Penselo
Council: Guimarães
Latitude: 412835 N
Longitude: 081748 W
Altitude: 386
Bibliography: Sarmento 1933a, 269-70. Silva 1986a, nº 273.

Nº: **88**
Name: Castro de Santo António
Parish: Pinheiro
Council: Guimarães
Latitude: (?)
Longitude: (?)
Altitude: (?)
Bibliography: Pina 1940, 513. Silva 1986a, nº 292.

Nº: **89**
Name: Castro
Parish: Polvoreira
Council: Guimarães
Latitude: 412429 N
Longitude: 081810 W
Altitude: 347
Bibliography: Sarmento 1933a, 211-5. Silva 1986a, nº 291.

Nº: **90**
Name: (?)
Parish: Ponte (S. João)
Council: Guimarães
Latitude: (?)
Longitude: (?)
Altitude: (?)
Bibliography: Sarmento 1933a, 264-6. Silva 1986a, nº 276.

Nº: **91**
Name: Cidade de Santa Eufémia
Parish: Prazins (Santa Eufémia)
Council: Guimarães
Latitude: 412904 N
Longitude: 081841 W
Altitude: 236
Bibliography: Silva 1986a, nº 272.

Nº: **92**
Name: Monte Castro
Parish: Prazins (Sto. Tirso)
Council: Guimarães
Latitude: 412920 N
Longitude: 081813 W
Altitude: 288
Bibliography: Sarmento 1933a, 266-8. Silva 1986a, nº 270.

Nº: **93**
Name: (?)
Parish: Ronfe
Council: Guimarães
Latitude: (?)

Longitude: (?)
Altitude: (?)
Bibliography: Silva 1986a, nº 278. Vasconcellos 1905 (2), 64, 329-31.

Nº: **94**
Name: Castro de S. Torcato
Parish: S. Torcato
Council: Guimarães
Latitude: 412906 N
Longitude: 081527 W
Altitude: 299
Bibliography: Cunha 1975, 527-28. Sarmento 1933a, 225.

Nº: **95**
Name: Penedo da Cabeça
Parish: S. Torcato
Council: Guimarães
Latitude: (?)
Longitude: (?)
Altitude: (?)
Bibliography: Cunha 1946a, 3. Sarmento 1933a, 223-7. Silva 1986a, nº 269.

Nº: **96**
Name: Sever
Parish: Sande (S. Clemente)
Council: Guimarães
Latitude: 412925 N
Longitude: 082200 W
Altitude: 160
Bibliography: Sarmento 1933a, 119-20. Silva 1986a, nº 274.

Nº: **97**
Name: Monte da Forca
Parish: Sande (Vila Nova)
Council: Guimarães
Latitude: 412759 N
Longitude: 082204 W
Altitude: 266
Bibliography: Sarmento 1933a, 166. Silva 1986a, nº 275.

Nº: **98**
Name: Senhora do Monte
Parish: Selho (S. Cristóvão)
Council: Guimarães
Latitude: 412422 N
Longitude: 082042 W
Altitude: 410
Bibliography: Pina 1940, 513. Sarmento 1933a, 239-40. Silva 1986a, nº 289.

Nº: **99**
Name: Monte da Santa
Parish: Selho (S. Jorge)
Council: Guimarães
Latitude: 412531 N
Longitude: 082145 W
Altitude: 249
Bibliography: Kalb 1980b, nº 19, Abb. 4. Sarmento 1933a, 256-9. Silva 1986a, nº 279.

Nº: **100**
Name: Penedo da Senhora
Parish: Serzedelo
Council: Guimarães
Latitude: 412455 N
Longitude: 082042 W
Altitude: 210
Bibliography: Sarmento 1933a, 246-7. Silva 1986a, nº 294.

Nº: **101**
Name: Penedo do Lobo
Parish: Silvares
Council: Guimarães
Latitude: 412605 N
Longitude: 082055 W
Altitude: 225
Bibliography: Sarmento 1933a, 259-62. Silva 1986a, nº 280.

Nº: **102**
Name: Sapadelo
Parish: Souto (S. Salvador)
Council: Guimarães
Latitude: (?)
Longitude: (?)
Altitude: (?)
Bibliography: Sarmento 1933a, 270-3. Silva 1986a, nº 268.

Nº: **103**
Name: Castro dos Mouros
Parish: Souto (Santa Maria)
Council: Guimarães
Latitude: 412938 N
Longitude: 081654 W
Altitude: 421
Bibliography: Silva 1986a, nº 266. Sarmento 1933, 270-3.

Nº: **104**
Name: Castro
Parish: Tagilde
Council: Guimarães
Latitude: (?)
Longitude: (?)
Altitude: (?)
Bibliography: Silva 1986a, nº 301.

Nº: **105**
Name: Monte da Forca
Parish: Urgeses
Council: Guimarães
Latitude: (?)
Longitude: (?)
Altitude: (?)
Bibliography: Sarmento 1933a, 236-7. Silva 1986a, nº 288.

Nº: **106**
Name: Santa Iria
Parish: Ajude/Verim
Council: Póvoa de Lanhoso
Latitude: 413820N
Longitude: 081847 W
Altitude: 110
Bibliography: Martins 1987, 173. Silva 1986a, nº 242.

Nº: **107**
Name: Monte Crasto
Parish: Brunhais
Council: Póvoa de Lanhoso
Latitude: 413522 N
Longitude: 081055 W
Altitude: 447
Bibliography: Cunha 1975, 523-4. Silva 1986a, nº 250.

Nº: **108**
Name: Bouça do Alto
Parish: Calvos
Council: Póvoa de Lanhoso
Latitude: (?)
Longitude: (?)
Altitude: (?)
Bibliography: Silva 1986a, nº 248.

Nº: **109**
Name: Santa Iria
Parish: Campos/Louredo
Council: Póvoa de Lanhoso
Latitude: 413219 N
Longitude: 081741 W
Altitude: 306
Bibliography: Sarmento 1933a, 166;1902, 27-33. Silva 1986a, nº 251, nº 263. Pina 1940, 506-7.

Nº: **110**
Name: Castro de S. Mamede
Parish: Frades
Council: Póvoa de Lanhoso
Latitude: 413738 N
Longitude: 081442 W
Altitude: 742
Bibliography: Silva 1986a, nº 244, nº 253. Cunha 1975, 553.

Nº: **111**
Name: Castelo
Parish: Garfe
Council: Póvoa de Lanhoso
Latitude: 413200 N
Longitude: 081447 W
Altitude: 230
Bibliography: Cunha 1975. Silva 1986a, nº 252.

Nº: **112**
Name: Castelo (Monte Pilar)
Parish: Lanhoso
Council: Póvoa de Lanhoso
Latitude: 413500 N
Longitude: O81647 W
Altitude: 385
Bibliography: Abasolo-Perez 1980, 107. BDGEMN 1942. Cunha 1975, 517-8. Garcia y Bellido 1946b, 356-8. Lopez Cuevillas 1946-47, 573, fig. 6; 1951c, 29-30. Martins 1987, 169. Raddatz 1969, 196, nº 33. Silva 1986a, nº 247. Teixeira 1939, 245-52; 1940b, 117-25; 1940c, 515-29; 1941, 138-9.

Nº: **113**
Name: Província
Parish: Lanhoso
Council: Póvoa de Lanhoso
Latitude: 413451 N
Longitude: 081802 W
Altitude: 496
Bibliography: Sarmento 1933a, 166. Silva 1986a, nº 246.

Nº: **114**
Name: Monte de Santo Tirso
Parish: Rendufinho
Council: Póvoa de Lanhoso
Latitude: 413645 N
Longitude: 081643 W
Altitude: 461
Bibliography: Cunha 1975. Martins 1987, 167. Silva 1986a, nº 245.

Nº: **115**
Name: Monte Vermelho - Monte Castro - Castro de S. João do Rei
Parish: S. João do Rei
Council: Póvoa de Lanhoso
Latitude: 413705 N
Longitude: 081740 W
Altitude: 202
Bibliography: Bellino 1909, 1-28. Cardozo 1938, 82-3. Cortez 1946a, 20. Kalb 1980b, nº 14, Abb. 2. Martins 1987, 171-2.

Monteagudo 1977, 268, nº 1795. Silva 1986a, nº 243. Pinto 1929b, 425. Cortez 1946, 20. Guimarães 1978, 415.

Nº: **116**
Name: Castro de Santinho
Parish: Taíde
Council: Póvoa de Lanhoso
Latitude: (?)
Longitude: (?)
Altitude: (?)
Bibliography: Silva 1986a, nº 249.

Nº: **117**
Name: Calcedónia
Parish: Campo do Gerês
Council: Terras de Bouro
Latitude: 414012 N
Longitude: 081158 W
Altitude: 910
Bibliography: Azevedo, PA, 1897, 104-5, 238. Corrêa-Teixeira 1946, 212-22. Cunha 1950c, 2, 4. Silva 1986a, nº 231.

Nº: **118**
Name: Castro da Perra da Moura
Parish: Chamoim
Council: Terras de Bouro
Latitude: (?)
Longitude: (?)
Altitude: (?)
Bibliography: Silva 1986a, nº 230. Silva, DM, 1958.

Nº: **119**
Name: Castro Mau
Parish: Chorense
Council: Terras de Bouro
Latitude: (?)
Longitude: (?)
Altitude: (?)
Bibliography: Silva 1986a, nº 229. Silva, DM, 1985, 135-48.

Nº: **120**
Name: Saim
Parish: Chorense
Council: Terras de Bouro
Latitude: (?)
Longitude: (?)
Altitude: (?)
Bibliography: Silva 1986a, nº 228. Silva, DM, 1985, 135-48.

Nº: **121**
Name: Monte do Facho
Parish: Calendário
Council: V. N. Famalicão
Latitude: 412454 N
Longitude: 083147 W
Altitude: 268
Bibliography: Teixeira et al. 1965, 46.

Nº: **122**
Name: S. Miguel-O-Anjo
Parish: Calendário
Council: V. N. Famalicão
Latitude: 412326 N
Longitude: 083155W
Altitude: 194

Nº: **123**
Name: S. Miguel-O-Anjo
Parish: Delães/Ruivães
Council: V. N. Famalicão
Latitude: 412355 N
Longitude: 082528 W
Altitude: 291
Bibliography: Lima 1926-28, 289-98. Sarmento 1933a, 420-1, 428-30. Silva 1986a, nº 260. Vasconcellos 1905 (2), 327-9.

Nº: **124**
Name: Penices
Parish: Gondifelos
Council: V. N. Famalicão
Latitude: 412419 N
Longitude: 083640 W
Altitude: 99
Bibliography: Queiroga 1987

Nº: **125**
Name: Santa Tecla
Parish: Oliveira (Sta. Maria)
Council: V. N. Famalicão
Latitude: 412501 N
Longitude: 082406 W
Altitude: 314
Bibliography: Sarmento 1901, 14-7. Silva 1986a, nº 259.

Nº: **126**
Name: Ermidas
Parish: Jesufrei
Council: V. N. Famalicão
Latitude: 412740 N
Longitude: 083120 W
Altitude: 209
Bibliography: Queiroga 1985a; 1985c, 56-7. Silva 1986a, nº 257.

Nº: **127**
Name: Eiras
Parish: Pousada de Saramagos
Council: V. N. Famalicão
Latitude: 412623 N
Longitude: 082630 W

Altitude: 384

Nº: **128**
Name: Santa Cristina
Parish: Requião
Council: V. N. Famalicão
Latitude: 412541 N
Longitude: 082748 W
Altitude: 300

Nº: **129**
Name: Bóca
Parish: Vale (S. Cosme)
Council: V. N. Famalicão
Latitude: 412701 N
Longitude: 082824 W
Altitude: 289

Nº: **130**
Name: Cruito
Parish: Vale S. Martinho
Council: V. N. Famalicão
Latitude: 412546 N
Longitude: 082811 W
Altitude: 210

Nº: **131**
Name: Castelo
Parish: Vermoim
Council: V. N. Famalicão
Latitude: 412543 N
Longitude: 082656 W
Altitude: 356
Bibliography: Almeida 1974c, 163. Cardozo 1972b, 173, nº 104, 175, nº 106. Queiroga 1985b, 56. Rosário 1973, 59. Sarmento 1933a, 81. Silva 1986a, nº 258.

Nº: **132**
Name: Castro
Parish: Anissó
Council: Vieira do Minho
Latitude: 413632 N
Longitude: 080947 W
Altitude: 733
Bibliography: Silva 1986a, nº 255.

Nº: **133**
Name: Monte do Castelo
Parish: Rossas
Council: Vieira do Minho
Latitude: 413437 N
Longitude: 080339 W
Altitude: 730
Bibliography: Cunha 1950b, 2, 4; 1975, 509-12. Silva 1986a, nº 256. Teixeira 1955-56, 17-38.

Nº: **134**
Name: Vila Seca
Parish: Vieira do Minho
Council: Vieira do Minho
Latitude: 413735 N
Longitude: 080737 W
Altitude: 485
Bibliography: Cunha 1975, 507-8. Silva 1986a, nº 254.

Nº: **135**
Name: S. Sebastião
Parish: Atiães
Council: Vila Verde
Latitude: 413742 N
Longitude: 082921 W
Altitude: 100
Bibliography: Regalo 1986, 87. Martins 1987, 179.

Nº: **136**
Name: Castelhão Barbudo
Parish: Barbudo
Council: Vila Verde
Latitude: 413902 N
Longitude: 082724 W
Altitude: 200
Bibliography: Martins 1987, 180. Cunha 1975, 530.

Nº: **137**
Name: Monte (Castro) do Barbudo ; Brito
Parish: Barbudo/Carreiras (S.Miguel)
Council: Vila Verde
Latitude: 413907N
Longitude: 082808 W
Altitude: 331
Bibliography: Azevedo, PA, 1897, 152. Cunha 1955b, 1, 3; 1975, 529-30. Hipólito 1960-61, 16-7. Martins 1985e; 1987, 179. Regalo 1986, 87. Silva 1986a, nº 217. Oliveira, M, 1908, 667-8.

Nº: **138**
Name: Lugar do Coto/ Póvoa/ Lugar do Reduto
Parish: Dossãos
Council: Vila Verde
Latitude: 414027 N
Longitude: 082802 W
Altitude: 250
Bibliography: Vieira 1886-87, 404. Martins 1987, 183-84. Regalo 1986, 88.

Nº: **139**
Name: Reduto das Coroas
Parish: Duas Igrejas
Council: Vila Verde

Latitude: 414235 N
Longitude: 082947 W
Altitude: 504
Bibliography: Regalo 1986, 88. Costa 1868, 235. Vieira 1886-87, 163. Leal - Ferreira 1873, 298. Azevedo, A P, 1896, 260.

Nº: **140**
Name: Castro (Montezelo) - Escariz
Parish: Escariz (S. Martinho - Isidoro)
Council: Vila Verde
Latitude: 413849 N
Longitude: 083200 W
Altitude: 235
Bibliography: Regalo 1986, 88. Silva 1986a, nº 219. Martins 1987, 193. Azevedo, PA, 1898, 143. Costa 1868, 304. Cunha, AC, 1975, 28. Oliveira 1908, 668.

Nº: **141**
Name: Castelo (Monte da Santa) Castro de Sta. Eulália
Parish: Geme
Council: Vila Verde
Latitude: 414002 N
Longitude: 082609 W
Altitude: 217
Bibliography: Silva 1986a, nº 216. Regalo 1986, 89. Martins 1987, 184.

Nº: **142**
Name: Penedos do Inferno
Parish: Gondomar
Council: Vila Verde
Latitude: 414504 N
Longitude: 082101W
Altitude: 700
Bibliography: Regalo 1986, 89.

Nº: **143**
Name: Castro de Vairão
Parish: Gormide (Oriz?)
Council: Vila Verde
Latitude: 414222 N
Longitude: 082247 W
Altitude: 462
Bibliography: Martins 1987, 194. Regalo 1986, 89. Silva 1986a, nº 212.

Nº: **144**
Name:
Parish: Lugar do Gaião - Cabanelas
Council: Vila Verde
Latitude: 413425 N
Longitude: 082930 W
Altitude: 60
Bibliography: Martins 1987, 181

Nº: **145**
Name:
Parish: Outeirinho
Council: Vila Verde
Latitude: 413834 N
Longitude: 082549 W
Altitude: 74
Bibliography: Martins 1987, 195

Nº: **146**
Name: Monte Sabroso
Parish: Pedregais
Council: Vila Verde
Latitude: 414216 N
Longitude: 082828 W
Altitude: 4
Bibliography: Silva 1986a, nº 214.

Nº: **147**
Name: Citânia de S. Julião
Parish: Ponte/Coucieiro
Council: Vila Verde
Latitude: 414115 N
Longitude: 082337 W
Altitude: 296
Bibliography: Alarcão, A, 1958, 281, 298, 300. Azevedo, PA, 1901, 50. Bellino 1909, 6. Cardozo 1947c, 65-70. Coffyn 1985, nº 138. Costa 1868-69, 216. Freitas, J, 1971, 133-5. Kalb 1980b, nº 13, Abb. 2. Martins 1984b; 1985a; 1985e; 1987, 183-88.. Martins-Silva 1984. Regalo 1986, 90-1. Rosário 1973, 8, 22-3. Silva 1986a, nº 215, p. 183, 148, 201-2, 204, 229, 305.

Nº: **148**
Name: Penedos da Portela
Parish: Portela das Cabras
Council: Vila Verde
Latitude: 414022 N
Longitude: 082922 W
Altitude: 453
Bibliography: Martins 1987. Regalo 1986, 91.

Nº: **149**
Name: Castelo dos Mouros/ Monte dos Francos
Parish: Rio Mau
Council: Vila Verde
Latitude: 414252 N
Longitude: 083038 W
Altitude: 590
Bibliography: Costa 1868, 235. Regalo 1986, 92.

Nº: **150**
Name: Moega

Parish: Rio Mau
Council: Vila Verde
Latitude: 414216 N
Longitude: 083020 W
Altitude: 300
Bibliography: Regalo 1986, 91.

Nº: **151**
Name: Sta. Helena / Sta. Cruz
Parish: Turiz
Council: Vila Verde
Latitude: 413753 N
Longitude: 082746 W
Altitude: 170
Bibliography: Martins 1987, 194. Regalo 1986, 92

Nº: **152**
Name: Castelo da Nobrega
Parish: Aboim da Nobrega / Sampriz
Council: Vila Verde/Ponte da Barca
Latitude: 414607 N
Longitude: 082221 W
Altitude: 773
Bibliography: Regalo 1986, 86. Cunha 1975, 503-06. Abreu, L, 1963, 15. Leal - Ferreira 1873. Costa 1868, 212.

2.2) District of Porto

Nº: **153**
Name: Monte do Crasto (Gião)
Parish: Anciães
Council: Amarante
Latitude: 411431 N
Longitude: 075820 W
Altitude: 403
Bibliography: Pinho 1905-08, 476. Silva 1986a, nº 412.

Nº: **154**
Name: Castelo
Parish: Carvalho de Rei
Council: Amarante
Latitude: 411406 N
Longitude: 080155 W
Altitude: 560
Bibliography: Silva 1986a, nº 413.

Nº: **155**
Name: Alto do Ladário (Paredes)
Parish: Gatão
Council: Amarante
Latitude: 411741 N
Longitude: 080359 W
Altitude: 273
Bibliography: Pinho 1905-08, 476. Silva 1986a, nº 410.

Nº: **156**
Name: Picoto (Outeiro das Chapas)
Parish: Salvador do Monte
Council: Amarante
Latitude: 411334 N
Longitude: 080548 W
Altitude: 330
Bibliography: Pinho 1929b. Lopez Cuevillas 1951c, 60. Silva 1986a, nº 407, 247, est. 109/2-2 A.

Nº: **157**
Name: Crasto
Parish: Várzea
Council: Amarante
Latitude: 411559 N
Longitude: 080036 W
Altitude: 294
Bibliography: Silva 1986a, nº 411.

Nº: **158**
Name: Castro de Coura
Parish: Vila Caiz
Council: Amarante
Latitude: 411334 N
Longitude: 080728 W
Altitude: 224
Bibliography: Silva 1986a, nº 408. Soeiro 1984, 39, 41-2.

Nº: **159**
Name: Igreja
Parish: Vila Garcia
Council: Amarante
Latitude: 411848 N
Longitude: 080418 W
Altitude: 219
Bibliography: Pinho 1905-08, 476. Silva 1986a, nº 409.

Nº: **160**
Name: Castro de Porto Manso
Parish: Ancede
Council: Baião
Latitude: 410544 N
Longitude: 080440 W
Altitude: 200
Bibliography: Jorge, VO, 1988, 12. Silva 1986a, nº 418. Vasconcelos, J, 1905-08, 669-70.

Nº: **161**
Name: Castro do Cruito
Parish: Gove
Council: Baião
Latitude: 410805 N
Longitude: 080203 W
Altitude: 516
Bibliography: Jorge, VO, 1988, 11-2. Pereira - González 1988. Silva 1986a, nº 416. Vasconcellos 1898, 177. Vasconcelos, J, 1905-08, 670-1.

Nº: **162**
Name: Castro de Pousada (Fiéis de Deus)
Parish: Santa Cruz do Douro
Council: Baião
Latitude: 410756 N
Longitude: 080101 W
Altitude: 716
Bibliography: Jorge, VO, 1988, 12. Vieira, A J, 1887, 443-73. Vasconcelos 1908, 671.

Nº: **163**
Name: Monte de Mantel
Parish: Santa Cruz do Douro
Council: Baião
Latitude: 410747 N
Longitude: 075944 W
Altitude: 515
Bibliography: Jorge, V O, 1988, 12. Silva 1986a, nº 420, nº 421. Vasconcellos 1898, 177. Vasconcelos, J, 1905-08, 671. Vieira 1887, 460.

Nº: **164**
Name: Crasto do Barreiro
Parish: Santa Marinha do Zêzere
Council: Baião
Latitude: 410825 N
Longitude: 075712 W
Altitude: 397
Bibliography: Cardozo 1985, 148-9. Jorge, V O, 1988, 12. Silva 1986a, nº 417. Vasconcellos 1889, 178-9; 1905 (2), 286. Vasconcelos, J, 1905-08, 672.

Nº: **165**
Name: Castelo (Pram)
Parish: Sta. Cruz do Douro
Council: Baião
Latitude: 410706 N
Longitude: 080059 W
Altitude: 404
Bibliography: Jorge, V O, 1988, 12. Silva 1986a, nº 419. Vasconcelos, J, 1908, 671.

Nº: **166**
Name: Castro
Parish: Teixeira
Council: Baião
Latitude: (?)
Longitude: (?)
Altitude: (?)
Bibliography: Jorge, V O , 1988, 12. Silva 1986a, nº 414. Vasconcelos, J, 1905-08, 672.

Nº: **167**
Name: Pedra Falada
Parish: Aião
Council: Felgueiras
Latitude: 411752N
Longitude: 081130 W
Altitude: 429
Bibliography: Silva 1986a, nº 367.

Nº: **168**
Name: Castro (Macieira)
Parish: Lagares
Council: Felgueiras
Latitude: (?)
Longitude: (?)
Altitude: (?)
Bibliography: Freitas 1960, 17. Silva 1986a, nº 362.

Nº: **169**
Name: Cristelo de S. Veríssimo
Parish: Lagares
Council: Felgueiras
Latitude: 412159 N
Longitude: 081438 W
Altitude: 454
Bibliography: Freitas 1960, 10. Sarmento 1933a, 204-6. Silva 1986a, nº 361.

Nº: **170**
Name: Santa Quitéria
Parish: Margaride
Council: Felgueiras
Latitude: 412214 N
Longitude: 081132 W
Altitude: 466
Bibliography: Freitas 1960, 16. Silva 1986a, nº 366.

Nº: **171**
Name: Sta. Maria da Pedreira
Parish: Pedreira
Council: Felgueiras
Latitude: 411922 N
Longitude: 081258 W

Altitude: 267
Bibliography: Fernandes, MA, 1989, 80.

Nº: **172**
Name: Senhor dos Perdidos
Parish: Penacova
Council: Felgueiras
Latitude: 412142 N
Longitude: 081448 W
Altitude: 459
Bibliography: Freitas 1960, 15. Silva 1986a, nº 363.

Nº: **173**
Name: Picoto (Monte do Castro)
Parish: Pombeiro de Ribavizela
Council: Felgueiras
Latitude: 412235 N
Longitude: 081247 W
Altitude: 331
Bibliography: Freitas 1960, 13. Silva 1986a, nº 364.

Nº: **174**
Name: Cividade
Parish: Refontoura
Council: Felgueiras
Latitude: 411941 N
Longitude: 081127 W
Altitude: 322
Bibliography: Fernandes, MA, 1989, 34, 80.

Nº: **175**
Name: Santa Comba (Cristelo)
Parish: Regilde
Council: Felgueiras
Latitude: 412208 N
Longitude: 081541 W
Altitude: 218
Bibliography: Freitas 1960, 12. Sarmento 1933a, 201-4. Silva 1986a, nº 360 = 300.

Nº: **176**
Name: Castro
Parish: Sendim
Council: Felgueiras
Latitude: 412253 N
Longitude: 081027 W
Altitude: 425
Bibliography: Almeida 1974a. Cardozo 1974, 1-4. Freitas 1960, 17. Pinto 1930b, 6; 1933c, 376-80. Silva 1986a, nº 365.

Nº: **177**
Name: Castro de Sto. Adrião de Vizela ou Senhora da Tocha
Parish: Vizela Sto Adrião
Council: Felgueiras
Latitude: 412141 N
Longitude: 081654 W
Altitude: 249
Bibliography: Sarmento 1933a, 198-200.

Nº: **178**
Name: Monte do Castelo
Parish: Covelo
Council: Gondomar
Latitude: 410538 N
Longitude: 082657 W
Altitude: 220
Bibliography: Oliveira, C, 1934, 22. Silva 1986a, nº 427. Soeiro 1984, 111.

Nº: **179**
Name: Alto do Castro
Parish: Medas
Council: Gondomar
Latitude: 410255 N
Longitude: 082612 W
Altitude: 121
Bibliography: Silva 1986a, nº 429. Soeiro 1984, 112, fig. 5.

Nº: **180**
Name: Castro (Broalhos)
Parish: Medas
Council: Gondomar
Latitude: 410410 N
Longitude: 082713 W
Altitude: 80
Bibliography: Silva 1986a, nº 428. Soeiro 1984, 112, fig. 5.

Nº: **181**
Name: Monte Crasto
Parish: São Cosme
Council: Gondomar
Latitude: 410826 N
Longitude: 082713 W
Altitude: 192
Bibliography: Oliveira, C, 1934, 18-9. Silva 1986a, nº 426.

Nº: **182**
Name: Monte de S. Domingos
Parish: Cristelos
Council: Lousada
Latitude: 411610 N
Longitude: 083732 W
Altitude: 246
Bibliography: Sarmento 1933a, 166. Silva 1986a, nº 357.

Nº: **183**
Name: (?)
Parish: Pias
Council: Lousada

Latitude: (?)
Longitude: (?)
Altitude: (?)
Bibliography: Silva 1986a, nº 358.

Nº: **184**
Name: Santa Agueda
Parish: Sousela
Council: Lousada
Latitude: 411830 N
Longitude: 081450 W
Altitude: 577
Bibliography: Silva 1986a, nº 356.

Nº: **185**
Name: Bacelo
Parish: Torno
Council: Lousada
Latitude: 411803 N
Longitude: 081242 W
Altitude: 218
Bibliography: Silva 1986a, nº 359.

Nº: **186**
Name: Alto da Maia
Parish: Aguas Santas
Council: Maia
Latitude: 411223 N
Longitude: 083410 W
Altitude: 146
Bibliography: Azevedo 1939, 102-5. Silva 1986a, nº 375.

Nº: **187**
Name: Castro de Avioso (Castelo da Maia)
Parish: Avioso (Sta. Maria)
Council: Maia
Latitude: 411623 N
Longitude: 083645 W
Altitude: 108
Bibliography: Almeida 1969, 33. Azevedo 1939, 102-5. Silva 1986a, nº 372.

Nº: **188**
Name: Santo Ovídio
Parish: Barca
Council: Maia
Latitude: (?)
Longitude: (?)
Altitude: (?)
Bibliography: Azevedo 1939, 102-5. Silva 1986a, nº 373.

Nº: **189**
Name: Monte do Faro
Parish: Gemunde
Council: Maia
Latitude: 411650 N
Longitude: 083928 W
Altitude: 144
Bibliography: Azevedo 1939, 102-5. Moreira 1969, 21. Silva 1986a, nº 371.

Nº: **190**
Name: Crasto
Parish: Vermoim
Council: Maia
Latitude: 411447 N
Longitude: 083608 W
Altitude: 99
Bibliography: Almeida 1969, 42. Azevedo 1939, 108-42. Silva 1986a, nº 374.

Nº: **191**
Name: Castro de Arados
Parish: Alpendurada/Magrelos
Council: Marco de Canaveses
Latitude: 410548 N
Longitude: 081337 W
Altitude: 480
Bibliography: Lanhas-Brandão 1967, 12-3. Ponte 1984, nº 71. Silva 1986a, nº 404. Vasconcellos 1905 (2), 79-80. Vasconcelos M, 1914-16.

Nº: **192**
Name: Freixo
Parish: Freixo
Council: Marco de Canaveses
Latitude: 410944 N
Longitude: 080837 W
Altitude: 274
Bibliography: Aguiar, MV, 1947, 210-6. Brandão 1973. Capela 1895, 247. Correia, JM, 1957, 22-3. Dias, LAT, 1980, 74-6; 1984, 86-90. Fonseca, C, 1940, 540. Monteiro 1955; 1960. Sarmento 1933a, 305-6. Silva 1986a, nº 402. Soares-Dias 1985. Vasconcellos 1900, 33; 1905 (2), 179; 1913, 222-24. Vasconcelos, M, 1914-16; 1935, 3. Vieira 1886-87, 496-7.

Nº: **193**
Name: Castro do Boi
Parish: Sande
Council: Marco de Canaveses
Latitude: (?)
Longitude: (?)
Altitude: (?)
Bibliography: Silva 1986a, nº 405. Vasconcelos, M, 1914-16.

Nº: **194**
Name: Crasto

Parish: Soalhães
Council: Marco de Canaveses
Latitude: (?)
Longitude: (?)
Altitude: (?)
Bibliography: Silva 1986a, nº 403. Vasconcelos, M, 1914-16, 327-9.

Nº: **195**
Name: Castro de S. Tiago
Parish: Tabuada
Council: Marco de Canaveses
Latitude: 411046 N
Longitude: 080447 W
Altitude: 499
Bibliography: Jorge, VO, 1988, 12.

Nº: **196**
Name: Castro de Quires
Parish: Vila Boa de Quires
Council: Marco de Canaveses
Latitude: 411220 N
Longitude: 081323 W
Altitude: 512
Bibliography: Silva 1986a, nº 401. Vasconcellos 1897 (1), 14. Vasconcelos, M, 1914-16.

Nº: **197**
Name: Monte do Castelo
Parish: Guifões
Council: Matosinhos
Latitude: 411155 N
Longitude: 084029 W
Altitude: 50
Bibliography: Almeida 1969, 33. Correia, JM, 1916, 336. Fortes 1905-8a - fibula de longo travessão sem espira. Felgueiras 1958, 3, 12-5. Sarmento 1933, 165. Corrêa 1935b; 1928; 1924 - (ceramica aretina). Pinto 1927, 24-5. Santos, JN, 1955, 25-60; 1962a, 111-6; 1962b. Silva 1986a, nº 369. Vasconcellos 1898, 270-2; 1901, 35; 1905 (2), 39.

Nº: **198**
Name: Monte Castro (Angeiras)
Parish: Lavra
Council: Matosinhos
Latitude: 411556 N
Longitude: 084249 W
Altitude: 25
Bibliography: Almeida 1969, 34. Felgueiras 1958, 3. Lanhas-Brandão 1969, 326. Ramos 1943, 13. Silva 1986a, nº 368.

Nº: **199**
Name: Recarei
Parish: Leça do Balio
Council: Matosinhos
Latitude: (?)
Longitude: (?)
Altitude: (?)
Bibliography: Silva 1986a, nº 370.

Nº: **200**
Name: Castro de S. Domingos
Parish: Carvalhosa/Ferreira/Paços de Ferreira
Council: Paços de Ferreira
Latitude: 411654 N
Longitude: 082125 W
Altitude: 383
Bibliography: Dinis 1971. Silva 1986a, nº 354; 1986b, 106.

Nº: **201**
Name: Castro do Monte do Socorro
Parish: Codeços
Council: Paços de Ferreira
Latitude: 411952 N
Longitude: 082115 W
Altitude: 423
Bibliography: Coelho, JF, 1962, 137; 1963, 179-80; 1965, 226-8. Dinis 1962, 223-91; 1971, 287-91. Silva 1986a, nº 348; 1986b, 105.

Nº: **202**
Name: Castro (S. Brás)
Parish: Frazão
Council: Paços de Ferreira
Latitude: 411601 N
Longitude: 082448 W
Altitude: 322
Bibliography: Dinis 1976, 217-21. Silva 1986a, nº 352; 1986b.

Nº: **203**
Name: Castro de Mortórios
Parish: Freamunde
Council: Paços de Ferreira
Latitude: 411714 N
Longitude: 081945 W
Altitude: 363
Bibliography: Dinis 1963, 94. Silva 1986a, nº 355, 141-3,149, 150, 152, 154-5, 157-8, 160-1, 193, 197-8, 201-2, 204, 213, 215-6, 220,226, 251, 308; 1986b, 106-7; 1983a , 144, 209, 227, 229. Paço-Jalhay 1948, 20; 1950a, 29-31; 1955, 194; 1968a,350-1; 1968b; 1974. Almeida 1973a(=1975, 495); 1974a, 145.

Ponte, 1984, nº 5, 10, 21, 30, 52. Coelho *et al.*, 1959, 277;. Silva-Centeno 1980, 68. Cuevillas 1951c, 38. Raddatz 1969, 197, nº37 (MACS, SA/Au1). Calo Lourido 1983, 1991.

Nº: **204**
Name: Castro de Busto
Parish: Meixomil/Penamaior
Council: Paços de Ferreira
Latitude: (?)
Longitude: (?)
Altitude: (?)
Bibliography: Dinis 1965, 189-95. Silva 1986a, nº 350; 1986, 106.

Nº: **205**
Name: Castelos
Parish: Paços de Ferreira
Council: Paços de Ferreira
Latitude: 411546 N
Longitude: 082238 W
Altitude: 339
Bibliography: Dinis 1978, 187-90. Silva 1986a, nº 353.

Nº: **206**
Name: Castro da Vila
Parish: Penamaior
Council: Paços de Ferreira
Latitude: 411650 N
Longitude: 082512 W
Altitude: 419
Bibliography: Dinis 1966, 512-18; 1971, 287-91; 1980, 389-93. Hipólito 1960-61, 45-6. Silva 1986a, nº 351; 1986, 106.

Nº: **207**
Name: Castro de S. Gonçalo
Parish: Raimonda
Council: Paços de Ferreira
Latitude: 411940 N
Longitude: 081935 W
Altitude: 460
Bibliography: Dinis 1971. Silva 1986a, nº 349; 1986, 105-6.

Nº: **208**
Name: Citânia de Sanfins
Parish: Sanfins de Ferreira
Council: Paços de Ferreira
Latitude: 411922 N
Longitude: 082305 W
Altitude: 570
Bibliography: Almeida 1973a (=1975, 495, 497, Fig. 1), 1974a; 1974c, 149-72. Argote 1734, 465-7. Brandão 1985. Coelho *et al.* 1959, 271-9. Garcia y Bellido 1944, 297-9. Dinis 1954, 63-73. Jalhay 1944, 414-35; 1949, 300-3; 1950a, 29-31; 1950b, 38-49; 1951, 5-9. Jalhay-Paço 1948, 699-717. Lopez Cuevillas 1951c, 38. Maluquer de Motes 1950, 60-1. Paço 1952, 375-84; 1953a, 489-94; 1953b, 673-89; 1953c, 683-97; 1954a, 139-53; 1954b, 197-212; 1954c, 139-53; 1956, 943-46; 1957, 353; 1959, 261-69; 1960, 19-20; 1961a, 106-19; 1963, 71-7; 1965b, 167-71; 1968a, 329-54; 1968b, 710-25; 1968c, 45-59; 1974, 3-9. Paço-Jalhay 1955, 189-257. Ponte 1984. Raddatz 1969, 197, nº 37. Sarmento 1933a, 168, 423-7. Silva 1980b, 80-2; 1983c; 1986a, nº 346; 1986b, 107-17. Silva-Centeno 1980, 57-78. Tavares 1970, 347-57. Vasconcellos 1895b; 1905 (2), 188-9.

Nº: **209**
Name: Castro
Parish: Sanfins de Ferreira/Lamoso/Codeços
Council: Paços de Ferreira
Latitude: 411950 N
Longitude: 082141W
Altitude: 480
Bibliography: Dinis 1971, 288. Paço 1952, 382-4. Proença Júnior 1908, 2, 18, 36, 40. Silva 1986a, nº 347; 1986, 105.

Nº: **210**
Name: (?)
Parish: Cristelo
Council: Paredes
Latitude: 411342 N
Longitude: 082126 W
Altitude: 333
Bibliography: Silva 1986a, nº 380.

Nº: **211**
Name: Monte do Picoto
Parish: Rebordosa
Council: Paredes
Latitude: (?)
Longitude: (?)
Altitude: (?)
Bibliography: Silva 1986a, nº 378.

Nº: **212**
Name: Serra do Muro (Monte)

Parish: Vandoma
Council: Paredes
Latitude: 411148 N
Longitude: 082330 W
Altitude: 485
Bibliography: Silva 1986a, nº 381. Silva, J, 1963, 126-30.

Nº: **213**
Name: Monte da Costa Figueira
Parish: Vilela
Council: Paredes
Latitude: (?)
Longitude: (?)
Altitude: (?)
Bibliography: Cabré Aguiló 1924, 82-3. Cardozo 1946b, 1-28. Coffyn 1983, 184; 1985, 271. Cuadrado 1955, 129. Freitas 1923. Kalb 1980b, nº 23, Abb. 4. Piggott 1983, 193. Silva 1986a, nº 379.

Nº: **214**
Name: Castro da Penha Grande
Parish: Abragão
Council: Penafiel
Latitude: 410935 N
Longitude: 081354 W
Altitude: 248
Bibliography: Silva 1986a, nº 394. Soeiro 1984, 50-1.

Nº: **215**
Name: Castro de Abragão
Parish: Abragão
Council: Penafiel
Latitude: 410921 N
Longitude: 081427 W
Altitude: 200
Bibliography: Silva 1986a, nº 395. Soeiro 1984, 50.

Nº: **216**
Name: Coto da Cividade
Parish: Eja
Council: Penafiel
Latitude: 410505 N
Longitude: 081803 W
Altitude: 207
Bibliography: Aguiar 1943, 13; 1945, 21. Almeida-Lopes 1981-82. Andrade 1918. Azevedo, PA, 1898. Miranda 1960. Pinho 1929a. Silva 1986a, nº 400. Soeiro 1984, 76.

Nº: **217**
Name: Castro de Bojefa (Abujefa)
Parish: Galegos
Council: Penafiel
Latitude: 410941 N
Longitude: 081914 W
Altitude: 330
Bibliography: Machado, FSL, 1920, 8. Silva 1986a, nº 393. Soeiro 1984, 96-8. Sousa, EF, 1962, 126-32.

Nº: **218**
Name: Castro de Irivo
Parish: Irivo
Council: Penafiel
Latitude: 411045 N
Longitude: 081959 W
Altitude: 157
Bibliography: Miranda, 1959. Silva 1986a, nº 388. Soeiro 1984, 100.

Nº: **219**
Name: Alto do Castro
Parish: Lagares
Council: Penafiel
Latitude: 410743 N
Longitude: 082116 W
Altitude: 284
Bibliography: Silva 1986a, nº 397. Soeiro 1984, 104. Sousa, EF, 1962, 128.

Nº: **220**
Name: Castelo de Ordins
Parish: Lagares
Council: Penafiel
Latitude: 410749 N
Longitude: 082130 W
Altitude: 275
Bibliography: Silva 1986a, nº 396. Soeiro 1984, 104.

Nº: **221**
Name: Castro
Parish: Marecos
Council: Penafiel
Latitude: 411047 N
Longitude: 081832 W
Altitude: 210
Bibliography: Silva 1986a, nº 389. Soeiro 1984, 95-6. Vasconcellos 1895c, 15-6.

Nº: **222**
Name: (?)
Parish: Novelas
Council: Penafiel
Latitude: (?)
Longitude: (?)
Altitude: (?)
Bibliography: Silva 1986a, nº 382.

Nº: **223**
Name: Castelo (Monte do castelo)

Parish: Oldrões
Council: Penafiel
Latitude: 410932 N
Longitude: 081712 W
Altitude: 324
Bibliography: Silva 1986a, nº 391. Soeiro 1984, 104, 160.

Nº: **224**
Name: Monte Mozinho
Parish: Oldrões
Council: Penafiel
Latitude: 410849 N
Longitude: 081840 W
Altitude: 408
Bibliography: Almeida 1974b; 1977, 16, 23, 26; 1980c, 51-6. Aguiar 1933, 32. Andrade 1920, 275-7. Brandão-Sousa 1966. Corrêa 1928, 194. Cortez 1948a, 269-81; 1951b, 161-84. Machado, FSL, 1920, 12-29. Miranda 1954. Paço 1966c, 555-60. Pinho 1931. Pinto 1928-29. Ponte 1984, nºs 25, 29, 44, 64, 74, 82, 83. Silva 1986a, 220, 226-7, 306, nº 390. Soeiro 1981-82, 97-120; 1984, 60, 125-299. Sousa, EF, 1954, 136-49; 1962, 126-32; 1965, 249-69.

Nº: **225**
Name: Cristelo
Parish: Paço de Sousa
Council: Penafiel
Latitude: 410937 N
Longitude: 082017 W
Altitude: 209
Bibliography: Silva 1986a, nº 392. Soeiro 1984, 100.

Nº: **226**
Name: Castro Louredo
Parish: Penafiel
Council: Penafiel
Latitude: 411239 N
Longitude: 081644 W
Altitude: 299
Bibliography: Silva 1986a, nº 387. Soeiro 1984, 86-7.

Nº: **227**
Name: Monte do Sameiro
Parish: Penafiel
Council: Penafiel
Latitude: 411235 N
Longitude: 081620 W
Altitude: 350
Bibliography: Silva 1986a, nº 386. Soeiro 1984, 86.

Nº: **228**
Name: Outeiro do Dino
Parish: Pinheiro
Council: Penafiel
Latitude: 410728 N
Longitude: 081828 W
Altitude: 274
Bibliography: Silva 1986a, nº 398. Soeiro 1984, 60.

Nº: **229**
Name: Suvidade
Parish: Recesinhos (S. Mamede)
Council: Penafiel
Latitude: 411358 N
Longitude: 081336 W
Altitude: 285
Bibliography: Silva 1986a, nº 383. Soeiro 1984, 46-8.

Nº: **230**
Name: Castro
Parish: Recesinhos (S. Martinho)
Council: Penafiel
Latitude: 411306 N
Longitude: 081300 W
Altitude: 310
Bibliography: Silva 1986a, nº 385. Soeiro 1984, 44-6.

Nº: **231**
Name: Crasto da Quintela
Parish: Rio de Moinhos
Council: Penafiel
Latitude: 410555 N
Longitude: 081629 W
Altitude: 119
Bibliography: Lanhas-Brandão 1967, 36-7. Silva 1986a, nº 399. Soeiro 1984, 55.

Nº: **232**
Name: Castro
Parish: Santa Marta
Council: Penafiel
Latitude: 411254 N
Longitude: 081551 W
Altitude: 403
Bibliography: Silva 1986a, nº 384. Soeiro 1984, 85.

Nº: **233**
Name: Castro de Nicola
Parish: Campanhã
Council: Porto
Latitude: (?)
Longitude: (?)
Altitude: (?)

Bibliography: Corrêa 1933-34; 1935b; 1936. Silva 1986a, nº 425.

Nº: **234**
Name: Castro
Parish: Foz do Douro
Council: Porto
Latitude: (?)
Longitude: (?)
Altitude: (?)
Bibliography: Corrêa 1933-34; 1935b; 1936. Silva 1986a, nº 422.

Nº: **235**
Name: Cristelo
Parish: Massarelos
Council: Porto
Latitude: (?)
Longitude: (?)
Altitude: (?)
Bibliography: Azevedo, R, 1960, 113. Bastos 1932, 51-2. Corrêa 1933-34; 1935b; 1936. Silva 1986a, nº 423.

Nº: **236**
Name: Cividade (Morro da Sé)
Parish: Sé
Council: Porto
Latitude: 410828 N
Longitude: 083635 W
Altitude: 76
Bibliography: Abrantes 1945; 1966. Azevedo, R, 1960; 1965. Basto 1963a, 137-41; 1963b, 143-7. Brandão 1963b, 236-45; 1965. Costa, AR, 1945. Corrêa 1933-34; 1935a; 1935b; 1936; 1940. Ferreira, JAP, 1959, 367-68. Machado 1955, 344-79; 1956; 1962, 213-16; 1968, 52-74. Pina 1965, 65-74. Pinto 1927, 24-5. Real 1984, 75-9. Silva 1986a, nº 424. Vasconcellos 1897 (1), 68; 1905 (2), 29; 1938a, 28-38; 1938b, 1394-1400.

Nº: **237**
Name: Monte de S. Félix
Parish: Laundos
Council: Póvoa de Varzim
Latitude: 412655 N
Longitude: 084255 W
Altitude: 208
Bibliography: Almeida 1972, 12-5. Araújo 1980, 103. Blanco Freijeiro 1957a, 288-9. Lopez Cuevillas 1951c, 79-80. Peixoto 1905-08b, 284. Sarmento 1933a, 166. Severo 1905-08e, 403-12. Silva 1986a, 263 (MNSR, CMP 148), 142 (MEHP), nº 322.

Nº: **238**
Name: Crasto
Parish: Navais
Council: Póvoa de Varzim
Latitude: 412630 N
Longitude: 084632 W
Altitude: 84
Bibliography: Sarmento 1933a, 165. Silva 1986a, nº 321.

Nº: **239**
Name: Monte da Cividade
Parish: Terroso
Council: Póvoa de Varzim
Latitude: 412457 N
Longitude: 084350 W
Altitude: 76
Bibliography: Almeida 1972, 20-3. Araújo 1980, 105. Fortes 1905-08j. Freire 1965, 207-9. Garcia y Bellido 1966a, 9-10. Gonçalves, F, 1964, 317-22. Peixoto 1905-08b, 284. Pinto 1932, 81-91. Ponte 1984, nº 14 = IAFCUP, 92 (IAFCUP107/6). Sarmento 1933a, 166. Silva 1980a, 305-15; 1986a, 142, 145-8, 150, 156, 158-62, 164, 193, 208-9, 218, 222, 225, nº 323.

Nº: **240**
Name: Castro de S. Marçal
Parish: Alvarelhos
Council: Santo Tirso
Latitude: 411753 N
Longitude: 083708 W
Altitude: 222
Bibliography: Almeida 1969, 31-2, 45; 1973, 15. Azevedo 1939, 101. Fortes 1899, 7-28. Santarém 1954; 1977, 161-70. Sarmento 1970, 48-52. Silva 1980b, 79-90; 1986a, 205, nº 335. Soeiro 1980, 237-43. Torres 1978-79, 15-250.

Nº: **241**
Name: Torre Alta
Parish: Areias
Council: Santo Tirso
Latitude: 412111 N
Longitude: 082830 W
Altitude: 63
Bibliography: Lima 1956, 222-3. Pimentel 1902, 64. Silva 1986a, nº 337.

Nº: 242
Name: Monte Padrão
Parish: Monte Córdova
Council: Santo Tirso
Latitude: 411853 N
Longitude: 082651 W
Altitude: 413
Bibliography: Ponte 1984, nº 15. Santarém 1951, 49-66; 1955a, 397-429. Silva 1986a,217, nº 344.

Nº: 243
Name: Senhora da Assunção (Môrro da Vela)
Parish: Monte Córdova
Council: Santo Tirso
Latitude: 411924 N
Longitude: 082630 W
Altitude: 472
Bibliography: Santarém 1954; 1955a, 397-429. Silva 1986a, nº 343.

Nº: 244
Name: Castro (Sta. Margarida)
Parish: Negrelos (S. Tomé)
Council: Santo Tirso
Latitude: 412105 N
Longitude: 082334 W
Altitude: 281
Bibliography: Pimentel 1902, 342. Silva 1986a, nº 345.

Nº: 245
Name: Castro de S. Pedro
Parish: Vilarinho
Council: Santo Tirso
Latitude: 412145 N
Longitude: 081931 W
Altitude: 346
Bibliography: Sarmento 1933a, 195-96. Silva 1986a, nº 299

Nº: 246
Name: Couce
Parish: Valongo
Council: Valongo
Latitude: (?)
Longitude: (?)
Altitude: (?)
Bibliography: Reis, JAL, 1904, 69. Silva 1986a, nº 377.

Nº: 247
Name: Santa Justa (Cavada dos Crastos)
Parish: Valongo
Council: Valongo
Latitude: 410136 N
Longitude: 082949 W
Altitude: 248
Bibliography: Kalb 1980b, nº 24, Abb. 4. Reis, JAL, 1904, 57. Silva 1986a, nº 376.

Nº: 248
Name: Castro de Argifonso (Alto dos Castelos)
Parish: Arcos
Council: Vila do Conde
Latitude: 412321 N
Longitude: 083918 W
Altitude: 157
Bibliography: Cortez 1946a, 149-55; 1949d, 270-8. Silva 1986a, nº 329.

Nº: 249
Name: Cividade
Parish: Bagunte
Council: Vila do Conde
Latitude: 412301 N
Longitude: 083910 W
Altitude: 206
Bibliography: Almeida 1974a, 195. Cortez 1948a, 269-81. Kalb 1980b, nº 22, Abb. 4. Lopez Cuevillas 1951c, 38. Silva 1986a, 156 (IAFCUP) nº 330.

Nº: 250
Name: Santagões
Parish: Bagunte
Council: Vila do Conde
Latitude: 412136 N
Longitude: 084100 W
Altitude: 54
Bibliography: Severo-Cardozo 1899-1903, 180 (Ref. Nom.). Silva 1986a, nº 328.

Nº: 251
Name: (?)
Parish: Ferreiró
Council: Vila do Conde
Latitude: 412102 N
Longitude: 084818 W
Altitude: 60
Bibliography: Silva 1986a, nº 332. Severo-Cardozo 1899-1903, 180.

Nº: 252
Name: Castro de S. Paio
Parish: Labruje
Council: Vila do Conde
Latitude: 411624 N
Longitude: 084342 W
Altitude: 19
Bibliography: Almeida 1969, 34. Lanhas-Brandão 1969, 296, 328, 333. Silva 1986a, nº 334.

Nº: **253**
Name: Castro
Parish: Retorta
Council: Vila do Conde
Latitude: 412120 N
Longitude: 084320 W
Altitude: 40
Bibliography: Alarcão-Alarcão 1963, 197-9. Almeida 1969, 33. Freitas, EAC, 1949, 20-7. Silva 1986a, nº 325. Severo-Cardozo 1899-1903, 180.

Nº: **254**
Name: Castro do Boi
Parish: Vairão
Council: Vila do Conde
Latitude: 411926 N
Longitude: 084030 W
Altitude: 123
Bibliography: Almeida 1969, 33-4. Azevedo, A, 1939, 102. Silva 1986a, nº 333. Freitas 1949, 20-1.

Nº: **255**
Name: (?)
Parish: Vila do Conde
Council: Vila do Conde
Latitude: 412117 N
Longitude: 084417 W
Altitude: 19
Bibliography: Silva 1986a, nº 324. Severo-Cardozo 1899-1903, 180.

Nº: **256**
Name: Castro de Casais
Parish: Arcos
Council: Vila do Conde
Latitude: 412332 N
Longitude: 083939 W
Altitude: 75
Bibliography: Silva 1986a, nº 324. Severo-Cardozo 1899-1903, 180.

2.3) District of Viana do Castelo

Nº: **257**
Name: Monte do Crasto (Cresto, Vilar)
Parish: Aboim das Choças
Council: Arcos de Valdevez
Latitude: 415550 N
Longitude: 082732 W
Altitude: 285
Bibliography: Brito 1908, 96-7. Gomes, CA, 1979, 2. Pereira 1904, 214-19. Silva 1986a, nº 148. Viana 1932, 159.

Nº: **258**
Name: Castro de Alvora
Parish: Alvora
Council: Arcos de Valdevez
Latitude: 415610 N
Longitude: 082655 W
Altitude: 269
Bibliography: Baptista 1986, 7, fig. 7. Gomes, CA, 1979. Silva 1986a, nº 150. Viana 1932, 159. Pereira 1924; 1929; 1934.

Nº: **259**
Name: Castelo de S. Miguel-O-Anjo
Parish: Azere
Council: Arcos de Valdevez
Latitude: 415124 N
Longitude: 082410 W
Altitude: 180
Bibliography: Baptista 1986, 8. Kalb 1980b, nº 10, Abb. 2. Pereira 1895, 161-75; 1898a, 231-8; 1898c; 1909, 310-14. Rocha 1895, 262-4. Silva 1986a, nº 162. Vasconcellos 1900. Viana 1932, 159.

Nº: **260**
Name: Coto das Portas
Parish: Cabreiro
Council: Arcos de Valdevez
Latitude: (?)
Longitude: (?)
Altitude: (?)
Bibliography: Silva 1986a, nº 153.

Nº: **261**
Name: Crasto das Necessidades
Parish: Cabreiro
Council: Arcos de Valdevez
Latitude: 415653 N
Longitude: 082402 W
Altitude: 453
Bibliography: Pereira 1898b, 289-303; 1904, 214-19. Silva 1986a, nº 152. Viana 1932, 160.

Nº: **262**
Name: Castro (Castelo do Mau Vizinho)
Parish: Cendufe
Council: Arcos de Valdevez

Latitude: 414838 N
Longitude: 082909 W
Altitude: 164
Bibliography: Gomes, CA, 1979, 6. Pereira 1908; 1910; 1915a, 1-16; 1933, 27. Silva 1986a, 305, nº 166. Viana 1932, 161. Paço 1966c, 559.

Nº: **263**
Name: Alto de Modorrão
Parish: Eiras
Council: Arcos de Valdevez
Latitude: (?)
Longitude: (?)
Altitude: (?)
Bibliography: Gomes, CA, 1979, 2. Pereira 1934. Silva 1986a, nº 147. Viana 1932, 161.

Nº: **264**
Name: Castro
Parish: Ermelo
Council: Arcos de Valdevez
Latitude: (?)
Longitude: (?)
Altitude: (?)
Bibliography: Pereira 1934. Silva 1986a, nº 171. Vasconcellos 1900. Viana 1932, 161.

Nº: **265**
Name: (?)
Parish: Giela
Council: Arcos de Valdevez
Latitude: 415051 N
Longitude: 082329 W
Altitude: 352
Bibliography: Pereira 1898a, 232-3; 1908, 202. Silva 1986a, nº 164. Viana 1932, 162.

Nº: **266**
Name: Crasto (Eiró)
Parish: Gondoriz
Council: Arcos de Valdevez
Latitude: 415259 N
Longitude: 082447 W
Altitude: 125
Bibliography: Silva 1986a, nº 157. Viana 1932, 162.

Nº: **267**
Name: Alto da Pedrada (Castro de Grade)
Parish: Grade
Council: Arcos de Valdevez
Latitude: 415215 N
Longitude: 082202 W
Altitude: 212
Bibliography: Pereira 1933, 227-8. Silva 1986a, nº 163. Viana 1932, 162.

Nº: **268**
Name: Crasto
Parish: Louredo
Council: Arcos de Valdevez
Latitude: (?)
Longitude: (?)
Altitude: (?)
Bibliography: Gomes, CA, 1979, 2. Silva 1986a, nº 151. Viana 1932, 162.

Nº: **269**
Name: (?)
Parish: Monte Redondo
Council: Arcos de Valdevez
Latitude: 414929 N
Longitude: 082642 W
Altitude: 271
Bibliography: Silva 1986a, nº 168.

Nº: **270**
Name: Castro de Telheira
Parish: Paçô
Council: Arcos de Valdevez
Latitude: (?)
Longitude: (?)
Altitude: (?)
Bibliography: Silva 1986a, nº 169.

Nº: **271**
Name: Coto da Lama
Parish: Portela
Council: Arcos de Valdevez
Latitude: (?)
Longitude: (?)
Altitude: (?)
Bibliography: Silva 1986a, nº 149. Viana 1932, 163.

Nº: **272**
Name: Castro da Cova do Ouro
Parish: Prozelo
Council: Arcos de Valdevez
Latitude: (?)
Longitude: (?)
Altitude: (?)
Bibliography: Silva 1986a, nº 159. Viana 1932, 163.

Nº: **273**
Name: Coto da Cividade
Parish: Prozelo
Council: Arcos de Valdevez
Latitude: (?)
Longitude: (?)

Altitude: (?)
Bibliography: Pereira 1934. Silva 1986a, nº 158.

Nº: **274**
Name: Castro de Anhó
Parish: Rio de Moinhos
Council: Arcos de Valdevez
Latitude: 415241 N
Longitude: 082742 W
Altitude: 300
Bibliography: Pereira 1934. Silva 1986a, nº 156. Vasconcellos 1900. Viana 1932, 163.

Nº: **275**
Name: Alto da Pena Cova
Parish: Salvador
Council: Arcos de Valdevez
Latitude: 415128 N
Longitude: 082302 W
Altitude: 447
Bibliography: Pereira 1915b, 1-51. Silva 1986a, nº 161.

Nº: **276**
Name: Castro
Parish: Santa Cristina
Council: Arcos de Valdevez
Latitude: 414906 N
Longitude: 082929 W
Altitude: 411
Bibliography: Silva 1986a, nº 165.

Nº: **277**
Name: Castro (Eirinha)
Parish: Senharei
Council: Arcos de Valdevez
Latitude: (?)
Longitude: (?)
Altitude: (?)
Bibliography: Silva 1986a, nº 155. Viana 1932, 164.

Nº: **278**
Name: Crasto
Parish: Távora (S. Vicente)
Council: Arcos de Valdevez
Latitude: 414905 N
Longitude: 082859 W
Altitude: 289
Bibliography: Silva 1986a, nº 167.

Nº: **279**
Name: Coto da Pena
Parish: Vale
Council: Arcos de Valdevez
Latitude: (?)
Longitude: (?)
Altitude: (?)
Bibliography: Baptista 1987, 7-8. Kalb 1980b, nº 11, Abb. 2. Pereira 1915b. Silva 1986a, nº 170. Viana 1932, 164.

Nº: **280**
Name: Eira
Parish: Vila Fonche
Council: Arcos de Valdevez
Latitude: 415057 N
Longitude: 082541 W
Altitude: 110
Bibliography: Silva 1986a, nº 160. Viana 1932, 164.

Nº: **281**
Name: Castelo
Parish: Arga de Baixo
Council: Caminha
Latitude: 415121 N
Longitude: 084239 W
Altitude: 364
Bibliography: Silva 1986a, nº 9. Viana 1926, 90; 1932, 159.

Nº: **282**
Name: (?)
Parish: Argela
Council: Caminha
Latitude: (?)
Longitude: (?)
Altitude: (?)
Bibliography: Silva 1986a, nº 10. Viana 1932, 159.

Nº: **283**
Name: Castro do Germano
Parish: Dem
Council: Caminha
Latitude: 414942 N
Longitude: 084601 W
Altitude: 318
Bibliography: Alves, L, 1980, 13. Silva 1986a, nº 8.

Nº: **284**
Name: Monte do facho
Parish: Moledo
Council: Caminha
Latitude: (?)
Longitude: (?)
Altitude: (?)
Bibliography: Silva 1986a, nº 3. Viana 1926, 90; 1932, 162.

Nº: **285**
Name: Monte de Sto. Amaro
Parish: Riba de Ancora
Council: Caminha
Latitude: 414832 N

Longitude: 084930 W
Altitude: 115
Bibliography: Sarmento, 1980=1933a, 63. Silva 1986a, nº 5.

Nº: **286**
Name: Gurito
Parish: Seixas
Council: Caminha
Latitude: 415338 N
Longitude: 084817 W
Altitude: 160
Bibliography: Silva 1986a, nº 12. Viana 1926, 90; 1932, 163.

Nº: **287**
Name: Crasto
Parish: Vilar de Mouros
Council: Caminha
Latitude: 415348 N
Longitude: 084744 W
Altitude: 120
Bibliography: Fortes 1905. Kalb 1980b, nº 2, Abb. 1. Pinto 1928. Sarmento, 1883-4=1933a, 165; 1888=1933a, 323-24.. Silva 1986a, nº 11. Viana 1926, 90; 1932, 164; 1955d, 529-34.

Nº: **288**
Name: Coto da Pena
Parish: Vilarelho
Council: Caminha
Latitude: 415153 N
Longitude: 085001 W
Altitude: 79
Bibliography: Fortes 1902. Silva 1986a, 141-42, 144, 147, 150-52, 154-57, 159, 162-64, 193-94, 196, 199, 202, 212, 215, 223, 229. Viana 1932, 164. Kalb 1980b, nº 1 (Abb1)

Nº: **289**
Name: Picoto dos Mouros
Parish: Vile/Riba de Ancora
Council: Caminha
Latitude: (?)
Longitude: (?)
Altitude: (?)
Bibliography: MHOP-DGPU 1978, nº 74. Sarmento 1980; 1933a, 63; 1883-84b; 1933a, 166. Silva 1986a, nº 4. Viana 1926, 90; 1932, 162.

Nº: **290**
Name: Cividade
Parish: Ancora/Afife
Council: Caminha/Viana do Castelo
Latitude: 414709 N
Longitude: 085110 W
Altitude: 177
Bibliography: Araújo, J, 1980, 105. Direcção Geral do Planeamento Urbanístico 1978, nº47. Cardozo 1959a; 1959b; 1972, 178-9, 184. Hawkes 1971a; 1984. Hawkes *et al.* 1959, 521-46. Meira 1945. Sarmento 1880=1933a, 63. Silva 1983-84, 128-9; 1986a, 142, 145, 149, 151, 153, 155-6, 158-61, 223, nº 6. Silva - Centeno 1980, 76. Viana 1926, 90; 1932, 159, 162; 1960-61; 1963.

Nº: **291**
Name: Cútero
Parish: Ancora/Afife
Council: Caminha/Viana do Castelo
Latitude: 414648 N
Longitude: 085037 W
Altitude: 232
Bibliography: Alves, L, 1980. Cartaillac 1886, 420. Cardozo 1930, 18; 1935a, 71; 1959b, 235-48. Corrêa 1928, 143. Figueiredo 1895, 143. Hawkes 1971a. Meira 1945, 29-31. Moreira, MAF, 1982, 54. Paço - Quesado 1956, 169-72. Sarmento 1880 (=1933a, 63-4); 1882-3 (1933a, 81); 1895, 23-95, 146; 1933a, 63-4, 81. Silva 1986a, nº 7. Viana 1932, 159; 1955c; 1961; 1962, 79.

Nº: **292**
Name: Cividade
Parish: Paderne
Council: Melgaço
Latitude: 420558 N
Longitude: 081649 W
Altitude: 159
Bibliography: Araújo 1980, 105. Pintor 1975, 123. Silva 1986a, nº 50. Viana 1926, 89; 1932, 163. Vasconcellos 1934, 45-9. Pintor 1975, 123.

Nº: **293**
Name: Crasto
Parish: Penso
Council: Melgaço
Latitude: 420448 N
Longitude: 081905 W
Altitude: 252
Bibliography: Silva 1986a, nº 49. Viana 1926, 89; 1932, 163.

Nº: **294**
Name: Crasto (Eira)
Parish: Roussas
Council: Melgaço
Latitude: 420623 N
Longitude: 081431 W
Altitude: 386
Bibliography: Coffyn 1983; 1985, nº 141. Kalb 1980b. Monteagudo 1977, nºs 1092, 1097, 1132, 1490, 1567a. Neves 1962, 94-99. Silva 1986a, nº 51. Viana 1926, 89; 1932, 163.

Nº: **295**
Name: Castelo da Pena da Rainha
Parish: Abedim
Council: Monção
Latitude: 415924 N
Longitude: 083100 W
Altitude: 472
Bibliography: Guerra 1900b, 134-6; 1900c. Marques, JATM, 1984a;1984b, nº 11;1987a, 95. Silva 1986a, nº 35. Veiga 1941, 9-186. Viana 1932, 158.

Nº: **296**
Name: Cividade
Parish: Anhões
Council: Monção
Latitude: 415953 N
Longitude: 082644 W
Altitude: 453
Bibliography: Araújo 1980, 103. Marques, JATM, 1984a nº 7; 1984b, nº 1; 1987a, 86-90 Silva 1986a, nº 42. Viana 1926, 89; 1932, 159.

Nº: **297**
Name: Mendoiro
Parish: Anhões
Council: Monção
Latitude: 415852 N
Longitude: 082430 W
Altitude: 794
Bibliography: Marques, JATM, 1984a; 1984b, nº 2; 1987a, 97-8. Moita 1971, 271-81. Silva 1986a, nº 43. Viana 1926, 89; 1932, 159.

Nº: **298**
Name: Castro da Senhora da Graça
Parish: Badil/Sá
Council: Monção
Latitude: 420345 N
Longitude: 082134 W
Altitude: 315
Bibliography: Marques, JATM, 1984a; 1984b, nº 3; 1985b; 1987a, 87-8. Silva 1986a, nº 45. Viana 1926, 89; 1932, 163.

Nº: **299**
Name: Monte da Assunção
Parish: Barbeita
Council: Monção
Latitude: 420321 N
Longitude: 082423 W
Altitude: 265
Bibliography: Marques, JATM, 1984a; 1984b, nº 4; 1984c; 1985a.; 1987a, 88. Nunes 1973. Silva 1986a, nº 44.

Nº: **300**
Name: Monte Crasto
Parish: Cambeses/Sago
Council: Monção
Latitude: 420245 N
Longitude: 082735 W
Altitude: 247
Bibliography: Marques, JATM, 1984a; 1984b, nº 5; 1987a, 86-7. Silva 1986a, nº 39. Viana 1926, 89; 1932, 163.

Nº: **301**
Name: Cotos Aguçados
Parish: Granja de Cima - Riba de Mouro
Council: Monção
Latitude: 415900 N
Longitude: 082130 W
Altitude: 585
Bibliography: Marques, JATM, 1987a, 99.

Nº: **302**
Name: Monte Crasto
Parish: Lara
Council: Monção
Latitude: 420115 N
Longitude: 083122 W
Altitude: 209
Bibliography: Marques, JATM, 1984a; 1984b, nº 10; 1987a, 94-5. Silva 1986a, nº 35.

Nº: **303**
Name: Castro de S. Caetano
Parish: Longos Vales
Council: Monção
Latitude: 420215 N
Longitude: 082633 W
Altitude: 342
Bibliography: Marques, JATM, 1984b, nº 6; 1987a, 88-9. Silva 1986a, nº 40. Vasconcellos 1902c. Viana 1926, 89; 1932, 162. Viana-

Neves 1959, 389-96. Caldas 1975, 22-3.

Nº: **304**
Name: Bustavade
Parish: Merufe
Council: Monção
Latitude: 420105 N
Longitude: 082515 W
Altitude: 754
Bibliography: Marques, JATM, 1984a; 1984b, nº 14; 1987a, 96-7. Silva 1986a, nº 41.

Nº: **305**
Name: Monte Redondo
Parish: Monção/Troviscoso
Council: Monção
Latitude: 420412 N
Longitude: 082735 W
Altitude: 73
Bibliography: Cortez 1945, 12-5. Dias, MH, 1984, 88-92. Marques, JATM, 1984a; 1984b, nº 13; 1987a, 91-2. Silva 1986a, nº 37. Viana 1932, 164.

Nº: **306**
Name: Castro do Outeiro da Torre
Parish: Moreira
Council: Monção
Latitude: 420100 N
Longitude: 082911 W
Altitude: 131
Bibliography: Marques, JATM, 1987a, 95-6.

Nº: **307**
Name: Monte dos Penedos
Parish: Pias
Council: Monção
Latitude: 420024 N
Longitude: 083004 W
Altitude: 198
Bibliography: Marques, JATM, 1987, 92.

Nº: **308**
Name: Monte da Senhora da Vista
Parish: Podame/Tangil
Council: Monção
Latitude: 420140 N
Longitude: 082221 W
Altitude: 244
Bibliography: Marques, JATM, 1984a; 1984b, nº 9; 1987a, 90-1. Silva 1986a, nº 46. Viana 1926, 89; 1932, 164.

Nº: **309**
Name: Coroa (Riba de Moura)
Parish: Riba de Moura
Council: Monção
Latitude: 420051 N
Longitude: 081834 W
Altitude: 925
Bibliography: Marques, JATM, 1987a, 98.

Nº: **310**
Name: Cividade
Parish: Riba de Mouro
Council: Monção
Latitude: 420214 N
Longitude: 082001 W
Altitude: 259
Bibliography: Marques, JATM, 1984a; 1984b, nº 7; 1987, 89-90. Silva 1986a, nº 48.

Nº: **311**
Name: Quintenda
Parish: Riba de Mouro
Council: Monção
Latitude: 420251 N
Longitude: 082034 W
Altitude: 500
Bibliography: Marques, JATM, 1987a, 98-9.

Nº: **312**
Name: Monte das Barrocas
Parish: Sago de Cima/Sago
Council: Monção
Latitude: 420154 N
Longitude: 082643 W
Altitude: 352
Bibliography: Marques, JATM, 1987a, 93.

Nº: **313**
Name: Coroas
Parish: Sta. Marinha (Riba de Moura)
Council: Monção
Latitude: 420000 N
Longitude: 084843 W
Altitude: 721
Bibliography: Marques, JATM, 1987a, 100.

Nº: **314**
Name: Crastelo
Parish: Tangil
Council: Monção
Latitude: 420124 N
Longitude: 082122 W
Altitude: 128
Bibliography: Marques, JATM, 1984a; 1984b, nº 8; 1987a, 93-4. Silva 1986a, nº 47. Viana 1926, 89; 1932, 164.

Nº: **315**
Name: Cristelo
Parish: Troviscoso
Council: Monção

Latitude: 420402 N
Longitude: 082655 W
Altitude: 78
Bibliography: Marques, JATM, 1984a; 1984b, nº 12; 1987a, 91. Silva 1986a, nº 38. Viana 1926, 89; 1932, 164.

Nº: **316**
Name: Couto da Trute
Parish: Trute
Council: Monção
Latitude: 415945 N
Longitude: 082742 W
Altitude: 425
Bibliography: Marques, JATM, 1987a,96.

Nº: **317**
Name: Castro
Parish: Bico
Council: Paredes de Coura
Latitude: 415327 N
Longitude: 082120 W
Altitude: 540
Bibliography: Oliveira 1976, 18. Silva 1986a, nº 65. Viana 1926, 89.

Nº: **318**
Name: Monte da Cividade
Parish: Cossourado/ Sta. Maria de Cossourado
Council: Paredes de Coura
Latitude: 415452 N
Longitude: 083805 W
Altitude: 376
Bibliography: Cunha 1956, 1; 1975, 497. Cunha, NCA, 1979, 131-2. MHOP-DGPU 1978, nº 81. Oliveira 1976. Pereira 1929, 1-51. Silva 1986a, nº 55. Viana 1926, 89; 1932, 161.

Nº: **319**
Name: Castro de S. Martinho
Parish: Coura
Council: Paredes de Coura
Latitude: (?)
Longitude: (?)
Altitude: (?)
Bibliography: Cunha, NCA, 1979, 131. MHOP-DGPU 1978, nº 76. Oliveira 1976, 22-3. Silva 1986a, nº 62.

Nº: **320**
Name: Madorra
Parish: Coura
Council: Paredes de Coura
Latitude: 415326 N
Longitude: 083822 W
Altitude: 201
Bibliography: Oliveira 1976, 23. Pereira 1929, 1. Silva 1986a, nº 63.

Nº: **321**
Name: Castro
Parish: Cristelo
Council: Paredes de Coura
Latitude: 415421 N
Longitude: 083214 W
Altitude: 491
Bibliography: Oliveira 1976, 23. Silva 1986a, nº 64. Viana 1926, 89; 1932, 161.

Nº: **322**
Name: Cidade
Parish: Cunha
Council: Paredes de Coura
Latitude: 415323 N
Longitude: 083626 W
Altitude: 409
Bibliography: Oliveira 1976, 24. Silva 1986a, nº 61.

Nº: **323**
Name: Castelo
Parish: Formariz
Council: Paredes de Coura
Latitude: (?)
Longitude: (?)
Altitude: (?)
Bibliography: Cunha, NCA, 1979, 131. Kalb 1980b, nº 4, Abb. 1. Oliveira 1976. Silva 1986a, nº 58. Viana 1926, 89; 1932, 161.

Nº: **324**
Name: Cidade de Pedrauca
Parish: Infesta
Council: Paredes de Coura
Latitude: 415426 N
Longitude: 083618 W
Altitude: 365
Bibliography: Oliveira, AL, 1976, 30-1. Silva 1986a, nº 50.

Nº: **325**
Name: Castelinho
Parish: Insalde
Council: Paredes de Coura
Latitude: (?)
Longitude: (?)
Altitude: (?)
Bibliography: Oliveira 1976, 32. Silva 1986a, nº 54.

Nº: **326**
Name: Castro
Parish: Insalde

Council: Paredes de Coura
Latitude: (?)
Longitude: (?)
Altitude: (?)
Bibliography: Oliveira 1976, 32. Silva 1986a, nº 53. Viana 1926, 89; 1932, 162.

Nº: **327**
Name: Castro de Bruzendes
Parish: Linhares
Council: Paredes de Coura
Latitude: 415539 N
Longitude: 083615 W
Altitude: 336
Bibliography: Cunha, NCA, 1979, 131, 133. Oliveira 1976. Silva 1986a, nº 56. Viana 1926, 89; 1932, 162.

Nº: **328**
Name: Monte da Modorra
Parish: Linhares
Council: Paredes de Coura
Latitude: 415555 N
Longitude: 083616 W
Altitude: 351
Bibliography: Oliveira 1976, 33. Silva 1986a, nº 57. Viana 1926, 89; 1932, 162.

Nº: **329**
Name: Eira do Cabo
Parish: Moselos
Council: Paredes de Coura
Latitude: (?)
Longitude: (?)
Altitude: (?)
Bibliography: Oliveira 1976, 34. Silva 1986a, nº 59. Viana 1926, 89; 1932, 162.

Nº: **330**
Name: Cidade Murada
Parish: Romarigães
Council: Paredes de Coura
Latitude: 415212 N
Longitude: 083616 W
Altitude: 280
Bibliography: Cunha, NCA, 1979, 131-3. Oliveira 1976, 42-3. Silva 1986a, nº 66. Viana 1926, 89; 1932, 163.

Nº: **331**
Name: Castro de Caçapedo
Parish: Azias
Council: Ponte da Barca
Latitude: 414640 N
Longitude: 082035 W
Altitude: 431

Nº: **332**
Name: Castro
Parish: Britelo
Council: Ponte da Barca
Latitude: 414937 N
Longitude: 081723 W
Altitude: 301
Bibliography: Silva 1986a, nº 177.

Nº: **333**
Name: Castro da Foz do Cabril
Parish: Castelo de Lindoso
Council: Ponte da Barca
Latitude: 415216 N
Longitude: 081048 W
Altitude: 375
Bibliography: Baptista 1986.

Nº: **334**
Name: Castro da Torre de Grade
Parish: Castelo Lindoso
Council: Ponte da Barca
Latitude: 415058 N
Longitude: 081025 W
Altitude: 669
Bibliography: Baptista 1986.

Nº: **335**
Name: Crasto (Cidadelhe)
Parish: Cidadelhe
Council: Ponte da Barca
Latitude: 415150 N
Longitude: 081432 W
Altitude: 215
Bibliography: Guerra 1882. Silva 1986a, nº 178. Vasconcellos 1913 (3), 43, 45.

Nº: **336**
Name: Castro de S. Martinho (Eiriz)
Parish: Crasto
Council: Ponte da Barca
Latitude: 414630 N
Longitude: 082601 W
Altitude: 180
Bibliography: Silva 1986a, nº 173. Almeida 1987, 125.

Nº: **337**
Name: S. Miguel
Parish: Entre Ambos os Rios
Council: Ponte da Barca
Latitude: 414906 N
Longitude: 081829 W
Altitude: 109

Nº: **338**
Name: Airó
Parish: Oleiros

Council: Ponte da Barca
Latitude: 414712 N
Longitude: 082600 W
Altitude: 180
Bibliography: Silva 1986a, nº 172. Viana 1932, 162.

Nº: **339**
Name: Castro de Leijó
Parish: Parada/Lindoso
Council: Ponte da Barca
Latitude: 415133 N
Longitude: 081226 W
Altitude: 505
Bibliography: Baptista 1986.

Nº: **340**
Name: Castro do Monte do Crasto
Parish: Parada/Lindoso
Council: Ponte da Barca
Latitude: 415204 N
Longitude: 081236 W
Altitude: 480
Bibliography: Baptista 1986.

Nº: **341**
Name: Airó
Parish: S. Sebastião/ S. Tomé
Council: Ponte da Barca
Latitude: 414654 N
Longitude: 082421 W
Altitude: 135
Bibliography: Silva 1986a, nº 174. Viana 1932, 164.

Nº: **342**
Name: Crasto
Parish: Sampriz
Council: Ponte da Barca
Latitude: 414712 N
Longitude: 082330 W
Altitude: 323
Bibliography: Brito 1908, 97-8. Centeno 1987. Silva 1986a, nº 176. Viana 1932, 163.

Nº: **343**
Name: Marialva
Parish: V. N. de Muia
Council: Ponte da Barca
Latitude: 414758 N
Longitude: 082246 W
Altitude: 140

Nº: **344**
Name: Monte Castro
Parish: Anais
Council: Ponte de Lima
Latitude: 404137 N
Longitude: 083242 W
Altitude: 365
Bibliography: Almeida-Baptista 1981, nº 13. Silva 1986a, nº 135. Viana 1926, 90; 1932, 159.

Nº: **345**
Name: Castro de Sto. Ovídio
Parish: Arcozelo
Council: Ponte de Lima
Latitude: 414643 N
Longitude: 083620 W
Altitude: 248
Bibliography: Almeida, CAB, 1990, 155-7. Almeida-Baptista 1981, nº 34. Brito 1908, 95. Hipólito 1960-61, 15-6. Reis 1978. Silva 1986a, nº 111. Viana 1962.

Nº: **346**
Name: Crasto
Parish: Bárrio
Council: Ponte de Lima
Latitude: 414950 N
Longitude: 083521 W
Altitude: 120
Bibliography: Almeida, CAB, 1990, 153. Almeida-Baptista 1981, nº 37. Silva 1986a, nº 105. Viana 1926.

Nº: **347**
Name: Quinta do Crasto (Castro de Vinhó)
Parish: Beiral do Lima
Council: Ponte de Lima
Latitude: 414604 N
Longitude: 082855 W
Altitude: 220
Bibliography: Almeida, CAB, 1990, 124-33. Almeida-Baptista 1981, nº 43. Silva 1986a, nº 130. Viana 1926, 90; 1932, 160.

Nº: **348**
Name: Vilela (Penouços)
Parish: Cabacos
Council: Ponte de Lima
Latitude: 414126 N
Longitude: 083406 W
Altitude: 300
Bibliography: Almeida-Baptista 1981, nº 3. Silva 1986a, nº 142.

Nº: **349**
Name: Castelo do Genso
Parish: Calheiros/Brandara
Council: Ponte de Lima
Latitude: 414755 N
Longitude: 083331 W
Altitude: 297

Bibliography: Almeida 1971, 293-6. Almeida, CAB, 1990, 154-5. Almeida-Baptista 1981, nº 35. Azevedo, PA, 1896, 318. Guerra 1900c, 175-7. Reis 1978, 9. Silva 1986a, nº 113. Viana 1932, 160.

Nº: **350**
Name: Castelo de Cadém
Parish: Calvelo
Council: Ponte de Lima
Latitude: 404101 N
Longitude: 083155 W
Altitude: 180
Bibliography: Almeida-Baptista 1981, nº 4. Silva 1986a, nº 143.

Nº: **351**
Name: Crasto
Parish: Cepões
Council: Ponte de Lima
Latitude: 414924 N
Longitude: 083524 W
Altitude: 100
Bibliography: Almeida, CAB, 1990a, 154. Almeida-Baptista 1981, nº 37. Silva 1986a, nº 106. Viana 1926, 90; 1932, 163.

Nº: **352**
Name: João do Monte (Cortes)
Parish: Correlhã
Council: Ponte de Lima
Latitude: 414410N
Longitude: 083550 W
Altitude: 250
Bibliography: Almeida, CAB, 1990, 106-7. Almeida-Baptista 1981, nº 20. Hipólito 1960-61, 15. Reis 1978, 9-36. Silva 1986a, nº 120. Viana 1926, 90; 1932, 161.

Nº: **353**
Name: Castro do Eirado
Parish: Correlhã
Council: Ponte de Lima
Latitude: 414411 N
Longitude: 083644 W
Altitude: 30
Bibliography: Almeida, CAB, 1990, 100-105. Almeida-Baptista 1981, nº 18. Silva 1986a, nº 119. Viana 1926, 90; 1932, 161.

Nº: **354**
Name: Tolocrasto (N.S. da Conceição)
Parish: Correlhã
Council: Ponte de Lima
Latitude: 414503 N
Longitude: 083555 W
Altitude: 54
Bibliography: Almeida-Baptista 1981, nº 21. Silva 1986a, nº 121. Almeida, CAB, 1990, 107-108.

Nº: **355**
Name: Monte dos Medos
Parish: Correlhã/Feitosa
Council: Ponte de Lima
Latitude: 414515 N
Longitude: 083533 W
Altitude: 25
Bibliography: Almeida, CAB, 1990, 116-7. Almeida-Baptista 1981, nº 22. Silva 1986a, nº 122.

Nº: **356**
Name: Bouça do Castro
Parish: Estorãos
Council: Ponte de Lima
Latitude: 414714 N
Longitude: 083823 W
Altitude: 25
Bibliography: Almeida, CAB, 1990, 167-72. Almeida-Baptista 1981, nº 32. Cortez 1948b. Pereira 1907b, 36-52. Reis 1978. Silva 1986a, nº 109. Vasconcellos 1913, 199-200. Viana 1932, 161.

Nº: **357**
Name: Socastro (Cerquido)
Parish: Estorãos
Council: Ponte de Lima
Latitude: 414820 N
Longitude: 084035 W
Altitude: 420
Bibliography: Almeida, CAB, 1990, 176.

Nº: **358**
Name: Castro de Sto. Estêvão
Parish: Facha
Council: Ponte de Lima
Latitude: 414312 N
Longitude: 083833 W
Altitude: 142
Bibliography: Almeida, CAB, 1990, 53-8. Almeida *et al.* 1981, 50-1, 70; 1982, 79. Almeida-Baptista 1981, nº 10. Silva 1986a,193, 203, 211, 213, nº 131. Soares - Cabral 1984, 179, 194. Viana 1926, 90; 1932, 161.

Nº: **359**
Name: Prazil
Parish: Facha
Council: Ponte de Lima

Latitude: 414303 N
Longitude: 083815 W
Altitude: 70
Bibliography: Almeida, CAB, 1990, 58-62. Almeida-Baptista 1981, nº 50. Silva 1986a, nº 132. Viana 1926, 90; 1932, 161.

Nº: **360**
Name: S. Cipriano
Parish: Facha
Council: Ponte de Lima
Latitude: 414315 N
Longitude: 083613 W
Altitude: 320
Bibliography: Almeida, CAB, 1990a, 53. Almeida-Baptista 1981, nº 49. Silva 1986a, nº 133. Viana 1926, 90; 1932, 161.

Nº: **361**
Name: Castro do Calvário
Parish: Fojo Lobal
Council: Ponte de Lima
Latitude: 414210 N
Longitude: 083500 W
Altitude: 332
Bibliography: Almeida, CAB, 1990, 113, 116. Almeida-Baptista 1981, nº 11. Silva 1986a, nº 140.

Nº: **362**
Name: Castro de S. Cristóvão
Parish: Fontão
Council: Ponte de Lima
Latitude: 414504 N
Longitude: 083855 W
Altitude: 88
Bibliography: Almeida, CAB, 1990, 172. Almeida-Baptista 1981, nº 31. Silva 1986a, nº 115.

Nº: **363**
Name: Quinta do Crasto
Parish: Fornelos
Council: Ponte de Lima
Latitude: 414324N
Longitude: 083204 W
Altitude: 163
Bibliography: Almeida, CAB, 1990, 113-4. Almeida-Baptista 1981,109, nº 47. Silva 1986a, nº 127. Viana 1932, 161.

Nº: **364**
Name: Santa Maria Madalena
Parish: Fornelos
Council: Ponte de Lima
Latitude: 414324 N
Longitude: 083204 W
Altitude: 256
Bibliography: Almeida-Baptista 1981, nº 23. Paço - Quesado 1956, 174-8. Silva 1986a, nº 126. Viana 1932, 161. Almeida, CAB, 1990, 113-4.

Nº: **365**
Name: Castro de Vilar
Parish: Gondufe
Council: Ponte de Lima
Latitude: 414601 N
Longitude: 083055 W
Altitude: 210
Bibliography: Almeida, CAB, 1990, 122-4. Almeida-Baptista 1981, nº 29. Silva 1986a, nº 128.

Nº: **366**
Name: Castro Mau (S. Lourenço da Armada)
Parish: Gondufe/Serdelo
Council: Ponte de Lima
Latitude: 414435 N
Longitude: 083004 W
Altitude: 609
Bibliography: Almeida, CAB, 1990, 109-12. Almeida-Baptista 1981, 301-2, nº 24. Paço - Quesado 1956, 176. Reis 1978, 9. Silva 1986a, nº 129. Viana 1926, 90; 1932, 163.

Nº: **367**
Name: Castro de Baixo
Parish: Labruja
Council: Ponte de Lima
Latitude: 415000 N
Longitude: 083531 W
Altitude: 96
Bibliography: Almeida, CAB, 1990, 151-2. Almeida-Baptista 1981, nº 39. Silva 1986a, nº 103.

Nº: **368**
Name: Castro de Cima
Parish: Labruja
Council: Ponte de Lima
Latitude: 415014 N
Longitude: 083522 W
Altitude: 133
Bibliography: Almeida, CAB, 1990, 153-4. Almeida-Baptista 1981, nº 40. Silva 1986a, nº 102.

Nº: **369**
Name: Cidade Murada
Parish: Labruja
Council: Ponte de Lima
Latitude: 415113N

Longitude: 083643 W
Altitude: 442
Bibliography: Almeida, CAB, 1990, 151. Almeida-Baptista 1981, nº 41. Cunha, NCA, 1979. Silva 1986a, nº 104. Viana 1932, 162.

Nº: **370**
Name: Bouça do Cavalo
Parish: Moreira de Lima
Council: Ponte de Lima
Latitude: 414640 N
Longitude: 083734 W
Altitude: 25
Bibliography: Almeida, CAB, 1990, 175.

Nº: **371**
Name: Cividade (Madorna)
Parish: Moreira de Lima
Council: Ponte de Lima
Latitude: 414732 N
Longitude: 083702 W
Altitude: 85
Bibliography: Almeida, CAB, 1990, 176.

Nº: **372**
Name: Sabugueiro
Parish: Poiares
Council: Ponte de Lima
Latitude: 413920 N
Longitude: 083734 W
Altitude: 81
Bibliography: Almeida-Baptista 1981, nº 44. Silva 1986a, nº 144.

Nº: **373**
Name: Castro de S. Simão
Parish: Refoios do Lima
Council: Ponte de Lima
Latitude: 414731 N
Longitude: 083330 W
Altitude: 180
Bibliography: Almeida, CAB, 1990, 144-5. Almeida-Baptista 1981, nº 45. Silva 1986a, nº 112. Viana 1926, 90; 1932, 163.

Nº: **374**
Name: Crasto do Soutinho (Pousada)
Parish: Refoios do Lima
Council: Ponte de Lima
Latitude: 414755N
Longitude: 083204 W
Altitude: 160
Bibliography: Almeida, CAB, 1990, 142-3. Almeida-Baptista 1981, nº 36. Calheiros 1910, 98. Silva 1986a, nº 114. Teixeira *et al.* 1972.

Nº: **375**
Name: Alto da Pena
Parish: Rendufe/Labrujó
Council: Ponte de Lima
Latitude: 415144 N
Longitude: 083255 W
Altitude: 746
Bibliography: Almeida, CAB, 1990, , 150. Almeida-Baptista 1981, nº 42. Silva 1986a, nº 101.

Nº: **376**
Name: Castelinho
Parish: Ribeira
Council: Ponte de Lima
Latitude: 414603 N
Longitude: 083440 W
Altitude: 56
Bibliography: Almeida, CAB, 1990, 113-5. Almeida-Baptista 1981, nº 26. Silva 1986a, nº 123.

Nº: **377**
Name: Castelo da Ermida
Parish: Ribeira
Council: Ponte de Lima
Latitude: 414534 N
Longitude: 083144 W
Altitude: 443
Bibliography: Almeida, CAB, 1990, 109-12. Almeida-Baptista 1981, nº 25. Cardoso, AP, 1966. Reis 1978, 9. Silva 1986a, nº 125.

Nº: **378**
Name: Monte de Sta. Catarina
Parish: Ribeira
Council: Ponte de Lima
Latitude: 414555 N
Longitude: 083242 W
Altitude: 225
Bibliography: Almeida, CAB, 1990, 112-13. Almeida-Baptista 1981, nº 28. Cardoso, AP, 1966. Silva 1986a, nº 124. Teixeira *et al.* 1972, 40.

Nº: **379**
Name: Castro de S. Cristóvão
Parish: S. Julião do Freixo
Council: Ponte de Lima
Latitude: 414000 N
Longitude: 083553 W
Altitude: 280
Bibliography: Almeida-Baptista 1981, nº 2. Azevedo, PA, 1898, 193-221. López Cuevillas 1953c. Silva 1986a, nº 145.

Nº: **380**
Name: Vilarinho
Parish: Seara
Council: Ponte de Lima
Latitude: 414404 N
Longitude: 083652 W
Altitude: 28
Bibliography: Almeida, CAB, 1990, 98. Almeida-Baptista 1981, nº 17. Silva 1986a, nº 118. Viana 1926, 90; 1932, 163.

Nº: **381**
Name: Cividade
Parish: Souto
Council: Ponte de Lima
Latitude: 414253 N
Longitude: 083352 W
Altitude: 160
Bibliography: Almeida, CAB, 1990, 118. Almeida-Baptista 1981, nº 14. Silva 1986a, nº 134. Viana 1932, 164.

Nº: **382**
Name: Castro de Sta. Comba
Parish: Sta. Comba
Council: Ponte de Lima
Latitude: 414542 N
Longitude: 083704 W
Altitude: 13
Bibliography: Almeida, CAB, 1990, 166-7.

Nº: **383**
Name: Castro Seco (Outeirais)
Parish: Vilar das Almas
Council: Ponte de Lima
Latitude: 413859 N
Longitude: 083205 W
Altitude: 235
Bibliography: Almeida-Baptista 1981, nº 1. Silva 1986a, nº 146.

Nº: **384**
Name: Castro dos Penedos do Castelo
Parish: Vilar do Monte
Council: Ponte de Lima
Latitude: 415016 N
Longitude: 083144 W
Altitude: 836
Bibliography: Almeida, CAB, 1990, 142.

Nº: **385**
Name: Castro da Senhora do Barco
Parish: Vitorino das Donas
Council: Ponte de Lima
Latitude: 414400N
Longitude: 083820 W
Altitude: 30
Bibliography: Almeida, CAB, 1990, 80. Almeida-Baptista 1981, nº 15. Silva 1986a, nº 116.

Nº: **386**
Name: Cividade
Parish: Vitorino das Donas
Council: Ponte de Lima
Latitude: 414400N
Longitude: 083755 W
Altitude: 30
Bibliography: Almeida, CAB, 1985c, 136-7; 1990, 82-98. Almeida-Baptista 1981, 309, nº 16. Silva 1986a, nº 117.

Nº: **387**
Name: Alto de Valadas
Parish: Vitorino dos Piães
Council: Ponte de Lima
Latitude: 414133 N
Longitude: 083613W
Altitude: 347
Bibliography: Almeida, CAB, 1990, 51. Almeida-Baptista 1981, 294, nº 7. Silva 1986a, nº 138. Viana 1932, 164.

Nº: **388**
Name: Castro de S. Simão
Parish: Vitorino dos Piães
Council: Ponte de Lima
Latitude: 414030 N
Longitude: 083704 W
Altitude: 179
Bibliography: Almeida-Baptista 1981, nº 5. Almeida, CAB, 1990b, 146. Silva 1986a, nº 137. Viana 1932, 164.

Nº: **389**
Name: Cresto
Parish: Vitorino dos Piães
Council: Ponte de Lima
Latitude: 414110 N
Longitude: 083703 W
Altitude: 220
Bibliography: Almeida, CAB, 1990, 48-51. Almeida-Baptista 1981, 294-4, nº 6. Reis 1978, 9. Silva 1986a, nº 136. Viana 1926, 90; 1932, 164.

Nº: **390**
Name: Trás-Cidade
Parish: Vitorino dos Piães
Council: Ponte de Lima
Latitude: 414151 N
Longitude: 083622 W
Altitude: 450

Bibliography: Almeida, CAB, 1990, , 51-3. Almeida-Baptista 1981, 294, nº 8. Silva 1986a, nº 139. Viana 1932, 164.

Nº: **391**
Name: Monte do Castro
Parish: Candemil
Council: V. N. Cerveira
Latitude: (?)
Longitude: (?)
Altitude: (?)
Bibliography: MHOP-DGPU, 1978, nº 84. Silva 1986a, nº 22.

Nº: **392**
Name: Cividade
Parish: Cornes
Council: V. N. Cerveira
Latitude: 415708 N
Longitude: 084039 W
Altitude: 70
Bibliography: Araújo 1980, 103. Silva 1986a, nº 21. Viana 1926, 90; 1932, 161.

Nº: **393**
Name: Castro da Torre
Parish: Covas
Council: V. N. Cerveira
Latitude: 415343 N
Longitude: 084055 W
Altitude: 250
Bibliography: MHOP-DGPU 1978, nº 75. Silva 1986a, nº 14. Viana 1932, 161.

Nº: **394**
Name: Monte de Góis
Parish: Gondarém
Council: V. N. Cerveira
Latitude: 415425 N
Longitude: 084530 W
Altitude: 270
Bibliography: Kalb 1980b, nº 3, Abb. 1. Silva 1986a, nº 15.

Nº: **395**
Name: Monte do Crasto
Parish: Gondarém
Council: V. N. Cerveira
Latitude: 415448 N
Longitude: 084513 W
Altitude: 110
Bibliography: Pereira 1930, 372-3. MHOP-DGPU 1978, nº 74. Silva 1986a, nº 16. Viana 1932, 162.

Nº: **396**
Name: Monte do Forte
Parish: Lovelhe
Council: V. N. Cerveira
Latitude: 415701 N
Longitude: 084424 W
Altitude: 46
Bibliography: MHOP-DGPU 1978, nº 88. Silva 1986a, nº 18. Viana 1932, 162.

Nº: **397**
Name: Crasto
Parish: Mentrestido
Council: V. N. Cerveira
Latitude: 415408 N
Longitude: 083956 W
Altitude: 253
Bibliography: MHOP-DGPU 1978, nº 77. Silva 1986a, nº 23. Viana 1926, 90; 1932, 162.

Nº: **398**
Name: Castro
Parish: Reboreda
Council: V. N. Cerveira
Latitude: 415722 N
Longitude: 084339 W
Altitude: 50
Bibliography: Silva 1986a, nº 19. Viana 1926, 90; 1932, 163.

Nº: **399**
Name: Mata dos Crastos
Parish: Sopo
Council: V. N. Cerveira
Latitude: 415353 N
Longitude: 084352 W
Altitude: 258
Bibliography: Silva 1986a, nº 13.

Nº: **400**
Name: Castro de Cerveira
Parish: V. N. Cerveira
Council: V. N. Cerveira
Latitude: 415625 N
Longitude: 084353 W
Altitude: 224
Bibliography: Pereira 1956. Silva 1986a, nº 17.

Nº: **401**
Name: Eirado
Parish: Arão
Council: Valença
Latitude: (?)
Longitude: (?)
Altitude: (?)

Bibliography: Oliveira 1978, 26. Silva 1986a, nº 28. Viana 1926, 89; 1932, 159.

Nº: **402**
Name: Castelo da Furna
Parish: Boivão
Council: Valença
Latitude: 415922 N
Longitude: 083257 W
Altitude: 608
Bibliography: Guerra 1900c, 176; 1900b. Kalb 1980b, nº 5, Abb. 2. Monteagudo 1977, nº 1740. Oliveira 1978, 23, 28. Silva 1986a, nº 34. Viana 1926, 89; 1932, 160.

Nº: **403**
Name: Crasto
Parish: Cerdal
Council: Valença
Latitude: (?)
Longitude: (?)
Altitude: (?)
Bibliography: Oliveira 1978, 23-4, 30. Silva 1986a, nº 27. Viana 1926, 89; 1932, 161.

Nº: **404**
Name: Monte do Grove
Parish: Fontoura
Council: Valença
Latitude: 415706 N
Longitude: 083759 W
Altitude: 80
Bibliography: Oliveira 1978, 24, 32. Silva 1986a, nº 26. Viana 1932, 161.

Nº: **405**
Name: Eirado
Parish: Friestas
Council: Valença
Latitude: 420244 N
Longitude: 083405 W
Altitude: 45
Bibliography: Oliveira 1978, 23. Silva 1986a, nº 32. Viana 1926, 89; 1932, 161.

Nº: **406**
Name: Crasto (Cristelo)
Parish: Ganfei
Council: Valença
Latitude: (?)
Longitude: (?)
Altitude: (?)
Bibliography: Oliveira 1978, 23-34. Silva 1986a, nº 29. Viana 1926, 89; 1932, 161.

Nº: **407**
Name: Crasto
Parish: Gondomil
Council: Valença
Latitude: 420140 N
Longitude: 083305 W
Altitude: 56
Bibliography: Almeida, CAB, 1984. Oliveira 1978, 23-4, 35, 40-1. Silva 1986a, nº 33. Viana 1926, 89; 1932, 162. Almeida - Abreu 1987, 198.

Nº: **408**
Name: Crastos
Parish: S. Pedro da Torre
Council: Valença
Latitude: (?)
Longitude: (?)
Altitude: (?)
Bibliography: Oliveira 1978, 23, 40-1. Silva 1986a, nº 25. Viana 1926, 89; 1932, 163.

Nº: **409**
Name: Eiras
Parish: Sanfins
Council: Valença
Latitude: 420132 N
Longitude: 083426 W
Altitude: 175
Bibliography: Oliveira 1978, 24. Silva 1986a, nº 31. Viana 1926, 89; 1932, 163.

Nº: **410**
Name: Outeiro de Madorra
Parish: Silva
Council: Valença
Latitude: 415717 N
Longitude: 083910 W
Altitude: 50
Bibliography: Oliveira 1978, 24, 41, 43. Silva 1986a, nº 24. Viana 1926, 89; 1932, 164.

Nº: **411**
Name: Castro de S. Sebastião
Parish: Silva
Council: Valença
Latitude: 415627 N
Longitude: 083951 W
Altitude: 250
Bibliography: Almeida, CAB, pers. comm.

Nº: **412**
Name: Eirado
Parish: Verdoejo
Council: Valença

Latitude: (?)
Longitude: (?)
Altitude: (?)
Bibliography: Oliveira 1978, 47. Silva 1986a, nº 30. Viana 1926, 89; 1932, 164.

Nº: **413**
Name: Agrichousa
Parish: Afife
Council: Viana do Castelo
Latitude: 414618 N
Longitude: 085020 W
Altitude: 130
Bibliography: Meira 1945, 29-30. Neves 1965, nº 3. Paço-Quesado 1956, 172. Quesado 1955, 4. Silva 1986a, nº 68. Viana 1932, 159.

Nº: **414**
Name: Castro de Santo António
Parish: Afife
Council: Viana do Castelo
Latitude: 414650 N
Longitude: 085141 W
Altitude: 71
Bibliography: Meira 1945, 26-9. MHOP-DGPU 1978, nº 46. Moreira, MAF, 1982, 54. Neves 1965, nº 4. Paço-Quesado 1956, 173-4. Silva 1983a, 144; 1983, 134; 1986a, 149-50, 194, nº 67. Viana 1932, 159; 1937, 75-83; 1955c, 88; 1963.

Nº: **415**
Name: Monte do Castro
Parish: Afife
Council: Viana do Castelo
Latitude: (?)
Longitude: (?)
Altitude: (?)
Bibliography: Meira 1945, 29-30. Paço-Quesado 1956, 173. Silva 1986a, nº 69.

Nº: **416**
Name: Outeiro
Parish: Alvarães
Council: Viana do Castelo
Latitude: (?)
Longitude: (?)
Altitude: (?)
Bibliography: Cepa 1939, 19-20. Silva 1986a, nº 99.

Nº: **417**
Name: Coroa
Parish: Amonde
Council: Viana do Castelo
Latitude: 414719 N
Longitude: 084523 W
Altitude: 212
Bibliography: Almeida, CAB, 1990a, 182. Silva 1986a, nº 84.

Nº: **418**
Name: Monte do Faro (Alto do Galeão)
Parish: Anha/Darque
Council: Viana do Castelo
Latitude: 414032 N
Longitude: 084735W
Altitude: 166
Bibliography: Almeida, CAB, 1990, 30-1. Neves 1965, nº 22. Paço-Quesado 1956, 80-90. Silva 1986a, nº 91.

Nº: **419**
Name: Castro do Pego
Parish: Areosa
Council: Viana do Castelo
Latitude: 414324 N
Longitude: 085045 W
Altitude: 160
Bibliography: Almeida, CAB, 1990a, 223. Guerra 1909. MHOP-DGPU 1978, nº 29. Neves 1965, nº 9. Paço-Quesado 1956, ??. Silva 1986a, nº 72. Viana 1926, 90; 1932, 159; 1962, 81.

Nº: **420**
Name: S. Silvestre
Parish: Cardielos
Council: Viana do Castelo
Latitude: 414314 N
Longitude: 084440 W
Altitude: 281
Bibliography: Almeida, CAB, 1990a, 183-4. MHOP-DGPU 1978, nº 17. Moreira, MAF, 1982, 54. Neves 1965, nº 17. Paço - Quesado 1956, 88. Silva 1986a, nº 80. Viana 1926, 90.

Nº: **421**
Name: Castro da Coroa (Castros Velhos)
Parish: Carreço
Council: Viana do Castelo
Latitude: (?)
Longitude: (?)
Altitude: (?)
Bibliography: Neves 1965, nº 8. Silva 1986a, nº 71. Viana 1932, 160.

Nº: **422**
Name: Montedor (Monte da Gândara)
Parish: Carreço
Council: Viana do Castelo

Latitude: 414459 N
Longitude: 085219 W
Altitude: 75
Bibliography: Silva 1986a, nº 70. Viana 1926, 90; 1932, 160; 1961, 9.

Nº: **423**
Name: Castro de Carmona
Parish: Carvoeiro
Council: Viana do Castelo
Latitude: 413850 N
Longitude: 083842 W
Altitude: 300
Bibliography: Almeida 1978b, 33. Almeida, CAB, 1990b, 134. Fonseca 1987, 138. Guerra 1900a, 6. MHOP-DGPU 1978, nº 5. Neves 1965, nº 28. Paço-Quesado 1956. 88-9. Silva 1986a, nº 100. Viana 1926, 90; 1932, 161.

Nº: **424**
Name: Castro de Moldes/Monte. do Castelo
Parish: Castelo de Neiva
Council: Viana do Castelo
Latitude: 413644 N
Longitude: 084658 W
Altitude: 128
Bibliography: Almeida 1980a; 1980b. Almeida, CAB, 1982b; 1985b. Azevedo, PA, 1897, 194; 1903, 74-5. Marques-Silva 1985. Moreira, MAF, 1982, 54-6. Neves 1965, nº 24. Paço-Quesado 1956. Silva 1986a, 197, 199, nº 97. Silva, EJL, 1982. Silva-Marques 1984. Viana 1932, 161. Almeida 1980, 250-1.

Nº: **425**
Name: Subidade
Parish: Deião
Council: Viana do Castelo
Latitude: 414213 N
Longitude: 084130 W
Altitude: 21
Bibliography: Almeida, CAB, 1990, 38. Viana 1926, 90; 1932, 161. Silva 1986a, nº 95.

Nº: **426**
Name: Senhora do Crasto
Parish: Deocriste
Council: Viana do Castelo
Latitude: 414101 N
Longitude: 084325W
Altitude: 248
Bibliography: Almeida, CAB, 1990, 36. Moreira, MAF, 1982, 54. Neves 1954, 13; 1965, 180. Paço-Quesado 1956, 80-90. Silva 1986a, nº 94. Teixeira *et al.* 1972. Viana 1926, 90.

Nº: **427**
Name: Castro do Peso
Parish: Geraz do Lima (S. Leocádia)
Council: Viana do Castelo
Latitude: 414122 N
Longitude: 083924 W
Altitude: 183
Bibliography: Almeida, CAB, 1990, 42-4. Almeida, CAB, *et al.* 1985. Moreira, MAF, 1982, 54. Silva 1986a, nº 96. Soeiro 1981, 99-102.

Nº: **428**
Name: Paço
Parish: Geraz do Lima (S. Maria
Council: Viana do Castelo
Latitude: 414212 N
Longitude: 084024 W
Altitude: 41
Bibliography: Almeida, CAB, 1990, 46-8.

Nº: **429**
Name: Cividade
Parish: Lanheses
Council: Viana do Castelo
Latitude: 414420 N
Longitude: 084010W
Altitude: 45
Bibliography: Almeida, CAB, 1990a, 189-203. MHOP-DGPU 1978, nº 15. Silva 1986a, nº 88. Viana 1932, 162.

Nº: **430**
Name: Outeiro
Parish: Lanheses
Council: Viana do Castelo
Latitude: (?)
Longitude: (?)
Altitude: (?)
Bibliography: Silva 1986a, nº 89. Viana 1932, 162. Almeida, CAB, 1990

Nº: **431**
Name: Castelão
Parish: Montaria
Council: Viana do Castelo
Latitude: (?)
Longitude: (?)
Altitude: (?)
Bibliography: Neves 1965, nº 19. Silva 1986a, nº 85. Viana 1932, 162.

Nº: **432**
Name: Castro Mau
Parish: Neiva (S. Romão)
Council: Viana do Castelo
Latitude: 413816 N
Longitude: 084635 W
Altitude: 112
Bibliography: MHOP-DGPU 1978, nº 3. Silva 1986a, nº 98. Viana 1926, 90; 1932, 163.

Nº: **433**
Name: Castro de S. Martinho
Parish: Nogueira
Council: Viana do Castelo
Latitude: 414505 N
Longitude: 084501 W
Altitude: 474
Bibliography: Almeida, CAB, 1990a, 182-3. MHOP-DGPU 1978, nº 16. Neves 1965, nº 18. Paço-Quesado 1956, 88-9. Silva 1986a, nº 83. Viana 1926, 90; 1932, 163.

Nº: **434**
Name: Rego do Castro
Parish: Outeiro
Council: Viana do Castelo
Latitude: 414604 N
Longitude: 084625 W
Altitude: 234
Bibliography: Almeida, CAB, 1990a, 217. Neves 1965, nº 7. Paço-Quesado 1956, 90. Silva 1986a, nº 82.

Nº: **435**
Name: Castro de Vieito
Parish: Perre
Council: Viana do Castelo
Latitude: 414334 N
Longitude: 084740 W
Altitude: 122
Bibliography: Almeida, CAB, 1990a, 217-8. Moreira, MAF, 1982, 54. MHOP-DGPU 1978, nº 20. Neves 1965, nº 15. Paço 1954; 1955, 510-15. Paço-Quesado 1956, 84; 1957, 387-93. Silva 1986a, nº 76. Viana 1926, 90; 1932, 163.

Nº: **436**
Name: Monte do Calvário
Parish: Perre
Council: Viana do Castelo
Latitude: 414323 N
Longitude: 084631 W
Altitude: 53
Bibliography: Almeida, CAB, 1990a, 212-5. Araújo, JR, 1948, 42-3. Azevedo, PA, 1901, 110. Cardozo 1934-36. Corrêa 1928, 188. Moreira, MAF, 1982, 54. Neves 1965, nº 12. Paço-Quesado 1956, 82. Silva 1986a, nº 74. Viana 1926, 90; 1932, 163.

Nº: **437**
Name: Monte dos Castelos
Parish: Perre
Council: Viana do Castelo
Latitude: 414420 N
Longitude: 084604 W
Altitude: 177
Bibliography: Almeida, CAB, 1990a, 215. Neves 1965, nº 11. Paço - Quesado 1956, 88. Silva 1986a, nº 78. Viana 1926, 90; 1932, 163.

Nº: **438**
Name: Miradoiros do Galo
Parish: Perre/Meadela
Council: Viana do Castelo
Latitude: 4144001N
Longitude: 084843 W
Altitude: 312
Bibliography: Almeida, CAB, 1990a, 221-2. Neves 1965, nº 13. Paço-Quesado 1956, 86-7, 168-9. Silva 1986a, nº 77. Viana 1926, 90; 1932, 163.

Nº: **439**
Name: Citânia (Cidade Velha)
Parish: Santa Maria Maior
Council: Viana do Castelo
Latitude: 414211 N
Longitude: 085045 W
Altitude: 226
Bibliography: Almeida, CAB, 1990a, 224-39. Bellino 1902. Caldas 1884, 333-5. Fortes 1904, 3; 1905-8a, 16-7. Kalb 1980b, nº 7, Abb. 2. MHOP-DGPU 1978, nº 25. Moreira, MAF, 1982, 54. Pereira 1910, 14; 1914, 11-4. Ponte 1981, 149-52. Silva 1986a, nº 73. Vasconcellos 1897 (1), 280, 380; 1902a; 1903a, 15-23; 1913 (3), 507-8. Viana 1932, 164; 1955a; 1962. Viana-Oliveira 1954, 40-72; 1955, 541-51. Vitorino 1909, 448.

Nº: **440**
Name: Castro de Serreleis
Parish: Serreleis
Council: Viana do Castelo
Latitude: 414225 N
Longitude: 084503 W
Altitude: 30
Bibliography: Almeida, CAB, 1990, 241. Viana 1938, 7-14.

Nº: **441**
Name: Capela das Almas
Parish: Viana do Castelo
Council: Viana do Castelo
Latitude: 414104 N
Longitude: 084945 W
Altitude: 5
Bibliography: Almeida, CAB, 1990, 241-2.

Nº: **442**
Name: Castro do Santinho (Roques)
Parish: Vila Franca/Mujães/ Sub-portela
Council: Viana do Castelo
Latitude: 414023 N
Longitude: 084325 W
Altitude: 269
Bibliography: Almeida 1978b, 40. Almeida, CAB, 1990, 32-5. Guerra 1910, 88. Moreira, MAF, 1982, 54. Neves 1942; 1959, 381-87; 1965, nº 26. Nunes 1979, 18-26. Sarmento 1933a, 209. Silva 1986a, nº 93. Viana 1926, 90; 1932, 162.

Nº: **443**
Name: Castro de Sabariz
Parish: Vila Fria
Council: Viana do Castelo
Latitude: 414015 N
Longitude: 084503W
Altitude: 110
Bibliography: Almeida, CAB, 1990, 31-2. Neves 1965, nº 25. Paço 1933, 272-6. Silva 1986a, nº 92. Viana 1932, 164.

Nº: **444**
Name: Cividade
Parish: Vila Mou
Council: Viana do Castelo
Latitude: 414351 N
Longitude: 084135W
Altitude: 55
Bibliography: Almeida, CAB, 1990, 186-7. Araújo, JR, 1962, 5-6. Moreira, MAF, 1981a; 1981b; 1982, 45-54. Silva 1986a, nº 90. Viana 1926, 90; 1932, 164.

Nº: **445**
Name: Crasto
Parish: Vilar de Murteda
Council: Viana do Castelo
Latitude: 414622 N
Longitude: 084300 W
Altitude: 250
Bibliography: Almeida, CAB, 1990a, 179. MHOP-DGPU 1978, nº 39. Neves 1965, nº 20. Silva 1986a, nº 86. Viana 1926, 90; 1932, 164. Vieira 1886-87, 229.

2.4) District of Vila Real

Nº: **446**
Name: Cornos das Alturas
Parish: Alturas de Barroso
Council: Boticas
Latitude: 414247 N
Longitude: 074931 W
Altitude: 1150
Bibliography: Martins, JB, 1984, nº 25. Santos Júnior 1982, 261. Silva 1986a, nº 480. Miranda Junior *et al.* 1983, 403, nº24.

Nº: **447**
Name: Vilarinho Seco (Coto dos Mouros)
Parish: Alturas de Barroso
Council: Boticas
Latitude: 414040 N
Longitude: 074804 W
Altitude: 973
Bibliography: Martins, JB, 1984, nº 23. Silva 1986a, nº 482. Miranda Junior *et al.* 1983, 403, nº22.

Nº: **448**
Name: Castelo de S. Romão
Parish: Alturas do Barroso
Council: Boticas
Latitude: 414318 N
Longitude: 074852 W
Altitude: 1111
Bibliography: Martins, JB, 1984, nº 24. Santos Júnior 1982, 260. Silva 1986a, nº 481. Miranda Junior *et al.* 1983, 403, nº23.

Nº: **449**
Name: Amurada (Murada)
Parish: Ardãos
Council: Boticas
Latitude: (?)
Longitude: (?)
Altitude: (?)
Bibliography: Martins, JB, 1984, nº 2. Miranda Júnior *et al.* 1983. Santos Júnior 1982, 260. Santos, MC, 1969. Silva 1986a, nº 484.

Nº: **450**
Name: Castro da Gorda
Parish: Ardãos
Council: Boticas
Latitude: 414636 N
Longitude: 073731 W
Altitude: 850
Bibliography: Martins, JB, 1984, nº 27. Miranda Júnior *et al.* 1983, 403, nº2, 408-11. Santos Júnior 1982, 260.

Nº: **451**
Name: Castro de Cunhas
Parish: Ardãos
Council: Boticas
Latitude: 414558 N
Longitude: 073508 W
Altitude: 700
Bibliography: Martins, JB, 1984, nº 1, 38. Santos Júnior 1982, 260. Silva 1986a, nº 485, 517. Miranda Junior *et al.* 1983, 403, nº1.

Nº: **452**
Name: Castro de Malhó (Amalhó)
Parish: Ardãos
Council: Boticas
Latitude: 414524 N
Longitude: 073724 W
Altitude: 689
Bibliography: Martins, JB, 1984, nº 3. Santos Júnior 1982, 260. Silva 1986a, nº 486. Montalvão 1971a, 34-38. Miranda Junior *et al.* 1983, 403, nº3.

Nº: **453**
Name: Castro de Carvalhelhos (Castelo dos Mouros)
Parish: Beça
Council: Boticas
Latitude: 414145 N
Longitude: 074357 W
Altitude: 850
Bibliography: Esparza 1980, nº 14. Martins, JB, 1984, nº 15. Miranda Júnior *et al.* 1983, 403, nº14, 430-5. Ponte 1984, nº 73. Santos Júnior 1957, 25-62; 1959-60, 361-8; 1963a,1963c, 187-93; 1964, 360-5; 1966b, 181-90; 1971a, 72-5; 1973a. 207-19; 1975b, 559-66; 1977b, 161-5; 1978c, 323-33; 1980b, 607-19; 1981b, 140-7; 1982, 249-63; 1983a, 511-9; 1984a, 673-82; 1984c, 411-24. Silva 1986a, 204, nº 483.

Nº: **454**
Name: Calhau dos Corvos
Parish: Bobadela
Council: Boticas
Latitude: (?)
Longitude: (?)
Altitude: (?)
Bibliography: Martins, JB, 1984, nº 6. Silva 1986a, nº 489.

Nº: **455**
Name: Castro da Cêrca
Parish: Bobadela
Council: Boticas
Latitude: 414358 N
Longitude: 073522 W
Altitude: 564
Bibliography: Martins, JB, 1984. Santos Júnior 1982, 260. Junior *et al.* 1983, 403, nº 5, 435. Montalvão 1971, 96.

Nº: **456**
Name: Castro da Cigadonha
Parish: Bobadela
Council: Boticas
Latitude: (?)
Longitude: (?)
Altitude: (?)
Bibliography: Martins, JB, 1984, nº 8. Santos Júnior 1982, 260. Silva 1986a, nº 491.

Nº: **457**
Name: Castro de Nogueira
Parish: Bobadela
Council: Boticas
Latitude: (?)
Longitude: (?)
Altitude: (?)
Bibliography: Kalb 1980b, nº 30. Martins, JB, 1984, nº 4. Ponte 1980, nº 51. Silva 1986a, nº 488. Santos Junior 1982, 260. Miranda Junior *et al.* 1983, 403, nº4.

Nº: **458**
Name: Castro do Brêjo
Parish: Bobadela
Council: Boticas
Latitude: 414357 N
Longitude: 073728 W
Altitude: 600
Bibliography: Martins, JB, 1984, nº 5. Miranda Júnior *et al.* 1983, 403, nº 5, 435-43. Santos Júnior 1982, 260. Silva 1986a, nº 490.

Nº: **459**
Name: Couto dos Mouros
Parish: Boticas
Council: Boticas
Latitude: 414021 N
Longitude: 073848 W
Altitude: 625
Bibliography: Martins, JB, 1984, nº 11. Miranda Júnior *et al.* 1983, 403, nº10, 425-6. Montalvão 1971, 96. Santos Júnior 1982, 260 (referred to as Côto da Moura). Silva 1986a, nº 496.

Nº: **460**
Name: Outeiro do Pardo
Parish: Boticas
Council: Boticas
Latitude: 410445 N
Longitude: 073858 W
Altitude: 500
Bibliography: Santos Júnior 1982, 260. Miranda Júnior *et al.* 1983, 403, nº 11, 423-5.

Nº: **461**
Name: Castro da Naia
Parish: Codeçoso
Council: Boticas
Latitude: 413912 N
Longitude: 074139 W
Altitude: 902
Bibliography: Martins, JB, 1984, nº 16. Santos Júnior 1982, 260. Silva 1986a, nº 503. Miranda Junior *et al.* 1983, 403, nº15.

Nº: **462**
Name: Secerigo
Parish: Codeçoso
Council: Boticas
Latitude: 413904 N
Longitude: 074314 W
Altitude: 750
Bibliography: Martins, JB, 1984, nº 17. Silva 1986a, nº 504. Santos Junior 1982, 260. Miranda Junior *et al.* 1983, 403.

Nº: **463**
Name: Castro de Lousas (Ervas Ruivas)
Parish: Couto de Dornelas
Council: Boticas
Latitude: 413635 N
Longitude: 074932 W
Altitude: 578
Bibliography: Martins, JB, 1984, nº 22. Miranda Júnior *et al.* 1983, 403, nº 21, 447-9. Santos Júnior 1982, 261. Silva 1986a, nº 499.

Nº: **464**
Name: Castro do Lesenho
Parish: Covas do Barroso
Council: Boticas
Latitude: 413840 N
Longitude: 084509 W
Altitude: 1074
Bibliography: Guerra 1982, 28-9, 31. Martins, JB, 1984, nº 19. Montalvão 1971, 96. Paço 1966c. Paris 1903, 1-8. Pereira 1908; 1915a. Santos Júnior 1982, 260. Sarmento 1880/81=1886=1896. Silva 1986a, nº 501. Vasconcellos 1913, 48-9. Miranda Junior *et al.* 1983, 403, nº 18.

Nº: **465**
Name: Castro do Poio
Parish: Covas do Barroso
Council: Boticas
Latitude: 413812 N
Longitude: 074745 W
Altitude: 695
Bibliography: Martins, JB, 1984, nº 20. Santos Júnior 1982, 260. Silva 1986a, nº 500. Miranda Junior *et al.* 1983, 403, nº 19.

Nº: **466**
Name: Souto da Lama
Parish: Gestosa
Council: Boticas
Latitude: 413926N
Longitude: 075000W
Altitude: 777
Bibliography: Martins, JB, 1984, nº 21. Miranda Júnior *et al.* 1983, 403, Nº 20, 443-7. Santos Júnior 1982, 260. Silva 1986a, nº 498.

Nº: **467**
Name: Cabeço
Parish: Granja
Council: Boticas
Latitude: 414225 N

Longitude: 073926 W
Altitude: 764
Bibliography: Martins, JB, 1984, nº 10. Montalvão 1971, 84-5. Miranda Júnior *et al.* 1983,403, Nº 9 411-9. Santos Júnior 1982, 260. Silva 1986a, nº 495.

Nº: **468**
Name: Castro de Mouril
Parish: Pinho
Council: Boticas
Latitude: 413922 N
Longitude: 073625 W
Altitude: 442
Bibliography: Martins, JB, 1984, nº 13. Miranda Júnior *et al.* 1983, 419-23. Santos Júnior 1982, 260. Silva 1986a, nº 506.

Nº: **469**
Name: Torre
Parish: Quintas (Seirrãos)
Council: Boticas
Latitude: (?)
Longitude: (?)
Altitude: (?)
Bibliography: Martins, JB, 1984, nº 14. Santos Júnior 1982, 260. Silva 1986a, nº 505. Miranda Junior *et al.* 1983, 403, nº 13.

Nº: **470**
Name: Castro de Sapelos
Parish: Sapiãos
Council: Boticas
Latitude: 414257 N
Longitude: 073607 W
Altitude: 600
Bibliography: Martins, JB, 1984, nº 7. Santos , MC, 1969. Miranda Júnior *et al.* 1983, 403, Nº 7, 426-30. Montalvão 1971, 42-6. Santos Júnior 1982, 260. Silva 1986a, nº 494.

Nº: **471**
Name: Muro
Parish: Sapiãos
Council: Boticas
Latitude: (?)
Longitude: (?)
Altitude: (?)
Bibliography: Martins, JB, 1984, nº 9. Silva 1986a, nº 493. Miranda Junior *et al.* 1983, 403, nº 8.

Nº: **472**
Name: Castro
Parish: Vilar
Council: Boticas
Latitude: 414029 N
Longitude: 074430 W
Altitude: 843
Bibliography: Martins, JB, 1984, nº 18. Santos Júnior 1982, 260. Silva 1986a, nº 502. Miranda Junior *et al.* 1983, 403, nº 17.

Nº: **473**
Name: Castro
Parish: Anelhe
Council: Chaves
Latitude: 414023 N
Longitude: 073507 W
Altitude: 485
Bibliography: Martins, JB, 1984, nº 6. Silva 1986a, nº 523.

Nº: **474**
Name: Atalaia
Parish: Arcossó
Council: Chaves
Latitude: (?)
Longitude: (?)
Altitude: (?)
Bibliography: Martins, JB, 1984, nº 7. Silva 1986a, nº 524.

Nº: **475**
Name: Castelões (Facho)
Parish: Calvão
Council: Chaves
Latitude: 414840 N
Longitude: 073337 W
Altitude: 929
Bibliography: Martins, JB, 1984, nº 11. Silva 1986a, nº 512.

Nº: **476**
Name: Outeiro dos mouros
Parish: Calvão
Council: Chaves
Latitude: 414738 N
Longitude: 073434 W
Altitude: 784
Bibliography: Martins, JB, 1984, nº 9. Silva 1986a, nº 513

Nº: **477**
Name: Castro (Castelo dos Mouros)
Parish: Curalha
Council: Chaves
Latitude: 414235 N
Longitude: 073138 W
Altitude: 431
Bibliography: Alves 1931. Esparza 1980, nº 15. Freitas-Santos 1980; 1981; 1982; 1983; 1984. Magalhães *et*

al. 1975; 1977; 1988. Martins, JB, 1984, nº 16. Montalvão 1971, 7-12. Santos Júnior 1984b. Silva 1986a, nº 522.

Nº: **478**
Name: Couto (Alto das Coroas)
Parish: Ervededo
Council: Chaves
Latitude: 414905 N
Longitude: 073157 W
Altitude: 775
Bibliography: Cardozo 1942a, 89-100. Martins, JB, 1984 nº 18. Ponte 1984, nº 75, 78, 85. Silva 1986a 228-9, nº 465, 277.

Nº: **479**
Name: Muro da Pastoria
Parish: Redondelo
Council: Chaves
Latitude: 414402 N
Longitude: 073344 W
Altitude: 602
Bibliography: Jorge 1985b. Martins, JB, 1984 nº 29. Montalvão 1971, 94. Silva 1986a nº 520. Soeiro 1985.

Nº: **480**
Name: Castro
Parish: Vidago
Council: Chaves
Latitude: 413801 N
Longitude: 073448 W
Altitude: 400
Bibliography: Martins, JB, 1984, nº 41. Montalvão 1971, 94. Silva 1986a, nº 525.

Nº: **481**
Name: Cambedo (Castelo de Vamba)
Parish: Vilarelho da Raia
Council: Chaves
Latitude: 415124 N
Longitude: 072900 W
Altitude: 525
Bibliography: Azevedo, M, 1895, 130. Martins, JB, 1984, nº 44. Montalvão 1971, 86-93. Silva 1986a, nº 507.

Nº: **482**
Name: Castelo
Parish: Cidadelhe
Council: Mesão Frio
Latitude: 411018 N
Longitude: 075035 W
Altitude: 443
Bibliography: Silva 1986a, 148, 151, 204, nº 633.

Nº: **483**
Name: Cebidaia
Parish: Atei
Council: Mondim de Basto
Latitude: 412722 N
Longitude: 075505 W
Altitude: 360
Bibliography: Silva 1986a, nº 592.

Nº: **484**
Name: Crastoeiro
Parish: Mondim de Basto
Council: Mondim de Basto
Latitude: 412451 N
Longitude: 075544 W
Altitude: 440
Bibliography: Silva 1986a, nº 596.

Nº: **485**
Name: Alto dos Palhaços
Parish: Vilar de Ferreiros
Council: Mondim de Basto
Latitude: 412502 N
Longitude: 075436 W
Altitude: 903
Bibliography: Silva 1986a, nº 594.

Nº: **486**
Name: Alto dos Palhacinhos
Parish: Vilar de Ferreiros
Council: Mondim de Basto
Latitude: 412508 N
Longitude: 075425 W
Altitude: 880
Bibliography: Silva 1986a, nº 595.

Nº: **487**
Name: Castro de Premurado
Parish: Vilar de Ferreiros
Council: Mondim de Basto
Latitude: 412522 N
Longitude: 075303 W
Altitude: 663

Nº: **488**
Name: Coroa do Coto
Parish: Cambeses
Council: Montalegre
Latitude: 414925 N
Longitude: 075145 W
Altitude: 1174
Bibliography: Barreiros 1915, 212, nº 27; 1919/20, 75. Silva 1986a, nº 445. Santos, MC, 1969. Costa, JG, 1968, 41.

Nº: **489**
Name: Crasto
Parish: Cambeses
Council: Montalegre
Latitude: 414915 N
Longitude: 075122 W
Altitude: 1020
Bibliography: Barreiros 1915, 212, nº 27; 1919/20, 75. Silva 1986a, nº 444.

Nº: **490**
Name: Alto do Crasto
Parish: Cervos
Council: Montalegre
Latitude: 414350 N
Longitude: 074051 W
Altitude: 886
Bibliography: Barreiros 1915, 211, nº 18; 1919-20, 72. Pinto 1928-29. Santos, MC, 1969. Silva 1986a, nº 474.

Nº: **491**
Name: Alto dos Fachos (Fonte da Moura)
Parish: Cervos
Council: Montalegre
Latitude: (?)
Longitude: (?)
Altitude: (?)
Bibliography: Barreiros 1919/20, 71. Silva 1986a, nº 475.

Nº: **492**
Name: Cabeço dos Mouros (Castro de Medeiros)
Parish: Chã
Council: Montalegre
Latitude: 414912 N
Longitude: 074615 W
Altitude: 1113
Bibliography: Barreiros 1915, 212, nº 32; 1919/20, 79-80. Costa, JG, 1968, 41. Kalb 1980b, nº 28, Abb. 5. Santos, MC, 1969. Silva 1986a, nº 448. Vasconcellos 1897 (RL 1), 55, 125.

Nº: **493**
Name: Castro de S. Vicente
Parish: Chã
Council: Montalegre
Latitude: 414627 N
Longitude: 074651 W
Altitude: 880
Bibliography: Barreiros 1915, 212, nº 33; 1919/20, 76. Costa, JG, 1968, 41. Freire 1967/68, 381-82. Santos Júnior 1963b. Santos - Freire 1964. Santos - Isidoro 1963. Silva 1986a, nº 447.

Nº: **494**
Name: Eiras (Bicancas do Castro)
Parish: Donões
Council: Montalegre
Latitude: 414922 N
Longitude: 074940 W
Altitude: 983
Bibliography: Barreiros 1915, 212, nº 29; 1919/20, 86. Costa, JG, 1968, 41. Santos, MC, 1969. Silva 1986a, nº 446.

Nº: **495**
Name: Alto da Ceada
Parish: Gralhas
Council: Montalegre
Latitude: 415038 N
Longitude: 074141 W
Altitude: 952
Bibliography: Silva 1986a, nº 455.

Nº: **496**
Name: Castelo de Romão
Parish: Gralhas
Council: Montalegre
Latitude: 415157 N
Longitude: 074310 W
Altitude: 1098
Bibliography: Barreiros 1915, 211, nº 6. Silva 1986a, nº 454.

Nº: **497**
Name: Castro do Soutelo
Parish: Gralhas
Council: Montalegre
Latitude: 415141 N
Longitude: 074035 W
Altitude: 825
Bibliography: Silva 1986, 88, nº 456. Barreiros 1920, 69.

Nº: **498**
Name: Castelo de Montalegre
Parish: Montalegre
Council: Montalegre
Latitude: 414933 N
Longitude: 074730 W
Altitude: 1000
Bibliography: Costa, JG, 1968. Silva 1986a, nº 449.

Nº: **499**
Name: Castro da Veiga (Monte do Castro)
Parish: Montalegre
Council: Montalegre

Latitude: 415002 N
Longitude: 074646 W
Altitude: 975
Bibliography: Costa, JG, 1968, 40-1. Barreiros 1915, 211, nº 1; 1920, 58. Silva 1986a, nº 450.

Nº: **500**
Name: Castro (Coroa do Castro)
Parish: Morgade
Council: Montalegre
Latitude: 414459 N
Longitude: 074531 W
Altitude: 993
Bibliography: Barreiros 1915, 212, nº 19. Santos, MC, 1969. Silva 1986a, nº 473.

Nº: **501**
Name: Castro de Lama Chão
Parish: Negrões
Council: Montalegre
Latitude: 414300 N
Longitude: 074522 W
Altitude: 1010
Bibliography: Barreiros 1920, 87. Santos Júnior 1968, 339-47.

Nº: **502**
Name: Castro de Negrões
Parish: Negrões
Council: Montalegre
Latitude: 414407 N
Longitude: 074601 W
Altitude: 1000
Bibliography: Barreiros 1915, 212, nº 22; 1919/20, 87. Santos, MC, 1969. Silva 1986a, nº 472.

Nº: **503**
Name: Castro do Mau Vizinho (de Vilarinho)
Parish: Negrões
Council: Montalegre
Latitude: 414345 N
Longitude: 074901 W
Altitude: 1018
Bibliography: Barreiros 1915, 212, nº 20; 1920, 87. Santos, MC, 1969. Silva 1986a, nº 470.

Nº: **504**
Name: Crasto (Porto das Antas)
Parish: Negrões
Council: Montalegre
Latitude: (?)
Longitude: (?)
Altitude: (?)
Bibliography: Silva 1986a, nº 471.

Nº: **505**
Name: Cabeça do Castro
Parish: Outeiro
Council: Montalegre
Latitude: 414733 N
Longitude: 075720 W
Altitude: 891
Bibliography: Barreiros 1915, 212, nº 31; 1919/20, 81. Silva 1986a, nº 441.

Nº: **506**
Name: Castro
Parish: Outeiro
Council: Montalegre
Latitude: (?)
Longitude: (?)
Altitude: (?)
Bibliography: Barreiros 1915, 212, nº 30; 1919/20, ?81. Silva 1986a, nº 442.

Nº: **507**
Name: Coto de Sendim (Castelo do Portelo)
Parish: Padornelos
Council: Montalegre
Latitude: 415340 N
Longitude: 074550 W
Altitude: 1200
Bibliography: Costa, JG, 1968. Silva 1986a, nº 452.

Nº: **508**
Name: Pedregalho
Parish: Padornelos
Council: Montalegre
Latitude: 415116 N
Longitude: 074435 W
Altitude: 1194
Bibliography: Barreiros 1915, 211, nº 5; 1919/20, 86. Silva 1986a, nº 453.

Nº: **509**
Name: Cigadonha
Parish: Padroso
Council: Montalegre
Latitude: 415113 N
Longitude: 074701 W
Altitude: 1060
Bibliography: Barreiros 1915, 211, nº 3; 1919/20, 83. Silva 1986a, nº 451.

Nº: **510**
Name: Castro (Alto do castro)
Parish: Paradela
Council: Montalegre
Latitude: 414512 N

Longitude: 075658 W
Altitude: 864
Bibliography: Barreiros 1915, 213, nº 51; 1919/20, 73-74. Silva 1986a, 247-49, nº 443. Cardozo 1959d. Parreira *et al.* 1980, nº129. MNAE, 566, 568, 569.

Nº: **511**
Name: Igreja do Castelo (Aldeia do Juriz)
Parish: Pitões das Júnias
Council: Montalegre
Latitude: 415018 N
Longitude: 075736 W
Altitude: 1055
Bibliography: Barreiros 1919/20, 85. Costa, JG, 1968. Silva 1986a, nº 440.

Nº: **512**
Name: Castelo da Lomba
Parish: Reigoso
Council: Montalegre
Latitude: 414413 N
Longitude: 075311 W
Altitude: 890
Bibliography: Silva 1986a, nº 465.

Nº: **513**
Name: Ponte dos Mouros (Espartida)
Parish: Reigoso
Council: Montalegre
Latitude: (?)
Longitude: (?)
Altitude: (?)
Bibliography: Silva 1986a, nº 466.

Nº: **514**
Name: Castelos
Parish: Salto
Council: Montalegre
Latitude: 413610 N
Longitude: 075611 W
Altitude: 1034
Bibliography: Barreiros 1915, 213, nº 45. Silva 1986a, nº 476.

Nº: **515**
Name: Cividades
Parish: Salto
Council: Montalegre
Latitude: (?)
Longitude: (?)
Altitude: (?)
Bibliography: Barreiros 1915, 213, nº 46. Silva 1986a, nº 477.

Nº: **516**
Name: Crasto
Parish: Salto
Council: Montalegre
Latitude: (?)
Longitude: (?)
Altitude: (?)
Bibliography: Barreiros 1915, 213, nº 48. Silva 1986a, nº 478.

Nº: **517**
Name: Cristelo
Parish: Salto
Council: Montalegre
Latitude: 413638 N
Longitude: 075259 W
Altitude: 850
Bibliography: Barreiros 1915, 213, nº 49. Silva 1986a, nº 479.

Nº: **518**
Name: Castelo de Carvalhosa
Parish: Santo André
Council: Montalegre
Latitude: (?)
Longitude: (?)
Altitude: (?)
Bibliography: Silva 1986a, nº 457.

Nº: **519**
Name: Cidade de Grou
Parish: Santo André
Council: Montalegre
Latitude: 415342 N
Longitude: 074100 W
Altitude: 1000
Bibliography: Barreiros 1915, 211, nº 9; 1919-20, 82. Silva 1986a, nº 458. Taboada Chivite 1955, 333, 352.

Nº: **520**
Name: Castro de Lamego
Parish: Sarraquinhos
Council: Montalegre
Latitude: 414830 N
Longitude: 073749 W
Altitude: 975
Bibliography: Silva 1986a, nº 463. Barreiros, 1920, 67.

Nº: **521**
Name: Fernão Mouro
Parish: Sarraquinhos
Council: Montalegre
Latitude: (?)
Longitude: (?)
Altitude: (?)
Bibliography: Barreiros 1915, 211, nº 13; 1919/20, 68, 69. Costa, JG, 1963. Silva 1986a, nº 464.

Nº: **522**
Name: Castro

Parish: Solveira
Council: Montalegre
Latitude: (?)
Longitude: (?)
Altitude: (?)
Bibliography: Barreiros 1915, 211, nº 15; 1919/20, 69. Coffyn 1985, nº 35. Costa, JG, 1963, 119-25. Kalb 1980b, nº 29, Abb. 6. Silva 1986a, nº 462.

Nº: **523**
Name: Castrelos
Parish: Tourém
Council: Montalegre
Latitude: 415402 N
Longitude: 075419 W
Altitude: 956
Bibliography: Barreiros 1915, 212, nº 25; 1919/20, 75. Costa, JG, 1968, 41. Fortes 1908, 665-86. Lema 1978. Silva 1986a, 250, nº 439. Cardozo 1943a, 109. López Cuevillas 1951c, 36. Raddatz 1969, 197, nº38.

Nº: **524**
Name: Castelo de S. Romão
Parish: Viade de Baixo
Council: Montalegre
Latitude: (?)
Longitude: (?)
Altitude: (?)
Bibliography: Barreiros 1915, 212, nº 24; 1919/20, 73. Costa, JG, 1968, 41. Silva 1986a, nº 469.

Nº: **525**
Name: Crasto (Boca dos Infernos)
Parish: Vila da Ponte
Council: Montalegre
Latitude: 414218 N
Longitude: 075323 W
Altitude: 1050
Bibliography: Barreiros 1915, 213, nº 41. Silva 1986a, nº 467.

Nº: **526**
Name: Ponte dos Mouros
Parish: Vila da Ponte
Council: Montalegre
Latitude: (?)
Longitude: (?)
Altitude: (?)
Bibliography: Silva 1986a, nº 468.

Nº: **527**
Name: Castelar
Parish: Vilar de Perdizes
Council: Montalegre
Latitude: (?)
Longitude: (?)
Altitude: (?)
Bibliography: Silva 1986a, nº 459.

Nº: **528**
Name: Mina
Parish: Vilar de Perdizes
Council: Montalegre
Latitude: 415044 N
Longitude: 074035 W
Altitude: 802
Bibliography: Esparza 1980, nº 22. Silva 1986a, nº 460. Barreiros 1919-20, 65.

Nº: **529**
Name: Praça de Crastelos
Parish: Vilar de Perdizes
Council: Montalegre
Latitude: (?)
Longitude: (?)
Altitude: (?)
Bibliography: Barreiros 1915, 211, nº 11; 1919/20, 62. Costa, JG, 1968, 41. Silva 1986a, nº 461.

Nº: **530**
Name: Monte dos Mouros
Parish: Cerva (Asnela)
Council: Ribeira da Pena
Latitude: 412821 N
Longitude: 075248 W
Altitude: 315
Bibliography: Silva 1986a, nº 559.

Nº: **531**
Name: Castro
Parish: Cerva (Couçadouro)
Council: Ribeira da Pena
Latitude: 412730 N
Longitude: 074817 W
Altitude: 495
Bibliography: Meneses, M, 1929, 34. Silva, 1986a, nº 560.

Nº: **532**
Name: Monte do Facho (Castro do Facho)
Parish: Cerva (Formoselos)
Council: Ribeira da Pena
Latitude: 412847 N
Longitude: 075236 W
Altitude: 528
Bibliography: Meneses, M, 1929, 42-3. Silva 1986a, nº 556.

Nº: **533**
Name: Outeiro dos Mouros
Parish: Cerva (Mourão)

Council: Ribeira da Pena
Latitude: 412838 N
Longitude: 074831 W
Altitude: 692
Bibliography: Meneses, M, 1929, 34, 41. Martins, MJC, 1981, 16-8. Silva 1986a, nº 558.

Nº: **534**
Name: Castro
Parish: Cerva (Seixinhos)
Council: Ribeira da Pena
Latitude: 412907 N
Longitude: 075038 W
Altitude: 480
Bibliography: Meneses, M, 1929, 34. Silva 1986a, nº 557.

Nº: **535**
Name: Monte Calvo (Fraga de Sulajes)
Parish: Cerva (Vilarinho)
Council: Ribeira da Pena
Latitude: 413005 N
Longitude: 074854 W
Altitude: 806
Bibliography: Meneses, M, 1929, 36-40. Silva 1986a, nº 555.

Nº: **536**
Name: Outeiro dos Mouros
Parish: Ribeira de Pena
Council: Ribeira da Pena
Latitude: 413038 N
Longitude: 074739 W
Altitude: 586
Bibliography: Silva 1986a, nº 553. Meneses, M, 1929, 34.

Nº: **537**
Name: Outeiro dos Mouros
Parish: Ribeira de Pena (Daivões)
Council: Ribeira de Pena
Latitude: 413123 N
Longitude: 075138 w
Altitude: 260
Bibliography: Meneses, M, 1929, 32, 34, 44. Silva 1986a, nº 554.

Nº: **538**
Name: Moradias
Parish: Cumieira
Council: Santa Marta de Penaguião
Latitude: (?)
Longitude: (?)
Altitude: (?)
Bibliography: Botelho 1902, 152-4. Silva 1986a, nº 627.

Nº: **539**
Name: Castro de Fontes
Parish: Fontes
Council: Santa Marta de Penaguião
Latitude: 411337 N
Longitude: 074805 W
Altitude: 658
Bibliography: Santos Júnior 1980c. Silva 1986a, nº 628.

Nº: **540**
Name: Monte de S. Pedro
Parish: Lobrigos (S. João)
Council: Santa Marta de Penaguião
Latitude: 411059 N
Longitude: 074609 W
Altitude: 443
Bibliography: Silva 1986a, nº 629.

Nº: **541**
Name: Cidadelhe de Aguiar
Parish: Afonsim
Council: Vila Pouca de Aguiar
Latitude: 413055 N
Longitude: 073901 W
Altitude: 916
Bibliography: Martins, JB, 1984, nº 13. Silva 1986a, nº 564.

Nº: **542**
Name: Castro
Parish: Capeludos
Council: Vila Pouca de Aguiar
Latitude: 413739 N
Longitude: 073752 W
Altitude: 512
Bibliography: Martins, JB, 1984, nº 12. Silva 1986a, 309, nº 561. Vasconcellos 1902a, 23-26; 1913 (R.L., 3), 54-5, 55, 60-1.

Nº: **543**
Name: (?)
Parish: Parada de Monteiros
Council: Vila Pouca de Aguiar
Latitude: (?)
Longitude: (?)
Altitude: (?)
Bibliography: Azevedo, M, 1895, 130-6. Botelho 1904, 58. Silva 1986a, nº 562.

Nº: **544**
Name: Castro de Rebordochão (Castelos)
Parish: Pensalvos
Council: Vila Pouca de Aguiar
Latitude: 413236 N
Longitude: 073818 W
Altitude: 929
Bibliography: Silva 1986a, nº 563.

Nº: **545**
Name: Castelo de Aguiar
Parish: Telões
Council: Vila Pouca de Aguiar
Latitude: 412804 N
Longitude: 074045 W
Altitude: 850
Bibliography: Silva 1986a, 255, nº 566. Severo 1905-08c., 108-09 (IAFCUP)

Nº: **546**
Name: Castro de Arnadelo
Parish: Torgueda
Council: Vila Real
Latitude: 411537 N
Longitude: 074735 W
Altitude: 702
Bibliography: Azevedo, M, 1895. Silva 1986a, 228, nº 608.

Nº: **547**
Name: Monte da Murada
Parish: Vilarinho de Samardã
Council: Vila Real
Latitude: (?)
Longitude: (?)
Altitude: (?)
Bibliography: Silva 1986a, nº 601.

APPENDIX II

1) Introduction

A reliable absolute chronology based on radiocarbon dating is essential to a better understanding of the cultural phasing within the castros of northern Portugal. Before this can be achieved however, some of the outstanding problems and inconsistencies must be tackled. The application of the radiocarbon dating method came relatively late to the Northwestern castros. Its use within the region has been sporadic and a degree of suspicion towards the feasibility of this method for dating archaeological sequences within the castros has arisen. This is largely due to the scarcity of such dates, to samples being carelessly taken or thoughtlessly contaminated, and to misinterpreting the results.

The impact made by the first few dates is well illustrated by the case of the 'wheel from Catoira'[1]. Even today, this example is still quoted by some archaeologists, as evidence of the shortcomings of the method. The great advances of the eighties have, however, gone a long way towards increasing the archaeologist's confidence in the method. In particular, the publication of calibration curves and the reduction of the standard deviations by laboratories, have contributed to this. In fact, northwestern archaeologists took little notice of the first few calibration curves produced (Suess 1970, Damon *et al.* 1972, Ralph *et al.* 1973, and Clark 1975), until Klein *et al.*'s (1982) curve was applied to the radiocarbon dates from Castro das Ermidas (Queiroga 1985, 35-9).

The general attitudes of those engaged in castros' research towards the use of scientific methods fall into two groups. The first group tends to accept new methods and approaches almost unconditionally, without taking their limitations and range of valid applications into consideration. This position is typical of the recent generation of archaeologists, in which I am included. The second uses only traditional methods and regards all technical advances with suspicion.

Old attitudes die hard in Portugal, although the exemplary efforts of laboratories to present results as calibrated dates, and to establish a dialogue with archaeologists (Soares - Cabral 1984), are gradually making an impact. The recent creation of a radiocarbon laboratory in Portugal is significant, and marks the beginning of a new phase in chronological studies for northern Portuguese Prehistory.

Until the mid-eighties, radiocarbon dating was generally done on the basis of one (or very few) date per site. If this date fitted in with the archaeologist's proposed pottery-based relative dating sequence, then it was accepted; if not it was rejected. In only a few cases did radiocarbon dating stimulate discussion, and lead to any chronological revision (Almeida *et al.* 1982)[2]. Laboratory accuracy was often questioned, to such an extent that it became common practice to give the same sample to several different laboratories for dating. The well-known margins of error between laboratories (Aitken 1990, 79-81) did little to improve matters. Again, the common saying that "one date is no date" became a hindrance to an acceptance of the reliability of the method, by understanding its limitations. The mid-eighties saw a change in attitude towards this problem, particularly

[1] - The wooden wheel discovered in a marsh, in Catoira, referred to above in Chapter 7, was initially given a date of 3670±45bc. Such an early date for an important item provoked much enthusiasm. This later turned to delusion when, on closer inspection of the piece under more propitious conditions, it was found to be much more recent (Lorenzo Fernandez 1973), as an XVIIIth century inscription revealed.

[2] - In this case, surprisingly, too much credit was given to those uncalibrated dates which proved to be more *acceptable* after calibration.

amongst the younger generation of archaeologists. Recently, it became customary to present all dates in cal BC form[3], although standard deviation limits are still not presented graphically.

As more dates are taken from the same context, so reliability increases, and the establishment of an 'internal scale' for a site becomes a possibility. Such a strategy was attempted at the site of Castelo de Matos in the north of Portugal (Queiroga - Figueiral 1992) and the results were rewarding. The high costs of this procedure are an important consideration however, particularly in a country where archaeological research is quite often at the archaeologist's own financial risk. As an experiment in methodology, three samples of charred seeds from Castelo de Matos were taken for radiocarbon dating. The results are presented in the chart below.

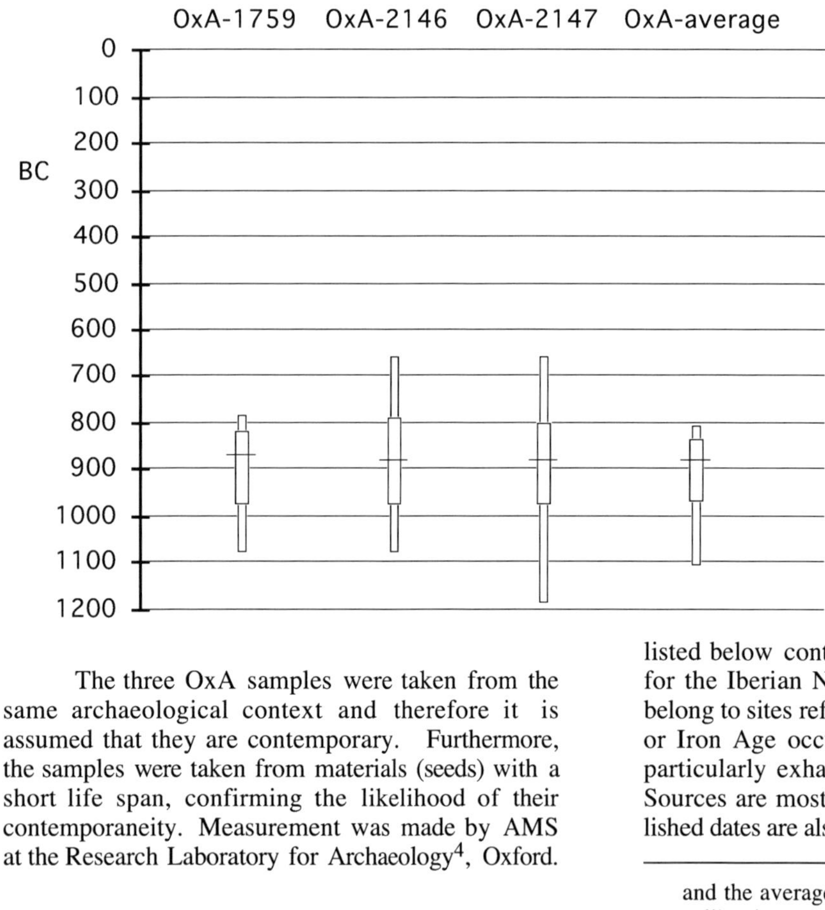

The three OxA samples were taken from the same archaeological context and therefore it is assumed that they are contemporary. Furthermore, the samples were taken from materials (seeds) with a short life span, confirming the likelihood of their contemporaneity. Measurement was made by AMS at the Research Laboratory for Archaeology[4], Oxford. The results, in radiocarbon years, were then calibrated according to the curve of Stuiver and Becker (1986). The probability distributions about a mean age in calendar years B.C. are shown graphically below:

The OxA-average for all three samples is also depicted. This graph shows just how much the age range in calendar years is reduced, as sample size increases, i.e. the lower value of two sigma in the mean of each individual date is roughly equivalent to a one sigma limit for the average of the dates combined.

Now these dates coincide with one of the *privileged* areas of the calibration curve, i.e. the first millennium. This good fortune should not be allowed to disguise the fact that the calibration curves are much less accurate for the period circa VIII to V centuries B.C. when, as illustrated below, each date in radiocarbon years B.P. may have several possible values in calendar years. The standard deviation in such cases will be correspondingly higher. Therefore, the allowable time-span of the dates, in calendar years, for a substantial part of the period of occupation of the castros, is rather large and of little help to archaeologists trying to construct reasonably accurate chronologies. This is not encouraging, as during the period from the Late Bronze Age to the end of the II century B.C. details of the Castros Culture are less well known, i.e. the accuracy of cultural dating is normally larger than one or two centuries.

The radiocarbon dataset listed below contains the radiocarbon dates produced for the Iberian Northwest until early 1992. They belong to sites referred to as having either Bronze Age or Iron Age occupation and, obviously, this list is particularly exhaustive for the north of Portugal. Sources are mostly bibliographical, although unpublished dates are also quoted[5].

---

3 - It is fair to point out that this is also due to the fact that, by then, laboratories started presenting their results in a time span corresponding to the 2 sigma limits of the calibrated age. However, efforts from archaeologists to calibrate dates also occurred (Eiroa 1980; Martins 1987, 742; Queiroga 1985, 35-9).

4 - The results provided by the Laboratory are:

OxA-1759	$d^{13}C$=-26‰	2730±70BP	(cal. BC 850-710)
OxA-2146	$d^{13}C$=-21,4‰	2700±90BP	(cal. BC 840-660)
OxA-2147	$d^{13}C$=-23,9‰	2710±90BP	(cal. BC 850-670)

and the average calibrated age ranges (using the calibration program of C.I.O. Groningen and the curve of Pearson and Stuiver 1986) are:

one sigma	900 - 820 (cal. BC)
two sigma	990 - 800 (cal. BC)

Thanks are due to Rupert Housley for all his kind help and advice regarding the careful selection of the samples and interpretation of the results.

5 - Information on unpublished dates was provided by A.P. Dinis (LY-4936, ICEN-45), in addition to dates from a joint research project with myself, at Castro de Penices. I am particularly grateful to Xulio

Complementary information for each site is also listed in this Appendix under the headings of site, location, type, cultural phase, references, sample ID, radiocarbon age BP, for easy comparison.

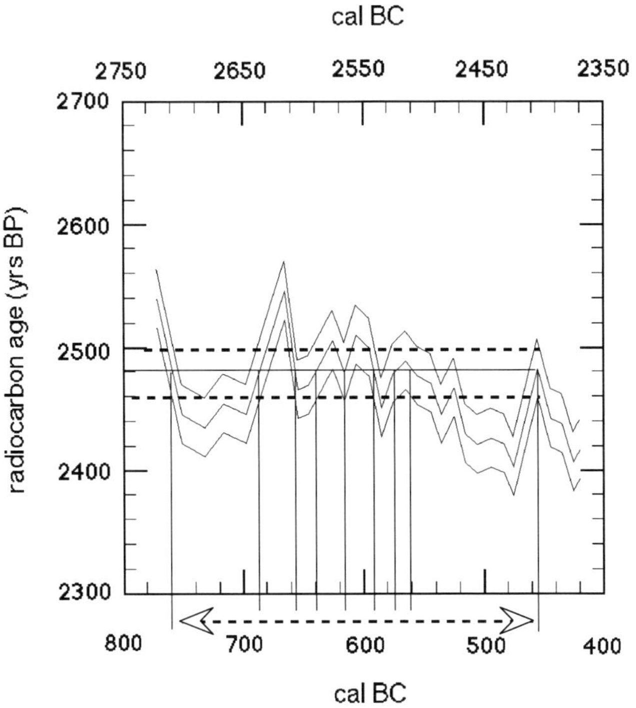

Intercepts obtained by the projection of a date of 2480 radiocarbon years BP on the calibration curve. Standard deviation is represented by dotted lines, but corresponding intercepts have been neglected in this particular example.

Site and location references give the most common designation of the site and also any administrative references. Dates are those from the north of Portugal and Galicia, although there are a few exceptions, and the reasons for these are provided. For types of site, a distinction is made between defended and open (not defended) settlements. Although this may not be necessary for the Iron Age, as all dated sites are defended, it may prove to be relevant to the Bronze Age where the contemporanity between open and defended sites is still open to discussion. Only the broadest cultural phasing is given, i.e. a distinction is made only between Bronze Age and Iron Age sites, for the sake of uniformity. Bibliographic references are restricted here to the articles in which radiocarbon dates or a proposed cultural dating are mentioned. Further references to the northern Portuguese sites can be found in the corresponding records of Appendix I.

Most of the references to dates in the dataset have been taken either from archaeological reports, or from articles in which a brief mention occurs. In most cases accordingly, only the date in radiocarbon years BP, together with its corresponding standard deviation, is available. Further information, such as the ‰ value of $d^{13}C$ and a measure of the half-life, is available in only a few cases. Therefore, all dates have been standardized on the Libby half-life of 5568 years, although it is known that at least some of the dates from the Gakushuin University Laboratory (GAK) used a half-life of 5570 years.

The radiocarbon calibration program CALIB (1987), produced by the Quaternary Isotope Laboratory of the University of Washington, and based on the calibration curve of Stuiver and Becker (1986), was used for the calibration of the dates presented in my dataset below. As all these dates fall below 3950 radiocarbon years BP, the decadal calibration file ATM10.14C was used for the whole set. No laboratory error multiplier was used for the conversion of the standard error of the radiocarbon age into a range of calendar ages.

Dates are listed in decreasing numerical order of their radiocarbon age BP. Figures 1.1 to 1.5 are

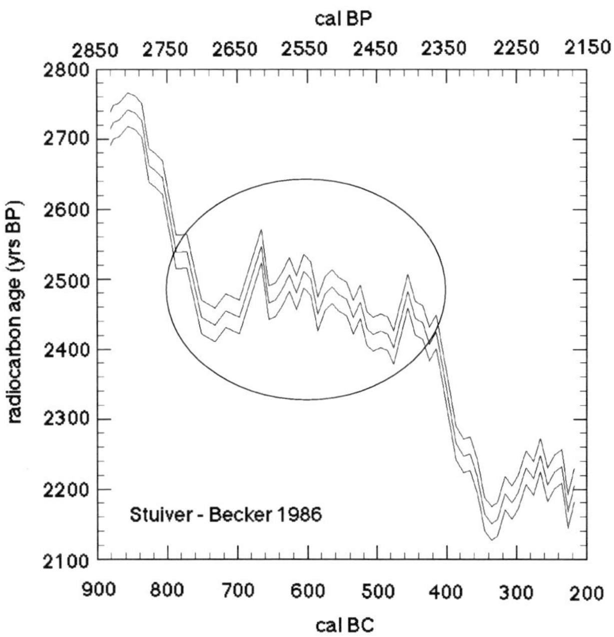

Carballo Arceo and Ramón Fábregas Valcarce for kindly permitting me to use their as yet unpublished article (Carballo - Fábregas 1992).

arranged in the same manner. In this way, a clearer idea can be obtained as to which areas were more densely covered by radiocarbon dating in the first millennium.

This straightforward chart is complemented by an additional representation of date range. In figures 2.1 to 2.4 dates are grouped by site, with the intention of showing the degree of overlap in age range within each individual site. Accordingly, only sites with more than one coherent radiocarbon date are represented in this set of figures. It should also be pointed out that this latter analysis is relatively broad, and therefore limited, as no discussion is presented as to the context of these dates. Occasionally, some dates within a coherent set from the same site are considered to be at variance with their cultural dating or contextual origins. Discussion of this problem would enlarge the scope of this Appendix beyond that intended.

It is now becoming clear that a substantial, and ever increasing, number of radiocarbon dates is available for research on the Castros Culture. This dataset should prove to be of considerable value in the future, as a means of resolving some of the more persistent doubts concerning absolute and relative chronologies in the first millennium.

2) Radiocarbon Dataset

Site: Castro de Troña
Location: Ponteareas, Pontevedra, Galicia
Type: Defended
Cultural phase: Iron Age
References: Carballo - Fábregas 1992, nº 21
Sample ID: GAK-12224
Radiocarbon Age BP: 3470 ± 110≈

Calibrated age (s)	cal BC	1808, 1804, 1771, 1760, 1754		
	cal BP	3757, 3753, 3720, 3709, 3703		

cal AD/BC (cal BP) age ranges obtained from intercepts (Method A) :

one Sigma**	cal BC	2008-2002 (3957-3951)	1920-1690 (3869-3639)	1671-1660 (3620-3609)
		1652-1642 (3601-3591)		
two Sigma**	cal BC	2134-2081 (4083-4030)	2071-2059 (4020-4008)	2040-1520 (3989-3469)

Summary of above
minimum of cal age ranges (cal ages) maximum of cal age ranges:

one sigma	cal BC	2008 (1808, 1804, 1771, 1760, 1754) 1642
	cal BP	3957 (3757, 3753, 3720, 3709, 3703) 3591
two sigma	cal BC	2134 (1808, 1804, 1771, 1760, 1754) 1520
	cal BP	4083 (3757, 3753, 3720, 3709, 3703) 3469

cal AD/BC age ranges (cal ages as above) from probability distribution (Method B) :

% area enclosed	cal BC (cal BP) age ranges		relative area under probability distribution
68.3 (one sigma)	cal BC	2009-1997 (3958-3946)	.03
		1940-1933 (3889-3882)	.02
		1930-1690 (3879-3639)	.88
		1672-1658 (3621-3607)	.04
		1653-1640 (3602-3589)	.03
95.4 (two sigma)	cal BC	2132-2082 (4081-4031)	.03
		2040-1520 (3989-3469)	.96

Site: Castro de Penalba
Location: Campo Lameiro, Pontevedra, Galicia
Type: Defended
Cultural phase: Iron Age
References: Carballo - Fábregas 1992, nº 15
Sample ID: GAK-12157
Radiocarbon Age BP: 3350 ± 160≈

Calibrated age (s)	cal BC	1686, 1672, 1658, 654, 1639
	cal BP	3635, 3621, 3607, 3603, 3588

cal AD/BC (cal BP) age ranges obtained from intercepts (Method A) :

one Sigma**	cal BC	1880-1841 (3829-3790)	1830-1490 (3779-3439)	1482-1457 (3431-3406)
two Sigma**	cal BC	2128-2123 (4077-4072)	2113-2099 (4062-4048)	2094-2083 (4043-4032)
		2068-2064 (4017-4013)	2040-1370 (3989-3319)	1366-1317 (3315-3266)

Summary of above
minimum of cal age ranges (cal ages) maximum of cal age ranges:

one sigma	cal BC	1880 (1686, 1672, 1658, 1654, 1639) 1457
	cal BP	3829 (3635, 3621, 3607, 3603, 3588) 3406
two sigma	cal BC	2128 (1686, 1672, 1658, 1654, 1639) 1317
	cal BP	4077 (3635, 3621, 3607, 3603, 3588) 3266

cal AD/BC age ranges (cal ages as above) from probability distribution (Method B) :

% area enclosed	cal BC (cal BP) age ranges		relative area under probability distribution
68.3 (one sigma)	cal BC	1878-1842 (3827-3791)	.08
		1830-1790 (3779-3739)	.09
		1780-1490 (3729-3439)	.80
		1478-1463 (3427-3412)	.03
95.4 (two sigma)	cal BC	2132-2082 (4081-4031)	.02

Site: Castro de Torroso
Location: Mos, Pontevedra, Galicia
Type: Defended
Cultural phase: Bronze Age-Iron Age
References: Carballo - Fábregas 1992, nº 19
Sample ID: GAK-12158
Radiocarbon Age BP: 3350 ± 90≈
Calibrated age (s) cal BC 1686, 1672, 1658, 1654, 1639
cal BP 3635, 3621, 3607, 3603, 3588
cal AD/BC (cal BP) age ranges obtained from intercepts (Method A) :

one Sigma**	cal BC	1743-1525 (3692-3474)	
two Sigma**	cal BC	1885-1837 (3834-3786)	1840-1440 (3789-3389)

Summary of above
minimum of cal age ranges (cal ages) maximum of cal age ranges:

one sigma	cal BC	1743 (1686, 1672, 1658, 1654, 1639) 1525
	cal BP	3692 (3635, 3621, 3607, 3603, 3588) 3474
two sigma	cal BC	1885 (1686, 1672, 1658, 1654, 1639) 1440
	cal BP	3834 (3635, 3621, 3607, 3603, 3588) 3389

cal AD/BC age ranges (cal ages as above) from probability distribution (Method B) :

% area enclosed	cal BC (cal BP) age ranges		relative area under probability distribution
68.3 (one sigma)	cal BC	1743-1525 (3692-3474)	1.00
95.4 (two sigma)	cal BC	1882-1839 (3831-3788)	.06
		1834-1487 (3783-3436)	.91
		1485-1452 (3434-3401)	.03

Site: Castro de Torroso
Location: Mos, Pontevedra, Galicia
Type: Defended
Cultural phase: Bronze Age-Iron Age
References: Carballo - Fábregas 1992, nº 19; Peña Santos 1987, 131-2
Sample ID: GAK-12159
Radiocarbon Age BP: 3340 ± 80≈
Calibrated age (s) cal BC 1684, 1673, 1637
cal BP 3633, 3622, 3586
cal AD/BC (cal BP) age ranges obtained from intercepts (Method A):

one Sigma**	cal BC	1737-1524 (3686-3473)	
two Sigma**	cal BC	1880-1841 (3829-3790)	1830-1450 (3779-3399)

Summary of above
minimum of cal age ranges (cal ages) maximum of cal age ranges:
one sigma cal BC 1737 (1684, 1673, 1637) 1524

	cal BP 3686 (3633, 3622, 3586) 3473			
two sigma	cal BC 1880 (1684, 1673, 1637) 1450			
	cal BP 3829 (3633, 3622, 3586) 3399			

cal AD/BC age ranges (cal ages as above) from probability distribution (Method B):

% area enclosed	cal BC (cal BP) age ranges		relative area under probability distribution
68.3 (one sigma)	cal BC	1732-1705 (3681-3654)	.13
		1695-1579 (3644-3528)	.63
		1576-1528 (3525-3477)	.24
95.4 (two sigma)	cal BC	1878-1843 (3827-3792)	.04
		1830-1817 (3779-3766)	.01
		1815-1793 (3764-3742)	.02
		1782-1488 (3731-3437)	.90
		1483-1455 (3432-3404)	.03

Site: Castro de Torroso
Location: Mos, Pontevedra, Galicia
Type: Defended
Cultural phase: Bronze Age-Iron Age
References: Carballo - Fábregas 1992, nº 19; Peña Santos 1987, 131-2
Sample ID: GAK-12160
Radiocarbon Age BP: 3340 ± 120≈

Calibrated age (s)	cal BC	1684, 1673, 1637	
	cal BP	3633, 3622, 3586	

cal AD/BC (cal BP) age ranges obtained from intercepts (Method A):

one Sigma**	cal BC	1769-1763 (3718-3712)	1750-1500 (3699-3449)
two Sigma**	cal BC	2009-2000 (3958-3949)	1920-1410 (3869-3359)

Summary of above

minimum of cal age ranges (cal ages) maximum of cal age ranges:

one sigma	cal BC	1769 (1684, 1673, 1637) 1500
	cal BP	3718 (3633, 3622, 3586) 3449
two sigma	cal BC	2009 (1684, 1673, 1637) 1410
	cal BP	3958 (3633, 3622, 3586) 3359

cal AD/BC age ranges (cal ages as above) from probability distribution (Method B):

% area enclosed	cal BC (cal BP) age ranges		relative area under probability distribution
68.3 (one sigma)	cal BC	1854-1847 (3803-3796)	.02
		1809-1804 (3758-3753)	.01
		1772-1759 (3721-3708)	.03
		1750-1490 (3699-3439)	.94
95.4 (two sigma)		BEYOND CALCULABLE RANGE	

Site: Tapado da Caldeira
Location: Baião, Portugal
Type: Open
Cultural phase: Bronze Age
References: Jorge, S.O., 1983; 1985, 160
Sample ID: KN-2769
Radiocarbon Age BP: 3290 ± 55≈

Calibrated age (s)	cal BC	1601, 1561, 1536	
	cal BP	3550, 3510, 3485	

cal AD/BC (cal BP) age ranges obtained from intercepts (Method A):

one Sigma**	cal BC	1686-1672 (3635-3621)	1658-1654 (3607-3603)	1640-1502 (3589-3451)
two Sigma**	cal BC	1732-1705 (3681-3654)	1700-1450 (3649-3399)	

Summary of above

minimum of cal age ranges (cal ages) maximum of cal age ranges:

one sigma	cal BC	1686 (1601, 1561, 1536) 1502
	cal BP	3635 (3550, 3510, 3485) 3451
two sigma	cal BC	1732 (1601, 1561, 1536) 1450
	cal BP	3681 (3550, 3510, 3485) 3399

cal AD/BC age ranges (cal ages as above) from probability distribution (Method B):

% area enclosed	cal BC (cal BP) age ranges		relative area under probability distribution
68.3 (one sigma)	cal BC	1684-1673 (3633-3622)	.08
		1637-1517 (3586-3466)	.91
95.4 (two sigma)	cal BC	1728-1710 (3677-3659)	.02
		1693-1488 (3642-3437)	.92
		1484-1454 (3433-3403)	.05

Site: Tapado da Caldeira
Location: Baião, Portugal
Type: Open
Cultural phase: Bronze Age
References: Jorge, S.O., 1983; 1985, 160
Sample ID: KN-2770
Radiocarbon Age BP: 3210 ± 55≈
Calibrated age (s) cal BC 1494
cal BP 3443

cal AD/BC (cal BP) age ranges obtained from intercepts (Method A):

one Sigma**	cal BC	1583-1573 (3532-3522)	1529-1434 (3478-3383)	
two Sigma**	cal BC	1681-1674 (3630-3623)	1620-1410 (3569-3359)	1406-1394 (3355-3343)

Summary of above
minimum of cal age ranges (cal ages) maximum of cal age ranges:

one sigma	cal BC	1583 (1494) 1434
	cal BP	3532 (3443) 3383
two sigma	cal BC	1681 (1494) 1394
	cal BP	3630 (3443) 3343

cal AD/BC age ranges (cal ages as above) from probability distribution (Method B):

% area enclosed	cal BC (cal BP) age ranges		relative area under probability distribution
68.3 (one sigma)	cal BC	1596-1592 (3545-3541)	.03
		1587-1569 (3536-3518)	.12
		1530-1433 (3479-3382)	.85
95.4 (two sigma)	cal BC	1683-1673 (3632-3622)	.01
		1625-1391 (3574-3340)	.98

Site: Castro de Penalba
Location: Campo Lameiro, Pontevedra, Galicia
Type: Defended
Cultural phase: Iron Age
References: Carballo - Fábregas 1992, nº 15
Sample ID: GAK-12971
Radiocarbon Age BP: 3120 ± 100≈
Calibrated age (s) cal BC 1428, 1424, 1413
cal BP 3377, 3373, 3362

cal AD/BC (cal BP) age ranges obtained from intercepts (Method A) :

one Sigma**	cal BC	1500-1310 (3449-3259)	1300-1296 (3249-3245)	1277-1265 (3226-3214)
two Sigma**	cal BC	1680-1675 (3629-3624)	1620-1210 (3569-3159)	1205-1189 (3154-3138)
		1180-1149 (3129-3098)	1146-1129 (3095-3078)	1123-1109 (3072-3058)

Summary of above
minimum of cal age ranges (cal ages) maximum of cal age ranges:

one sigma	cal BC	1500 (1428, 1424, 1413) 1265
	cal BP	3449 (3377, 3373, 3362) 3214
two sigma	cal BC	1680 (1428, 1424, 1413) 1109
	cal BP	3629 (3377, 3373, 3362) 3058

cal AD/BC age ranges (cal ages as above) from probability distribution (Method B) :

% area enclosed	cal BC (cal BP) age ranges		relative area under probability distribution
68.3 (one sigma)	cal BC	1520-1290 (3469-3239)	.94

		1282-1262 (3231-3211)	.06
95.4 (two sigma)	cal BC	1620-1210 (3569-3159)	.91
		1207-1188 (3156-3137)	.02
		1183-1128 (3132-3077)	.04

Site: Castro de Penalba
Location: Campo Lameiro, Pontevedra, Galicia
Type: Defended
Cultural phase: Iron Age
References: Carballo - Fábregas 1992, nº 15
Sample ID: GAK-12155
Radiocarbon Age BP: 3110 ± 110≈
Calibrated age (s) cal BC 1411
cal BP 3360
cal AD/BC (cal BP) age ranges obtained from intercepts (Method A) :
one Sigma** cal BC 1500-1290 (3449-3239) 1282-1262 (3231-3211)
two Sigma** cal BC 1682-1674 (3631-3623) 1620-1080 (3569-3029) 1062-1052 (3011-3001)

Summary of above
minimum of cal age ranges (cal ages) maximum of cal age ranges:

one sigma	cal BC	1500 (1411) 1262
	cal BP	3449 (3360) 3211
two sigma	cal BC	1682 (1411) 1052
	cal BP	3631 (3360) 3001

cal AD/BC age ranges (cal ages as above) from probability distribution (Method B) :

% area enclosed	cal BC (cal BP) age ranges		relative area under probability distribution
68.3 (one sigma)	cal BC	1520-1260 (3469-3209)	.95
		1230-1217 (3179-3166)	.03
		1198-1195 (3147-3144)	.01
95.4 (two sigma)	BEYOND CALCULABLE RANGE		

Site: Castro de Troña
Location: Ponteareas, Pontevedra, Galicia
Type: Defended
Cultural phase: Iron Age
References: Carballo - Fábregas 1992, nº 21
Sample ID: GAK-12223
Radiocarbon Age BP: 3100 ± 90≈
Calibrated age (s) cal BC 1409
cal BP 3358
cal AD/BC (cal BP) age ranges obtained from intercepts (Method A) :
one Sigma** cal BC 1491-1480 (3440-3429) 1459-1293 (3408-3242) 1280-1263 (3229-3212)
two Sigma** cal BC 1600-1567 (3549-3516) 1530-1210 (3479-3159) 1205-1189 (3154-3138)
1180-1148 (3129-3097) 1146-1129 (3095-3078) 1123-1109 (3072-3058)

Summary of above
minimum of cal age ranges (cal ages) maximum of cal age ranges:

one sigma	cal BC	1491 (1409) 1263
	cal BP	3440 (3358) 3212
two sigma	cal BC	1600 (1409) 1109
	cal BP	3549 (3358) 3058

cal AD/BC age ranges (cal ages as above) from probability distribution (Method B) :

% area enclosed	cal BC (cal BP) age ranges		relative area under probability distribution
68.3 (one sigma)	cal BC	1494-1471 (3443-3420)	.08
		1466-1290 (3415-3239)	.83
		1283-1261 (3232-3210)	.08
95.4 (two sigma)	cal BC	1601-1553 (3550-3502)	.03
		1542-1188 (3491-3137)	.89
		1183-1128 (3132-3077)	.05

	1125-1108 (3074-3057)		.01

Site: Castro de Penalba
Location: Pontevedra, Galicia
Type: Defended
Cultural phase: Iron Age
References: Alvarez Nuñez 1986, 22, 61; Carballo - Fábregas 1992, nº 15; Eiroa 1988, 118
Sample ID: GAK-11333
Radiocarbon Age BP: 3080 ± 120≈

Calibrated age (s)	cal BC	1388, 1330, 1325		
	cal BP	3337, 3279, 3274		

cal AD/BC (cal BP) age ranges obtained from intercepts (Method A):

one Sigma**	cal BC	1493-1477 (3442-3426)	1460-1240 (3409-3189)	1236-1213 (3185-3162)
		1201-1193 (3150-3142)	1141-1133 (3090-3082)	
two Sigma**	cal BC	1680-1675 (3629-3624)	1620-1010 (3569-2959)	

Summary of above
minimum of cal age ranges (cal ages) maximum of cal age ranges:

one sigma	cal BC	1493 (1388, 1330, 1325) 1133
	cal BP	3442 (3337, 3279, 3274) 3082
two sigma	cal BC	1680 (1388, 1330, 1325) 1010
	cal BP	3629 (3337, 3279, 3274) 2959

cal AD/BC age ranges (cal ages as above) from probability distribution (Method B):

% area enclosed	cal BC (cal BP) age ranges		relative area under probability distribution
68.3 (one sigma)	cal BC	1500-1210 (3449-3159)	.93
		1202-1192 (3151-3141)	.03
		1142-1132 (3091-3081)	.03
		1118-1114 (3067-3063)	.01
95.4 (two sigma)	cal BC	1620-1010 (3569-2959)	1.00

Site: Castro de Penalba
Location: Campo Lameiro, Pontevedra, Galicia
Type: Defended
Cultural phase: Iron Age
References: Carballo - Fábregas 1992, nº 15
Sample ID: GAK-12156
Radiocarbon Age BP: 3050 ± 110≈

Calibrated age (s)	cal BC	1372, 1359, 1354, 1337, 1320
	cal BP	3321, 3308, 3303, 3286, 3269

cal AD/BC (cal BP) age ranges obtained from intercepts (Method A) :

one Sigma**	cal BC	1440-1210 (3389-3159)	1203-1191 (3152-3140)	1163-1155 (3112-3104)
		1143-1131 (3092-3080)	1120-1113 (3069-3062)	
two Sigma**	cal BC	1597-1571 (3546-3520)	1530-1000 (3479-2949)	987-975 (2936-2924)

Summary of above
minimum of cal age ranges (cal ages) maximum of cal age ranges:

one sigma	cal BC	1440 (1372, 1359, 1354, 1337, 1320) 1113
	cal BP	3389 (3321, 3308, 3303, 3286, 3269) 3062
two sigma	cal BC	1597 (1372, 1359, 1354, 1337, 1320) 975
	cal BP	3546 (3321, 3308, 3303, 3286, 3269) 2924

cal AD/BC age ranges (cal ages as above) from probability distribution (Method B) :

% area enclosed	cal BC (cal BP) age ranges		relative area under probability distribution
68.3 (one sigma)	cal BC	1440-1210 (3389-3159)	.84
		1203-1191 (3152-3140)	.04
		1176-1154 (3125-3103)	.06
		1143-1131 (3092-3080)	.04
		1120-1112 (3069-3061)	.02
95.4 (two sigma)	cal BC	1597-1567 (3546-3516)	.01
		1530-1000 (3479-2949)	.98

Site:	Citânia de S. Julião			
Location:	Ponte, Vila Verde, Portugal			
Type:	Defended			
Cultural phase:	Bronze Age-Iron Age			
References:	Carballo - Fábregas 1992, nº 30; Martins 1987a, 44			
Sample ID:	ICEN-54			
Radiocarbon Age BP:	3030 ± 80≈			
Calibrated age (s)	cal BC	1368, 1365, 1347, 1346, 1317		
	cal BP	3317, 3314, 3296, 3295, 3266		

cal AD/BC (cal BP) age ranges obtained from intercepts (Method A) :

one Sigma**	cal BC	1411-1212 (3360-3161)	1202-1192 (3151-3141)	1142-1132 (3091-3081)
		1118-1114 (3067-3063)		
two Sigma**	cal BC	1492-1479 (3441-3428)	1460-1070 (3409-3019)	1066-1049 (3015-2998)
		1038-1017 (2987-2966)		

Summary of above

minimum of cal age ranges (cal ages) maximum of cal age ranges:

one sigma	cal BC	1411 (1368, 1365, 1347, 1346, 1317) 1114
	cal BP	3360 (3317, 3314, 3296, 3295, 3266) 3063
two sigma	cal BC	1492 (1368, 1365, 1347, 1346, 1317) 1017
	cal BP	3441 (3317, 3314, 3296, 3295, 3266) 2966

cal AD/BC age ranges (cal ages as above) from probability distribution (Method B) :

% area enclosed	cal BC (cal BP) age ranges		relative area under probability distribution
68.3 (one sigma)	cal BC	1410-1213 (3359-3162)	.92
		1201-1192 (3150-3141)	.04
		1141-1133 (3090-3082)	.04
95.4 (two sigma)	cal BC	1457-1070 (3406-3019)	.97

Site:	Citânia de S. Julião			
Location:	Ponte, Vila Verde, Portugal			
Type:	Defended			
Cultural phase:	Bronze Age-Iron Age			
References:	Carballo - Fábregas 1992, nº 30; Martins 1988c, 126-7			
Sample ID:	ICEN-25			
Radiocarbon Age BP:	3010 ± 35≈			
Calibrated age (s)	cal BC	1294, 1279, 1264		
	cal BP	3243, 3228, 3213		

cal AD/BC (cal BP) age ranges obtained from intercepts (Method A):

one Sigma**	cal BC	1373-1358 (3322-3307)	1355-1336 (3304-3285)	1321-1242 (3270-3191)
		1234-1215 (3183-3164)	1200-1194 (3149-3143)	1139-1134 (3088-3083)
two Sigma**	cal BC	1408-1209 (3357-3158)	1204-1190 (3153-3139)	1179-1151 (3128-3100)
		1145-1129 (3094-3078)	1122-1111 (3071-3060)	

Summary of above

minimum of cal age ranges (cal ages) maximum of cal age ranges:

one sigma	cal BC	1373 (1294, 1279, 1264) 1134
	cal BP	3322 (3243, 3228, 3213) 3083
two sigma	cal BC	1408 (1294, 1279, 1264) 1111
	cal BP	3357 (3243, 3228, 3213) 3060

cal AD/BC age ranges (cal ages as above) from probability distribution (Method B):

% area enclosed	cal BC (cal BP) age ranges		relative area under probability distribution
68.3 (one sigma)	cal BC	1372-1359 (3321-3308)	.10
		1354-1337 (3303-3286)	.12
		1320-1257 (3269-3206)	.52
		1253-1244 (3202-3193)	.06
		1233-1215 (3182-3164)	.13
		1199-1194 (3148-3143)	.04
		1139-1135 (3088-3084)	.03
95.4 (two sigma)	cal BC	1393-1210 (3342-3159)	.84
		1204-1190 (3153-3139)	.05

	1179-1151 (3128-3100)	.04
	1144-1130 (3093-3079)	.05
	1122-1112 (3071-3061)	.02

Site: Tapado da Caldeira
Location: Baião, Portugal
Type: Open
Cultural phase: Bronze Age
References: Jorge, S.O., 1985, 160
Sample ID: CSIC-597
Radiocarbon Age BP: 2990 ± 50≈
Calibrated age (s) cal BC 1289, 1284, 1260, 1227, 1225, 1197, 1196, 1137, 1136
cal BP 3238, 3233, 3209, 3176, 3174, 3146, 3145, 3086, 3085
cal AD/BC (cal BP) age ranges obtained from intercepts (Method A):

one Sigma**	cal BC	1371-1360 (3320-3309)	1353-1339 (3302-3288)	1319-1210 (3268-3159)
		1203-1191 (3152-3140)	1169-1154 (3118-3103)	1143-1131 (3092-3080)
		1120-1112 (3069-3061)		
two Sigma**	cal BC	1410-1080 (3359-3029)	1063-1051 (3012-3000)	

Summary of above
minimum of cal age ranges (cal ages) maximum of cal age ranges:

one sigma	cal BC	1371 (1289, 1284, 1260, 1227, 1225, 1197, 1196, 1137, 1136) 1112
	cal BP	3320 (3238, 3233, 3209, 3176, 3174, 3146, 3145, 3086, 3085) 3061
two sigma	cal BC	1410 (1289, 1284, 1260, 1227, 1225, 1197, 1196, 1137, 1136) 1051
	cal BP	3359 (3238, 3233, 3209, 3176, 3174, 3146, 3145, 3086, 3085) 3000

cal AD/BC age ranges (cal ages as above) from probability distribution (Method B):

% area enclosed	cal BC (cal BP) age ranges		relative area under probability distribution
68.3 (one sigma)	cal BC	1370-1360 (3319-3309)	.05
		1352-1340 (3301-3289)	.06
		1319-1211 (3268-3160)	.65
		1203-1191 (3152-3140)	.07
		1165-1155 (3114-3104)	.05
		1143-1131 (3092-3080)	.07
		1120-1113 (3069-3062)	.04
95.4 (two sigma)	cal BC	1393-1079 (3342-3028)	.99
		1062-1052 (3011-3001)	.01

Site: Lavra
Location: Marco de Canaveses, Portugal
Type: Open
Cultural phase: Bronze Age
References:
Sample ID: ICEN-414
Radiocarbon Age BP: 2980 ± 70≈
Calibrated age (s) cal BC 1287, 1286, 1258, 1230, 1216, 1198, 1195, 1138, 1135
cal BP 3236, 3235, 3207, 3179, 3165, 3147, 3144, 3087, 3084
cal AD/BC (cal BP) age ranges obtained from intercepts (Method A):

one Sigma**	cal BC	1373-1357 (3322-3306)	1356-1336 (3305-3285)	1321-1207 (3270-3156)
		1206-1188 (3155-3137)	1182-1128 (3131-3077)	1125-1108 (3074-3057)
		1095-1084 (3044-3033)	1058-1055 (3007-3004)	
two Sigma**	cal BC	1430-1421 (3379-3370)	1410-1010 (3359-2959)	980- 976 (2929-2925)

Summary of above
minimum of cal age ranges (cal ages) maximum of cal age ranges:

one sigma	cal BC	1373 (1287, 1286, 1258, 1230, 1216, 1198, 1195, 1138, 1135) 1055
	cal BP	3322 (3236, 3235, 3207, 3179, 3165, 3147, 3144, 3087, 3084) 3004
two sigma	cal BC	1430 (1287, 1286, 1258, 1230, 1216, 1198, 1195, 1138, 1135) 976
	cal BP	3379 (3236, 3235, 3207, 3179, 3165, 3147, 3144, 3087, 3084) 2925

cal AD/BC age ranges (cal ages as above) from probability distribution (Method B):

% area enclosed	cal BC (cal BP) age ranges	relative area under probability distribution

68.3 (one sigma)	cal BC	1371-1360 (3320-3309)	.05
		1353-1339 (3302-3288)	.06
		1320-1208 (3269-3157)	.53
		1205-1189 (3154-3138)	.08
		1181-1148 (3130-3097)	.14
		1146-1129 (3095-3078)	.08
		1123-1109 (3072-3058)	.06
95.4 (two sigma)	cal BC	1410-1047 (3359-2996)	.96
		1044-1013 (2993-2962)	.04

Site: Castro de Torroso
Location: Mos, Pontevedra, Galicia
Type: Defended
Cultural phase: Bronze Age-Iron Age
References: Carballo - Fábregas 1992, nº 19; Peña Santos 1987, 131-2
Sample ID: GAK-12161
Radiocarbon Age BP: 2980 ± 110≈
Calibrated age (s) cal BC 1287, 1286, 1258, 1230, 1216, 1198, 1195, 1138, 1135
cal BP 3236, 3235, 3207, 3179, 3165, 3147, 3144, 3087, 3084
cal AD/BC (cal BP) age ranges obtained from intercepts (Method A):
one Sigma** cal BC 1407-1401 (3356-3350) 1400-1070 (3349-3019)
1065-1050 (3014-2999) 1035-1021 (2984-2970)
two Sigma** cal BC 1493-1475 (3442-3424) 1470- 900 (3419-2849)

Summary of above
minimum of cal age ranges (cal ages) maximum of cal age ranges:
one sigma cal BC 1407 (1287, 1286, 1258, 1230, 1216, 1198, 1195, 1138, 1135) 1021
cal BP 3356 (3236, 3235, 3207, 3179, 3165, 3147, 3144, 3087, 3084) 2970
two sigma cal BC 1493 (1287, 1286, 1258, 1230, 1216, 1198, 1195, 1138, 1135) 900
cal BP 3442 (3236, 3235, 3207, 3179, 3165, 3147, 3144, 3087, 3084) 2849

cal AD/BC age ranges (cal ages as above) from probability distribution (Method B):

% area enclosed	cal BC (cal BP) age ranges		relative area under probability distribution
68.3 (one sigma)	cal BC	1378-1331 (3327-3280)	.15
		1320-1080 (3269-3029)	.82
		1062-1052 (3011-3001)	.03
95.4 (two sigma)	cal BC	1460- 970 (3409-2919)	.96

Site: Castro de Troña
Location: Ponteareas, Pontevedra, Galicia
Type: Defended
Cultural phase: Iron Age
References: Carballo - Fábregas 1992, nº 21
Sample ID: GAK-12222
Radiocarbon Age BP: 2960 ± 120≈
Calibrated age (s) cal BC 1238, 1236, 1214, 1200, 1193, 1140, 1133
cal BP 3187, 3185, 3163, 3149, 3142, 3089, 3082
cal AD/BC (cal BP) age ranges obtained from intercepts (Method A) :
one Sigma** cal BC 1390-1329 (3339-3278) 1330-1010 (3279-2959)
two Sigma** cal BC 1493-1477 (3442-3426) 1470-890 (3419-2839) 879-835 (2828-2784)

Summary of above
minimum of cal age ranges (cal ages) maximum of cal age ranges:
one sigma cal BC 1390 (1238, 1236, 1214, 1200, 1193, 1140, 1133) 1010
cal BP 3339 (3187, 3185, 3163, 3149, 3142, 3089, 3082) 2959
two sigma cal BC 1493 (1238, 1236, 1214, 1200, 1193, 1140, 1133) 835
cal BP 3442 (3187, 3185, 3163, 3149, 3142, 3089, 3082) 2784

cal AD/BC age ranges (cal ages as above) from probability distribution (Method B) :

% area enclosed	cal BC (cal BP) age ranges		relative area under probability distribution
68.3 (one sigma)	cal BC	1373-1357 (3322-3306)	.05
		1356-1336 (3305-3285)	.06

		1320-1070 (3269-3019)	.80
		1066-1049 (3015-2998)	.05
		1037-1019 (2986-2968)	.05
95.4 (two sigma)	BEYOND CALCULABLE RANGE		

Site: Coto da Pena
Location: Vilarelho, Caminha, Portugal
Type: Defended
Cultural phase: Bronze Age-Iron Age
References: Carballo - Fábregas 1992, nº 24; Silva 1986a, 34
Sample ID: UGRA-200
Radiocarbon Age BP: 2930 ± 100≈

Calibrated age(s)	cal BC	1210, 1204, 1190, 1177, 1153, 1144 , 1130, 1121, 1112		
	cal BP	3159, 3153, 3139, 3126, 3102, 3093 , 3079, 3070, 3061		

cal AD/BC (cal BP) age ranges obtained from intercepts (Method A):

one Sigma**	cal BC	1368-1364 (3317-3313)	1348-1344 (3297-3293)	1320-1010 (3269-2959)
		984- 975 (2933-2924)		
two Sigma**	cal BC	1431-1418 (3380-3367)	1420- 890 (3369-2839)	875-836 (2824-2785)

Summary of above

minimum of cal age ranges (cal ages) maximum of cal age ranges:

one sigma	cal BC	1368 (1210, 1204, 1190, 1177, 1153, 1144, 1130, 1121, 1112) 975
	cal BP	3317 (3159, 3153, 3139, 3126, 3102, 3093, 3079, 3070, 3061) 2924
two sigma	cal BC	1431 (1210, 1204, 1190, 1177, 1153, 1144, 1130, 1121, 1112) 836
	cal BP	3380 (3159, 3153, 3139, 3126, 3102, 3093, 3079, 3070, 3061) 2785

cal AD/BC age ranges (cal ages as above) from probability distribution (Method B):

% area enclosed	cal BC (cal BP) age ranges		relative area under probability distribution
68.3 (one sigma)	cal BC	1312-1298 (3261-3247)	.04
		1296-1277 (3245-3226)	.06
		1270-1010 (3219-2959)	.90
95.4 (two sigma)	cal BC	1410- 900 (3359-2849)	1.00

Site: Coto da Pena
Location: Vilarelho, Caminha, Portugal
Type: Defended
Cultural phase: Bronze Age-Iron Age
References: Carballo - Fábregas 1992, nº 24; Silva 1986a, 34, 173
Sample ID: UGRA-220
Radiocarbon Age BP: 2920 ± 110≈

Calibrated age(s)	cal BC	1209, 1205, 1190, 1179, 1150, 1145, 1129, 1122, 1110
	cal BP	3158, 3154, 3139, 3128, 3099, 3094, 3078, 3071, 3059

cal AD/BC (cal BP) age ranges obtained from intercepts (Method A):

one Sigma**	cal BC	1368-1364 (3317-3313)	1348-1345 (3297-3294)	1320- 970 (3269-2919)
		957- 939 (2906-2888)		
two Sigma**	cal BC	1430- 830 (3379-2779)		

Summary of above

minimum of cal age ranges (cal ages) maximum of cal age ranges:

one sigma	cal BC	1368 (1209, 1205, 1190, 1179, 1150, 1145, 1129, 1122, 1110) 939
	cal BP	3317 (3158, 3154, 3139, 3128, 3099, 3094, 3078, 3071, 3059) 2888
two sigma	cal BC	1430 (1209, 1205, 1190, 1179, 1150, 1145, 1129, 1122, 1110) 830
	cal BP	3379 (3158, 3154, 3139, 3128, 3099, 3094, 3078, 3071, 3059) 2779

cal AD/BC age ranges (cal ages as above) from probability distribution (Method B):

% area enclosed	cal BC (cal BP) age ranges		relative area under probability distribution
68.3 (one sigma)	cal BC	1312-1299 (3261-3248)	.03
		1296-1277 (3245-3226)	.05
		1270-1000 (3219-2949)	.88
		988-975 (2937-2924)	.04
95.4 (two sigma)	cal BC	1410-900 (3359-2849)	.98
		867-840 (2816-2789)	.02

Site: Castro de Vermoim
Location: Vermoim, Vila Nova de Famalicão, Portugal
Type: Defended
Cultural phase: Iron Age
References: (unpublished)
Sample ID: GAK-11463
Radiocarbon Age BP: 2920 ± 160≈

Calibrated age (s)	cal BC	1209, 1205, 1190, 1179, 1150, 1145, 1129, 1122, 1110
	cal BP	3158, 3154, 3139, 3128, 3099, 3094, 3078, 3071, 3059

cal AD/BC (cal BP) age ranges obtained from intercepts (Method A) :

one Sigma**	cal BC	1389-1329 (3338-3278)	1320-970 (3269-2919)	966-904 (2915-2853)
two Sigma**	cal BC	1520-800 (3469-2749)		

Summary of above

minimum of cal age ranges (cal ages) maximum of cal age ranges:

one sigma	cal BC	1389 (1209, 1205, 1190, 1179, 1150, 1145, 1129, 1122, 1110) 904
	cal BP	3338 (3158, 3154, 3139, 3128, 3099, 3094, 3078, 3071, 3059) 2853
two sigma	cal BC	1520 (1209, 1205, 1190, 1179, 1150, 1145, 1129, 1122, 1110) 800
	cal BP	3469 (3158, 3154, 3139, 3128, 3099, 3094, 3078, 3071, 3059) 2749

cal AD/BC age ranges (cal ages as above) from probability distribution (Method B) :

% area enclosed	cal BC (cal BP) age ranges		relative area under probability distribution
68.3 (one sigma)	cal BC	1370-1361 (3319-3310)	.02
		1352-1340 (3301-3289)	.03
		1320-970 (3269-2919)	.90
		959-935 (2908-2884)	.05
95.4 (two sigma)	cal BC	1500-800 (3449-2749)	1.00

Site: Citânia de S. Julião
Location: Ponte, Vila Verde, Portugal
Type: Defended
Cultural phase: Bronze Age-Iron Age
References: Carballo - Fábregas 1992, nº 30; Martins 1988c, 130
Sample ID: CSIC-734
Radiocarbon Age BP: 2900 ± 50≈

Calibrated age (s)	cal BC	1188, 1184, 1127, 1126, 1107, 1105, 1083, 1059, 1054
	cal BP	3137, 3133, 3076, 3075, 3056, 3054, 3032, 3008, 3003

cal AD/BC (cal BP) age ranges obtained from intercepts (Method A) :

one Sigma**	cal BC	1213-1201 (3162-3150)	1193-1141 (3142-3090)	1133-1117 (3082-3066)
		1116-1047 (3065-2996)	1046-1011 (2995-2960)	
two Sigma**	cal BC	1294-1278 (3243-3227)	1260-970 (3209-2919)	961-931 (2910-2880)

Summary of above

minimum of cal age ranges (cal ages) maximum of cal age ranges:

one sigma	cal BC	1213 (1188, 1184, 1127, 1126, 1107, 1105, 1083, 1059, 1054) 1011
	cal BP	3162 (3137, 3133, 3076, 3075, 3056, 3054, 3032, 3008, 3003) 2960
two sigma	cal BC	1294 (1188, 1184, 1127, 1126, 1107, 1105, 1083, 1059, 1054) 931
	cal BP	3243 (3137, 3133, 3076, 3075, 3056, 3054, 3032, 3008, 3003) 2880

cal AD/BC age ranges (cal ages as above) from probability distribution (Method B) :

% area enclosed	cal BC (cal BP) age ranges		relative area under probability distribution
68.3 (one sigma)	cal BC	1212-1202 (3161-3151)	.06
		1192-1142 (3141-3091)	.29
		1132-1118 (3081-3067)	.09
		1115-1048 (3064-2997)	.41
		1043-1014 (2992-2963)	.16
95.4 (two sigma)	cal BC	1294-1279 (3243-3228)	.02
		1263-971 (3212-2920)	.95
		960-933 (2909-2882)	.03

Site:	Citânia de S. Julião			
Location:	Ponte, Vila Verde, Portugal			
Type:	Defended			
Cultural phase:	Bronze Age-Iron Age			
References:	Carballo - Fábregas 1992, nº 30; Martins 1988c, 126-7			
Sample ID:	ICEN-27			
Radiocarbon Age BP:	2890 ± 45≈			
Calibrated age (s)	cal BC	1187, 1186, 1080, 1061, 1053		
	cal BP	3136, 3135, 3029,3010, 3002		

cal AD/BC (cal BP) age ranges obtained from intercepts (Method A):

one Sigma**	cal BC	1211-1202 (3160-3151)	1192-1142 (3141-3091)
		1131-1119 (3080-3068)	1114-1009 (3063-2958)
two Sigma**	cal BC	1290-1282 (3239-3231)	1261-1197 (3210-3146)
		1200- 970 (3149-2919)	961- 929 (2910-2878)

Summary of above

minimum of cal age ranges (cal ages) maximum of cal age ranges:

one sigma	cal BC	1211 (1187, 1186, 1080, 1061, 1053) 1009
	cal BP	3160 (3136, 3135, 3029, 3010, 3002) 2958
two sigma	cal BC	1290 (1187, 1186, 1080, 1061, 1053) 929
	cal BP	3239 (3136, 3135, 3029, 3010, 3002) 2878

cal AD/BC age ranges (cal ages as above) from probability distribution (Method B):

% area enclosed	cal BC (cal BP) age ranges		relative area under probability distribution
68.3 (one sigma)	cal BC	1210-1203 (3159-3152)	.05
		1191-1164 (3140-3113)	.16
		1155-1143 (3104-3092)	.07
		1131-1120 (3080-3069)	.07
		1113-1011 (3062-2960)	.65
95.4 (two sigma)	cal BC	1260-1227 (3209-3176)	.05
		1219-1197 (3168-3146)	.06
		1196-1137 (3145-3086)	.21
		1136- 971 (3085-2920)	.63
		960- 933 (2909-2882)	.04

Site:	Castro de Sta. Catarina,	
Location:	Penha, Guimarães, Portugal	
Type:	Open (?)	
Cultural phase:	Bronze Age	
References:	Cardozo 1970, 91-5	
Sample ID:	GRN-5568	
Radiocarbon Age BP:	2880 ± 65≈	
Calibrated age (s)	cal BC	1078, 1063, 1052
	cal BP	3027, 3012, 3001

cal AD/BC (cal BP) age ranges obtained from intercepts (Method A):

one Sigma**	cal BC	1212-1201 (3161-3150)	1192-1141 (3141-3090)	1132-1118 (3081-3067)
		1115- 997 (3064-2946)	995- 973 (2944-2922)	957- 941 (2906-2890)
two Sigma**	cal BC	1313-1297 (3262-3246)	1296-1275 (3245-3224)	1270- 900 (3219-2849)
		859- 848 (2808-2797)		

Summary of above

minimum of cal age ranges (cal ages) maximum of cal age ranges:

one sigma	cal BC	1212 (1078, 1063, 1052) 941
	cal BP	3161 (3027, 3012, 3001) 2890
two sigma	cal BC	1313 (1078, 1063, 1052) 848
	cal BP	3262 (3027, 3012, 3001) 2797

cal AD/BC age ranges (cal ages as above) from probability distribution (Method B):

% area enclosed	cal BC (cal BP) age ranges		relative area under probability distribution
68.3 (one sigma)	cal BC	1212-1202 (3161-3151)	.05
		1192-1142 (3141-3091)	.22
		1132-1119 (3081-3068)	.06
		1114-1000 (3063-2949)	.57

		992- 974 (2941-2923)	.07
		951- 945 (2900-2894)	.02
95.4 (two sigma)	cal BC	1296-1277 (3245-3226)	.02
		1265- 899 (3214-2848)	.97

Site: San Estebán del Rio Sil
Location: Galicia
Type: (??)
Cultural phase: Bronze Age
References: Carballo - Fábregas 1992, foot. 3; M. Almagro Gorbea 1977, 522
Sample ID: CSIC-215
Radiocarbon Age BP: 2880 ± 70≈
Calibrated age (s) cal BC 1078, 1063, 1052
cal BP 3027, 3012, 3001

cal AD/BC (cal BP) age ranges obtained from intercepts (Method A) :

one Sigma**	cal BC	1213-1201 (3162-3150)	1193-1141 (3142-3090)	1133-1117 (3082-3066)
		1116-973 (3065-2922)	958-938 (2907-2887)	
two Sigma**	cal BC	1368-1365 (3317-3314)	1320-900 (3269-2849)	866-839 (2815-2788)

Summary of above
minimum of cal age ranges (cal ages) maximum of cal age ranges:

one sigma	cal BC	1213 (1078, 1063, 1052) 938
	cal BP	3162 (3027, 3012, 3001) 2887
two sigma	cal BC	1368 (1078, 1063, 1052) 839
	cal BP	3317 (3027, 3012, 3001) 2788

cal AD/BC age ranges (cal ages as above) from probability distribution (Method B) :

% area enclosed	cal BC (cal BP) age ranges		relative area under probability distribution
68.3 (one sigma)	cal BC	1212-1202 (3161-3151)	.05
		1192-1142 (3141-3091)	.21
		1132-1118 (3081-3067)	.06
		1115-999 (3064-2948)	.56
		993-974 (2942-2923)	.08
		955-943 (2904-2892)	.05
95.4 (two sigma)	cal BC	1315-1271 (3264-3220)	.04
		1266-897 (3215-2846)	.95

Site: Citânia de S. Julião
Location: Ponte, Vila Verde, Portugal
Type: Defended
Cultural phase: Bronze Age-Iron Age
References: Carballo - Fábregas 1992, nº 30; Martins 1985, 216; 1986a, 159-60; 1988c, 126-7
Sample ID: GIF-6993
Radiocarbon Age BP: 2840 ± 80≈
Calibrated age (s) cal BC 1010
cal BP 2959

cal AD/BC (cal BP) age ranges obtained from intercepts (Method A):

one Sigma**	cal BC	1209-1204 (3158-3153)	1190-1178 (3139-3127)	1151-1145 (3100-3094)
		1130-1122 (3079-3071)	1111- 967 (3060-2916)	966- 903 (2915-2852)
two Sigma**	cal BC	1294-1279 (3243-3228)	1260- 830 (3209-2779)	

Summary of above
minimum of cal age ranges (cal ages) maximum of cal age ranges:

one sigma	cal BC	1209 (1010) 903
	cal BP	3158 (2959) 2852
two sigma	cal BC	1294 (1010) 830
	cal BP	3243 (2959) 2779

cal AD/BC age ranges (cal ages as above) from probability distribution (Method B):

% area enclosed	cal BC (cal BP) age ranges		relative area under probability distribution
68.3 (one sigma)	cal BC	1209-1205 (3158-3154)	.02
		1190-1179 (3139-3128)	.04

		1151-1145 (3100-3094)	.02
		1129-1122 (3078-3071)	.03
		1110- 967 (3059-2916)	.63
		966- 904 (2915-2853)	.25
95.4 (two sigma)	cal BC	1260-1227 (3209-3176)	.04
		1223-1197 (3172-3146)	.04
		1196- 831 (3145-2780)	.92

Site: Citânia de S. Julião
Location: Ponte, Vila Verde, Portugal
Type: Defended
Cultural phase: Bronze Age-Iron Age
References: Carballo - Fábregas 1992, nº 30; Martins 1988c, 126-7
Sample ID: ICEN-28
Radiocarbon Age BP: 2820 ± 40≈

Calibrated age (s)	cal BC	1002, 989, 975, 947		
	cal BP	2951, 2938, 2924, 2896		

cal AD/BC (cal BP) age ranges obtained from intercepts (Method A):

one Sigma**	cal BC	1070-1066 (3019-3015)	1050-1037 (2999-2986)	1020- 969 (2969-2918)
		963- 923 (2912-2872)		
two Sigma**	cal BC	1208-1206 (3157-3155)	1189-1181 (3138-3130)	1148-1146 (3097-3095)
		1128-1124 (3077-3073)	1109-1089 (3058-3038)	1086-1057 (3035-3006)
		1056- 895 (3005-2844)	869- 837 (2818-2786)	

Summary of above
minimum of cal age ranges (cal ages) maximum of cal age ranges:

one sigma	cal BC	1070 (1002, 989, 975, 947) 923
	cal BP	3019 (2951, 2938, 2924, 2896) 2872
two sigma	cal BC	1208 (1002, 989, 975, 947) 837
	cal BP	3157 (2951, 2938, 2924, 2896) 2786

cal AD/BC age ranges (cal ages as above) from probability distribution (Method B):

% area enclosed	cal BC (cal BP) age ranges		relative area under probability distribution
68.3 (one sigma)	cal BC	1073-1065 (3022-3014)	.05
		1050-1031 (2999-2980)	.13
		1023- 969 (2972-2918)	.46
		964- 920 (2913-2869)	.36
95.4 (two sigma)	cal BC	1189-1179 (3138-3128)	.02
		1128-1123 (3077-3072)	.01
		1109-1089 (3058-3038)	.02
		1087- 894 (3036-2843)	.90
		874- 840 (2823-2789)	.04

Site: Castro de Penalba
Location: Pontevedra, Galicia
Type: Defended
Cultural phase: Iron Age
References: Alvarez Nuñez 1986, 61; Carballo - Fábregas 1992, nº 15; Eiroa 1988, 118
Sample ID: GAK-11331
Radiocarbon Age BP: 2820 ± 120≈

Calibrated age (s)	cal BC	1002, 989, 975, 947		
	cal BP	2951, 2938, 2924, 2896		

cal AD/BC (cal BP) age ranges obtained from intercepts (Method A):

one Sigma**	cal BC	1211-1202 (3160-3151)	1192-1142 (3141-3091)	1131-1119 (3080-3068)
		1110-890 (3059-2839)	885-832 (2834-2781)	
two Sigma**	cal BC	1375-1333 (3324-3282)	1320-790 (3269-2739)	

Summary of above
minimum of cal age ranges (cal ages) maximum of cal age ranges:

one sigma	cal BC	1211 (1002, 989, 975, 947) 832
	cal BP	3160 (2951, 2938, 2924, 2896) 2781
two sigma	cal BC	1375 (1002, 989, 975, 947) 790
	cal BP	3324 (2951, 2938, 2924, 2896) 2739

cal AD/BC age ranges (cal ages as above) from probability distribution (Method B):

% area enclosed	cal BC (cal BP) age ranges		relative area under probability distribution
68.3 (one sigma)	cal BC	1210-1204 (3159-3153)	.02
		1190-1177 (3139-3126)	.04
		1153-1144 (3102-3093)	.03
		1130-1121 (3079-3070)	.03
		1110- 890 (3059-2839)	.75
		880- 834 (2829-2783)	.13
95.4 (two sigma)	cal BC	1353-1339 (3302-3288)	.01
		1320- 800 (3269-2749)	.98

Site: Castro do Barbudo
Location: Barbudo, Vila Verde, Portugal
Type: Defended
Cultural phase: Bronze Age-Iron Age
References: Carballo - Fábregas 1992, nº 29; Martins 1989, 66
Sample ID: CSIC-735
Radiocarbon Age BP: 2750 ± 60≈
Calibrated age (s) cal BC 901
cal BP 2850

cal AD/BC (cal BP) age ranges obtained from intercepts (Method A) :

one Sigma**	cal BC	1000-992 (2949-2941)	974-953 (2923-2902)	944-830 (2893-2779)
two Sigma**	cal BC	1078-1063 (3027-3012)	1050-800 (2999-2749)	

Summary of above
minimum of cal age ranges (cal ages) maximum of cal age ranges:

one sigma	cal BC	1000 (901) 830
	cal BP	2949 (2850) 2779
two sigma	cal BC	1078 (901) 800
	cal BP	3027 (2850) 2749

cal AD/BC age ranges (cal ages as above) from probability distribution (Method B) :

% area enclosed	cal BC (cal BP) age ranges		relative area under probability distribution
68.3 (one sigma)	cal BC	998-994 (2947-2943)	.03
		973-957 (2922-2906)	.12
		941-831 (2890-2780)	.85
95.4 (two sigma)	cal BC	1075-1064 (3024-3013)	.01
		1051-1027 (3000-2976)	.04
		1026-804 (2975-2753)	.95

Site: Citânia de S. Julião
Location: Ponte, Vila Verde, Portugal
Type: Defended
Cultural phase: Bronze Age-Iron Age
References: Carballo - Fábregas 1992, nº 30; Martins 1985, 216; 1986a, 159-60; 1988c, 126
Sample ID: GIF-7013
Radiocarbon Age BP: 2750 ± 60≈
Calibrated age (s) cal BC 901
cal BP 2850

cal AD/BC (cal BP) age ranges obtained from intercepts (Method A):

one Sigma**	cal BC	1000- 992 (2949-2941)	974- 953 (2923-2902)	944- 830 (2893-2779)
two Sigma**	cal BC	1078-1063 (3027-3012)	1050- 800 (2999-2749)	

Summary of above
minimum of cal age ranges (cal ages) maximum of cal age ranges:

one sigma	cal BC	1000 (901) 830
	cal BP	2949 (2850) 2779
two sigma	cal BC	1078 (901) 800
	cal BP	3027 (2850) 2749

cal AD/BC age ranges (cal ages as above) from probability distribution (Method B):

% area enclosed	cal BC (cal BP) age ranges		relative area under probability distribution
68.3 (one sigma)	cal BC	998- 994 (2947-2943)	.03
		973- 957 (2922-2906)	.12
		941- 831 (2890-2780)	.85
95.4 (two sigma)	cal BC	1075-1064 (3024-3013)	.01
		1051-1027 (3000-2976)	.04
		1026- 804 (2975-2753)	.95

Site: Castro do Barbudo
Location: Barbudo, Vila Verde, Portugal
Type: Defended
Cultural phase: Bronze Age-Iron Age
References: Carballo - Fábregas 1992, nº 29; Martins 1989, 66
Sample ID: CSIC-735R
Radiocarbon Age BP: 2740 ± 50≈
Calibrated age (s) cal BC 898, 858, 850
cal BP 2847, 2807, 2799
cal AD/BC (cal BP) age ranges obtained from intercepts (Method A) :
one Sigma** cal BC 972-959 (2921-2908) 934-830 (2883-2779)
two Sigma** cal BC 1048-1043 (2997-2992) 1010-800 (2959-2749)

Summary of above
minimum of cal age ranges (cal ages) maximum of cal age ranges:

one sigma	cal BC	972 (898, 858, 850) 830
	cal BP	2921 (2847, 2807, 2799) 2779
two sigma	cal BC	1048 (898, 858, 850) 800
	cal BP	2997 (2847, 2807, 2799) 2749

cal AD/BC age ranges (cal ages as above) from probability distribution (Method B) :

% area enclosed	cal BC (cal BP) age ranges		relative area under probability distribution
68.3 (one sigma)	cal BC	970-962 (2919-2911)	.08
		929-832 (2878-2781)	.92
95.4 (two sigma)	cal BC	1012-805 (2961-2754)	1.00

Site: Castelo de Matos
Location: Matos, Baião, Portugal
Type: Defended
Cultural phase: Bronze Age
References: Figueiral - Queiroga 1988; Hedges et al 1990, 225-6; Queiroga 1984; Queiroga - Figueiral 1989
Sample ID: OXA-1759
Radiocarbon Age BP: 2730 ± 70≈
Calibrated age (s) cal BC 895, 867, 839
cal BP 2844, 2816, 2788
cal AD/BC (cal BP) age ranges obtained from intercepts (Method A):
one Sigma** cal BC 973- 958 (2922-2907) 938- 817 (2887-2766)
two Sigma** cal BC 1077-1063 (3026-3012) 1050- 790 (2999-2739)

Summary of above
minimum of cal age ranges (cal ages) maximum of cal age ranges:

one sigma	cal BC	973 (895, 867, 839) 817
	cal BP	2922 (2844, 2816, 2788) 2766
two sigma	cal BC	1077 (895, 867, 839) 790
	cal BP	3026 (2844, 2816, 2788) 2739

cal AD/BC age ranges (cal ages as above) from probability distribution (Method B):

% area enclosed	cal BC (cal BP) age ranges		relative area under probability distribution
68.3 (one sigma)	cal BC	973- 958 (2922-2907)	.10
		938- 816 (2887-2765)	.90
95.4 (two sigma)	cal BC	1079-1062 (3028-3011)	.02
		1052- 793 (3001-2742)	.98

Site:	Bouça do Frade		
Location:	Campelo, Baião, Portugal		
Type:	Open		
Cultural phase:	Bronze Age		
References:	Jorge, S.O., 1985, 160		
Sample ID:	CSIC-631		
Radiocarbon Age BP:	2720 ± 50≈		
Calibrated age (s)	cal BC	893, 878, 835	
	cal BP	2842, 2827, 2784	

cal AD/BC (cal BP) age ranges obtained from intercepts (Method A):

one Sigma**	cal BC	969- 963 (2918-2912)	924- 827 (2873-2776)
two Sigma**	cal BC	1007- 981 (2956-2930)	980- 800 (2929-2749)

Summary of above

minimum of cal age ranges (cal ages) maximum of cal age ranges:

one sigma	cal BC	969 (893, 878, 835) 827
	cal BP	2918 (2842, 2827, 2784) 2776
two sigma	cal BC	1007 (893, 878, 835) 800
	cal BP	2956 (2842, 2827, 2784) 2749

cal AD/BC age ranges (cal ages as above) from probability distribution (Method B):

% area enclosed	cal BC (cal BP) age ranges		relative area under probability distribution
68.3 (one sigma)	cal BC	968- 964 (2917-2913)	.04
		920- 828 (2869-2777)	.96
95.4 (two sigma)	cal BC	1005- 987 (2954-2936)	.03
		975- 801 (2924-2750)	.97

Site:	Bouça do Frade		
Location:	Campelo, Baião, Portugal		
Type:	Open		
Cultural phase:	Bronze Age		
References:	Jorge, S.O., 1985, 160		
Sample ID:	CSIC-630		
Radiocarbon Age BP:	2720 ± 50≈		
Calibrated age (s) cal BC	893, 878, 835		
	cal BP	2842, 2827, 2784	

cal AD/BC (cal BP) age ranges obtained from intercepts (Method A):

one Sigma**	cal BC	969- 963 (2918-2912)	924- 827 (2873-2776)
two Sigma**	cal BC	1007- 981 (2956-2930)	980- 800 (2929-2749)

Summary of above

minimum of cal age ranges (cal ages) maximum of cal age ranges:

one sigma	cal BC	969 (893, 878, 835) 827
	cal BP	2918 (2842, 2827, 2784) 2776
two sigma	cal BC	1007 (893, 878, 835) 800
	cal BP	2956 (2842, 2827, 2784) 2749

cal AD/BC age ranges (cal ages as above) from probability distribution (Method B):

% area enclosed	cal BC (cal BP) age ranges		relative area under probability distribution
68.3 (one sigma)	cal BC	968- 964 (2917-2913)	.04
		920- 828 (2869-2777)	.96
95.4 (two sigma)	cal BC	1005- 987 (2954-2936)	.03
		975- 801 (2924-2750)	.97

Site:	Castelo de Matos
Location:	Matos, Baião, Portugal
Type:	Defended
Cultural phase:	Bronze Age
References:	Figueiral - Queiroga 1988; Hedges et al 1990, 225-6; Queiroga 1984; Queiroga - Figueiral 1989
Sample ID:	OXA-2147
Radiocarbon Age BP:	2710 ± 90≈

Calibrated age (s)	cal BC	890, 881, 834	
	cal BP	2839, 2830, 2783	

cal AD/BC (cal BP) age ranges obtained from intercepts (Method A):

one Sigma**	cal BC	972- 958 (2921-2907)	937- 800 (2886-2749)
two Sigma**	cal BC	1187-1184 (3136-3133)	1082-1060 (3031-3009)
		1050- 770 (2999-2719)	673- 663 (2622-2612)

Summary of above

minimum of cal age ranges (cal ages) maximum of cal age ranges:

one sigma	cal BC	972 (890, 881, 834) 800
	cal BP	2921 (2839, 2830, 2783) 2749
two sigma	cal BC	1187 (890, 881, 834) 663
	cal BP	3136 (2839, 2830, 2783) 2612

cal AD/BC age ranges (cal ages as above) from probability distribution (Method B):

% area enclosed	cal BC (cal BP) age ranges		relative area under probability distribution
68.3 (one sigma)	cal BC	1000- 992 (2949-2941)	.03
		974- 955 (2923-2904)	.10
		943- 798 (2892-2747)	.87
95.4 (two sigma)	cal BC	1191-1166 (3140-3115)	.01
		1130-1120 (3079-3069)	.00

Site: Bouça do Frade
Location: Campelo, Baião, Portugal
Type: Open
Cultural phase: Bronze Age
References: Jorge, S.O., 1985, 160
Sample ID: CSIC-632
Radiocarbon Age BP: 2710 ± 50≈

Calibrated age (s)	cal BC	890, 881, 834	
	cal BP	2839, 2830, 2783	

cal AD/BC (cal BP) age ranges obtained from intercepts (Method A):

one Sigma**	cal BC	968- 965 (2917-2914)	906- 815 (2855-2764)
two Sigma**	cal BC	1002- 989 (2951-2938)	970- 800 (2919-2749)

Summary of above

minimum of cal age ranges (cal ages) maximum of cal age ranges:

one sigma	cal BC	968 (890, 881, 834) 815
	cal BP	2917 (2839, 2830, 2783) 2764
two sigma	cal BC	1002 (890, 881, 834) 800
	cal BP	2951 (2839, 2830, 2783) 2749

cal AD/BC age ranges (cal ages as above) from probability distribution (Method B):

% area enclosed	cal BC (cal BP) age ranges		relative area under probability distribution
68.3 (one sigma)	cal BC	968- 965 (2917-2914)	.03
		906- 816 (2855-2765)	.96
95.4 (two sigma)	cal BC	1002- 989 (2951-2938)	.02
		974- 951 (2923-2900)	.07
		944- 798 (2893-2747)	.91

Site: Citânia de S. Julião
Location: Ponte, Vila Verde, Portugal
Type: Defended
Cultural phase: Bronze Age-Iron Age
References: Carballo - Fábregas 1992, nº 30; Martins 1988c, 126
Sample ID: ICEN-23
Radiocarbon Age BP: 2700 ± 40≈

Calibrated age (s)	cal BC	888, 885, 832	
	cal BP	2837, 2834, 2781	

cal AD/BC (cal BP) age ranges obtained from intercepts (Method A):

one Sigma**	cal BC	900- 814 (2849-2763)	
two Sigma**	cal BC	971- 960 (2920-2909)	933- 798 (2882-2747)

Summary of above
minimum of cal age ranges (cal ages) maximum of cal age ranges:

one sigma	cal BC	900 (888, 885, 832) 814
	cal BP	2849 (2837, 2834, 2781) 2763
two sigma	cal BC	971 (888, 885, 832) 798
	cal BP	2920 (2837, 2834, 2781) 2747

cal AD/BC age ranges (cal ages as above) from probability distribution (Method B):

% area enclosed	cal BC (cal BP) age ranges		relative area under probability distribution
68.3 (one sigma)	cal BC	899- 856 (2848-2805)	.58
		855- 820 (2804-2769)	.42
95.4 (two sigma)	cal BC	970- 961 (2919-2910)	.03
		930- 800 (2879-2749)	.97

Site: Castelo de Matos
Location: Matos, Baião, Portugal
Type: Defended
Cultural phase: Bronze Age
References: Figueiral - Queiroga 1988; Hedges *et al* 1990, 225-6; Queiroga 1984; Queiroga - Figueiral 1989
Sample ID: OXA-2146
Radiocarbon Age BP: 2700 ± 90≈

Calibrated age (s)	cal BC	888, 885, 832		
	cal BP	2837, 2834, 2781		

cal AD/BC (cal BP) age ranges obtained from intercepts (Method A):

one Sigma**	cal BC	971- 960 (2920-2909)	933- 798 (2882-2747)	
two Sigma**	cal BC	1079-1062 (3028-3011)	1050- 770 (2999-2719)	676- 661 (2625-2610)

Summary of above
minimum of cal age ranges (cal ages) maximum of cal age ranges:

one sigma	cal BC	971 (888, 885, 832) 798
	cal BP	2920 (2837, 2834, 2781) 2747
two sigma	cal BC	1079 (888, 885, 832) 661
	cal BP	3028 (2837, 2834, 2781) 2610

cal AD/BC age ranges (cal ages as above) from probability distribution (Method B):

% area enclosed	cal BC (cal BP) age ranges		relative area under probability distribution
68.3 (one sigma)	cal BC	999- 993 (2948-2942)	.02
		974- 956 (2923-2905)	.09
		942- 795 (2891-2744)	.89
95.4 (two sigma)	cal BC	1190-1177 (3139-3126)	.01
		1130-1121 (3079-3070)	.00
		1111- 757 (3060-2706)	.93

Site: Lavra
Location: Marco de Canaveses, Porto, Portugal
Type: Open
Cultural phase: Bronze Age
References:
Sample ID: CSIC-824
Radiocarbon Age BP: 2665 ± 60≈

Calibrated age (s)	cal BC	827	
	cal BP	2776	

cal AD/BC (cal BP) age ranges obtained from intercepts (Method A):

one Sigma**	cal BC	895- 869 (2844-2818)	837- 797 (2786-2746)
two Sigma**	cal BC	971- 960 (2920-2909)	930- 790 (2879-2739)
		778- 771 (2727-2720)	668- 664 (2617-2613)

Summary of above
minimum of cal age ranges (cal ages) maximum of cal age ranges:

one sigma	cal BC	895 (827) 797
	cal BP	2844 (2776) 2746

two sigma	cal BC	971 (827) 664	
	cal BP	2920 (2776) 2613	

cal AD/BC age ranges (cal ages as above) from probability distribution (Method B):

% area enclosed	cal BC (cal BP) age ranges		relative area under probability distribution
68.3 (one sigma)	cal BC	898- 857 (2847-2806)	.39
		849- 795 (2798-2744)	.61
95.4 (two sigma)	cal BC	1004- 987 (2953-2936)	.01
		975- 764 (2924-2713)	.96
		680- 661 (2629-2610)	.02

Site: Castro do Barbudo
Location: Barbudo, Vila Verde, Portugal
Type: Defended
Cultural phase: Bronze Age-Iron Age
References: Carballo - Fábregas 1992, nº 29; Martins 1989, 66
Sample ID: ICEN-21
Radiocarbon Age BP: 2650 ± 40≈

Calibrated age (s)	cal BC	811	
	cal BP	2760	

cal AD/BC (cal BP) age ranges obtained from intercepts (Method A) :

one Sigma**	cal BC	832-798 (2781-2747)	
two Sigma**	cal BC	899-857 (2848-2806)	856-790 (2805-2739)

Summary of above
minimum of cal age ranges (cal ages) maximum of cal age ranges:

one sigma	cal BC	832 (811) 798
	cal BP	2781 (2760) 2747
two sigma	cal BC	899 (811) 790
	cal BP	2848 (2760) 2739

cal AD/BC age ranges (cal ages as above) from probability distribution (Method B) :

% area enclosed	cal BC (cal BP) age ranges		relative area under probability distribution
68.3 (one sigma)	cal BC	891-880 (2840-2829)	.15
		834-795 (2783-2744)	.85
95.4 (two sigma)	cal BC	906-787 (2855-2736)	.98
		779-770 (2728-2719)	.01
		671-665 (2620-2614)	.01

Site: Castro das Ermidas
Location: Jesufrei, V.N. Famalicão, Portugal
Type: Defended
Cultural phase: Iron Age
References: Carballo - Fábregas 1992, nº 32; Queiroga 1985, 35-7
Sample ID: GAK-11461
Radiocarbon Age BP: 2650 ± 120≈

Calibrated age (s)	cal BC	811
	cal BP	2760

cal AD/BC (cal BP) age ranges obtained from intercepts (Method A):

one Sigma**	cal BC	969- 964 (2918-2913)	920- 770 (2869-2719)	672- 663 (2621-2612)
two Sigma**	cal BC	1187-1185 (3136-3134)	1081-1060 (3030-3009)	1050- 480 (2999-2429)
		474- 426 (2423-2375)	425- 412 (2374-2361)	

Summary of above
minimum of cal age ranges (cal ages) maximum of cal age ranges:

one sigma	cal BC	969 (811) 663
	cal BP	2918 (2760) 2612
two sigma	cal BC	1187 (811) 412
	cal BP	3136 (2760) 2361

cal AD/BC age ranges (cal ages as above) from probability distribution (Method B):

% area enclosed	cal BC (cal BP) age ranges	relative area under probability distribution

68.3 (one sigma)	cal BC	1003- 987 (2952-2936)	.03
		970- 760 (2919-2709)	.73
		688- 657 (2637-2606)	.08
		640- 591 (2589-2540)	.11
		575- 555 (2524-2504)	.04
95.4 (two sigma)	cal BC	1127-1125 (3076-3074)	.00
		1083-1059 (3032-3008)	.02
		1050- 410 (2999-2359)	.98

Site: Castro da Senhora da Guia
Location: Baiões, S. Pedro do Sul, Portugal
Type: Defended
Cultural phase: Bronze Age
References: Carballo - Fábregas 1992, nº 34; Eiroa 1980, 78; Kalb 1974-77; Soares - Cabral 1984, 179, 191
Sample ID: GRN-7484
Radiocarbon Age BP: 2650 ± 130≈
Calibrated age(s) cal BC 811
cal BP 2760
cal AD/BC (cal BP) age ranges obtained from intercepts (Method A):
one Sigma** cal BC 970-962 (2919-2911) 930- 770 (2879-2719) 675-662 (2624-2611)

two Sigma** cal BC 1208-1205 (3157-3154) 1189-1181 (3138-3130) 1148-1146 (3097-3095)
1129-1123 (3078-3072) 1109-1088 (3058-3037) 1086-1057 (3035-3006)
1060-410 (3009-2359)

Summary of above
minimum of cal age ranges (cal ages) maximum of cal age ranges:

one sigma	cal BC	970 (811) 662
	cal BP	2919 (2760) 2611
two sigma	cal BC	1208 (811) 410
	cal BP	3157 (2760) 2359

cal AD/BC age ranges (cal ages as above) from probability distribution (Method B):

% area enclosed	cal BC (cal BP) age ranges		relative area under probability distribution
68.3 (one sigma)	cal BC	1006-983 (2955-2932)	.04
		980-760 (2929-2709)	.69
		689-657 (2638-2606)	.08
		643-590 (2592-2539)	.11
		577-547 (2526-2496)	.06
		458-452 (2407-2401)	.01
95.4 (two sigma)	cal BC	1189-1181 (3138-3130)	.00
		1128-1124 (3077-3073)	.00
		1086-1057 (3035-3006)	.02
		1060-410 (3009-2359)	.96

Site: Castro de A Graña
Location: Capela, Toques, A Coruña, Galicia
Type: Defended
Cultural phase: Iron Age
References: Carballo - Fábregas 1992, nº 7
Sample ID: GD-6074
Radiocarbon Age BP: 2650 ± 100≈
Calibrated age (s) cal BC 811
cal BP 2760
cal AD/BC (cal BP) age ranges obtained from intercepts (Method A) :
one Sigma** cal BC 900-790 (2849-2739)
two Sigma** cal BC 1048-1042 (2997-2991) 1010-750 (2959-2699) 720-540 (2669-2489)
535-519 (2484-2468) 464-446 (2413-2395)

Summary of above
minimum of cal age ranges (cal ages) maximum of cal age ranges:

one sigma	cal BC	900 (811) 790
	cal BP	2849 (2760) 2739

two sigma	cal BC	1048 (811) 446
	cal BP	2997 (2760) 2395

cal AD/BC age ranges (cal ages as above) from probability distribution (Method B) :

% area enclosed	cal BC (cal BP) age ranges		relative area under probability distribution
68.3 (one sigma)	cal BC	973-957 (2922-2906)	.05
		940-760 (2889-2709)	.80
		684-659 (2633-2608)	.07
		633-620 (2582-2569)	.03
		612-593 (2561-2542)	.04
95.4 (two sigma)	cal BC	1050-510 (2999-2459)	.96
		469-431 (2418-2380)	.03

Site: Castro de Penices
Location: Gondifelos, VN Famalicão, Portugal
Type: Defended
Cultural phase: Iron Age
References: (unpublished)
Sample ID: ICEN-467
Radiocarbon Age BP: 2640 ± 60≈

Calibrated age (s)	cal BC	805		
	cal BP	2754		

cal AD/BC (cal BP) age ranges obtained from intercepts (Method A):

one Sigma**	cal BC	888- 884 (2837-2833)	833- 793 (2782-2742)	
two Sigma**	cal BC	968- 965 (2917-2914)	910- 770 (2859-2719)	676- 661 (2625-2610)

Summary of above

minimum of cal age ranges (cal ages) maximum of cal age ranges:

one sigma	cal BC	888 (805) 793
	cal BP	2837 (2754) 2742
two sigma	cal BC	968 (805) 661
	cal BP	2917 (2754) 2610

cal AD/BC age ranges (cal ages as above) from probability distribution (Method B):

% area enclosed	cal BC (cal BP) age ranges		relative area under probability distribution
68.3 (one sigma)	cal BC	898- 857 (2847-2806)	.29
		849- 847 (2798-2796)	.01
		845- 787 (2794-2736)	.67
		669- 665 (2618-2614)	.03
95.4 (two sigma)	cal BC	973- 957 (2922-2906)	.02
		937- 757 (2886-2706)	.86
		690- 657 (2639-2606)	.06
		642- 590 (2591-2539)	.04

Site: Castro de Torroso
Location: Mos, Pontevedra, Galicia
Type: Defended
Cultural phase: Bronze Age-Iron Age
References: Carballo - Fábregas 1992, nº 19
Sample ID: GRN-14589
Radiocarbon Age BP: 2635 ± 30≈

Calibrated age (s)	cal BC	804	
	cal BP	2753	

cal AD/BC (cal BP) age ranges obtained from intercepts (Method A) :

one Sigma**	cal BC	828-797 (2777-2746)	
two Sigma**	cal BC	889-883 (2838-2832)	833-791 (2782-2740)

Summary of above

minimum of cal age ranges (cal ages) maximum of cal age ranges:

one sigma	cal BC	828 (804) 797
	cal BP	2777 (2753) 2746
two sigma	cal BC	889 (804) 791

	cal BP	2838 (2753) 2740		

cal AD/BC age ranges (cal ages as above) from probability distribution (Method B) :

% area enclosed		cal BC (cal BP) age ranges	relative area under probability distribution
68.3 (one sigma)	cal BC	827-798 (2776-2747)	1.00
95.4 (two sigma)	cal BC	897-862 (2846-2811)	.12
		837-788 (2786-2737)	.86
		672-665 (2621-2614)	.01

Site: Castro de A Graña
Location: Capela, Toques, A Coruña, Galicia
Type: Defended
Cultural phase: Iron Age
References: Carballo - Fábregas 1992, nº 7
Sample ID: GD-6068
Radiocarbon Age BP: 2610 ± 70≈

Calibrated age (s)	cal BC	799		
	cal BP	2748		

cal AD/BC (cal BP) age ranges obtained from intercepts (Method A) :

one Sigma**	cal BC	830-771 (2779-2720)	669-664 (2618-2613)	
two Sigma**	cal BC	900-760 (2849-2709)	691-656 (2640-2605)	653-589 (2602-2538)
		579-543 (2528-2492)	527-525 (2476-2474)	459-451 (2408-2400)

Summary of above

minimum of cal age ranges (cal ages) maximum of cal age ranges:

one sigma	cal BC	830 (799) 664
	cal BP	2779 (2748) 2613
two sigma	cal BC	900 (799) 451
	cal BP	2849 (2748) 2400

cal AD/BC age ranges (cal ages as above) from probability distribution (Method B) :

% area enclosed		cal BC (cal BP) age ranges	relative area under probability distribution
68.3 (one sigma)	cal BC	896-863 (2845-2812)	.13
		840-760 (2789-2709)	.59
		685-658 (2634-2607)	.14
		634-619 (2583-2568)	.05
		612-593 (2561-2542)	.07
		568-563 (2517-2512)	.02
95.4 (two sigma)	cal BC	927-745 (2876-2694)	.60
		730-537 (2679-2486)	.35
		535-519 (2484-2468)	.01
		465-435 (2414-2384)	.03

Site: Castro de Vermoim
Location: Vermoim, Vila Nova de Famalicão, Portugal
Type: Defended
Cultural phase: Iron Age
References: (unpublished)
Sample ID: GAK-11462
Radiocarbon Age BP: 2600 ± 120≈

Calibrated age (s)	cal BC	797		
	cal BP	2746		

cal AD/BC (cal BP) age ranges obtained from intercepts (Method A) :

one Sigma**	cal BC	893-877 (2842-2826)	836-758 (2785-2707)	687-657 (2636-2606)
		639-616 (2588-2565)	615-591 (2564-2540)	573-556 (2522-2505)
		456-454 (2405-2403)		
two Sigma**	cal BC	1010-400 (2959-2349)		

Summary of above

minimum of cal age ranges (cal ages) maximum of cal age ranges:

one sigma	cal BC	893 (797) 454
	cal BP	2842 (2746) 2403

two sigma	cal BC	1010 (797) 400		
	cal BP	2959 (2746) 2349		

cal AD/BC age ranges (cal ages as above) from probability distribution (Method B) :

% area enclosed	cal BC (cal BP) age ranges		relative area under probability distribution
68.3 (one sigma)	cal BC	900-750 (2849-2699)	.48
		719-706 (2668-2655)	.03
		690-590 (2639-2539)	.31
		584-540 (2533-2489)	.12
		531-522 (2480-2471)	.02
		462-447 (2411-2396)	.04
95.4 (two sigma)	cal BC	1007-981 (2956-2930)	.02
		980-410 (2929-2359)	.98

Site: Castro de Torroso
Location: Mos, Pontevedra, Galicia
Type: Defended
Cultural phase: Bronze Age-Iron Age
References: Carballo - Fábregas 1992, nº 19
Sample ID: GRN-14588
Radiocarbon Age BP: 2580 ± 30≈

Calibrated age (s)	cal BC	793		
	cal BP	2742		

cal AD/BC (cal BP) age ranges obtained from intercepts (Method A) :

one Sigma**	cal BC	800-788 (2749-2737)	667-665 (2616-2614)	
two Sigma**	cal BC	812-765 (2761-2714)	677-661 (2626-2610)	606-604 (2555-2553)

Summary of above
minimum of cal age ranges (cal ages) maximum of cal age ranges:

one sigma	cal BC	800 (793) 665
	cal BP	2749 (2742) 2614
two sigma	cal BC	812 (793) 604
	cal BP	2761 (2742) 2553

cal AD/BC age ranges (cal ages as above) from probability distribution (Method B) :

% area enclosed	cal BC (cal BP) age ranges		relative area under probability distribution
68.3 (one sigma)	cal BC	804-767 (2753-2716)	.77
		676-663 (2625-2612)	.23
95.4 (two sigma)	cal BC	828-758 (2777-2707)	.65
		689-658 (2638-2607)	.21
		636-618 (2585-2567)	.04
		613-592 (2562-2541)	.06
		575-556 (2524-2505)	.02
		458-450 (2407-2399)	.01

Site: Castrovite
Location: A Estrada, Pontevedra, Galicia
Type: Defended
Cultural phase: Iron Age
References: Carballo - Fábregas 1992, nº 9
Sample ID: ICEN-412
Radiocarbon Age BP: 2570 ± 40≈

Calibrated age (s)	cal BC	792		
	cal BP	2741		

cal AD/BC (cal BP) age ranges obtained from intercepts (Method A) :

one Sigma**	cal BC	800-768 (2749-2717)	672-663 (2621-2612)	
two Sigma**	cal BC	820-759 (2769-2708)	686-658 (2635-2607)	637-617 (2586-2566)
		614-592 (2563-2541)	571-559 (2520-2508)	

Summary of above
minimum of cal age ranges (cal ages) maximum of cal age ranges:

one sigma	cal BC	800 (792) 663

	cal BP	2749 (2741) 2612
two sigma	cal BC	820 (792) 559
	cal BP	2769 (2741) 2508

cal AD/BC age ranges (cal ages as above) from probability distribution (Method B) :

% area enclosed	cal BC (cal BP) age ranges		relative area under probability distribution
68.3 (one sigma)	cal BC	808-761 (2757-2710)	.57
		684-659 (2633-2608)	.26
		632-621 (2581-2570)	.07
		610-594 (2559-2543)	.10
95.4 (two sigma)	cal BC	829-754 (2778-2703)	.47
		719-696 (2668-2645)	.03
		695-655 (2644-2604)	.21
		651-587 (2600-2536)	.19
		583-544 (2532-2493)	.07
		462-446 (2411-2395)	.03

Site: Coto do Mosteiro
Location: Lobás, Carballiño, Galicia
Type: Defended
Cultural phase: Iron Age
References: Carballo - Fábregas 1992, nº 16
Sample ID: ICEN-411
Radiocarbon Age BP: 2560 ± 50≈

Calibrated age (s)	cal BC	790
	cal BP	2739

cal AD/BC (cal BP) age ranges obtained from intercepts (Method A) :

one Sigma**	cal BC	800-764 (2749-2713)	678-661 (2627-2610)	607-601 (2556-2550)
two Sigma**	cal BC	827-753 (2776-2702)	718-713 (2667-2662)	690-590 (2639-2539)
		584-539 (2533-2488)	532-522 (2481-2471)	462-448 (2411-2397)

Summary of above
minimum of cal age ranges (cal ages) maximum of cal age ranges:

one sigma	cal BC	800 (790) 601
	cal BP	2749 (2739) 2550
two sigma	cal BC	827 (790) 448
	cal BP	2776 (2739) 2397

cal AD/BC age ranges (cal ages as above) from probability distribution (Method B) :

% area enclosed	cal BC (cal BP) age ranges		relative area under probability distribution
68.3 (one sigma)	cal BC	805-758 (2754-2707)	.40
		688-658 (2637-2607)	.23
		638-616 (2587-2565)	.12
		615-591 (2564-2540)	.14
		575-556 (2524-2505)	.08
		457-452 (2406-2401)	.02
95.4 (two sigma)	cal BC	828-745 (2777-2694)	.35
		731-537 (2680-2486)	.58
		535-519 (2484-2468)	.02
		465-435 (2414-2384)	.05

Site: Castro de Torroso
Location: Mos, Pontevedra, Galicia
Type: Defended
Cultural phase: Bronze Age-Iron Age
References: Carballo - Fábregas 1992, nº 19; Peña Santos 1987, 132
Sample ID: GRN-13706
Radiocarbon Age BP: 2555 ± 30≈

Calibrated age (s)	cal BC	789
	cal BP	2738

cal AD/BC (cal BP) age ranges obtained from intercepts (Method A):

one Sigma**	cal BC	795- 767 (2744-2716)	674- 662 (2623-2611)

two Sigma**	cal BC	802- 760 (2751-2709)	685- 658 (2634-2607)	636- 618 (2585-2567)
		614- 592 (2563-2541)	569- 561 (2518-2510)	

Summary of above
minimum of cal age ranges (cal ages) maximum of cal age ranges:

one sigma	cal BC	795 (789) 662
	cal BP	2744 (2738) 2611
two sigma	cal BC	802 (789) 561
	cal BP	2751 (2738) 2510

cal AD/BC age ranges (cal ages as above) from probability distribution (Method B):

% area enclosed	cal BC (cal BP) age ranges		relative area under probability distribution
68.3 (one sigma)	cal BC	799- 762 (2748-2711)	.54
		683- 660 (2632-2609)	.30
		629- 623 (2578-2572)	.05
		609- 595 (2558-2544)	.11
95.4 (two sigma)	cal BC	805- 755 (2754-2704)	.42
		717- 696 (2666-2645)	.03
		694- 656 (2643-2605)	.25
		645- 588 (2594-2537)	.20
		581- 546 (2530-2495)	.07
		461- 446 (2410-2395)	.03

Site: Castro de Romariz
Location: Romariz, Vila da Feira, Portugal
Type: Defended
Cultural phase: Iron Age
References: Carballo - Fábregas 1992, nº 33; Silva 1986a, 40
Sample ID: UGRA-203
Radiocarbon Age BP: 2550 ± 100≈

Calibrated age(s)	cal BC	788
	cal BP	2737

cal AD/BC (cal BP) age ranges obtained from intercepts (Method A):

one Sigma**	cal BC	813- 752 (2762-2701)	721- 704 (2670-2653)	700- 590(2649-2539)
		585- 537 (2534-2486)	534- 520(2483-2469)	463- 447 (2412-2396)
two Sigma**	cal BC	900- 400 (2849-2349)		

Summary of above
minimum of cal age ranges (cal ages) maximum of cal age ranges:

one sigma	cal BC	813 (788) 447
	cal BP	2762 (2737) 2396
two sigma	cal BC	900 (788) 400
	cal BP	2849 (2737) 2349

cal AD/BC age ranges (cal ages as above) from probability distribution (Method B):

% area enclosed	cal BC (cal BP) age ranges		relative area under probability distribution
68.3 (one sigma)	cal BC	812-752 (2761-2701)	.24
		720-540 (2669-2489)	.66
		533-521 (2482-2470)	.04
		463-446 (2412-2395)	.06
95.4 (two sigma)	cal BC	897-863 (2846-2812)	.03
		840-410 (2789-2359)	.97

Site: Castro de Penalba
Location: Campo Lameiro, Pontevedra, Galicia
Type: Defended
Cultural phase: Iron Age
References: Aira *et al.* 1990, 82; Carballo - Fábregas 1992, nº 15
Sample ID: PA-88
Radiocarbon Age BP: 2548 ± 126≈

Calibrated age (s)	cal BC	788, 666
	cal BP	2737, 2615

cal AD/BC (cal BP) age ranges obtained from intercepts (Method A):

one Sigma**	cal BC	830-510 (2779-2459)	505-485 (2454-2434)	470-430 (2419-2379)
		418-415 (2367-2364)		
two Sigma**	cal BC	972-958 (2921-2907)	940-390 (2889-2339)	

Summary of above

minimum of cal age ranges (cal ages) maximum of cal age ranges:

one sigma	cal BC	830 (788, 666) 415
	cal BP	2779 (2737, 2615) 2364
two sigma	cal BC	972 (788,666) 390
	cal BP	2921 (2737, 2615) 2339

cal AD/BC age ranges (cal ages as above) from probability distribution (Method B) :

% area enclosed	cal BC (cal BP) age ranges		relative area under probability distribution
68.3 (one sigma)	cal BC	818-740 (2767-2689)	.25
		730-520 (2679-2469)	.66
		465-436 (2414-2385)	.09
95.4 (two sigma)	cal BC	940-390 (2889-2339)	.99

Site: Castro de Torroso
Location: Mos, Pontevedra, Galicia
Type: Defended
Cultural phase: Bronze Age-Iron Age
References: Carballo - Fábregas 1992, nº 19; Peña Santos 1987, 132
Sample ID: GRN-13705
Radiocarbon Age BP: 2540 ± 30≈

Calibrated age (s)	cal BC	787, 772, 668, 665
	cal BP	2736, 2721, 2617, 2614

cal AD/BC (cal BP) age ranges obtained from intercepts (Method A):

one Sigma**	cal BC	793- 763 (2742-2712)	680- 660 (2629-2609)	628- 624 (2577-2573)
		608- 597 (2557-2546)		
two Sigma**	cal BC	800- 755 (2749-2704)	691- 589 (2640-2538)	580- 543 (2529-2492)
		527- 524 (2476-2473)	460- 451 (2409-2400)	

Summary of above

minimum of cal age ranges (cal ages) maximum of cal age ranges:

one sigma	cal BC	793 (787, 772, 668, 665) 597
	cal BP	2742 (2736, 2721, 2617, 2614) 2546
two sigma	cal BC	800 (787, 772, 668, 665) 451
	cal BP	2749 (2736, 2721, 2617, 2614) 2400

cal AD/BC age ranges (cal ages as above) from probability distribution (Method B):

% area enclosed	cal BC (cal BP) age ranges		relative area under probability distribution
68.3 (one sigma)	cal BC	795- 760 (2744-2709)	.40
		686- 659 (2635-2608)	.28
		634- 619 (2583-2568)	.11
		612- 593 (2561-2542)	.15
		569- 563 (2518-2512)	.04
		456- 454 (2405-2403)	.02
95.4 (two sigma)	cal BC	800- 753 (2749-2702)	.32
		721- 655 (2670-2604)	.28
		653- 587 (2602-2536)	.26
		584- 544 (2533-2493)	.10
		463- 445 (2412-2394)	.04

Site: Castro de Penices
Location: Gondifelos, V.N. Famalicão, Portugal
Type: Defended
Cultural phase: Iron Age
References: (unpublished)
Sample ID: ICEN-832
Radiocarbon Age BP: 2530 ± 45≈

Calibrated age(s)	cal BC	769, 671, 663		
	cal BP	2718, 2620, 2612		

cal AD/BC (cal BP) age ranges obtained from intercepts (Method A):

one Sigma**	cal BC	794-758 (2743-2707)	687-657 (2636-2606)	639-616 (2588-2565)
		615-591 (2564-2540)	573-556 (2522-2505)	456-454 (2405-2403)
two Sigma**	cal BC	800-510 (2749-2459)	468-432 (2417-2381)	

Summary of above

minimum of cal age ranges (cal ages) maximum of cal age ranges:

one sigma	cal BC	794 (769, 671, 663) 454
	cal BP	2743 (2718, 2620, 2612) 2403
two sigma	cal BC	800 (769, 671, 663) 432
	cal BP	2749 (2718, 2620, 2612) 2381

cal AD/BC age ranges (cal ages as above) from probability distribution (Method B):

% area enclosed	cal BC (cal BP) age ranges		relative area under probability distribution
68.3 (one sigma)	cal BC	795-756 (2744-2705)	.27
		690-657 (2639-2606)	.22
		644-590 (2593-2539)	.32
		578- 548 (2527-2497)	.15
		459- 451(2408-2400)	.04
95.4 (two sigma)	cal BC	802- 517 (2751-2466)	.93
		467- 433 (2416-2382)	.07

Average of samples ICEN 831 and ICEN-832

Sample ID: ICEN-831
Radiocarbon Age BP: 2420 ± 45≈
Sample ID: ICEN-832
Radiocarbon Age BP: 2530 ± 45≈

Weighted average: ICEN-831/832 (average)
Radiocarbon Age BP: 2475 ± 31.8≈

Calibrated age(s)	cal BC	758, 688, 657, 641, 591, 575, 551, 457, 453
	cal BP	2707, 2637, 2606, 2590, 2540, 2524, 2500, 2406, 2402

cal AD/BC (cal BP) age ranges obtained from intercepts (Method A):

one Sigma**	cal BC	766- 732 (2715-2681)	731- 675 (2680-2624)	662- 517 (2611-2466)
		466- 434 (2415-2383)		
two Sigma**	cal BC	790- 410 (2739-2359)		

Summary of above

minimum of cal age ranges (cal ages) maximum of cal age ranges:

one sigma	cal BC	766 (758, 688, 657, 641, 591, 575, 551, 457, 453) 434
	cal BP	2715 (2707, 2637, 2606, 2590, 2540, 2524, 2500, 2406, 2402) 2383
two sigma	cal BC	790 (758, 688, 657, 641, 591, 575, 551, 457, 453) 410
	cal BP	2739 (2707, 2637, 2606, 2590, 2540, 2524, 2500, 2406, 2402) 2359

cal AD/BC age ranges (cal ages as above) from probability distribution (Method B):

% area enclosed	cal BC (cal BP) age ranges		relative area under probability distribution
68.3 (one sigma)	cal BC	764- 751 (2713-2700)	.07
		724- 678 (2673-2627)	.22
		660- 627 (2609-2576)	.17
		624- 608 (2573-2557)	.08
		598- 538 (2547-2487)	.30
		533- 520 (2482-2469)	.06
		464- 442 (2413-2391)	.11
95.4 (two sigma)	cal BC	770- 668 (2719-2617)	.31
		663- 484 (2612-2433)	.56
		471- 429 (2420-2378)	.12

Site:	Castro de Torroso			
Location:	Mos, Pontevedra, Galicia			
Type:	Defended			
Cultural phase:	Bronze Age-Iron Age			
References:	Carballo - Fábregas 1992, nº 19; Peña Santos 1987, 131-2			
Sample ID:	GRN-13678			
Radiocarbon Age BP:	2515 ± 30≈			
Calibrated age (s)	cal BC	766, 676, 662		
	cal BP	2715, 2625, 2611		

cal AD/BC (cal BP) age ranges obtained from intercepts (Method A):

one Sigma**	cal BC	789- 758 (2738-2707)	687- 657 (2636-2606)	641- 591 (2590-2540)
		575- 552 (2524-2501)	457- 454 (2406-2403)	
two Sigma**	cal BC	796- 736 (2745-2685)	729- 517 (2678-2466)	466- 435 (2415-2384)

Summary of above

minimum of cal age ranges (cal ages) maximum of cal age ranges:

one sigma	cal BC	789 (766, 676, 662) 454
	cal BP	2738 (2715, 2625, 2611) 2403
two sigma	cal BC	796 (766, 676, 662) 435
	cal BP	2745 (2715, 2625, 2611) 2384

cal AD/BC age ranges (cal ages as above) from probability distribution (Method B):

% area enclosed	cal BC (cal BP) age ranges		relative area under probability distribution
68.3 (one sigma)	cal BC	789- 757 (2738-2706)	.22
		690- 657 (2639-2606)	.24
		642- 590 (2591-2539)	.36
		577- 554 (2526-2503)	.14
		458- 451 (2407-2400)	.04
95.4 (two sigma)	cal BC	794- 750 (2743-2699)	.19
		730- 539 (2679-2488)	.72
		533- 520 (2482-2469)	.02
		465- 435 (2414-2384)	.07

Site:	Castro de Penarrubia			
Location:	Orbazai, Lugo			
Type:	Defended			
Cultural phase:	Iron Age			
References:	Aira - Guitián 1985-86, 193; Arias Vilas 1979, 617; Carballo - Fábregas 1992, nº 8; Eiroa 1980, 77; 1988, 118			
Sample ID:	CSIC-358			
Radiocarbon Age BP:	2510 ± 50≈			
Calibrated age (s)	cal BC	765, 677, 661, 606,605		
	cal BP	2714, 2626, 2610, 2555, 2554		

cal AD/BC (cal BP) age ranges obtained from intercepts (Method A) :

one Sigma**	cal BC	791-753 (2740-2702)	694-587 (2643-2536)	583-539 (2532-2488)
		531-522 (2480-2471)	462-448 (2411-2397)	
two Sigma**	cal BC	800-410 (2749-2359)		

Summary of above

minimum of cal age ranges (cal ages) maximum of cal age ranges:

one sigma	cal BC	791 (765, 677, 661, 606, 605) 448
	cal BP	2740 (2714, 2626, 2610, 2555, 2554) 2397
two sigma	cal BC	800 (765, 677, 661, 606, 605) 410
	cal BP	2749 (2714, 2626, 2610, 2555, 2554) 2359

cal AD/BC age ranges (cal ages as above) from probability distribution (Method B) :

% area enclosed	cal BC (cal BP) age ranges		relative area under probability distribution
68.3 (one sigma)	cal BC	790-754 (2739-2703)	.18
		716-712 (2665-2661)	.02
		693-588 (2642-2537)	.53
		582-542 (2531-2491)	.19
		528-524 (2477-2473)	.02
		461-448 (2410-2397)	.06

95.4 (two sigma)	cal BC	796-508 (2745-2457)	.88
		501-487 (2450-2436)	.02
		470-430 (2419-2379)	.09

Site: Castrovite
Location: A Estrada, Pontevedra, Galicia
Type: Defended
Cultural phase: Iron Age
References: Carballo - Fábregas 1992, nº9
Sample ID: CSIC-816
Radiocarbon Age BP: 2500 ± 50≈
Calibrated age (s) cal BC 763, 680, 660, 629, 623, 609, 596
cal BP 2712, 2629, 2609, 2578, 2572, 2558, 2545
cal AD/BC (cal BP) age ranges obtained from intercepts (Method A) :
one Sigma** cal BC 789-750 (2738-2699) 724-536 (2673-2485) 535-519 (2484-2468)
464-446 (2413-2395)
two Sigma** cal BC 800-410 (2749-2359)

Summary of above
minimum of cal age ranges (cal ages) maximum of cal age ranges:

one sigma	cal BC	789 (763, 680, 660, 629, 623, 609, 596) 446
	cal BP	2738 (2712, 2629, 2609, 2578, 2572, 2558, 2545) 2395
two sigma	cal BC	800 (763, 680, 660, 629, 623, 609, 596) 410
	cal BP	2749 (2712, 2629, 2609, 2578, 2572, 2558, 2545) 2359

cal AD/BC age ranges (cal ages as above) from probability distribution (Method B) :

% area enclosed	cal BC (cal BP) age ranges		relative area under probability distribution
68.3 (one sigma)	cal BC	787-754 (2736-2703)	.15
		717-707 (2666-2656)	.04
		694-666 (2643-2615)	.13
		665-587 (2614-2536)	.38
		583-541 (2532-2490)	.20
		530-523 (2479-2472)	.03
		462-448 (2411-2397)	.07
95.4 (two sigma)	cal BC	793-507 (2742-2456)	.87
		503-486 (2452-2435)	.03
		470-430 (2419-2379)	.10

Site: Castro de Penalba
Location: Campo Lameiro, Pontevedra, Galicia
Type: Defended
Cultural phase: Iron Age
References: Aira *et al.* 1990, 82; Carballo - Fábregas 1992, nº 15
Sample ID: GRN-14134
Radiocarbon Age BP: 2490 ± 35≈
Calibrated age (s) cal BC 761, 683, 659, 633, 620, 612, 594
cal BP 2710, 2632, 2608, 2582, 2569, 2561, 2543
cal AD/BC (cal BP) age ranges obtained from intercepts (Method A) :
one Sigma** cal BC 770-752 (2719-2701) 722-702 (2671-2651) 696-670 (2645-2619)
664-537 (2613-2486) 534-520 (2483-2469) 463-446 (2412-2395)
two Sigma** cal BC 792-478 (2741-2427) 474-426 (2423-2375) 425-412 (2374-2361)

Summary of above
minimum of cal age ranges (cal ages) maximum of cal age ranges:

one sigma	cal BC	770 (761, 683, 659, 633, 620, 612, 594) 446
	cal BP	2719 (2710, 2632, 2608, 2582, 2569, 2561, 2543) 2395
two sigma	cal BC	792 (761, 683, 659, 633, 620, 612, 594) 412
	cal BP	2741 (2710, 2632, 2608, 2582, 2569, 2561, 2543) 2361

Site: Castro de Penalba
Location: Campo Lameiro, Pontevedra, Galicia
Type: Defended

Cultural phase: Iron Age
References: Aira *et al.* 1990, 82; Carballo - Fábregas 1992, nº 15
Sample ID: GRN-14132
Radiocarbon Age BP: 2485 ± 35≈

Calibrated age (s)	cal BC	760, 685, 658, 635, 618, 613, 593, 568, 563
	cal BP	2709, 2634, 2607, 2584, 2567, 2562, 2542, 2517, 2512

cal AD/BC (cal BP) age ranges obtained from intercepts (Method A) :

one Sigma**	cal BC	769-745 (2718-2694) 726-672 (2675-2621) 663-519 (2612-2468) 464-443 (2413-2392)
two Sigma**	cal BC	792-411 (2741-2360)

Summary of above
minimum of cal age ranges (cal ages) maximum of cal age ranges:

one sigma	cal BC	769 (760, 685, 658, 635, 618, 613, 593, 568, 563) 443
	cal BP	2718 (2709, 2634, 2607, 2584, 2567, 2562, 2542, 2517, 2512) 2392
two sigma	cal BC	792 (760, 685, 658, 635, 618, 613, 593, 568, 563) 411
	cal BP	2741 (2709, 2634, 2607, 2584, 2567, 2562, 2542, 2517, 2512) 2360

cal AD/BC age ranges (cal ages as above) from probability distribution (Method B) :

% area enclosed	cal BC (cal BP) age ranges		relative area under probability distribution
68.3 (one sigma)	cal BC	767-753 (2716-2702)	.07
		719-697 (2668-2646)	.10
		695-674 (2644-2623)	.11
		661-587 (2610-2536)	.38
		584-540 (2533-2489)	.23
		530-522 (2479-2471)	.04
		462-447 (2411-2396)	.08
95.4 (two sigma)	cal BC	788-509 (2737-2458)	.87
		500-488 (2449-2437)	.02
		470-430 (2419-2379)	.11

Site: Castro das Ermidas
Location: Jesufrei, V.N. Famalicão, Portugal
Type: Defended
Cultural phase: Iron Age
References: Carballo - Fábregas 1992, nº 32; Queiroga 1985, 35-7
Sample ID: GAK-11460
Radiocarbon Age BP: 2480 ± 120≈

Calibrated age (s)	cal BC	759, 686, 657, 638, 616, 615, 592, 572, 558, 456, 455
	cal BP	2708, 2635, 2606, 2587, 2565, 2564, 2541, 2521, 2507, 2405, 2404

cal AD/BC (cal BP) age ranges obtained from intercepts (Method A):

one Sigma**	cal BC	800- 400 (2749-2349)
two Sigma**	cal BC	894- 873 (2843-2822) 840- 360 (2789-2309) 270- 262 (2219-2211)

Summary of above
minimum of cal age ranges (cal ages) maximum of cal age ranges:

one sigma	cal BC	800 (759, 686, 657, 638, 616, 615, 592, 572, 558, 456, 455) 400
	cal BP	2749 (2708, 2635, 2606, 2587, 2565, 2564, 2541, 2521, 2507, 2405, 2404) 2349
two sigma	cal BC	894 (759, 686, 657, 638, 616, 615, 592, 572, 558, 456, 455) 262
	cal BP	2843 (2708, 2635, 2606, 2587, 2565, 2564, 2541, 2521,2507, 2405, 2404) 2211

cal AD/BC age ranges (cal ages as above) from probability distribution (Method B):

% area enclosed	cal BC (cal BP) age ranges		relative area under probability distribution
68.3 (one sigma)	cal BC	770- 670 (2719-2619)	.34
		660- 510 (2609-2459)	.51
		498- 492 (2447-2441)	.02
		469- 431 (2418-2380)	.13
95.4 (two sigma)	cal BC	896- 866 (2845-2815)	.01
		840- 360 (2789-2309)	.98

Site: Castro de Borneiro
Location: Cabana, A Coruña, Galicia
Type: Defended
Cultural phase: Iron Age
References: Carballo - Fábregas 1992, nº 3; Eiroa 1973b, 61; 1980, 76-7; 1988, 112-3, 118
Sample ID: CSIC-83
Radiocarbon Age BP: 2470 ± 100≈

Calibrated age (s)	cal BC	757, 689, 656, 645, 590, 577, 545, 458, 452
	cal BP	2706, 2638, 2605, 2594, 2539, 2526, 2494, 2407, 2401

cal AD/BC (cal BP) age ranges obtained from intercepts (Method A):

one Sigma**	cal BC	790- 410 (2739-2359)
two Sigma**	cal BC	830- 380 (2779-2329)

Summary of above

minimum of cal age ranges (cal ages) maximum of cal age ranges:

one sigma	cal BC	790 (757, 689, 656, 645, 590, 577, 545, 458, 452) 410
	cal BP	2739 (2706, 2638, 2605, 2594, 2539, 2526, 2494, 2407, 2401) 2359
two sigma	cal BC	830 (757, 689, 656, 645, 590, 577, 545, 458, 452) 380
	cal BP	2779 (2706, 2638, 2605, 2594, 2539, 2526, 2494, 2407, 2401) 2329

cal AD/BC age ranges (cal ages as above) from probability distribution (Method B):

% area enclosed	cal BC (cal BP) age ranges		relative area under probability distribution
68.3 (one sigma)	cal BC	766- 676 (2715-2625)	.32
		660- 510 (2609-2459)	.54
		468- 432 (2417-2381)	.13
95.4 (two sigma)	cal BC	810- 390 (2759-2339)	1.00

Site: Castro de Penalba
Location: Campo Lameiro, Pontevedra, Galicia
Type: Defended
Cultural phase: Iron Age
References: Aira *et al.* 1990, 82; Carballo - Fábregas 1992, nº 15
Sample ID: GRN-14133
Radiocarbon Age BP: 2445 ± 30≈

Calibrated age (s)	cal BC	750, 724, 536, 535, 519, 464, 446
	cal BP	2699, 2673, 2485, 2484, 2468, 2413, 2395

cal AD/BC (cal BP) age ranges obtained from intercepts (Method A) :

one Sigma**	cal BC	759-685 (2708-2634)	658-636 (2607-2585)	617-614 (2566-2563)
		592-569 (2541-2518)	561-478 (2510-2427)	474-426 (2423-2375)
		425-412 (2374-2361)		
two Sigma**	cal BC	768-674 (2717-2623)	662-406 (2611-2355)	

Summary of above

minimum of cal age ranges (cal ages) maximum of cal age ranges:

one sigma	cal BC	759 (750, 724, 536, 535, 519, 464, 446) 412
	cal BP	2708 (2699, 2673, 2485, 2484, 2468, 2413, 2395) 2361
two sigma	cal BC	768 (750, 724, 536, 535, 519, 464, 446) 406
	cal BP	2717 (2699, 2673, 2485, 2484, 2468, 2413, 2395) 2355

cal AD/BC age ranges (cal ages as above) from probability distribution (Method B) :

% area enclosed	cal BC (cal BP) age ranges		relative area under probability distribution
68.3 (one sigma)	cal BC	757-688 (2706-2637)	.36
		656-644 (2605-2593)	.06
		590-576 (2539-2525)	.07
		544-484 (2493-2433)	.29
		471-458 (2420-2407)	.07
		454-429 (2403-2378)	.13
		419-415 (2368-2364)	.02
95.4 (two sigma)	cal BC	763-679 (2712-2628)	.30
		659-629 (2608-2578)	.08
		622-610 (2571-2559)	.03
		595-410 (2544-2359)	.59

Site: Cidade de San Cibrán das Lás
Location: San Amaro, Ourense, Galicia
Type: Defended
Cultural phase: Iron Age
References: Carballo - Fábregas 1992, nº 18
Sample ID: ICEN-410
Radiocarbon Age BP: 2440 ± 60≈

Calibrated age (s)	cal BC	742, 727, 518, 465, 439		
	cal BP	2691, 2676, 2467, 2414, 2388		

cal AD/BC (cal BP) age ranges obtained from intercepts (Method A) :

one Sigma**	cal BC	764-678 (2713-2627)	660-626 (2609-2575)	625-607 (2574-2556)
		601-407 (2550-2356)		
two Sigma**	cal BC	790-390 (2739-2339)		

Summary of above
minimum of cal age ranges (cal ages) maximum of cal age ranges:

one sigma	cal BC	764 (742, 727, 518, 465, 439) 407
	cal BP	2713 (2691, 2676, 2467, 2414, 2388) 2356
two sigma	cal BC	790 (742, 727, 518, 465, 439) 390
	cal BP	2739 (2691, 2676, 2467, 2414, 2388) 2339

cal AD/BC age ranges (cal ages as above) from probability distribution (Method B) :

% area enclosed	cal BC (cal BP) age ranges		relative area under probability distribution
68.3 (one sigma)	cal BC	758-687 (2707-2636)	.31
		657-642 (2606-2591)	.06
		591-575 (2540-2524)	.07
		548-478 (2497-2427)	.30
		474-457 (2423-2406)	.08
		454-412 (2403-2361)	.18
95.4 (two sigma)	cal BC	768-673 (2717-2622)	.28
		662-405 (2611-2354)	.72

Site: Castro de Torroso
Location: Mos, Pontevedra, Galicia
Type: Defended
Cultural phase: Bronze Age-Iron Age
References: Carballo - Fábregas 1992, nº 19
Sample ID: GRN-14587
Radiocarbon Age BP: 2435 ± 30≈

Calibrated age (s)	cal BC	733, 731, 517, 466, 435		
	cal BP	2682, 2680, 2466, 2415, 2384		

cal AD/BC (cal BP) age ranges obtained from intercepts (Method A) :

one Sigma**	cal BC	757-688 (2706-2637)	657-642 (2606-2591)	590-576 (2539-2525)
		548-458 (2497-2407)	453-411 (2402-2360)	
two Sigma**	cal BC	766-676 (2715-2625)	661-404 (2610-2353)	

Summary of above
minimum of cal age ranges (cal ages) maximum of cal age ranges:

one sigma	cal BC	757 (733, 731, 517, 466, 435) 411
	cal BP	2706 (2682, 2680, 2466, 2415, 2384) 2360
two sigma	cal BC	766 (733, 731, 517, 466, 435) 404
	cal BP	2715 (2682, 2680, 2466, 2415, 2384) 2353

cal AD/BC age ranges (cal ages as above) from probability distribution (Method B) :

% area enclosed	cal BC (cal BP) age ranges		relative area under probability distribution
68.3 (one sigma)	cal BC	755-691 (2704-2640)	.35
		588-580 (2537-2529)	.04
		541-529 (2490-2478)	.06
		524-479 (2473-2428)	.25
		474-459 (2423-2408)	.08
		452-426 (2401-2375)	.14
		425-413 (2374-2362)	.06

95.4 (two sigma)   cal BC   762-681 (2711-2630)   .30
658-632 (2607-2581)   .06
619-612 (2568-2561)   .01
594-408 (2543-2357)   .62

Site: Castro das Ermidas
Location: Jesufrei, VN Famalicão, Portugal
Type: Defended
Cultural phase: Iron Age
References: (unpublished)
Sample ID: ICEN-469
Radiocarbon Age BP: 2420 ± 50≈
Calibrated age (s)   cal BC   506, 485, 470, 430, 418, 415
cal BP   2455, 2434, 2419, 2379, 2367, 2364
cal AD/BC (cal BP) age ranges obtained from intercepts (Method A):
one Sigma**   cal BC   758- 688 (2707-2637)   657- 641 (2606-2590)   591- 575 (2540-2524)
551- 457 (2500-2406)   454- 405 (2403-2354)
two Sigma**   cal BC   770- 671 (2719-2620)   660- 390 (2609-2339)

Summary of above
minimum of cal age ranges (cal ages) maximum of cal age ranges:
one sigma   cal BC   758 (506, 485, 470, 430, 418, 415) 405
cal BP   2707 (2455, 2434, 2419, 2379, 2367, 2364) 2354
two sigma   cal BC   770 (506, 485, 470, 430, 418, 415) 390
cal BP   2719 (2455, 2434, 2419, 2379, 2367, 2364) 2339

cal AD/BC age ranges (cal ages as above) from probability distribution (Method B):

% area enclosed	cal BC (cal BP) age ranges		relative area under probability distribution
68.3 (one sigma)	cal BC	756- 690 (2705-2639)	.32
		589- 579 (2538-2528)	.05
		542- 528 (2491-2477)	.06
		524- 459 (2473-2408)	.34
		452- 407 (2401-2356)	.23
95.4 (two sigma)	cal BC	764- 678 (2713-2627)	.27
		660- 627 (2609-2576)	.07
		624- 608 (2573-2557)	.03
		598- 401 (2547-2350)	.63

Site: Castro de Penices
Location: Gondifelos, V.N. Famalicão, Portugal
Type: Defended
Cultural phase: Iron Age
References: (unpublished)
Sample ID: ICEN-831
Radiocarbon Age BP: 2420 ± 45≈
Calibrated age(s)   cal BC   506, 485, 470, 430, 418, 415
cal BP   2455, 2434, 2419, 2379, 2367, 2364
cal AD/BC (cal BP) age ranges obtained from intercepts (Method A):
one Sigma**   cal BC   757- 689 (2706-2638)   656- 644 (2605-2593)   590- 577 (2539-2526)
545- 458 (2494-2407)   452- 406 (2401-2355)
two Sigma**   cal BC   768- 673 (2717-2622)   660- 400 (2609-2349)

Summary of above
minimum of cal age ranges (cal ages) maximum of cal age ranges:
one sigma   cal BC   757 (506, 485, 470, 430, 418, 415) 406
cal BP   2706 (2455, 2434, 2419, 2379, 2367, 2364) 2355
two sigma   cal BC   768 (506, 485, 470, 430, 418, 415) 400
cal BP   2717 (2455, 2434, 2419, 2379, 2367, 2364) 2349

cal AD/BC age ranges (cal ages as above) from probability distribution (Method B):

% area enclosed	cal BC (cal BP) age ranges		relative area under probability distribution
68.3 (one sigma)	cal BC	755- 691 (2704-2640)	.33

		588- 580 (2537-2529)	.04
		540- 529 (2489-2478)	.05
		523- 460 (2472-2409)	.35
		451- 408 (2400-2357)	.23
95.4 (two sigma)	cal BC	763- 679 (2712-2628)	.28
		659- 629 (2608-2578)	.07
		622- 610 (2571-2559)	.02
		595- 402 (2544-2351)	.64

Site: Castro de Troña
Location: Ponteareas, Pontevedra, Galicia
Type: Defended
Cultural phase: Iron Age
References: Carballo - Fábregas 1992, nº 21; Eiroa 1988, 118; Hidalgo Cuñarro 1988-89, 90
Sample ID: CSIC-690
Radiocarbon Age BP: 2400 ± 50≈
Calibrated age (s) cal BC 411
cal BP 2360
cal AD/BC (cal BP) age ranges obtained from intercepts (Method A):

one Sigma**	cal BC	753- 717 (2702-2666)	714- 694 (2663-2643)	587- 583 (2536-2532)
		539- 531 (2488-2480)	522- 462 (2471-2411)	448- 402 (2397-2351)
two Sigma**	cal BC	765- 678 (2714-2627)	661- 607 (2610-2556)	600- 390 (2549-2339)

Summary of above
minimum of cal age ranges (cal ages) maximum of cal age ranges:

one sigma	cal BC	753 (411) 402
	cal BP	2702 (2360) 2351
two sigma	cal BC	765 (411) 390
	cal BP	2714 (2360) 2339

cal AD/BC age ranges (cal ages as above) from probability distribution (Method B):

% area enclosed	cal BC (cal BP) age ranges		relative area under probability distribution
68.3 (one sigma)	cal BC	754- 693 (2703-2642)	.30
		587- 583 (2536-2532)	.02
		538- 532 (2487-2481)	.03
		522- 461 (2471-2410)	.36
		450- 402 (2399-2351)	.28
95.4 (two sigma)	cal BC	763- 680 (2712-2629)	.26
		659- 630 (2608-2579)	.05
		621- 610 (2570-2559)	.02
		595- 394 (2544-2343)	.67

Site: Castro de Penalba
Location: Campo Lameiro, Pontevedra, Galicia
Type: Defended
Cultural phase: Iron Age
References: Carballo - Fábregas 1992, nº 15
Sample ID: CSIC-636
Radiocarbon Age BP: 2390 ± 50≈
Calibrated age (s) cal BC 409
cal BP 2358
cal AD/BC (cal BP) age ranges obtained from intercepts (Method A) :

one Sigma**	cal BC	748-725 (2697-2674)	519-464 (2468-2413)	445-400 (2394-2349)
two Sigma**	cal BC	762-681 (2711-2630)	659-630 (2608-2579)	622-610 (2571-2559)
		590-390 (2539-2339)		

Summary of above
minimum of cal age ranges (cal ages) maximum of cal age ranges:

one sigma	cal BC	748 (409) 400
	cal BP	2697 (2358) 2349
two sigma	cal BC	762 (409) 390
	cal BP	2711 (2358) 2339

cal AD/BC age ranges (cal ages as above) from probability distribution (Method B) :

% area enclosed		cal BC (cal BP) age ranges	relative area under probability distribution
68.3 (one sigma)	cal BC	753-718 (2702-2667)	.18
		717-694 (2666-2643)	.12
		586-584 (2535-2533)	.01
		536-534 (2485-2483)	.01
		520-462 (2469-2411)	.37
		449-399 (2398-2348)	.31
95.4 (two sigma)	cal BC	762-680 (2711-2629)	.25
		659-631 (2608-2580)	.05
		620-611 (2569-2560)	.01
		594-390 (2543-2339)	.69

Site: Castro de Penices
Location: Gondifelos, V.N. Famalicão, Portugal
Type: Defended
Cultural phase: Iron Age
References: (unpublished)
Sample ID: ICEN-833
Radiocarbon Age BP: 2380 ± 60≈
Calibrated age(s) cal BC 408
cal BP 2357

cal AD/BC (cal BP) age ranges obtained from intercepts (Method A):

one Sigma**	cal BC	747- 725 (2696-2674)	519- 464 (2468-2413)	444- 396(2393-2345)
two Sigma**	cal BC	764- 678 (2713-2627)	661- 626 (2610-2575)	625- 607 (2574-2556)
		600- 380 (2549-2329)		

Summary of above
minimum of cal age ranges (cal ages) maximum of cal age ranges:

one sigma	cal BC	747 (408) 396
	cal BP	2696 (2357) 2345
two sigma	cal BC	764 (408) 380
	cal BP	2713 (2357) 2329

cal AD/BC age ranges (cal ages as above) from probability distribution (Method B):

% area enclosed		cal BC (cal BP) age ranges	relative area under probability distribution
68.3 (one sigma)	cal BC	754- 693 (2703-2642)	.28
		587- 582 (2536-2531)	.02
		538- 532 (2487-2481)	.03
		522- 461 (2471-2410)	.35
		450- 394 (2399-2343)	.32
95.4 (two sigma)	cal BC	766- 675 (2715-2624)	.24
		661- 606 (2610-2555)	.08
		604- 378 (2553-2327)	.68

Site: Castromao
Location: Celanova, Ourense, Galicia
Type: Defended
Cultural phase: Iron Age
References: Carballo - Fábregas 1992, n° 22; Fariña Busto 1991, 54
Sample ID: CSIC-638
Radiocarbon Age BP: 2370 ± 50≈
Calibrated age (s) cal BC 406
cal BP 2355

cal AD/BC (cal BP) age ranges obtained from intercepts (Method A) :

one Sigma**	cal BC	509-499 (2458-2448)	490-469 (2439-2418)	431-416 (2380-2365)
		415-396 (2364-2345)		
two Sigma**	cal BC	758-687 (2707-2636)	657-640 (2606-2589)	591-574 (2540-2523)
		554-457 (2503-2406)	454-384 (2403-2333)	

Summary of above
minimum of cal age ranges (cal ages) maximum of cal age ranges:

one sigma	cal BC	509 (406) 396
	cal BP	2458 (2355) 2345
two sigma	cal BC	758 (406) 384
	cal BP	2707 (2355) 2333

cal AD/BC age ranges (cal ages as above) from probability distribution (Method B) :

% area enclosed		cal BC (cal BP) age ranges	relative area under probability distribution
68.3 (one sigma)	cal BC	747-722 (2696-2671)	.12
		716-696 (2665-2645)	.10
		519-463 (2468-2412)	.37
		447-392 (2396-2341)	.40
95.4 (two sigma)	cal BC	762-680 (2711-2629)	.23
		658-632 (2607-2581)	.03

Site: El Cerco
Location: Sejas de Aliste, Zamora
Type: Defended
Cultural phase: Iron Age
References: Carballo - Fábregas 1992, nº 31; Esparza 1986, 401
Sample ID: GRN-10339
Radiocarbon Age BP: 2360 ± 80≈

Calibrated age (s)	cal BC	404		
	cal BP	2353		

cal AD/BC (cal BP) age ranges obtained from intercepts (Method A) :

one Sigma**	cal BC	744-726 (2693-2675)	519-465 (2468-2414)	441-388 (2390-2337)
two Sigma**	cal BC	768-673 (2717-2622)	660-350 (2609-2299)	296-230 (2245-2179)
		220-211 (2169-2160)		

Summary of above
minimum of cal age ranges (cal ages) maximum of cal age ranges:

one sigma	cal BC	744 (404) 388
	cal BP	2693 (2353) 2337
two sigma	cal BC	768 (404) 211
	cal BP	2717 (2353) 2160

cal AD/BC age ranges (cal ages as above) from probability distribution (Method B) :

% area enclosed		cal BC (cal BP) age ranges	relative area under probability distribution
68.3 (one sigma)	cal BC	757-688 (2706-2637)	.25
		656-645 (2605-2594)	.03
		590-577 (2539-2526)	.04
		544-458 (2493-2407)	.34
		453-377 (2402-2326)	.33
95.4 (two sigma)	cal BC	769-671 (2718-2620)	.21
		663-351 (2612-2300)	.71

Site: Castelo de Matos
Location: Matos, Baião, Portugal
Type: Defended
Cultural phase: Bronze Age
References: Figueiral - Queiroga 1988; Hedges et al 1990, 226; Queiroga 1984; Queiroga - Figueiral 1989
Sample ID: GIF-7718
Radiocarbon Age BP: 2340 ± 60≈

Calibrated age (s)	cal BC	400		
	cal BP	2349		

cal AD/BC (cal BP) age ranges obtained from intercepts (Method A):

one Sigma**	cal BC	411- 388 (2360-2337)		
two Sigma**	cal BC	755- 691 (2704-2640)	589- 580 (2538-2529)	543- 527 (2492-2476)
		524- 460 (2473-2409)	451- 355 (2400-2304)	290- 276 (2239-2225)
		275- 258 (2224-2207)		

Summary of above
minimum of cal age ranges (cal ages) maximum of cal age ranges:

one sigma	cal BC	411 (400) 388	
	cal BP	2360 (2349) 2337	
two sigma	cal BC	755 (400) 258	
	cal BP	2704 (2349) 2207	

cal AD/BC age ranges (cal ages as above) from probability distribution (Method B):

% area enclosed	cal BC (cal BP) age ranges		relative area under probability distribution
68.3 (one sigma)	cal BC	748- 721 (2697-2670)	.10
		716- 696 (2665-2645)	.07
		519- 463 (2468-2412)	.28
		448- 361 (2397-2310)	.52
		269- 262 (2218-2211)	.03
95.4 (two sigma)	cal BC	761- 682 (2710-2631)	.17
		658- 634 (2607-2583)	.02

Site: Castromao
Location: Celanova, Ourense, Galicia
Type: Defended
Cultural phase: Iron Age
References: Carballo - Fábregas 1992, nº 22; Fariña Busto 1991, 54
Sample ID: CSIC-639
Radiocarbon Age BP: 2330 ± 50≈

Calibrated age (s)	cal BC	399		
	cal BP	2348		

cal AD/BC (cal BP) age ranges obtained from intercepts (Method A) :

one Sigma**	cal BC	408-388 (2357-2337)		
two Sigma**	cal BC	733-731 (2682-2680)	517-466 (2466-2415)	435-358 (2384-2307)
		288-281 (2237-2230)	273-260 (2222-2209)	245-234 (2194-2183)

Summary of above
minimum of cal age ranges (cal ages) maximum of cal age ranges:

one sigma	cal BC	408 (399) 388
	cal BP	2357 (2348) 2337
two sigma	cal BC	733 (399) 234
	cal BP	2682 (2348) 2183

cal AD/BC age ranges (cal ages as above) from probability distribution (Method B) :

% area enclosed	cal BC (cal BP) age ranges		relative area under probability distribution
68.3 (one sigma)	cal BC	702-697 (2651-2646)	.02
		516-467 (2465-2416)	.23
		440-435 (2389-2384)	.02
		434-360 (2383-2309)	.64
		271-260 (2220-2209)	.05
		238-235 (2187-2184)	.01
95.4 (two sigma)	cal BC	757-687 (2706-2636)	.12
		590-575 (2539-2524)	.01
		542-527 (2491-2476)	.01
		525-457 (2474-2406)	.19

Site: Castro de Troña
Location: Ponteareas, Pontevedra, Galicia
Type: Defended
Cultural phase: Iron Age
References: Eiroa 1988, 118
Sample ID: CSIC-684
Radiocarbon Age BP: 2330 ± 50≈

Calibrated age (s)	cal BC	399		
	cal BP	2348		

cal AD/BC (cal BP) age ranges obtained from intercepts (Method A):

one Sigma**	cal BC	408- 388 (2357-2337)		
two Sigma**	cal BC	733- 731 (2682-2680)	517- 466 (2466-2415)	435- 358 (2384-2307)
		288- 281 (2237-2230)	273- 260 (2222-2209)	245- 234 (2194-2183)

Summary of above
minimum of cal age ranges (cal ages) maximum of cal age ranges:

one sigma	cal BC	408 (399) 388
	cal BP	2357 (2348) 2337
two sigma	cal BC	733 (399) 234
	cal BP	2682 (2348) 2183

cal AD/BC age ranges (cal ages as above) from probability distribution (Method B):

% area enclosed	cal BC (cal BP) age ranges		relative area under probability distribution
68.3 (one sigma)	cal BC	702- 697 (2651-2646)	.02
		516- 467 (2465-2416)	.23
		440- 435 (2389-2384)	.02
		434- 360 (2383-2309)	.64
		271- 260 (2220-2209)	.05
		238- 235 (2187-2184)	.01
95.4 (two sigma)	cal BC	757- 687 (2706-2636)	.12
		590- 575 (2539-2524)	.01
		542- 527 (2491-2476)	.01
		525- 457 (2474-2406)	.19

Site: Castro de Troña
Location: Ponteareas, Pontevedra, Galicia
Type: Defended
Cultural phase: Iron Age
References: Carballo - Fábregas 1992, nº 21; Hidalgo 1988-89, 90
Sample ID: CSIC-689
Radiocarbon Age BP: 2330 ± 50≈

Calibrated age (s)	cal BC	399		
	cal BP	2348		

cal AD/BC (cal BP) age ranges obtained from intercepts (Method A) :

one Sigma**	cal BC	408-388 (2357-2337)		
two Sigma**	cal BC	733-731 (2682-2680)	517-466 (2466-2415)	435-358 (2384-2307)
		288-281 (2237-2230)	273-260 (2222-2209)	245-234 (2194-2183)

Summary of above
minimum of cal age ranges (cal ages) maximum of cal age ranges:

one sigma	cal BC	408 (399) 388
	cal BP	2357 (2348) 2337
two sigma	cal BC	733 (399) 234
	cal BP	2682 (2348) 2183

cal AD/BC age ranges (cal ages as above) from probability distribution (Method B) :

% area enclosed	cal BC (cal BP) age ranges		relative area under probability distribution
68.3 (one sigma)	cal BC	702-697 (2651-2646)	.02
		516-467 (2465-2416)	.23
		440-435 (2389-2384)	.02
		434-360 (2383-2309)	.64
		271-260 (2220-2209)	.05
		238-235 (2187-2184)	.01
95.4 (two sigma)	cal BC	757-687 (2706-2636)	.12
		590-575 (2539-2524)	.01
		542-527 (2491-2476)	.01
		525-457 (2474-2406)	.19

Site: Castro de Penalba
Location: Campo Lameiro, Pontevedra, Galicia
Type: Defended
Cultural phase: Iron Age
References: Alvarez Nuñez 1986, 61; Carballo - Fábregas 1992, nº 15; Eiroa 1988, 118
Sample ID: GAK-11332
Radiocarbon Age BP: 2310 ± 120≈

Calibrated age (s)	cal BC	395		
	cal BP	2344		

cal AD/BC (cal BP) age ranges obtained from intercepts (Method A):

one Sigma**	cal BC	516- 467 (2465-2416)	434- 350 (2383-2299)	317- 311 (2266-2260)
		301- 228 (2250-2177)	222- 210 (2171-2159)	
two Sigma**	cal BC	790- 90 (2739-2039)		

Summary of above

minimum of cal age ranges (cal ages) maximum of cal age ranges:

one sigma	cal BC	516 (395) 210
	cal BP	2465 (2344) 2159
two sigma	cal BC	790 (395) 90
	cal BP	2739 (2344) 2039

cal AD/BC age ranges (cal ages as above) from probability distribution (Method B):

% area enclosed	cal BC (cal BP) age ranges		relative area under probability distribution
68.3 (one sigma)	cal BC	754- 692 (2703-2641)	.14
		588- 581 (2537-2530)	.02
		540- 530 (2489-2479)	.02
		523- 461 (2472-2410)	.16
		450- 340 (2399-2289)	.33
		320- 200 (2269-2149)	.33
95.4 (two sigma)	cal BC	770- 670 (2719-2619)	.14
		660- 160 (2609-2109)	.84
		152- 111 (2101-2060)	.03

Site: Castrovite
Location: A Estrada, Pontevedra, Galicia
Type: Defended
Cultural phase: Iron Age
References: Carballo - Fábregas 1992, nº 9
Sample ID: CSIC-815
Radiocarbon Age BP: 2300 ± 50≈

Calibrated age (s)	cal BC	393		
	cal BP	2342		

cal AD/BC (cal BP) age ranges obtained from intercepts (Method A) :

one Sigma**	cal BC	403-364 (2352-2313)	267-264 (2216-2213)	
two Sigma**	cal BC	479-473 (2428-2422)	427-424 (2376-2373)	413-350 (2362-2299)
		317-313 (2266-2262)	299-229 (2248-2178)	221-210 (2170-2159)

Summary of above

minimum of cal age ranges (cal ages) maximum of cal age ranges:

one sigma	cal BC	403 (393) 264
	cal BP	2352 (2342) 2213
two sigma	cal BC	479 (393) 210
	cal BP	2428 (2342) 2159

cal AD/BC age ranges (cal ages as above) from probability distribution (Method B) :

% area enclosed	cal BC (cal BP) age ranges		relative area under probability distribution
68.3 (one sigma)	cal BC	408-356 (2357-2305)	.63
		290-277 (2239-2226)	.09
		275-257 (2224-2206)	.15
		251-233 (2200-2182)	.13
95.4 (two sigma)	cal BC	716-696 (2665-2645)	.02
		517-464 (2466-2413)	.06
		448-343 (2397-2292)	.52
		324-206 (2273-2155)	.39

Site: Castro de Cortegada
Location: Silleda, Pontevedra, Galicia
Type: Defended
Cultural phase: Iron Age

References: Carballo - Fábregas 1992, nº 12
Sample ID: CSIC-783
Radiocarbon Age BP: 2280 ± 50≈
Calibrated age (s) cal BC 389
cal BP 2338
cal AD/BC (cal BP) age ranges obtained from intercepts (Method A) :

one Sigma**	cal BC	399-358 (2348-2307)	288-282 (2237-2231)	272-260 (2221-2209)
		245-234 (2194-2183)		
two Sigma**	cal BC	409-346 (2358-2295)	322-226 (2271-2175)	225-207 (2174-2156)

Summary of above
minimum of cal age ranges (cal ages) maximum of cal age ranges:

one sigma	cal BC	399 (389) 234
	cal BP	2348 (2338) 2183
two sigma	cal BC	409 (389) 207
	cal BP	2358 (2338) 2156

cal AD/BC age ranges (cal ages as above) from probability distribution (Method B) :

% area enclosed	cal BC (cal BP) age ranges		relative area under probability distribution
68.3 (one sigma)	cal BC	401-355 (2350-2304)	.53
		291-257 (2240-2206)	.30
		251-233 (2200-2182)	.16
			.01
95.4 (two sigma)	cal BC	428-421 (2377-2370)	.01
		413-336 (2362-2285)	.45

Site: Castro de Cortegada
Location: Silleda, Pontevedra, Galicia
Type: Defended
Cultural phase: Iron Age
References: Carballo Arceo 1989; Carballo - Fábregas 1992, nº 12
Sample ID: CSIC-780
Radiocarbon Age BP: 2260 ± 60≈
Calibrated age (s) cal BC 383
cal BP 2332
cal AD/BC (cal BP) age ranges obtained from intercepts (Method A):

one Sigma**	cal BC	397- 351 (2346-2300)	296- 230 (2245-2179)	220- 211 (2169-2160)
two Sigma**	cal BC	410- 200 (2359-2149)	194- 173 (2143-2122)	

Summary of above
minimum of cal age ranges (cal ages) maximum of cal age ranges:

one sigma	cal BC	397 (383) 211
	cal BP	2346 (2332) 2160
two sigma	cal BC	410 (383) 173
	cal BP	2359 (2332) 2122

cal AD/BC age ranges (cal ages as above) from probability distribution (Method B):

% area enclosed	cal BC (cal BP) age ranges		relative area under probability distribution
68.3 (one sigma)	cal BC	397- 351 (2346-2300)	.40
		297- 230 (2246-2179)	.53
		220- 211 (2169-2160)	.06
95.4 (two sigma)	cal BC	428- 421 (2377-2370)	1.00

Site: Castromao
Location: Celanova, Ourense, Galicia
Type: Defended
Cultural phase: Iron Age
References: Carballo - Fábregas 1992, 22; Fariña Busto 1991, 54-5
Sample ID: CSIC-640
Radiocarbon Age BP: 2250 ± 50≈
Calibrated age (s) cal BC 377, 370, 365, 266, 265
cal BP 2326, 2319, 2314, 2215, 2214

cal AD/BC (cal BP) age ranges obtained from intercepts (Method A) :

one Sigma**	cal BC	394-351 (2343-2300)	297-230 (2246-2179)	220-211 (2169-2160)
two Sigma**	cal BC	400-200 (2349-2149)	188-174 (2137-2123)	

Summary of above
minimum of cal age ranges (cal ages) maximum of cal age ranges:

one sigma	cal BC	394 (377, 370, 365, 266, 265) 211
	cal BP	2343 (2326, 2319, 2314, 2215, 2214) 2160
two sigma	cal BC	400 (377, 370, 365, 266, 265) 174
	cal BP	2349 (2326, 2319, 2314, 2215, 2214) 2123

cal AD/BC age ranges (cal ages as above) from probability distribution Method B) :

% area enclosed	cal BC (cal BP) age ranges		relative area under probability distribution
68.3 (one sigma)	cal BC	392-353 (2341-2302)	.38
		294-231 (2243-2180)	.57
		218-213 (2167-2162)	.05
95.4 (two sigma)	cal BC	401-336 (2350-2285)	.34
		334-202 (2283-2151)	.64
		184-176 (2133-2125)	.01

Site: Castro das Ermidas
Location: Jesufrei, VN Famalicão, Portugal
Type: Defended
Cultural phase: Iron Age
References: (unpublished)
Sample ID: ICEN-470
Radiocarbon Age BP: 2230 ± 50≈

Calibrated age (s)	cal BC	359, 286, 284, 271, 261, 241, 235
	cal BP	2308, 2235, 2233, 2220, 2210, 2190, 2184

cal AD/BC (cal BP) age ranges obtained from intercepts (Method A):

one Sigma**	cal BC	389- 347 (2338-2296)	321- 227 (2270-2176)	224- 208 (2173-2157)
two Sigma**	cal BC	400- 170 (2349-2119)	137- 131 (2086-2080)	

Summary of above
minimum of cal age ranges (cal ages) maximum of cal age ranges:

one sigma	cal BC	389 (359, 286, 284, 271, 261, 241, 235) 208
	cal BP	2338 (2308, 2235, 2233, 2220, 2210, 2190, 2184) 2157
two sigma	cal BC	400 (359, 286, 284, 271, 261, 241, 235) 131
	cal BP	2349 (2308, 2235, 2233, 2220, 2210, 2190, 2184) 2080

cal AD/BC age ranges (cal ages as above) from probability distribution (Method B):

% area enclosed	cal BC (cal BP) age ranges		relative area under probability distribution
68.3 (one sigma)	cal BC	387- 350 (2336-2299)	.29
		317- 312 (2266-2261)	.04
		300- 229 (2249-2178)	.58
		221- 210 (2170-2159)	.09
95.4 (two sigma)	cal BC	396- 198 (2345-2147)	.97
		190- 174 (2139-2123)	.03

Site: Castro de Troña
Location: Ponteareas, Pontevedra, Galicia
Type: Defended
Cultural phase: Iron Age
References: Carballo - Fábregas 1992, nº 21; Eiroa 1988, 118; Hidalgo Cuñarro 1988-89, 90
Sample ID: CSIC-688
Radiocarbon Age BP: 2230 ± 50≈

Calibrated age (s)	cal BC	359, 286, 284, 271, 261, 241, 235
	cal BP	2308, 2235, 2233, 2220, 2210, 2190, 2184

cal AD/BC (cal BP) age ranges obtained from intercepts (Method A):

one Sigma**	cal BC	389- 347 (2338-2296)	321- 227 (2270-2176)	224- 208 (2173-2157)
two Sigma**	cal BC	400- 170 (2349-2119)	137- 131 (2086-2080)	

Summary of above
minimum of cal age ranges (cal ages) maximum of cal age ranges:

one sigma	cal BC	389 (359, 286, 284, 271, 261, 241, 235) 208
	cal BP	2338 (2308, 2235, 2233, 2220, 2210, 2190, 2184) 2157
two sigma	cal BC	400 (359, 286, 284, 271, 261, 241, 235) 131
	cal BP	2349 (2308, 2235, 2233, 2220, 2210, 2190, 2184) 2080

cal AD/BC age ranges (cal ages as above) from probability distribution (Method B):

% area enclosed	cal BC (cal BP) age ranges		relative area under probability distribution
68.3 (one sigma)	cal BC	387- 350 (2336-2299)	.29
		317- 312 (2266-2261)	.04
		300- 229 (2249-2178)	.58
		221- 210 (2170-2159)	.09
95.4 (two sigma)	cal BC	396- 198 (2345-2147)	.97
		190- 174 (2139-2123)	.03

Site: Castro de Recarea
Location: Coiro, Mazaricos, A Coruña, Galicia
Type: Defended
Cultural phase: Iron Age
References: Carballo - Fábregas 1992, nº 5; Eiroa 1988, 118
Sample ID: CSIC-697
Radiocarbon Age BP: 2230 ± 50≈

Calibrated age (s)	cal BC	359, 286, 284, 271, 261, 241, 235
	cal BP	2308, 2235, 2233, 2220, 2210, 2190, 2184

cal AD/BC (cal BP) age ranges obtained from intercepts (Method A):

one Sigma**	cal BC	389- 347 (2338-2296)	321- 227 (2270-2176)	224- 208 (2173-2157)
two Sigma**	cal BC	400- 170 (2349-2119)	137- 131 (2086-2080)	

Summary of above
minimum of cal age ranges (cal ages) maximum of cal age ranges:

one sigma	cal BC	389 (359, 286, 284, 271, 261, 241, 235) 208
	cal BP	2338 (2308, 2235, 2233, 2220, 2210, 2190, 2184) 2157
two sigma	cal BC	400 (359, 286, 284, 271, 261, 241, 235) 131
	cal BP	2349 (2308, 2235, 2233, 2220, 2210, 2190, 2184) 2080

cal AD/BC age ranges (cal ages as above) from probability distribution (Method B):

% area enclosed	cal BC (cal BP) age ranges		relative area under probability distribution
68.3 (one sigma)	cal BC	387- 350 (2336-2299)	.29
		317- 312 (2266-2261)	.04
		300- 229 (2249-2178)	.58
		221- 210 (2170-2159)	.09
95.4 (two sigma)	cal BC	396- 198 (2345-2147)	.97
		190- 174 (2139-2123)	.03

Site: Castro das Orellas
Location: Silleda, Pontevedra, Galicia
Type: Defended
Cultural phase: Iron Age
References: Carballo Arceo 1989; Carballo - Fábregas 1992, nº 14
Sample ID: CSIC-787
Radiocarbon Age BP: 2230 ± 69≈

Calibrated age (s)	cal BC	359, 286, 284, 271, 261, 241, 235
	cal BP	2308, 2235, 2233, 2220, 2210, 2190, 2184

cal AD/BC (cal BP) age ranges obtained from intercepts (Method A):

one Sigma**	cal BC	393- 340 (2342-2289)	326- 204 (2275-2153)
two Sigma**	cal BC	410- 160 (2359-2109)	151- 110 (2100-2059)

Summary of above
minimum of cal age ranges (cal ages) maximum of cal age ranges:

one sigma	cal BC	393 (359, 286, 284, 271, 261, 241, 235) 204
	cal BP	2342 (2308, 2235, 2233, 2220, 2210, 2190, 2184) 2153

two sigma	cal BC	410 (359, 286, 284, 271, 261, 241, 235) 110	
	cal BP	2359 (2308, 2235, 2233, 2220, 2210, 2190, 2184) 2059	

cal AD/BC age ranges (cal ages as above) from probability distribution (Method B):

% area enclosed	cal BC (cal BP) age ranges		relative area under probability distribution
68.3 (one sigma)	cal BC	389- 348 (2338-2297)	.28
		320- 227 (2269-2176)	.62
		224- 208 (2173-2157)	.11
95.4 (two sigma)	cal BC	406- 156 (2355-2105)	.96
		152- 113 (2101-2062)	.04

Site: Castro de Cortegada,
Location: Silleda, Pontevedra, Galicia
Type: Defended
Cultural phase: Iron Age
References: Carballo Arceo 1989; Carballo - Fábregas 1992, nº 12
Sample ID: CSIC-781
Radiocarbon Age BP: 2220 ± 60≈

Calibrated age (s)	cal BC	356, 289, 279, 274, 259, 248, 233	
	cal BP	2305, 2238, 2228, 2223, 2208, 2197, 2182	

cal AD/BC (cal BP) age ranges obtained from intercepts (Method A):

one Sigma**	cal BC	389- 339 (2338-2288)	328- 203 (2277-2152)
two Sigma**	cal BC	400- 160 (2349-2109)	144- 116 (2093-2065)

Summary of above
minimum of cal age ranges (cal ages) maximum of cal age ranges:

one sigma	cal BC	389 (356, 289, 279, 274, 259, 248, 233) 203
	cal BP	2338 (2305, 2238, 2228, 2223, 2208, 2197 2182) 2152
two sigma	cal BC	400 (356, 289, 279, 274, 259, 248, 233) 116
	cal BP	2349 (2305, 2238, 2228, 2223, 2208, 2197, 2182) 2065

cal AD/BC age ranges (cal ages as above) from probability distribution (Method B):

% area enclosed	cal BC (cal BP) age ranges		relative area under probability distribution
68.3 (one sigma)	cal BC	383- 348 (2332-2297)	.24
		320- 227 (2269-2176)	.65
		223- 208 (2172-2157)	.11
95.4 (two sigma)	cal BC	399- 167 (2348-2116)	.97
		141- 123 (2090-2072)	.03

Site: Castrovite
Location: A Estrada, Pontevedra, Galicia
Type: Defended
Cultural phase: Iron Age
References: Carballo - Fábregas 1992, nº 9
Sample ID: CSIC-819
Radiocarbon Age BP: 2220 ± 60≈

Calibrated age (s)	cal BC	356, 289, 279, 274, 259, 248, 233	
	cal BP	2305, 2238, 2228, 2223, 2208, 2197, 2182	

cal AD/BC (cal BP) age ranges obtained from intercepts (Method A) :

one Sigma**	cal BC	389-339 (2338-2288)	328-203 (2277-2152)
two Sigma**	cal BC	400-160 (2349-2109)	144-116 (2093-2065)

Summary of above
minimum of cal age ranges (cal ages) maximum of cal age ranges:

one sigma	cal BC	389 (356, 289, 279, 274, 259, 248, 233) 203
	cal BP	2338 (2305, 2238, 2228, 2223, 2208, 2197, 2182) 2152
two sigma	cal BC	400 (356, 289, 279, 274, 259, 248, 233) 116
	cal BP	2349 (2305, 2238, 2228, 2223, 2208, 2197, 2182) 2065

cal AD/BC age ranges (cal ages as above) from probability distribution (Method B) :

% area enclosed	cal BC (cal BP) age ranges	relative area under probability distribution

68.3 (one sigma)	cal BC	383-348 (2332-2297)	.24
		320-227 (2269-2176)	.65
		223-208 (2172-2157)	.11
95.4 (two sigma)	cal BC	399-167 (2348-2116)	.97
		141-123 (2090-2072)	.03

Site: As Muradellas
Location: Lubian, Zamora
Type: Defended
Cultural phase: Iron Age
References: Carballo - Fábregas 1992, nº 25; Esparza Arroyo 1986, 401
Sample ID: I-13259
Radiocarbon Age BP: 2210 ± 80≈
Calibrated age (s) cal BC 354, 292, 257, 253, 232, 217, 214
cal BP 2303, 2241, 2206, 2202, 2181, 2166, 2163
cal AD/BC (cal BP) age ranges obtained from intercepts (Method A) :
one Sigma** cal BC 391-172 (2340-2121) 136-134 (2085-2083)
two Sigma** cal BC 410-90 (2359-2039) 78-72 (2027-2021)

Summary of above
minimum of cal age ranges (cal ages) maximum of cal age ranges:

one sigma	cal BC	391 (354, 292, 257, 253, 232, 217, 214) 134
	cal BP	2340 (2303, 2241, 2206, 2202, 2181, 2166, 2163) 2083
two sigma	cal BC	410 (354, 292, 257, 253, 232, 217, 214) 72
	cal BP	2359 (2303, 2241, 2206, 2202, 2181, 2166, 2163) 2021

cal AD/BC age ranges (cal ages as above) from probability distribution (Method B) :

% area enclosed	cal BC (cal BP) age ranges		relative area under probability distribution
68.3 (one sigma)	cal BC	386-337 (2335-2286)	.26
		333-202 (2282-2151)	.74
95.4 (two sigma)	cal BC	409-90 (2358-2039)	.99

Site: Castro de Montaz
Location: Oleiros, Silleda, Pontevedra, Galicia
Type: Defended
Cultural phase: Iron Age
References: Carballo - Fábregas 1992, nº 13
Sample ID: CSIC-788
Radiocarbon Age BP: 2210 ± 50≈
Calibrated age (s) cal BC 354, 292, 257, 253, 232, 217, 214
cal BP 2303, 2241, 2206, 2202, 2181, 2166, 2163
cal AD/BC (cal BP) age ranges obtained from intercepts (Method A) :
one Sigma** cal BC 385-339 (2334-2288) 328-203 (2277-2152)
two Sigma** cal BC 400-170 (2349-2119) 142-121 (2091-2070)

Summary of above
minimum of cal age ranges (cal ages) maximum of cal age ranges:

one sigma	cal BC	385 (354, 292, 257, 253, 232, 217, 214) 203
	cal BP	2334 (2303, 2241, 2206, 2202, 2181, 2166, 2163) 2152
two sigma	cal BC	400 (354, 292, 257, 253, 232, 217, 214) 121
	cal BP	2349 (2303, 2241, 2206, 2202, 2181, 2166, 2163) 2070

cal AD/BC age ranges (cal ages as above) from probability distribution (Method B) :

% area enclosed	cal BC (cal BP) age ranges		relative area under probability distribution
68.3 (one sigma)	cal BC	377-372 (2326-2321)	.04
		364-346 (2313-2295)	.13
		322-226 (2271-2175)	.70
		225-207 (2174-2156)	.13
95.4 (two sigma)	cal BC	393-170 (2342-2119)	.99
		137-131 (2086-2080)	.01

Site: Santo Estevão
Location: Facha, Ponte de Lima, Portugal
Type: Defended
Cultural phase: Iron Age
References: Almeida *et al.* 1982, 79; Carballo - Fábregas 1992, nº 28
Sample ID: CSIC-500
Radiocarbon Age BP: 2210 ± 50≈
Calibrated age (s) cal BC 354, 292, 257, 253, 232, 217, 214
cal BP 2303, 2241, 2206, 2202, 2181, 2166, 2163
cal AD/BC (cal BP) age ranges obtained from intercepts (Method A):
one Sigma** cal BC 385- 339 (2334-2288) 328- 203 (2277-2152)
two Sigma** cal BC 400- 170 (2349-2119) 142- 121 (2091-2070)

Summary of above
minimum of cal age ranges (cal ages) maximum of cal age ranges:
one sigma cal BC 385 (354, 292, 257, 253, 232, 217, 214) 203
cal BP 2334 (2303, 2241, 2206, 2202, 2181, 2166, 2163) 2152
two sigma cal BC 400 (354, 292, 257, 253, 232, 217, 214) 121
cal BP 2349 (2303, 2241, 2206, 2202, 2181, 2166, 2163) 2070

cal AD/BC age ranges (cal ages as above) from probability distribution (Method B):

% area enclosed	cal BC (cal BP) age ranges		relative area under probability distribution
68.3 (one sigma)	cal BC	377- 372 (2326-2321)	.04
		364- 346 (2313-2295)	.13
		322- 226 (2271-2175)	.70
		225- 207 (2174-2156)	.13
95.4 (two sigma)	cal BC	393- 170 (2342-2119)	.99
		137- 131 (2086-2080)	.01

Site: Citânia de S. Julião
Location: Ponte, Vila Verde, Portugal
Type: Defended
Cultural phase: Bronze Age-Iron Age
References: Carballo - Fábregas 1992, nº 30; Martins 1988c, 126-7
Sample ID: ICEN-30
Radiocarbon Age BP: 2210 ± 35≈
Calibrated age (s) cal BC 354, 292, 257, 253, 232, 217, 214
cal BP 2303, 2241, 2206, 2202, 2181, 2166, 2163
cal AD/BC (cal BP) age ranges obtained from intercepts (Method A):
one Sigma** cal BC 377- 370 (2326-2319) 365- 346 (2314-2295) 322- 266 (2271-2215)
265- 226 (2214-2175) 225- 207 (2174-2156)
two Sigma** cal BC 391- 196 (2340-2145) 195- 172 (2144-2121)

Summary of above
minimum of cal age ranges (cal ages) maximum of cal age ranges:
one sigma cal BC 377 (354, 292, 257, 253, 232, 217, 214) 207
cal BP 2326 (2303, 2241, 2206, 2202, 2181, 2166, 2163) 2156
two sigma cal BC 391 (354, 292, 257, 253, 232, 217, 214) 172
cal BP 2340 (2303, 2241, 2206, 2202, 2181, 2166, 2163) 2121

cal AD/BC age ranges (cal ages as above) from probability distribution (Method B):

% area enclosed	cal BC (cal BP) age ranges		relative area under probability distribution
68.3 (one sigma)	cal BC	362- 348 (2311-2297)	.12
		320- 268 (2269-2217)	.43
		264- 227 (2213-2176)	.32
		223- 208 (2172-2157)	.13
95.4 (two sigma)	cal BC	387- 201 (2336-2150)	.97
		186- 175 (2135-2124)	.03

Site: Crastoeiro
Location: Campos, Mondim de Basto, Portugal
Type: Defended

Cultural phase: Iron Age
References: (unpublished -Dinis pers. comm.)
Sample ID: ICEN-45
Radiocarbon Age BP: 2210 ± 45≈

Calibrated age (s)	cal BC	354, 292, 257, 253, 232, 217, 214	
	cal BP	2303, 2241, 2206, 2202, 2181, 2166, 2163	

cal AD/BC (cal BP) age ranges obtained from intercepts (Method A):

one Sigma**	cal BC	382- 342 (2331-2291)	325- 205 (2274-2154)
two Sigma**	cal BC	394- 169 (2343-2118)	140- 126 (2089-2075)

Summary of above
minimum of cal age ranges (cal ages) maximum of cal age ranges:

one sigma	cal BC	382 (354, 292, 257, 253, 232, 217, 214) 205
	cal BP	2331 (2303, 2241, 2206, 2202, 2181, 2166, 2163) 2154
two sigma	cal BC	394 (354, 292, 257, 253, 232, 217, 214) 126
	cal BP	2343 (2303, 2241, 2206, 2202, 2181, 2166, 2163) 2075

cal AD/BC age ranges (cal ages as above) from probability distribution (Method B):

% area enclosed	cal BC (cal BP) age ranges		relative area under probability distribution
68.3 (one sigma)	cal BC		.01
		364- 347 (2313-2296)	.13
		322- 266 (2271-2215)	.42
		265- 226 (2214-2175)	.30
		224- 207 (2173-2156)	.13
95.4 (two sigma)	cal BC	391- 172 (2340-2121)	1.00

Site: Castro de Vermoim
Location: Vermoim, Vila Nova de Famalicão, Portugal
Type: Defended
Cultural phase: Iron Age
References: (unpublished)
Sample ID: GAK-11459
Radiocarbon Age BP: 2210 ± 100≈

Calibrated age (s)	cal BC	354, 292,257, 253, 232, 217, 214
	cal BP	2303, 2241, 2206, 2202, 2181, 2166,2163

cal AD/BC (cal BP) age ranges obtained from intercepts (Method A) :

one Sigma**	cal BC	400-170 (2349-2119)	141-124 (2090-2073)	
two Sigma**	cal BC	482-472 (2431-2421)	428-421 (2377-2370)	410-90 (2359-2039)
		83-66 (2032-2015)	64-37 (2013-1986)	31-21 (1980-1970)
		12-cal AD1 (1961-1949)		

Summary of above
minimum of cal age ranges (cal ages) maximum of cal age ranges:

one sigma	cal BC	400 (354, 292, 257, 253, 232, 217, 214) 124
	cal BP	2349 (2303, 2241, 2206, 2202, 2181, 2166, 2163) 2073
two sigma	cal BC	482 (354, 292, 257, 253, 232, 217, 214) cal AD 0
	cal BP	2431 (2303, 2241, 2206, 2202, 2181, 2166, 2163) 1950

cal AD/BC age ranges (cal ages as above) from probability distribution (Method B) :

% area enclosed	cal BC (cal BP) age ranges		relative area under probability distribution
68.3 (one sigma)	cal BC	390-170 (2339-2119)	1.00
95.4 (two sigma)	cal BC	.00	

Site: Castro de Borneiro
Location: Cabana, A Coruña, Galicia
Type: Defended
Cultural phase: Iron Age
References: Carballo - Fábregas 1992, nº 3
Sample ID: GRN-16755
Radiocarbon Age BP: 2185 ± 20≈

Calibrated age (s)	cal BC	349, 318, 309, 302, 228, 222, 209
	cal BP	2298, 2267, 2258, 2251, 2177, 2171, 2158

cal AD/BC (cal BP) age ranges obtained from intercepts (Method A) :

one Sigma**	cal BC	354-340 (2303-2289)	327-291 (2276-2240)	257-252 (2206-2201)
		232-217 (2181-2166)	214-204 (2163-2153)	
two Sigma**	cal BC	362-268 (2311-2217)	263-196 (2212-2145)	195-172 (2144-2121)

Summary of above

minimum of cal age ranges (cal ages) maximum of cal age ranges:

one sigma	cal BC	354 (349, 318, 309, 302, 228, 222, 209) 204
	cal BP	2303 (2298, 2267, 2258, 2251, 2177, 2171, 2158) 2153
two sigma	cal BC	362 (349, 318, 309, 302, 228, 222, 209) 172
	cal BP	2311 (2298, 2267, 2258, 2251, 2177, 2171, 2158) 2121

cal AD/BC age ranges (cal ages as above) from probability distribution (Method B) :

% area enclosed	cal BC (cal BP) age ranges		relative area under probability distribution
68.3 (one sigma)	cal BC	354-338 (2303-2287)	.17
		330-291 (2279-2240)	.44
			.01
		258-250 (2207-2199)	.08
		232-217 (2181-2166)	.17
		215-203 (2164-2152)	.13
95.4 (two sigma)	cal BC	360-269 (2309-2218)	.57
		263-239 (2212-2188)	.10
		235-197 (2184-2146)	.25
		194-173 (2143-2122)	.08

Site: Castro de Cortegada
Location: Silleda, Pontevedra, Galicia
Type: Defended
Cultural phase: Iron Age
References: Carballo Arceo 1989; Carballo - Fábregas 1992, nº 12
Sample ID: CSIC-784
Radiocarbon Age BP: 2180 ± 60≈

Calibrated age (s)	cal BC	348, 320, 306, 305, 227, 223, 208
	cal BP	2297, 2269, 2255, 2254, 2176, 2172, 2157

cal AD/BC (cal BP) age ranges obtained from intercepts (Method A):

one Sigma**	cal BC	363- 268 (2312-2217)	264- 170 (2213-2119)	138- 128 (2087-2077)
two Sigma**	cal BC	390- 90 (2339-2039)	77- 73 (2026-2022)	

Summary of above

minimum of cal age ranges (cal ages) maximum of cal age ranges:

one sigma	cal BC	363 (348, 320, 306, 305, 227, 223, 208) 128
	cal BP	2312 (2297, 2269, 2255, 2254, 2176, 2172, 2157) 2077
two sigma	cal BC	390 (348, 320, 306, 305, 227, 223, 208) 73
	cal BP	2339 (2297, 2269, 2255, 2254, 2176, 2172, 2157) 2022

cal AD/BC age ranges (cal ages as above) from probability distribution (Method B):

% area enclosed probability distribution	cal BC (cal BP) age ranges		relative area under
68.3 (onc sigma)	cal BC	359- 286 (2308-2235)	.45
		284- 271 (2233-2220)	.07
		261- 242 (2210-2191)	.11
		235- 172 (2184-2121)	.37
95.4 (two sigma)	cal BC	388- 107 (2337-2056)	1.00

Site: Crastoeiro
Location: Campos, Mondim de Basto, Portugal
Type: Defended
Cultural phase: Iron Age
References: (unpublished -Dinis pers. comm.)
Sample ID: LY-4936
Radiocarbon Age BP: 2175 ± 40≈

Calibrated age (s)	cal BC	347, 321, 227, 224, 208
	cal BP	2296, 2270, 2176, 2173, 2157

cal AD/BC (cal BP) age ranges obtained from intercepts (Method A):

one Sigma**	cal BC	356- 290 (2305-2239)	278- 274 (2227-2223)	259- 249 (2208-2198)
		233- 216 (2182-2165)	215- 196 (2164-2145)	194- 172 (2143-2121)
two Sigma**	cal BC	384- 158 (2333-2107)	149- 111 (2098-2060)	

Summary of above

minimum of cal age ranges (cal ages) maximum of cal age ranges:

one sigma	cal BC	356 (347, 321, 227, 224, 208) 172
	cal BP	2305 (2296, 2270, 2176, 2173, 2157) 2121
two sigma	cal BC	384 (347, 321, 227, 224, 208) 111
	cal BP	2333 (2296, 2270, 2176, 2173, 2157) 2060

cal AD/BC age ranges (cal ages as above) from probability distribution (Method B):

% area enclosed	cal BC (cal BP) age ranges		relative area under probability distribution
68.3 (one sigma)	cal BC	355- 290 (2304-2239)	.51
		277- 275 (2226-2224)	.02
		258- 249 (2207-2198)	.06
		233- 216 (2182-2165)	.13
		215- 197 (2164-2146)	.14
		193- 173 (2142-2122)	.14
95.4 (two sigma)	cal BC	381- 163 (2330-2112)	.94
		143- 119 (2092-2068)	.06

Site: Castro de Cortegada
Location: Silleda, Pontevedra, Galicia
Type: Defended
Cultural phase: Iron Age
References: Carballo Arceo 1989; Carballo - Fábregas 1992, nº 12
Sample ID: CSIC-786
Radiocarbon Age BP: 2170 ± 60≈

Calibrated age (s)	cal BC	346, 322, 226, 225, 207
	cal BP	2295, 2271, 2175, 2174, 2156

cal AD/BC (cal BP) age ranges obtained from intercepts (Method A):

one Sigma**	cal BC	360- 286 (2309-2235)	285- 270 (2234-2219)	
		262- 238 (2211-2187)	235- 168 (2184-2117)	141- 123 (2090-2072)
two Sigma**	cal BC	390- 90 (2339-2039)	78- 72 (2027-2021)	

Summary of above

minimum of cal age ranges (cal ages) maximum of cal age ranges:

one sigma	cal BC	360 (346, 322, 226, 225, 207) 123
	cal BP	2309 (2295, 2271, 2175, 2174, 2156) 2072
two sigma	cal BC	390 (346, 322, 226, 225, 207) 72
	cal BP	2339 (2295, 2271, 2175, 2174, 2156) 2021

cal AD/BC age ranges (cal ages as above) from probability distribution (Method B):

% area enclosed	cal BC (cal BP) age ranges		relative area under probability distribution
68.3 (one sigma)	cal BC	358- 288 (2307-2237)	.43
		282- 272 (2231-2221)	.05
		260- 245 (2209-2194)	.08
		234- 169 (2183-2118)	.38
		139- 128 (2088-2077)	.06
95.4 (two sigma)	cal BC	383- 96 (2332-2045)	1.00

Site: Castro Real
Location: Boimorto, A Coruña, Galicia
Type: Defended
Cultural phase: Iron Age
References: Carballo - Fábregas 1992, nº 6
Sample ID: CSIC-731
Radiocarbon Age BP: 2170 ± 50≈

Calibrated age (s)	cal BC	346, 322, 226, 225, 207
	cal BP	2295, 2271, 2175, 2174, 2156

cal AD/BC (cal BP) age ranges obtained from intercepts (Method A) :

one Sigma**	cal BC	357-288 (2306-2237)	280-273 (2229-2222)	259-246 (2208-2195)
		234-170 (2183-2119)	138-128 (2087-2077)	
two Sigma**	cal BC	390-90 (2339-2039)	76-74 (2025-2023)	

Summary of above

minimum of cal age ranges (cal ages) maximum of cal age ranges:

one sigma	cal BC	357 (346, 322, 226, 225, 207) 128
	cal BP	2306 (2295, 2271, 2175, 2174, 2156) 2077
two sigma	cal BC	390 (346, 322, 226, 225, 207) 74
	cal BP	2339 (2295, 2271, 2175, 2174, 2156) 2023

cal AD/BC age ranges (cal ages as above) from probability distribution (Method B) :

% area enclosed	cal BC (cal BP) age ranges		relative area under probability distribution
68.3 (one sigma)	cal BC	356-289 (2305-2238)	.46
		278-274 (2227-2223)	.03
		259-248 (2208-2197)	.06
		233-170 (2182-2119)	.41
		137-131 (2086-2080)	.04
95.4 (two sigma)	cal BC	380-111 (2329-2060)	1.00

Site: Santo Estevão
Location: Facha, Ponte de Lima, Portugal
Type: Defended
Cultural phase: Iron Age
References: Almeida *et al.* 1982, 79; Carballo - Fábregas 1992, nº 28
Sample ID: CSIC-499
Radiocarbon Age BP: 2160 ± 50≈

Calibrated age (s)	cal BC	341, 325, 204
	cal BP	2290, 2274, 2153

cal AD/BC (cal BP) age ranges obtained from intercepts (Method A):

one Sigma**	cal BC	355-291 (2304-2240)	257-251 (2206-2200)	232-216 (2181-2165)
		215-168 (2164-2117)	141-123 (2090-2072)	
two Sigma**	cal BC	390-90 (2339-2039)	78-73 (2027-2022)	

Summary of above

minimum of cal age ranges (cal ages) maximum of cal age ranges:

one sigma	cal BC	355 (341, 325, 204) 123
	cal BP	2304 (2290, 2274, 2153) 2072
two sigma	cal BC	390 (341, 325, 204) 73
	cal BP	2339 (2290, 2274, 2153) 2022

cal AD/BC age ranges (cal ages as above) from probability distribution (Method B):

% area enclosed	cal BC (cal BP) age ranges		relative area under probability distribution
68.3 (one sigma)	cal BC	355- 291 (2304-2240)	.43
		258- 250 (2207-2199)	.04
		232- 217 (2181-2166)	.10
		215- 167 (2164-2116)	.32
		141- 123 (2090-2072)	.10
95.4 (two sigma)	cal BC	364-94 (2313-2043)	.99

Site: Castro de Troña
Location: Ponteareas, Pontevedra, Galicia
Type: Defended
Cultural phase: Iron Age
References: Carballo - Fábregas 1992, nº 21; Eiroa 1988, 118; Hidalgo Cuñarro 1985a, 7; 1987, 35; 1988-89, 90.
Sample ID: CSIC-602
Radiocarbon Age BP: 2160 ± 50≈

Calibrated age (s)	cal BC	341, 325, 204
	cal BP	2290, 2274, 2153

cal AD/BC (cal BP) age ranges obtained from intercepts (Method A):

one Sigma**	cal BC	355- 291 (2304-2240)	257- 251 (2206-2200)	232- 216 (2181-2165)
		215- 168 (2164-2117)	141- 123 (2090-2072)	
two Sigma**	cal BC	390- 90 (2339-2039)	78- 73 (2027-2022)	

Summary of above
minimum of cal age ranges (cal ages) maximum of cal age ranges:

one sigma	cal BC	355 (341, 325, 204) 123
	cal BP	2304 (2290, 2274, 2153) 2072
two sigma	cal BC	390 (341, 325, 204) 73
	cal BP	2339 (2290, 2274, 2153) 2022

cal AD/BC age ranges (cal ages as above) from probability distribution (Method B):

% area enclosed	cal BC (cal BP) age ranges		relative area under probability distribution
68.3 (one sigma)	cal BC	355- 291 (2304-2240)	.43
		258- 250 (2207-2199)	.04
		232- 217 (2181-2166)	.10
		215- 167 (2164-2116)	.32
		141- 123 (2090-2072)	.10
95.4 (two sigma)	cal BC	364- 94 (2313-2043)	.99

Site: Citânia de S. Julião
Location: Ponte, Vila Verde, Portugal
Type: Defended
Cultural phase: Bronze Age-Iron Age
References: Carballo - Fábregas 1992, nº 30; Martins 1988c, 126
Sample ID: ICEN-29
Radiocarbon Age BP: 2150 ± 45≈

Calibrated age (s)	cal BC	335, 202
	cal BP	2284, 2151

cal AD/BC (cal BP) age ranges obtained from intercepts (Method A):

one Sigma**	cal BC	352-295 (2301-2244)	230-219 (2179-2168)	212-166 (2161-2115)
		143-120 (2092-2069)		
two Sigma**	cal BC	377-369 (2326-2318)	365-266 (2314-2215)	270-90 (2219-2039)
		78-72 (2027-2021)		

Summary of above
minimum of cal age ranges (cal ages) maximum of cal age ranges:

one sigma	cal BC	352 (335, 202) 120
	cal BP	2301 (2284, 2151) 2069
two sigma	cal BC	377 (335, 202) 72
	cal BP	2326 (2284, 2151) 2021

cal AD/BC age ranges (cal ages as above) from probability distribution (Method B):

% area enclosed	cal BC (cal BP) age ranges		relative area under probability distribution
68.3 (one sigma)	cal BC	352- 294 (2301-2243)	.41
		230- 219 (2179-2168)	.08
		212- 166 (2161-2115)	.36
		143- 119 (2092-2068)	.15
95.4 (two sigma)	cal BC	360- 269 (2309-2218)	.37
		263- 237 (2212-2186)	.06
		236- 94 (2185-2043)	.57

Site: Castro de Montaz
Location: Oleiros, Silleda, Pontevedra, Galicia
Type: Defended
Cultural phase: Iron Age
References: Carballo - Fábregas 1992, nº 13
Sample ID: CSIC-789
Radiocarbon Age BP: 2120 ± 50≈

Calibrated age (s)	cal BC	170, 138, 130
	cal BP	2119, 2087, 2079

cal AD/BC (cal BP) age ranges obtained from intercepts (Method A) :

one Sigma**	cal BC	347-322 (2296-2271)	226-224 (2175-2173)	207-94 (2156-2043)
two Sigma**	cal BC	358-287 (2307-2236)	282-272 (2231-2221)	260-244 (2209-2193)
		230-90 (2179-2039)	82-67 (2031-2016)	62-39 (2011-1988)
		27-24 (1976-1973)	10-2 (1959-1951)	

Summary of above

minimum of cal age ranges (cal ages) maximum of cal age ranges:

one sigma	cal BC	347 (170, 138, 130) 94
	cal BP	2296 (2119, 2087, 2079) 2043
two sigma	cal BC	358 (170, 138, 130) 2
	cal BP	2307 (2119, 2087, 2079) 1951

cal AD/BC age ranges (cal ages as above) from probability distribution (Method B) :

% area enclosed	cal BC (cal BP) age ranges		relative area under probability distribution
68.3 (one sigma)	cal BC	346-322 (2295-2271)	.16
			.01
		207-94 (2156-2043)	.83
95.4 (two sigma)	cal BC	358-286 (2307-2235)	.22
		284-271 (2233-2220)	.01
		262-243 (2211-2192)	.02
		234-88 (2183-2037)	.67
		82-67 (2031-2016)	.03
		62-40 (2011-1989)	.03

Site: Castro de Fozara
Location: Ponteareas, Pontevedra, Galicia
Type: Defended
Cultural phase: Iron Age
References: Carballo - Fábregas 1992, nº 20; Hidalgo - Rodríguez 1988, 137
Sample ID: CSIC-693
Radiocarbon Age BP: 2120 ± 50≈

Calibrated age (s)	cal BC	170, 138, 130
	cal BP	2119, 2087, 2079

cal AD/BC (cal BP) age ranges obtained from intercepts (Method A) :

one Sigma**	cal BC	347-322 (2296-2271)	226-224 (2175-2173)	207-94 (2156-2043)
two Sigma**	cal BC	358-287 (2307-2236)	282-272 (2231-2221)	260-244 (2209-2193)
		230-90 (2179-2039)	82-67 (2031-2016)	62-39 (2011-1988)
		27-24 (1976-1973)	10-2 (1959-1951)	

Summary of above

minimum of cal age ranges (cal ages) maximum of cal age ranges:

one sigma	cal BC	347 (170, 138, 130) 94
	cal BP	2296 (2119, 2087, 2079) 2043
two sigma	cal BC	358 (170, 138, 130) 2
	cal BP	2307 (2119, 2087, 2079) 1951

cal AD/BC age ranges (cal ages as above) from probability distribution (Method B) :

% area enclosed	cal BC (cal BP) age ranges		relative area under probability distribution
68.3 (one sigma)	cal BC	346-322 (2295-2271)	.16
			.01
		207-94 (2156-2043)	.83
95.4 (two sigma)	cal BC	358-286 (2307-2235)	.22
		284-271 (2233-2220)	.01
		262-243 (2211-2192)	.02
		234-88 (2183-2037)	.67
		82-67 (2031-2016)	.03
		62-40 (2011-1989)	.03

Site: Castro de Fozara
Location: Ponteareas, Pontevedra, Galicia
Type: Defended
Cultural phase: Iron Age

References: Carballo - Fábregas 1992, nº 20; Hidalgo - Rodríguez 1988, 137
Sample ID: CSIC-691
Radiocarbon Age BP: 2110 ± 50≈
Calibrated age (s) cal BC 168, 140, 125
cal BP 2117, 2089, 2074
cal AD/BC (cal BP) age ranges obtained from intercepts (Method A) :

one Sigma**	cal BC	343-324 (2292-2273)	205-93 (2154-2042)	77-73 (2026-2022)
two Sigma**	cal BC	355-290 (2304-2239)	277-275 (2226-2224)	258-250 (2207-2199)
		233-216 (2182-2165)	210-90 (2159-2039)	83-66 (2032-2015)
		64-37 (2013-1986)	32-20 (1981-1969)	12-cal AD 1 (1961-1949)

Summary of above
minimum of cal age ranges (cal ages) maximum of cal age ranges:

one sigma	cal BC	343 (168, 140, 125) 73
	cal BP	2292 (2117, 2089, 2074) 2022
two sigma	cal BC	355 (168, 140, 125) cal AD 0
	cal BP	2304 (2117, 2089, 2074) 1950

cal AD/BC age ranges (cal ages as above) from probability distribution (Method B) :

% area enclosed	cal BC (cal BP) age ranges		relative area under probability distribution
68.3 (one sigma)	cal BC	344-324 (2293-2273)	.12
		205-93 (2154-2042)	.85
		77-73 (2026-2022)	.03
95.4 (two sigma)	cal BC	356-289 (2305-2238)	.18
		260-247 (2209-2196)	.01
		233-87 (2182-2036)	.68
		84-66 (2033-2015)	.04
		64-37 (2013-1986)	.05
		32-20 (1981-1969)	.01
		12-cal AD 1 (1961-1949)	.02

Site: Castro de Fozara
Location: Ponteareas, Pontevedra, Galicia
Type: Defended
Cultural phase: Iron Age
References: Carballo - Fábregas 1992, nº 20; Hidalgo - Rodríguez 1988, 137
Sample ID: CSIC-692
Radiocarbon Age BP: 2110 ± 50≈
Calibrated age (s) cal BC 168, 140, 125
cal BP 2117, 2089, 2074
cal AD/BC (cal BP) age ranges obtained from intercepts (Method A) :

one Sigma**	cal BC	343-324 (2292-2273)	205-93 (2154-2042)	77-73 (2026-2022)
two Sigma**	cal BC	355-290 (2304-2239)	277-275 (2226-2224)	258-250 (2207-2199)
		233-216 (2182-2165)	210-90 (2159-2039)	83-66 (2032-2015)
		64-37 (2013-1986)	32-20 (1981-1969)	12-cal AD 1 (1961-1949)

Summary of above
minimum of cal age ranges (cal ages) maximum of cal age ranges:

one sigma	cal BC	343 (168, 140, 125) 73
	cal BP	2292 (2117, 2089, 2074) 2022
two sigma	cal BC	355 (168, 140, 125) cal AD 0
	cal BP	2304 (2117, 2089, 2074) 1950

cal AD/BC age ranges (cal ages as above) from probability distribution (Method B) :

% area enclosed	cal BC (cal BP) age ranges		relative area under probability distribution
68.3 (one sigma)	cal BC	344-324 (2293-2273)	.12
		205-93 (2154-2042)	.85
		77-73 (2026-2022)	.03
95.4 (two sigma)	cal BC	356-289 (2305-2238)	.18
		260-247 (2209-2196)	.01
		233-87 (2182-2036)	.68
		84-66 (2033-2015)	.04
		64-37 (2013-1986)	.05

	32-20 (1981-1969)	.01
	12-cal AD 1 (1961-1949)	.02

Site: Castro de Borneiro
Location: Cabana, A Coruña, Galicia
Type: Defended
Cultural phase: Iron Age
References: Carballo - Fábregas 1992 nº3
Sample ID: GRN-15890
Radiocarbon Age BP: 2095 ± 20≈

Calibrated age (s)	cal BC	166, 144, 117		
	cal BP	2115, 2093, 2066		

cal AD/BC (cal BP) age ranges obtained from intercepts (Method A) :

one Sigma**	cal BC	170-138 (2119-2087)	130-94 (2079-2043)	
two Sigma**	cal BC	200-186 (2149-2135)	175-91 (2124-2040)	79-71 (2028-2020)

Summary of above
minimum of cal age ranges (cal ages) maximum of cal age ranges:

one sigma	cal BC	170 (166, 144, 117) 94
	cal BP	2119 (2115, 2093, 2066) 2043
two sigma	cal BC	200 (166, 144, 117) 71
	cal BP	2149 (2115, 2093, 2066) 2020

cal AD/BC age ranges (cal ages as above) from probability distribution (Method B) :

% area enclosed	cal BC (cal BP) age ranges		relative area under probability distribution
68.3 (one sigma)	cal BC	169-140 (2118-2089)	.49
		126-95 (2075-2044)	.51
95.4 (two sigma)	cal BC	202-91 (2151-2040)	.96
		79-71 (2028-2020)	.03

Site: Castro de Cortegada
Location: Silleda, Pontevedra, Galicia
Type: Defended
Cultural phase: Iron Age
References: Carballo - Fábregas 1992, nº 12
Sample ID: CSIC-782
Radiocarbon Age BP: 2090 ± 50≈

Calibrated age (s)	cal BC	161, 145, 114
	cal BP	2110, 2094, 2063

cal AD/BC (cal BP) age ranges obtained from intercepts (Method A) :

one Sigma**	cal BC	200-188 (2149-2137)	175-90 (2124-2039)
		79-70 (2028-2019)	56-48 (2005-1997)
two Sigma**	cal BC	351-296 (2300-2245)	230-220 (2179-2169)
		210-cal AD 1 (2159-1949)	9-21 (1941-1929)

Summary of above
minimum of cal age ranges (cal ages) maximum of cal age ranges:

one sigma	cal BC	200 (161, 145, 114) 48
	cal BP	2149 (2110, 2094, 2063) 1997
two sigma	cal BC	351 (161, 145, 114) cal AD 21
	cal BP	2300 (2110, 2094, 2063) 1929

cal AD/BC age ranges (cal ages as above) from probability distribution (Method B) :

% area enclosed	cal BC (cal BP) age ranges		relative area under probability distribution
68.3 (one sigma)	cal BC	202-90 (2151-2039)	.84
		80-70 (2029-2019)	.07
95.4 (two sigma)	cal BC	351-295 (2300-2244)	.11
		230-220 (2179-2169)	.01
		212-cal AD 3 (2161-1947)	.86
		10-21 (1940-1929)	.01

Site: Castrovite
Location: A Estrada, Pontevedra, Galicia
Type: Defended
Cultural phase: Iron Age
References: Carballo - Fábregas 1992, nº 9
Sample ID: CSIC-817
Radiocarbon Age BP: 2070 ± 45≈
Calibrated age (s) cal BC 94, 75
cal BP 2043, 2024
cal AD/BC (cal BP) age ranges obtained from intercepts (Method A) :
one Sigma** cal BC 170-138 (2119-2087) 129-89 (2078-2038) 81-68 (2030-2017)
60-42 (2009-1991) 8-3 (1957-1952)
two Sigma** cal BC 344-324 (2293-2273) 206-cal AD 25 (2155-1925) 43-46 (1907-1904)

Summary of above
minimum of cal age ranges (cal ages) maximum of cal age ranges:
one sigma cal BC 170 (94, 75) 3
cal BP 2119 (2043, 2024) 1952
two sigma cal BC 344 (94, 75) cal AD 46
cal BP 2293 (2043, 2024) 1904

cal AD/BC age ranges (cal ages as above) from probability distribution (Method B) :

% area enclosed	cal BC (cal BP) age ranges		relative area under probability distribution
68.3 (one sigma)	cal BC	172-136 (2121-2085)	.29
		133-88 (2082-2037)	.37
		82-67 (2031-2016)	.12
		61-40 (2010-1989)	.15
			.01
		10-2 (1959-1951)	.05
95.4 (two sigma)	cal BC	345-323 (2294-2272)	.03
		206-cal AD 25 (2155-1925)	.96

Site: Castro de Mohías
Location: Coaña, Asturias
Type: Defended
Cultural phase: Iron Age
References: Almagro *et al.* 1978, 169; Carballo - Fábregas 1992, nº2; Eiroa 1980, 76-7; 1988, 118
Sample ID: CSIC-110
Radiocarbon Age BP: 2050 ± 80≈
Calibrated age (s) cal BC 92, 78, 72
cal BP 2041, 2027, 2021
cal AD/BC (cal BP) age ranges obtained from intercepts (Method A):
one Sigma** cal BC 196- 194 (2145-2143) 172-cal AD 27 (2121-1923) 42- 48 (1908-1902)
two Sigma** cal BC 355- 291 (2304-2240) 257- 251 (2206-2200) 232- 217 (2181-2166)
210-cal AD 80 (2159-1870) 109- 126 (1841-1824)

Summary of above
minimum of cal age ranges (cal ages) maximum of cal age ranges:
one sigma cal BC 196 (92, 78, 72) cal AD 48
cal BP 2145 (2041, 2027, 2021) 1902
two sigma cal BC 355 (92, 78, 72) cal AD 126
cal BP 2304 (2041, 2027, 2021) 1824

cal AD/BC age ranges (cal ages as above) from probability distribution (Method B):

% area enclosed	cal BC (cal BP) age ranges		relative area under probability distribution
68.3 (one sigma)	cal BC	172- cal AD 26 (2121-1924)	.98
		43-47 (1907-1903)	.02
95.4 (two sigma)	cal BC	355- 291 (2304-2240)	.08
		232- 217 (2181-2166)	.02
		215- cal AD 82 (2164-1868)	.88
		93- 126 (1857-1824)	.02

Site: Castro do Marco
Location: Silleda, Pontevedra, Galicia
Type: Defended
Cultural phase: Iron Age
References: Carballo - Fábregas 1992, nº 10
Sample ID: CSIC-790
Radiocarbon Age BP: 2050 ± 50≈

Calibrated age (s)	cal BC	92, 78, 72		
	cal BP	2041, 2027, 2021		

cal AD/BC (cal BP) age ranges obtained from intercepts (Method A) :

one Sigma**	cal BC	167-142 (2116-2091)	121-86 (2070-2035)	84-18 (2033-1967)
		14-cal AD 1 (1963-1949)		
two Sigma**	cal BC	338-330 (2287-2279)	200-cal AD 30 (2149-1920)	37-54 (1913-1896)

Summary of above
minimum of cal age ranges (cal ages) maximum of cal age ranges:

one sigma	cal BC	167 (92, 78, 72) cal AD 1
	cal BP	2116 (2041, 2027, 2021) 1949
two sigma	cal BC	338 (92, 78, 72) cal AD 54
	cal BP	2287 (2041, 2027, 2021) 1896

cal AD/BC age ranges (cal ages as above) from probability distribution (Method B) :

% area enclosed	cal BC (cal BP) age ranges		relative area under probability distribution
68.3 (one sigma)	cal BC	167-142 (2116-2091)	.17
		120-86 (2069-2035)	.25
		84-36 (2033-1985)	.38
		34-19 (1983-1968)	.10
		13-cal AD 1 (1962-1949)	.10
95.4 (two sigma)	cal BC	339-328 (2288-2277)	.01
		203-cal AD 32 (2152-1918)	.96
		37-53 (1913-1897)	.03

Site: Castro de Cartimil
Location: Silleda, Pontevedra, Galicia
Type: Defended
Cultural phase: Iron Age
References: Carballo - Fábregas 1992, nº 11
Sample ID: CSIC-792
Radiocarbon Age BP: 2040 ± 40≈

Calibrated age (s)	cal BC	91, 79, 71		
	cal BP	2040, 2028, 2020		

cal AD/BC (cal BP) age ranges obtained from intercepts (Method A) :

one Sigma**	cal BC	105-86 (2054-2035)	84-18 (2033-1967)	14-cal AD 1 (1963-1949)
two Sigma**	cal BC	171-136 (2120-2085)	133-cal AD 30 (2082-1920)	39-52 (1911-1898)

Summary of above
minimum of cal age ranges (cal ages) maximum of cal age ranges:

one sigma	cal BC	105 (91, 79, 71) cal AD 1
	cal BP	2054 (2040, 2028, 2020) 1949
two sigma	cal BC	171 (91, 79, 71) cal AD 52
	cal BP	2120 (2040, 2028, 2020) 1898

cal AD/BC age ranges (cal ages as above) from probability distribution (Method B) :

% area enclosed	cal BC (cal BP) age ranges		relative area under probability distribution
68.3 (one sigma)	cal BC	158-150 (2107-2099)	.06
		108-86 (2057-2035)	.17
		85-17 (2034-1966)	.63
		14-cal AD 1 (1963-1949)	.14
95.4 (two sigma)	cal BC	172-136 (2121-2085)	.13
		133-cal AD 31 (2082-1919)	.85
		39-51 (1911-1899)	.02

Site:	Castro de Cameixa			
Location:	Boborás, Ourense, Galicia			
Type:	Defended			
Cultural phase:	Iron Age			
References:	Carballo - Fábregas 1992, nº 17; Fariña Busto 1991, 53			
Sample ID:	CSIC-743			
Radiocarbon Age BP:	2030 ± 40≈			
Calibrated age (s)	cal BC	90, 80, 70, 58, 44, 6, 4		
	cal BP	2039, 2029, 2019, 2007, 1993, 1955, 1953		

cal AD/BC (cal BP) age ranges obtained from intercepts (Method A) :

one Sigma**	cal BC	94-cal AD 3 (2043-1947)	11-19 (1939-1931)	
two Sigma**	cal BC	169-139 (2118-2088)	127-cal AD32 (2076-1918)	37-54 (1913-1896)

Summary of above
minimum of cal age ranges (cal ages) maximum of cal age ranges:

one sigma	cal BC	94 (90, 80, 70, 58, 44, 6, 4) cal AD 19
	cal BP	2043 (2039, 2029, 2019, 2007, 1993, 1955, 1953) 1931
two sigma	cal BC	169 (90, 80, 70, 58, 44, 6, 4) cal AD 54
	cal BP	2118 (2039, 2029, 2019, 2007, 1993, 1955, 1953) 1896

cal AD/BC age ranges (cal ages as above) from probability distribution (Method B) :

% area enclosed	cal BC (cal BP) age ranges		relative area under probability distribution
68.3 (one sigma)	cal BC	94-cal AD 3 (2043-1947)	.93
		10-20 (1940-1930)	.07
95.4 (two sigma)	cal BC	170-139 (2119-2088)	.09
		127-cal AD 33 (2076-1917)	.87
		37-54 (1913-1896)	.04

Site:	Castro de Troña			
Location:	Ponteareas, Pontevedra, Galicia			
Type:	Defended			
Cultural phase:	Iron Age			
References:	Carballo - Fábregas 1992, nº 21; Eiroa 1988, 118; Hidalgo Cuñarro 1988-89, 90			
Sample ID:	CSIC-719			
Radiocarbon Age BP:	2030 ± 50≈			
Calibrated age (s)	cal BC	90, 80, 70, 58, 44, 6, 4		
	cal BP	2039, 2029, 2019, 2007, 1993, 1955, 1953		

cal AD/BC (cal BP) age ranges obtained from intercepts (Method A):

one Sigma**	cal BC	95- cal AD 4 (2044-1946)	6-24 (1944-1926)	44-46 (1906-1904)
two Sigma**	cal BC	197- 193 (2146-2142)	170- cal AD 70 (2119-1880)	

Summary of above
minimum of cal age ranges (cal ages) maximum of cal age ranges:

one sigma	cal BC	95 (90, 80, 70, 58, 44, 6, 4) cal AD 46
	cal BP	2044 (2039, 2029, 2019, 2007, 1993, 1955, 1953) 1904
two sigma	cal BC	197 (90, 80, 70, 58, 44, 6, 4) cal AD 70
	cal BP	2146 (2039, 2029, 2019, 2007, 1993, 1955, 1953) 1880

cal AD/BC age ranges (cal ages as above) from probability distribution (Method B):

% area enclosed	cal BC (cal BP) age ranges		relative area under probability distribution
68.3 (one sigma)	cal BC	105-cal AD 5 (2054-1945)	.86
		6-25 (1944-1925)	.12
		44-46 (1906-1904)	.02
95.4 (two sigma)	cal BC	174-cal AD 68 (2123-1882)	.99

Site:	Castro Real
Location:	Boimorto, A Coruña, Galicia
Type:	Defended
Cultural phase:	Iron Age
References:	Carballo - Fábregas 1992, nº 6
Sample ID:	CSIC-729
Radiocarbon Age BP:	2030 ± 50≈

Calibrated age (s)	cal BC	90, 80, 70, 58, 44, 6, 4
	cal BP	2039, 2029, 2019, 2007, 1993, 1955, 1953

cal AD/BC (cal BP) age ranges obtained from intercepts (Method A) :

one Sigma**	cal BC	95-cal AD 4 (2044-1946)	6-24 (1944-1926)	44-46 (1906-1904)
two Sigma**	cal BC	197-193 (2146-2142)	170-cal AD 70 (2119-1880)	

Summary of above

minimum of cal age ranges (cal ages) maximum of cal age ranges:

one sigma	cal BC	95 (90, 80, 70, 58, 44, 6, 4) cal AD 46
	cal BP	2044 (2039, 2029, 2019, 2007, 1993, 1955, 1953) 1904
two sigma	cal BC	197 (90, 80, 70, 58, 44, 6, 4) cal AD 70
	cal BP	2146 (2039, 2029, 2019, 2007, 1993, 1955, 1953) 1880

cal AD/BC age ranges (cal ages as above) from probability distribution Method B) :

% area enclosed	cal BC (cal BP) age ranges		relative area under probability distribution
68.3 (one sigma)	cal BC	105-cal AD 5 (2054-1945)	.86
		6-25 (1944-1925)	.12
		44-46 (1906-1904)	.02
95.4 (two sigma)	cal BC	174-cal AD 68 (2123-1882)	.99

Site: Castro do Marco
Location: Silleda, Pontevedra, Galicia
Type: Defended
Cultural phase: Iron Age
References: Carballo - Fábregas 1992, nº 10
Sample ID: CSIC-791
Radiocarbon Age BP: 2030 ± 50≈

Calibrated age (s)	cal BC	90, 80, 70, 58, 44, 6, 4
	cal BP	2039, 2029, 2019, 2007, 1993, 1955, 1953

cal AD/BC (cal BP) age ranges obtained from intercepts (Method A) :

one Sigma**	cal BC	95-cal AD 4 (2044-1946)	6-24 (1944-1926)	44-46 (1906-1904)
two Sigma**	cal BC	197-193 (2146-2142)	170-cal AD 70 (2119-1880)	

Summary of above

minimum of cal age ranges (cal ages) maximum of cal age ranges:

one sigma	cal BC	95 (90, 80, 70, 58, 44, 6, 4) cal AD 46
	cal BP	2044 (2039, 2029, 2019, 2007, 1993, 1955, 1953) 1904
two sigma	cal BC	197 (90, 80, 70, 58, 44, 6, 4) cal AD 70
	cal BP	2146 (2039, 2029, 2019, 2007, 1993, 1955, 1953) 1880

cal AD/BC age ranges (cal ages as above) from probability distribution (Method B) :

% area enclosed	cal BC (cal BP) age ranges		relative area under probability distribution
68.3 (one sigma)	cal BC	105-cal AD 5 (2054-1945)	.86
		6-25 (1944-1925)	.12
		44-46 (1906-1904)	.02
95.4 (two sigma)	cal BC	174-cal AD 68 (2123-1882)	.99

Site: Corona de Corporales
Location: Truchas, León
Type: Defended
Cultural phase: Iron Age
References: Carballo - Fábregas 1992, nº 23; Sánchez - Fernández 1985, 277
Sample ID: CSIC-429
Radiocarbon Age BP: 2020 ± 60≈

Calibrated age (s)	cal BC	89, 82, 68, 60, 41, 9, 3
	cal BP	2038, 2031, 2017, 2009, 1990, 1958, 952

cal AD/BC (cal BP) age ranges obtained from intercepts (Method A) :

one Sigma**	cal BC	95-cal AD 29 (2044-1921)	40-51 (1910-1899)
two Sigma**	cal BC	200-188 (2149-2137)	170-cal AD 80 (2119-1870)

Summary of above

minimum of cal age ranges (cal ages) maximum of cal age ranges:

one sigma	cal BC	95 (89, 82, 68, 60, 41, 9, 3) cal AD 51
	cal BP	2044 (2038, 2031, 2017, 2009, 1990, 1958, 1952) 1899
two sigma	cal BC	200 (89, 82, 68, 60, 41, 9, 3) cal AD 80
	cal BP	2149 (2038, 2031, 2017, 2009, 1990, 1958, 1952) 1870

cal AD/BC age ranges (cal ages as above) from probability distribution (Method B) :

% area enclosed	cal BC (cal BP) age ranges		relative area under probability distribution
68.3 (one sigma)	cal BC	106-cal AD 30 (2055-1920)	.94
		40-51 (1910-1899)	.06
95.4 (two sigma)	cal BC	202-cal AD 81 (2151-1869)	.98
		106-126 (1844-1824)	.02

Site: Castro de Vixil
Location: Boizán, Vilalba, Lugo, Galicia
Type: Defended
Cultural phase: Iron Age
References: Aira - Guitián 1985-86, 193 and Fariña Busto 1991, 52 (with incorrect values); Carballo - Fábregas 1992, nº 4
Sample ID: CSIC-610
Radiocarbon Age BP: 2020 ± 50≈

Calibrated age (s)	cal BC	89, 82, 68, 60, 41, 9, 3
	cal BP	2038, 2031, 2017, 2009, 1990, 1958, 1952

cal AD/BC (cal BP) age ranges obtained from intercepts (Method A) :

one Sigma**	cal BC	94-cal AD 27 (2043-1923)	42-48 (1908-1902)
two Sigma**	cal BC	171-137 (2120-2086)	130-cal AD 70 (2079-1880)

Summary of above
minimum of cal age ranges (cal ages) maximum of cal age ranges:

one sigma	cal BC	94 (89, 82, 68, 60, 41, 9, 3) cal AD 48
	cal BP	2043 (2038, 2031, 2017, 2009, 1990, 1958, 1952) 1902
two sigma	cal BC	171 (89, 82, 68, 60, 41, 9, 3) cal AD 70
	cal BP	2120 (2038, 2031, 2017, 2009, 1990, 1958, 1952) 1880

cal AD/BC age ranges (cal ages as above) from probability distribution (Method B) :

% area enclosed	cal BC (cal BP) age ranges		relative area under probability distribution
68.3 (one sigma)	cal BC	94-cal AD 26 (2043-1924)	.97
		43-47 (1907-1903)	.03
95.4 (two sigma)	cal BC	172-136 (2121-2085)	.08
		134-cal AD 73 (2083-1877)	.91

Site: Castro de Vixil
Location: Boizán, Vilalba, Lugo, Galicia
Type: Defended
Cultural phase: Iron Age
References: Aira y Guitián 1985-86, 193 (with incorrect values and ID); Carballo - Fábregas 1992, nº4
Sample ID: CSIC-611
Radiocarbon Age BP: 2020 ± 50≈

Calibrated age (s)	cal BC	89, 82, 68, 60, 41, 9, 3
	cal BP	2038, 2031, 2017, 2009, 1990, 1958, 1952

cal AD/BC (cal BP) age ranges obtained from intercepts (Method A) :

one Sigma**	cal BC	94-cal AD 27 (2043-1923)	42-48 (1908-1902)
two Sigma**	cal BC	171-137 (2120-2086)	130-cal AD 70 (2079-1880)

Summary of above
minimum of cal age ranges (cal ages) maximum of cal age ranges:

one sigma	cal BC	94 (89, 82, 68, 60, 41, 9, 3) cal AD 48
	cal BP	2043 (2038, 2031, 2017, 2009, 1990, 1958, 1952) 1902
two sigma	cal BC	171 (89, 82, 68, 60, 41, 9, 3) cal AD 70
	cal BP	2120 (2038, 2031, 2017, 2009, 1990, 1958, 1952) 1880

cal AD/BC age ranges (cal ages as above) from probability distribution (Method B) :

% area enclosed	cal BC (cal BP) age ranges	relative area under

			probability distribution
68.3 (one sigma)	cal BC	94-cal AD 26 (2043-1924)	.97
		43-47 (1907-1903)	.03
95.4 (two sigma)	cal BC	172-136 (2121-2085)	.08
		134-cal AD 73 (2083-1877)	.91

Site: Corona de Corporales
Location: Truchas, León
Type: Defended
Cultural phase: Iron Age
References: Carballo - Fábregas 1992, nº 23; Sánchez - Fernández 1985, 277
Sample ID: CSIC-428
Radiocarbon Age BP: 2010 ± 60≈
Calibrated age (s) cal BC 88, 83, 67, 63, 38, 29, 23, 11, 1
cal BP 2037, 2032, 2016, 2012, 1987, 1978, 1972, 1960, 1950
cal AD/BC (cal BP) age ranges obtained from intercepts (Method A) :
one Sigma** cal BC 94-cal AD 32 (2043-1918) 38-53 (1912-1897)
two Sigma** cal BC 197-193 (2146-2142) 170-cal AD 80 (2119-1870) 109-126 (1841-1824)

Summary of above
minimum of cal age ranges (cal ages) maximum of cal age ranges:
one sigma cal BC 94 (88, 83, 67, 63, 38, 29, 23, 11, 1) cal AD 53
cal BP 2043 (2037, 2032, 2016, 2012, 1987, 1978, 1972, 1960, 1950) 1897
two sigma cal BC 197 (88, 83, 67, 63, 38, 29, 23, 11, 1) cal AD 126
cal BP 2146 (2037, 2032, 2016, 2012, 1987, 1978, 1972, 1960, 1950) 1824

cal AD/BC age ranges (cal ages as above) from probability distribution (Method B) :

% area enclosed	cal BC (cal BP) age ranges		relative area under probability distribution
68.3 (one sigma)	cal BC	94-cal AD 32 (2043-1918)	.91
		38-53 (1912-1897)	.09
95.4 (two sigma)	cal BC	173-cal AD 83 (2122-1867)	.96
		90-126 (1860-1824)	.03

Site: Castro de Labradas
Location: Arrabalde, Zamora
Type: Defended
Cultural phase: Iron Age
References: Carballo - Fábregas 1992, nº 27; Esparza Arroyo 1986, 401
Sample ID: I-12855
Radiocarbon Age BP: 2010 ± 90≈
Calibrated age (s) cal BC 88, 83, 67, 63, 38, 29, 23, 11, 1
cal BP 2037, 2032, 2016, 2012, 1987, 1978, 1972, 1960, 1950
cal AD/BC (cal BP) age ranges obtained from intercepts (Method A) :
one Sigma** cal BC 167-142 (2116-2091) 120-cal AD 69 (2069-1881)
two Sigma** cal BC 350-316 (2299-2265) 314-299 (2263-2248) 229-221 (2178-2170)
210-cal AD 130 (2159-1820) 149-171 (1801-1779) 200-212 (1750-1738)

Summary of above
minimum of cal age ranges (cal ages) maximum of cal age ranges:
one sigma cal BC 167 (88, 83, 67, 63, 38, 29, 23, 11, 1) cal AD 69
cal BP 2116 (2037, 2032, 2016, 2012, 1987, 1978, 1972, 1960, 1950) 1881
two sigma cal BC 350 (88, 83, 67, 63, 38, 29, 23, 11, 1) cal AD 212
cal BP 2299 (2037, 2032, 2016, 2012, 1987, 1978, 1972, 1960, 1950) 1738

cal AD/BC age ranges (cal ages as above) from probability distribution (Method B) :

% area enclosed	cal BC (cal BP) age ranges		relative area under probability distribution
68.3 (one sigma)	cal BC	167-142 (2116-2091)	.09
		121-cal AD 70 (2070-1880)	.91
95.4 (two sigma)	cal BC	351-296 (2300-2245)	.04
		212-cal AD 175 (2161-1775)	.94
		198-217 (1752-1733)	.01

Site: Castro de Troña
Location: Ponteareas, Pontevedra, Galicia,
Type: Defended
Cultural phase: Iron Age
References: Carballo - Fábregas 1992, nº 21; Eiroa 1988, 118; Hidalgo Cuñarro 1988-89, 90
Sample ID: CSIC-718
Radiocarbon Age BP: 2010 ± 50≈
Calibrated age (s) cal BC 88, 83, 67, 63, 38, 29, 23, 11, 1
cal BP 2037, 2032, 2016, 2012, 1987, 1978, 1972, 1960, 1950
cal AD/BC (cal BP) age ranges obtained from intercepts (Method A):
one Sigma** cal BC 93-76 (2042-2025) 74-cal AD 29 (2023-1921) 40-51 (1910-1899)
two Sigma** cal BC 169-139 (2118-2088) 130-cal AD 70 (2079-1880)

Summary of above
minimum of cal age ranges (cal ages) maximum of cal age ranges:
one sigma cal BC 93 (88, 83, 67, 63, 38, 29, 23, 11, 1) cal AD 51
cal BP 2042 (2037, 2032, 2016, 2012, 1987, 1978, 1972, 1960, 1950) 1899
two sigma cal BC 169 (88, 83, 67, 63, 38, 29, 23, 11, 1) cal AD 70
cal BP 2118 (2037, 2032, 2016, 2012, 1987, 1978, 1972, 1960, 1950) 1880

cal AD/BC age ranges (cal ages as above) from probability distribution (Method B):

% area enclosed	cal BC (cal BP) age ranges		relative area under probability distribution
68.3 (one sigma)	cal BC	92- 78 (2041-2027)	.12
		73-cal AD 27 (2022-1923)	.83
		42- 48 (1908-1902)	.05
95.4 (two sigma)	cal BC	171- 137 (2120-2086)	.06
		131-cal AD 79 (2080-1871)	.93

Site: Corona de Corporales
Location: Truchas, León
Type: Defended
Cultural phase: Iron Age
References: Carballo - Fábregas 1992, nº 23; Sánchez - Fernández 1985, 278
Sample ID: UGRA-48
Radiocarbon Age BP: 2000 ± 130≈
Calibrated age (s) cal BC 86, 84, 65, 36, 34, 19, 13, cal AD 1
cal BP 2035, 2033, 2014, 1985, 1983, 1968, 1962, 1949
cal AD/BC (cal BP) age ranges obtained from intercepts (Method A) :
one Sigma** cal BC 170-cal AD 130 (2119-1820)
two Sigma** cal BC 380-cal AD 260 (2329-1690) 282-289 (1668-1661) 300-322 (1650-1628)

Summary of above
minimum of cal age ranges (cal ages) maximum of cal age ranges:
one sigma cal BC 170 (86, 84, 65, 36, 34, 19, 13, cal AD 1) cal AD 130
cal BP 2119 (2035, 2033, 2014, 1985, 1983, 1968, 1962, 1949) 1820
two sigma cal BC 380 (86, 84, 65, 36, 34, 19, 13, cal AD 1) cal AD 322
cal BP 2329 (2035, 2033, 2014, 1985, 1983, 1968, 1962, 1949) 1628

cal AD/BC age ranges (cal ages as above) from probability distribution (Method B) :

% area enclosed	cal BC (cal BP) age ranges		relative area under probability distribution
68.3 (one sigma)	cal BC	200-185 (2149-2134)	.03
		180-cal AD 130 (2129-1820)	.97
95.4 (two sigma)	cal BC	364-266 (2313-2215)	.09
		270-cal AD 260 (2219-1690)	.90

Site: Castro de A Graña
Location: Capela, Toques, A Coruña, Galicia
Type: Defended
Cultural phase: Iron Age
References: Carballo - Fábregas 1992, nº 7
Sample ID: GD-5555

Radiocarbon Age BP: 1980 ± 50≈
Calibrated age (s) cal AD 4, 7, 23, 45
cal BP 1946, 1943, 1927, 1905
cal AD/BC (cal BP) age ranges obtained from intercepts (Method A) :
one Sigma** cal BC 90-80 (2039-2029) 70-57 (2019-2006)
44-6 (1993-1955) 5-cal AD 66 (1954-1884)
two Sigma** cal BC 110-cal AD 130 (2059-1820)

Summary of above
minimum of cal age ranges (cal ages) maximum of cal age ranges:
one sigma cal BC 90 (cal AD 4, 7, 23, 45) cal AD 66
cal BP 2039 (1946, 1943, 1927, 1905) 1884
two sigma cal BC 110 (cal AD 4, 7, 23, 45) cal AD 130
cal BP 2059 (1946, 1943, 1927, 1905) 1820

cal AD/BC age ranges (cal ages as above) from probability distribution (Method B) :

% area enclosed	cal BC (cal BP) age ranges		relative area under probability distribution
68.3 (one sigma)	cal BC	89-81 (2038-2030)	.07
		69-59 (2018-2008)	.08
		42-7 (1991-1956)	.29
		4-cal AD 62 (1953-1888)	.56
95.4 (two sigma)	cal BC	159-149 (2108-2098)	.01
		106-cal AD 128 (2055-1822)	.99

Site: Castro de Cameixa
Location: Boborás, Ourense, Galicia
Type: Defended
Cultural phase: Iron Age
References: Carballo - Fábregas 1992, nº 17; Fariña Busto 1991, 53
Sample ID: CSIC-742
Radiocarbon Age BP: 1970 ± 50≈
Calibrated age (s) cal AD 26, 42, 48
cal BP 1924, 1908, 1902
cal AD/BC (cal BP) age ranges obtained from intercepts (Method A) :
one Sigma** cal BC 89-81 (2038-2030) 68-60 (2017-2009)
42-8 (1991-1957) 3-cal AD 70 (1952-1880)
two Sigma** cal BC 90-cal AD 130 (2039-1820)

Summary of above
minimum of cal age ranges (cal ages) maximum of cal age ranges:
one sigma cal BC 89 (cal AD 26, 42, 48) cal AD 70
cal BP 2038 (1924, 1908, 1902) 1880
two sigma cal BC 90 (cal AD 26, 42, 48) cal AD 130
cal BP 2039 (1924, 1908, 1902) 1820

cal AD/BC age ranges (cal ages as above) from probability distribution (Method B) :

% area enclosed	cal BC (cal BP) age ranges		relative area under probability distribution
68.3 (one sigma)	cal BC	89-81 (2038 2030)	.06
		68-60 (2017-2009)	.06
		41-9 (1990-1958)	.24
		3-cal AD 70 (1952-1880)	.63
95.4 (two sigma)	cal BC	105-cal AD 130 (2054-1820)	1.00

Site: Castelo de Matos
Location: Matos, Baião, Portugal
Type: Defended
Cultural phase: Iron Age
References: Figueiral - Queiroga 1988; Hedges et al. 1990, 226; Queiroga 1984; Queiroga - Figueiral 1989
Sample ID: GIF-7719
Radiocarbon Age BP: 1970 ± 60≈
Calibrated age (s) cal AD 26, 42, 48
cal BP 1924, 1908, 1902

cal AD/BC (cal BP) age ranges obtained from intercepts (Method A):

one Sigma**	cal BC	90-80 (2039-2029)	70-57 (2019-2006)
		44-6 (1993-1955)	5-cal AD 74 (1954-1876)
two Sigma**	cal BC	165-144 (2114-2093)	120-cal AD 130 (2069-1820)

Summary of above

minimum of cal age ranges (cal ages) maximum of cal age ranges:

one sigma	cal BC	90 (cal AD, 26, 42, 48) cal AD 74
	cal BP	2039 (1924, 1908, 1902) 1876
two sigma	cal BC	165 (cal AD, 26, 42, 48) cal AD 130
	cal BP	2114 (1924, 1908, 1902) 1820

cal AD/BC age ranges (cal ages as above) from probability distribution (Method B):

% area enclosed	cal BC (cal BP) age ranges		relative area under probability distribution
68.3 (one sigma)	cal BC	90-80 (2039-2029)	.07
		70-57 (2019-2006)	.08
		44-cal AD 75 (1993-1875)	.85
95.4 (two sigma)	cal BC	168-142 (2117-2091)	.03
		121-cal AD 133 (2070-1817)	.96

Site: Castrovite
Location: A Estrada, Pontevedra, Galicia
Type: Defended
Cultural phase: Iron Age
References: Carballo - Fábregas 1992, nº 9
Sample ID: CSIC-820
Radiocarbon Age BP: 1965 ± 45≈

Calibrated age (s)	cal AD	28, 41, 49
	cal BP	1922, 1909, 1901

cal AD/BC (cal BP) age ranges obtained from intercepts (Method A) :

one Sigma**	cal BC	88-82 (2037-2031)	67-62 (2016-2011)	39-28 (1988-1977)
		24-10 (1973-1959)	1-cal AD 70 (1950-1880)	
two Sigma**	cal BC	93-77 (2042-2026)	74-cal AD 129 (2023-1821)	

Summary of above

minimum of cal age ranges (cal ages) maximum of cal age ranges:

one sigma	cal BC	88 (cal AD 28, 41, 49) cal AD 70
	cal BP	2037 (1922, 1909, 1901) 1880
two sigma	cal BC	93 (cal AD 28, 41, 49) cal AD 129
	cal BP	2042 (1922, 1909, 1901) 1821

cal AD/BC age ranges (cal ages as above) from probability distribution (Method B) :

% area enclosed	cal BC (cal BP) age ranges		relative area under probability distribution
68.3 (one sigma)	cal BC	88-82 (2037-2031)	.05
		67-62 (2016-2011)	.04
		39-27 (1988-1976)	.09
		24-10 (1973-1959)	.11
		2-cal AD 71 (1951-1879)	.70
95.4 (two sigma)	cal BC	92-78 (2041-2027)	.06
		72-cal AD 127 (2021-1823)	.94

Site: Castro Real
Location: Boimorto, A Coruña, Galicia
Type: Defended
Cultural phase: Iron Age
References: Carballo - Fábregas 1992, nº 6
Sample ID: CSIC-730
Radiocarbon Age BP: 1950 ± 50≈

Calibrated age (s)	cal AD	31, 38, 53
	cal BP	1919, 1912, 1897

cal AD/BC (cal BP) age ranges obtained from intercepts (Method A) :

one Sigma**	cal BC	87-84 (2036-2033)	36-33 (1985-1982)

		20-13 (1969-1962)	cal AD 1-78 (1949-1872)	
two Sigma**	cal BC	92-77 (2041-2026)	70-cal AD 130 (2019-1820)	

Summary of above
minimum of cal age ranges (cal ages) maximum of cal age ranges:

one sigma	cal BC	87 (cal AD 31, 38, 53) cal AD 78
	cal BP	2036 (1919, 1912, 1897) 1872
two sigma	cal BC	92 (cal AD 31, 38, 53) cal AD 130
	cal BP	2041 (1919, 1912, 1897) 1820

cal AD/BC age ranges (cal ages as above) from probability distribution (Method B) :

% area enclosed	cal BC (cal BP) age ranges		relative area under probability distribution
68.3 (one sigma)	cal BC	88-83 (2037-2032)	.04
		67-63 (2016-2012)	.03
		38-29 (1987-1978)	.06
		22-11 (1971-1960)	.07
		1-cal AD 81 (1950-1869)	.70
		107-126 (1843-1824)	.11
95.4 (two sigma)	cal BC	93-76 (2042-2025)	.05
		74-cal AD 133 (2023-1817)	.93
		155-171 (1795-1779)	.01

Site: Castro de Fazouro
Location: Foz, Lugo, Galicia
Type: Defended
Cultural phase: Iron Age
References: Barbi Alonso 1991, 321; Carballo - Fábregas 1992, nº1
Sample ID: UBAR-117
Radiocarbon Age BP: 1930 ± 60≈

Calibrated age (s)	cal AD	65
	cal BP	1885

cal AD/BC (cal BP) age ranges obtained from intercepts (Method A) :

one Sigma**	cal BC	cal AD 2-14 (1948-1936)	16-129 (1934-1821)	
two Sigma**	cal BC	92-77 (2041-2026)	70-cal AD 180 (2019-1770)	190-227 (1760-1723)

Summary of above
minimum of cal age ranges (cal ages) maximum of cal age ranges:

one sigma	cal BC	16 (cal AD 65) cal AD 129
	cal BP	1965 (1885) 1821
two sigma	cal BC	92 (cal AD 65) cal AD 227
	cal BP	2041 (1885) 1723

cal AD/BC age ranges (cal ages as above) from probability distribution (Method B) :

% area enclosed	cal BC (cal BP) age ranges		relative area under probability distribution
68.3 (one sigma)	cal BC	87-83 (2036-2032)	.02
		66-64 (2015-2013)	.01
		37-32 (1986-1981)	.03
		21-12 (1970-1961)	.04
		cal AD 1-131 (1949-1819)	.90
95.4 (two sigma)	cal BC	92-78 (2041-2027)	.03
		72-cal AD 181 (2021-1769)	.91
		191-226 (1759-1724)	.06

Site: Castro de Troña
Location: Ponteareas, Pontevedra, Galicia
Type: Defended
Cultural phase: Iron Age
References: Carballo - Fábregas 1992, nº 21; Hidalgo 1988-89, 91
Sample ID: CSIC-741
Radiocarbon Age BP: 1930 ± 45≈

Calibrated age (s)	cal AD	65
	cal BP	1885

cal AD/BC (cal BP) age ranges obtained from intercepts (Method A) :

one Sigma**	cal AD	4-6 (1946-1944) 24-44 (1926-1906) 45-83 (1905-1867) 90-127 (1860-1823)
two Sigma**	cal BC	89-81 (2038-2030) 69-58 (2018-2007) 43-7 (1992-1956) 4-cal AD 134 (1953-1816) 157-168 (1793-1782) 201-210 (1749-1740)

Summary of above
minimum of cal age ranges (cal ages) maximum of cal age ranges:

one sigma	cal AD	4 (65) 127
	cal BP	1946 (1885) 1823
two sigma	cal BC	89 (cal AD 65) cal AD 210
	cal BP	2038 (1885) 1740

cal AD/BC age ranges (cal ages as above) from probability distribution (Method B) :

% area enclosed	cal BC (cal BP) age ranges		relative area under probability distribution
68.3 (one sigma)	cal AD	4-7 (1946-1943)	.03
		22-128 (1928-1822)	.97
95.4 (two sigma)	cal BC	90-80 (2039-2029)	.02
		70-58 (2019-2007)	.02
		43-7 (1992-1956)	.08
		4-cal AD 134 (1953-1816)	.81
		144-175 (1806-1775)	.04
		199-216 (1751-1734)	.02

Site: Castro de Troña
Location: Ponteareas, Pontevedra, Galicia
Type: Defended
Cultural phase: Iron Age
References: Carballo - Fábregas 1992, nº 21; Hidalgo 1988-89, 90
Sample ID: CSIC-740
Radiocarbon Age BP: 1920 ± 45≈

Calibrated age (s)	cal AD	69
	cal BP	1881

cal AD/BC (cal BP) age ranges obtained from intercepts (Method A) :

one Sigma**	cal AD	27-42 (1923-1908) 48-129 (1902-1821)
two Sigma**	cal BC	88-82 (2037-2031) 68-61 (2017-2010) 40-9 (1989-1958) 2-cal AD 173 (1951-1777) 198-214 (1752-1736)

Summary of above
minimum of cal age ranges (cal ages) maximum of cal age ranges:

one sigma	cal AD	27 (69) 129
	cal BP	1923 (1881) 1821
two sigma	cal BC	88 (cal AD 69) cal AD 214
	cal BP	2037 (1881) 1736

cal AD/BC age ranges (cal ages as above) from probability distribution (Method B) :

% area enclosed	cal BC (cal BP) age ranges		relative area under probability distribution
68.3 (one sigma)	cal AD	24-44 (1926-1906)	.18
		46-130 (1904-1820)	.82
95.4 (two sigma)	cal BC	88-82 (2037-2031)	.01
		68-61 (2017-2010)	.01
		40-10 (1989-1959)	.05
		2-cal AD 177 (1951-1773)	.88
		197-222 (1753-1728)	.04

Site: Castro de Mohías
Location: Coaña, Asturias
Type: Defended
Cultural phase: Iron Age
References: Almagro *et al.* 1978, 169; Carballo - Fábregas 1992, nº2; Eiroa 1980, 76-7; 1988, 118
Sample ID: CSIC-109
Radiocarbon Age BP: 1870 ± 80≈

Calibrated age (s)	cal AD	129
	cal BP	1821

cal AD/BC (cal BP) age ranges obtained from intercepts (Method A):

one Sigma**	cal AD	31-38 (1919-1912)	53- 232 (1897-1718)	
two Sigma**	cal BC	90-80 (2039-2029)	70-57 (2019-2006)	44-6 (1993-1955)
		cal AD 1- 260 (1949-1690)	276- 340 (1674-1610)	

Summary of above

minimum of cal age ranges (cal ages) maximum of cal age ranges:

one sigma	cal AD	31 (129) 232
	cal BP	1919 (1821) 1718
two sigma	cal BC	90 (cal AD, 129) cal AD 340
	cal BP	2039 (1821) 1610

cal AD/BC age ranges (cal ages as above) from probability distribution (Method B):

% area enclosed	cal BC (cal BP) age ranges		relative area under probability distribution
68.3 (one sigma)	cal AD	30-39 (1920-1911)	.04
		52- 233 (1898-1717)	.96
95.4 (two sigma)	cal BC	90-80 (2039-2029)	.01
		70-57 (2019-2006)	.01
		44-cal AD 263 (1993-1687)	.89
		274- 341 (1676-1609)	.08

Site: Castro de Borneiro
Location: Cabana, A Coruña, Galicia
Type: Defended
Cultural phase: Iron Age
References: Carballo - Fábregas 1992, nº 3
Sample ID: GRN-17042
Radiocarbon Age BP: 1870 ± 35≈

Calibrated age (s)	cal AD	129
	cal BP	1821

cal AD/BC (cal BP) age ranges obtained from intercepts (Method A) :

one Sigma**	cal AD	74-134 (1876-1816)	155-168 (1795-1782)	201-211 (1749-1739)
two Sigma**	cal AD	32-37 (1918-1913)	54-230 (1896-1720)	

Summary of above

minimum of cal age ranges (cal ages) maximum of cal age ranges:

one sigma	cal AD	74 (129) 211
	cal BP	1876 (1821) 1739
two sigma	cal AD	32 (129) 230
	cal BP	1918 (1821) 1720

cal AD/BC age ranges (cal ages as above) from probability distribution (Method B) :

% area enclosed	cal AD (cal BP) age ranges		relative area under probability distribution
68.3 (one sigma)	cal AD	73-134 (1877-1816)	.68
		148-172 (1802-1778)	.21
		200-212 (1750-1738)	.11
95.4 (two sigma)	cal AD	58-229 (1892-1721)	1.00

Site: Corona de Corporales
Location: Truchas, León
Type: Defended
Cultural phase: Iron Age
References: Carballo - Fábregas 1992, nº 23
Sample ID: CSIC-486
Radiocarbon Age BP: 1850 ± 40≈

Calibrated age (s)	cal AD	132
	cal BP	1818

cal AD/BC (cal BP) age ranges obtained from intercepts (Method A) :

one Sigma**	cal AD	79-118 (1871-1832)	126-181 (1824-1769)	192-226 (1758-1724)
two Sigma**	cal AD	61-252 (1889-1698)		

Summary of above
minimum of cal age ranges (cal ages) maximum of cal age ranges:

one sigma	cal AD	79 (132) 226
	cal BP	1871 (1818) 1724
two sigma	cal AD	61 (132) 252
	cal BP	1889 (1818) 1698

cal AD/BC age ranges (cal ages as above) from probability distribution (Method B) :

% area enclosed	cal AD (cal BP) age ranges		relative area under probability distribution
68.3 (one sigma)	cal AD	81-95 (1869-1855)	.12
		96-112 (1854-1838)	.13
		127-179 (1823-1771)	.48
		195-225 (1755-1725)	.27
95.4 (two sigma)	cal AD	61-248 (1889-1702)	.98
		305-317 (1645-1633)	.02

Site: Castro de Fazouro
Location: Foz, Lugo, Galicia
Type: Defended
Cultural phase: Iron Age
References: Barbi Alonso 1991, 321; Carballo - Fábregas 1992, nº 1
Sample ID: UBAR-118 **(values referring to this sample were not plotted in charts)**
Radiocarbon Age BP: 1840 ± 160≈

Calibrated age (s)	cal AD	133, 203, 207
	cal BP	1817, 1747, 1743

cal AD/BC (cal BP) age ranges obtained from intercepts (Method A) :

one Sigma**	cal BC	87-84 (2036-2033)	36-34 (1985-1983)	19-13 (1968-1962)
		cal AD 1-360 (1949-1590)	368-384 (1582-1566)	
two Sigma**	cal BC	342-325 (2291-2274)	210-cal AD 540 (2159-1410)	

Summary of above
minimum of cal age ranges (cal ages) maximum of cal age ranges:

one sigma	cal BC	87 (cal AD 133, 203, 207) cal AD 384
	cal BP	2036 (1817, 1747, 1743) 1566
two sigma	cal BC	342 (cal AD 133, 203, 207) cal AD 540
	cal BP	2291 (1817, 1747, 1743) 1410

cal AD/BC age ranges (cal ages as above) from probability distribution (Method B) :

% area enclosed	cal BC (cal BP) age ranges		relative area under probability distribution
68.3 (one sigma)	cal BC	18-14 (1967-1963)	.01
	cal AD	1-360 (1949-1590)	.95
		369-383 (1581-1567)	.03
95.4 (two sigma)	cal BC	200-cal AD 540 (2149-1410)	1.00

Site: Castrovite
Location: A Estrada, Pontevedra, Galicia
Type: Defended
Cultural phase: Iron Age
References: Carballo - Fábregas 1992, nº 9
Sample ID: CSIC-818
Radiocarbon Age BP: 1830 ± 50≈

Calibrated age (s)	cal AD	134, 152, 169, 200, 211
	cal BP	1816, 1798, 1781, 1750, 1739

cal AD/BC (cal BP) age ranges obtained from intercepts (Method A) :

one Sigma**	cal AD	82-93 (1868-1857)	103-106 (1847-1844)	127-238 (1823-1712)
two Sigma**	cal AD	34-36 (1916-1914)	60-260 (1890-1690)	279-294 (1671-1656)
		296-336 (1654-1614)		

Summary of above
minimum of cal age ranges (cal ages) maximum of cal age ranges:

one sigma	cal AD	82 (134, 152, 169, 200, 211) 238

	cal BP	1868 (1816, 1798, 1781, 1750, 1739) 1712	
two sigma	cal AD	34 (134, 152, 169, 200, 211) 336	
	cal BP	1916 (1816, 1798, 1781, 1750, 1739) 1614	

cal AD/BC age ranges (cal ages as above) from probability distribution (Method B) :

% area enclosed	cal AD (cal BP) age ranges		relative area under probability distribution
68.3 (one sigma)	cal AD	82-93 (1868-1857)	.07
		101-109 (1849-1841)	.05
		127-236 (1823-1714)	.89
95.4 (two sigma)	cal AD	65-260 (1885-1690)	.90
		280-292 (1670-1658)	.02
		297-335 (1653-1615)	.08

3) Notes

The following dates were excluded from the dataset as their age in calendar years falls substantially outside the first millennium BC.

Site: Castromao
Location: Celanova, Ourense, Galicia
Type: Defended
Cultural phase: Iron Age
References: Carballo - Fábregas 1992, 22; Fariña Busto 1991, 54
Sample ID: CSIC-637-R
Radiocarbon Age BP: 7200 ± 60≈

Site: Castromao
Location: Celanova, Ourense, Galicia
Type: Defended
Cultural phase: Iron Age
References: Carballo - Fábregas 1992, 22; Fariña Busto 1991, 54
Sample ID: CSIC-637
Radiocarbon Age BP: 7180 ± 60≈

Site: Castro de Torroso
Location: Mos, Pontevedra, Galicia
Type: Defended
Cultural phase: Bronze Age-Iron Age
References: Carballo - Fábregas 1992, nº 19
Sample ID: GRN-14590
Radiocarbon Age BP: 6245 ± 35≈

Site: Castro de Penalba
Location: Campo Lameiro, Pontevedra, Galicia
Type: Defended
Cultural phase: Iron Age
References: Carballo - Fábregas 1992, nº 15
Sample ID: GAK-12970
Radiocarbon Age BP: 4010 ± 120≈

Site: As Muradellas
Location: Lubian, Zamora
Type: Defended
Cultural phase: Iron Age
References: Carballo - Fábregas 1992, nº 25; Esparza 1986, 401
Sample ID: GAK-9521
Radiocarbon Age BP: 3750 ± 290≈
Calibrated age (s) cal BC 2199, 2155, 2143

cal BP 4148, 4104, 4092

Site: Barrocal Alto,
Location: Peredo da Bemposta, Mogadouro, Portugal
Type: Open
Cultural phase: Bronze Age
References:
Sample ID: UGRA-223
Radiocarbon Age BP: 3230 ± 100≈
Calibrated age (s) cal BC 1502
cal BP 3451

Site: Pastoria
Location: Pastoria, Chaves
Type: Open
Cultural phase: Bronze Age
References: Jorge 1985
Sample ID: LY-3375
Radiocarbon Age BP: 3040 ± 140≈

Site: Pastoria
Location: Pastoria, Chaves, Portugal
Type: Open (??)
Cultural phase: Bronze Age (??)
References: Jorge 1985
Sample ID: LY-3376
Radiocarbon Age BP: 1940 ± 90≈

Site: Corno do Cunho
Location: Peredo da Bemposta, Mogadouro, Portugal
Type: Open
Cultural phase: Bronze Age
References:
Sample ID: UGRA-221
Radiocarbon Age BP: 2340 ± 100≈

Site: Citânia de S. Julião
Location: Ponte, Vila Verde, Portugal
Type: Defended
Cultural phase: Bronze Age-Iron Age
References: Carballo - Fábregas 1992, nº 30; Martins 1988c, 126
Sample ID: ICEN-31
Radiocarbon Age BP: 1550 ± 40≈

Site: Castro de Mohías
Location: Coaña, Asturias
Type: Defended
Cultural phase: Iron Age
References: Almagro Gorbea 1972, 234; Carballo - Fábregas 1992, nº2; Eiroa 1980, 76-7; 1988, 118; Martínez Fernandez 1971, 355
Sample ID: CSIC-38
Radiocarbon Age BP: 1380 ± 100≈

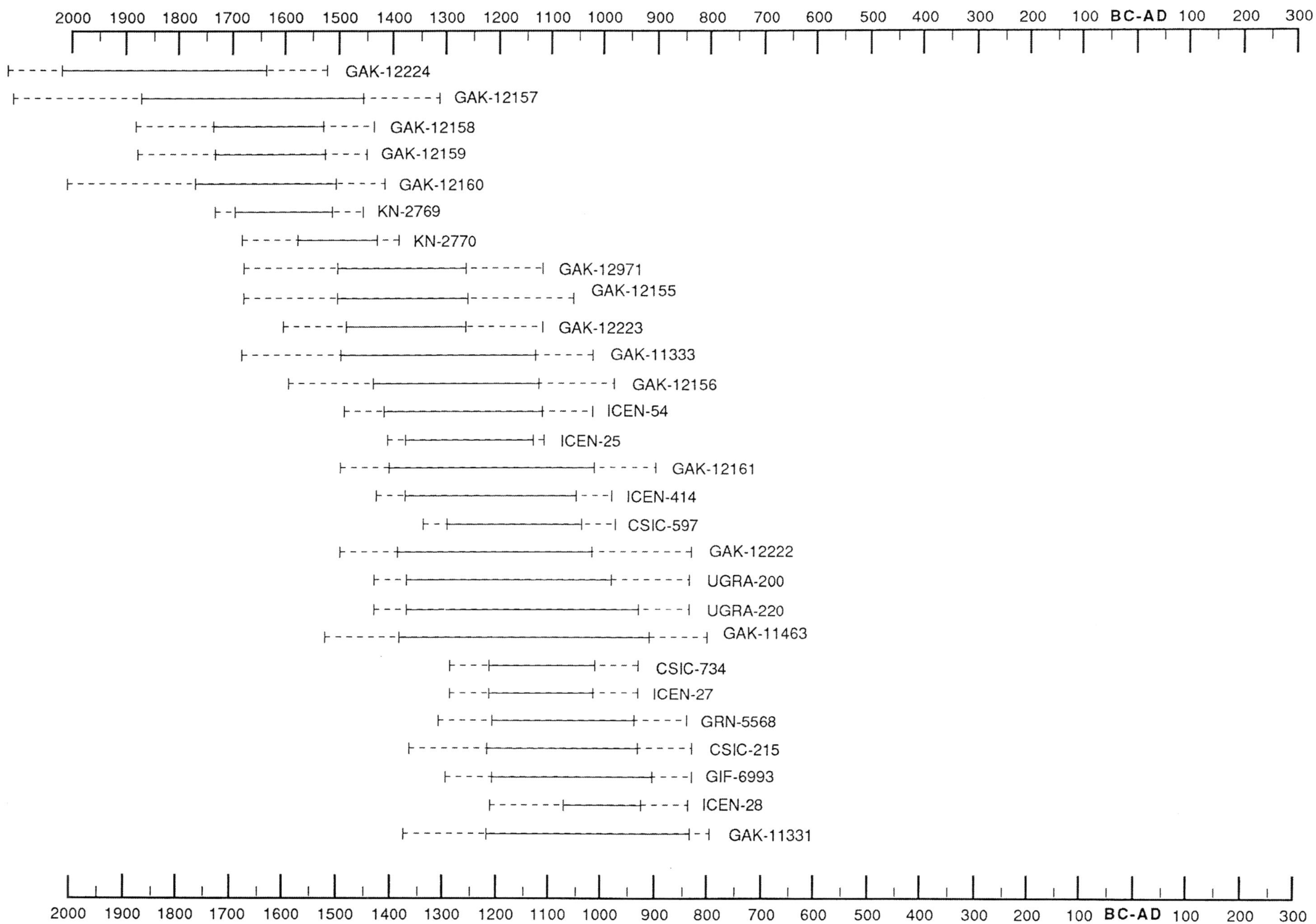

2000 1900 1800 1700 1600 1500 1400 1300 1200 1100 1000 900 800 700 600 500 400 300 200 100 BC-AD 100 200 300
GAK-12224
GAK-12157
GAK-12158
GAK-12159
GAK-12160
KN-2769
KN-2770
GAK-12971
GAK-12155
GAK-12223
GAK-11333
GAK-12156
ICEN-54
ICEN-25
GAK-12161
ICEN-414
CSIC-597
GAK-12222
UGRA-200
UGRA-220
GAK-11463
CSIC-734
ICEN-27
GRN-5568
CSIC-215
GIF-6993
ICEN-28
GAK-11331
2000 1900 1800 1700 1600 1500 1400 1300 1200 1100 1000 900 800 700 600 500 400 300 200 100 BC-AD 100 200 300

Figure 1.1

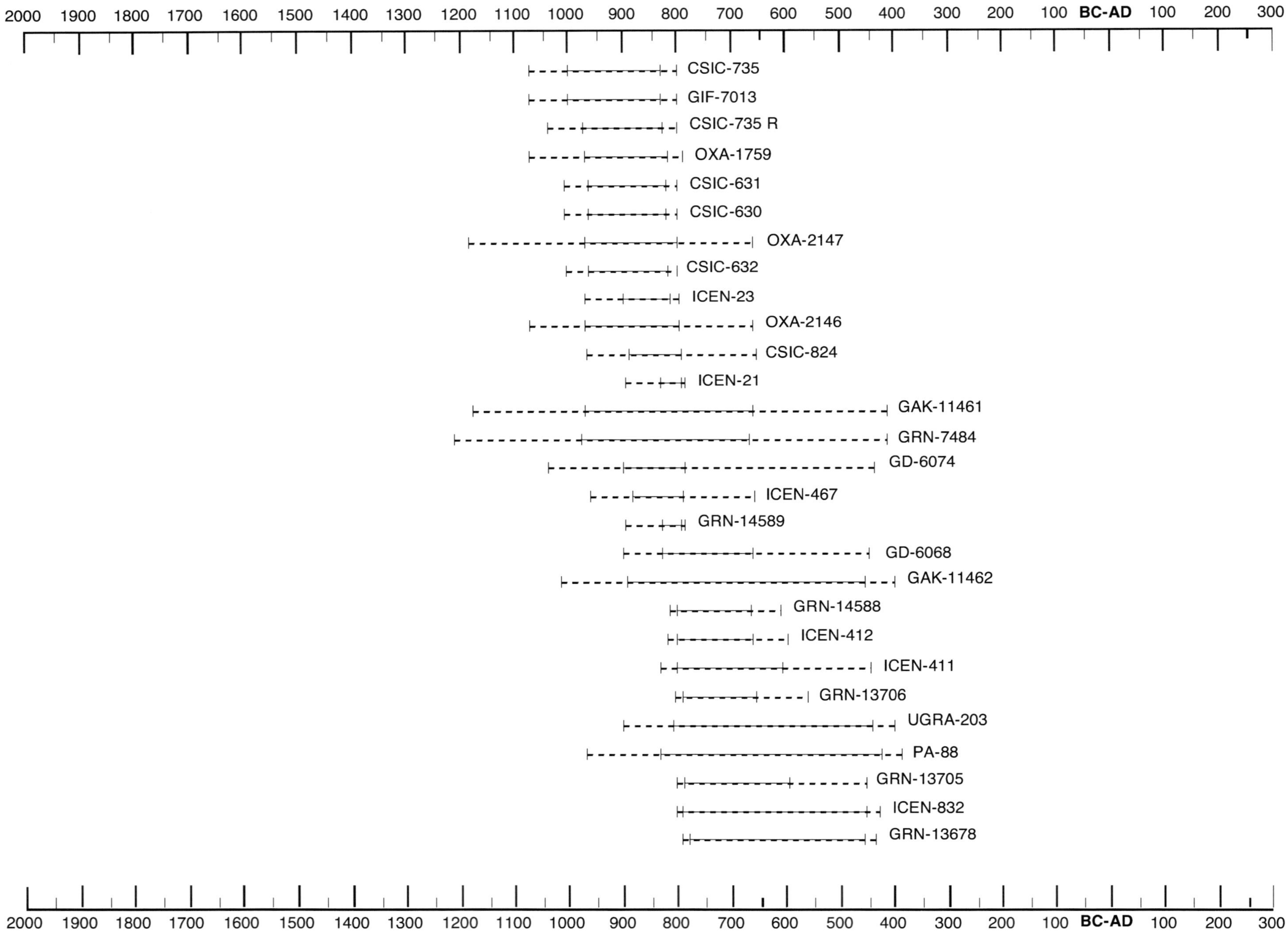

2000 1900 1800 1700 1600 1500 1400 1300 1200 1100 1000 900 800 700 600 500 400 300 200 100 BC-AD 100 200 300
CSIC-735
GIF-7013
CSIC-735 R
OXA-1759
CSIC-631
CSIC-630
OXA-2147
CSIC-632
ICEN-23
OXA-2146
CSIC-824
ICEN-21
GAK-11461
GRN-7484
GD-6074
ICEN-467
GRN-14589
GD-6068
GAK-11462
GRN-14588
ICEN-412
ICEN-411
GRN-13706
UGRA-203
PA-88
GRN-13705
ICEN-832
GRN-13678
2000 1900 1800 1700 1600 1500 1400 1300 1200 1100 1000 900 800 700 600 500 400 300 200 100 BC-AD 100 200 300

Figure 1.2

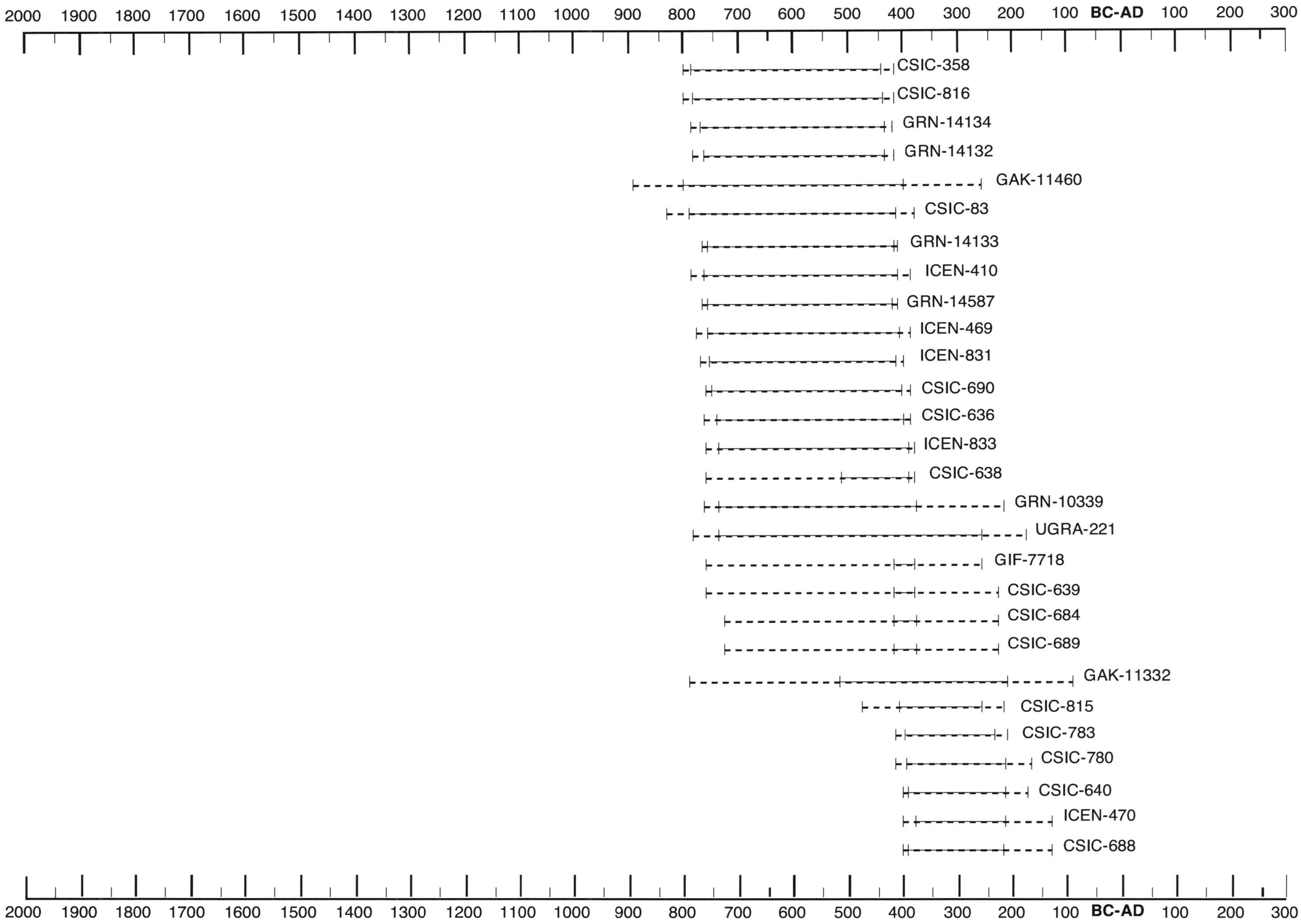
2000
1900
1800
1700
1600
1500
1400
1300
1200
1100
1000
900
800
700
600
500
400
300
200
100
BC-AD
100
200
300
CSIC-358
CSIC-816
GRN-14134
GRN-14132
GAK-11460
CSIC-83
GRN-14133
ICEN-410
GRN-14587
ICEN-469
ICEN-831
CSIC-690
CSIC-636
ICEN-833
CSIC-638
GRN-10339
UGRA-221
GIF-7718
CSIC-639
CSIC-684
CSIC-689
GAK-11332
CSIC-815
CSIC-783
CSIC-780
CSIC-640
ICEN-470
CSIC-688

Figure 1.3

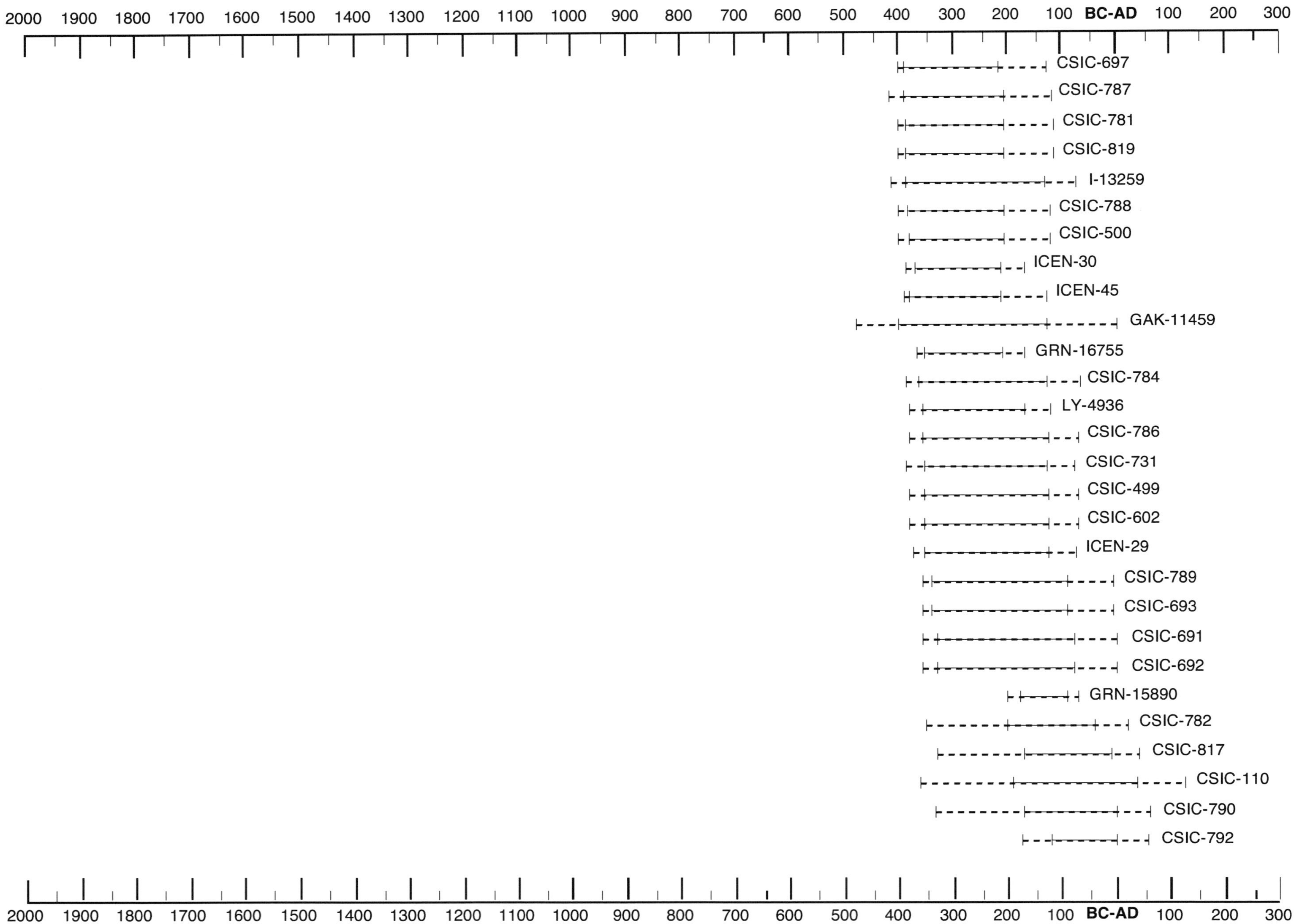

2000 1900 1800 1700 1600 1500 1400 1300 1200 1100 1000 900 800 700 600 500 400 300 200 100 BC-AD 100 200 300
CSIC-697
CSIC-787
CSIC-781
CSIC-819
I-13259
CSIC-788
CSIC-500
ICEN-30
ICEN-45
GAK-11459
GRN-16755
CSIC-784
LY-4936
CSIC-786
CSIC-731
CSIC-499
CSIC-602
ICEN-29
CSIC-789
CSIC-693
CSIC-691
CSIC-692
GRN-15890
CSIC-782
CSIC-817
CSIC-110
CSIC-790
CSIC-792
2000 1900 1800 1700 1600 1500 1400 1300 1200 1100 1000 900 800 700 600 500 400 300 200 100 BC-AD 100 200 300

Figure 1.4

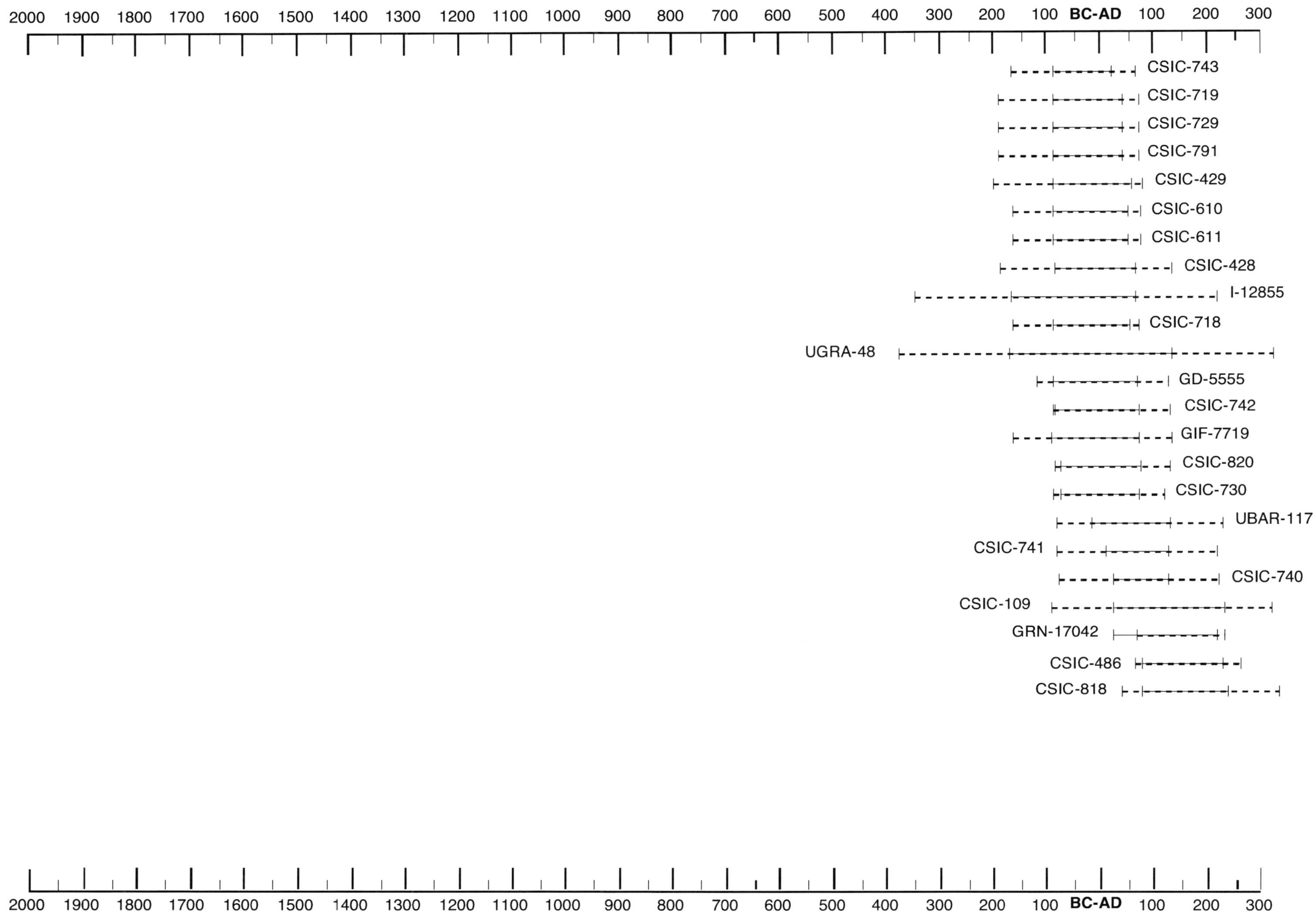

2000 1900 1800 1700 1600 1500 1400 1300 1200 1100 1000 900 800 700 600 500 400 300 200 100 BC-AD 100 200 300
CSIC-743
CSIC-719
CSIC-729
CSIC-791
CSIC-429
CSIC-610
CSIC-611
CSIC-428
I-12855
CSIC-718
UGRA-48
GD-5555
CSIC-742
GIF-7719
CSIC-820
CSIC-730
UBAR-117
CSIC-741
CSIC-740
CSIC-109
GRN-17042
CSIC-486
CSIC-818
2000 1900 1800 1700 1600 1500 1400 1300 1200 1100 1000 900 800 700 600 500 400 300 200 100 BC-AD 100 200 300

Figure 1.5

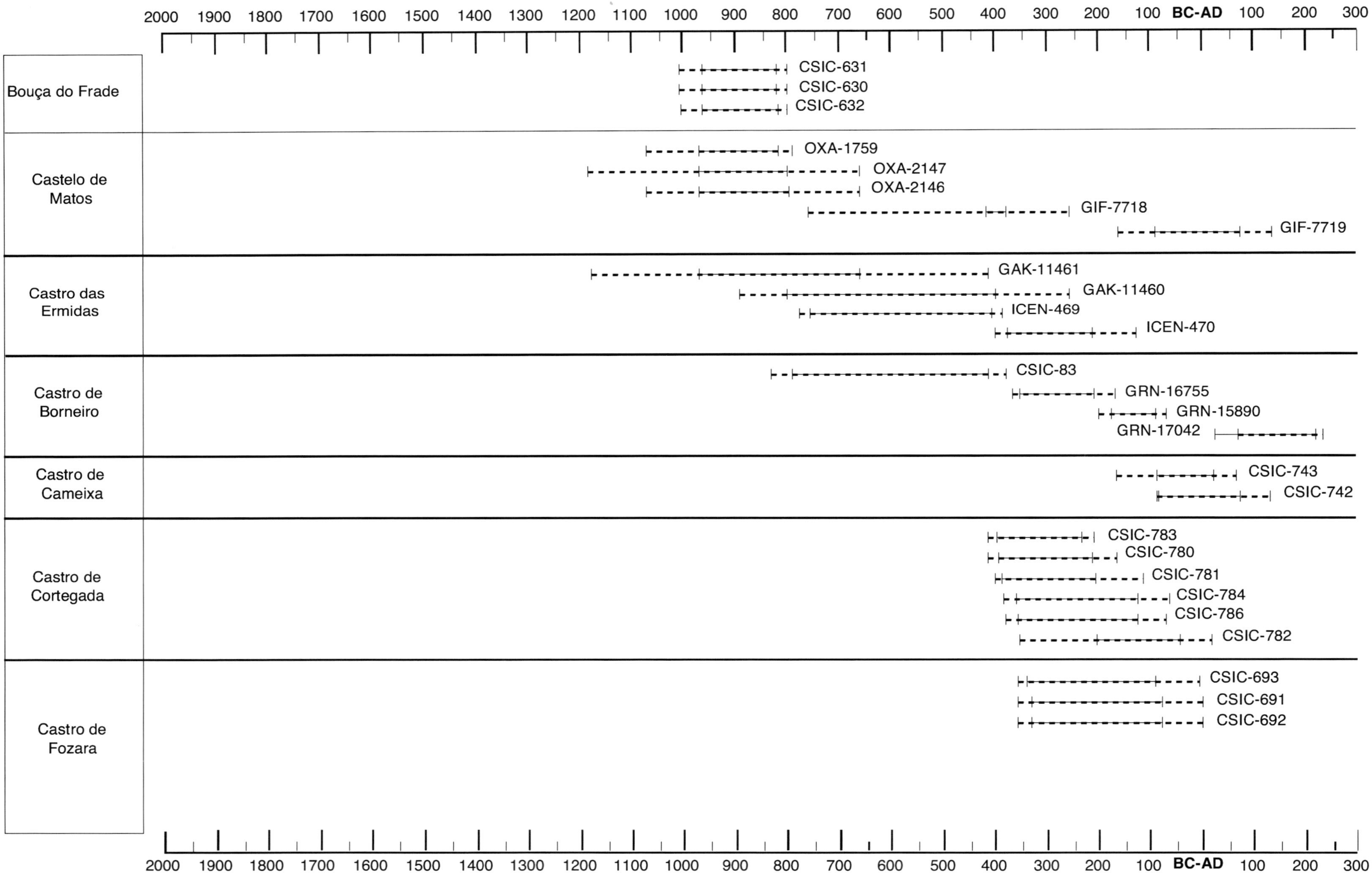
2000
1900
1800
1700
1600
1500
1400
1300
1200
1100
1000
900
800
700
600
500
400
300
200
100
BC-AD
100
200
300
Bouça do Frade
CSIC-631
CSIC-630
CSIC-632
Castelo de Matos
OXA-1759
OXA-2147
OXA-2146
GIF-7718
GIF-7719
Castro das Ermidas
GAK-11461
GAK-11460
ICEN-469
ICEN-470
Castro de Borneiro
CSIC-83
GRN-16755
GRN-15890
GRN-17042
Castro de Cameixa
CSIC-743
CSIC-742
Castro de Cortegada
CSIC-783
CSIC-780
CSIC-781
CSIC-784
CSIC-786
CSIC-782
Castro de Fozara
CSIC-693
CSIC-691
CSIC-692

Figure 2.1

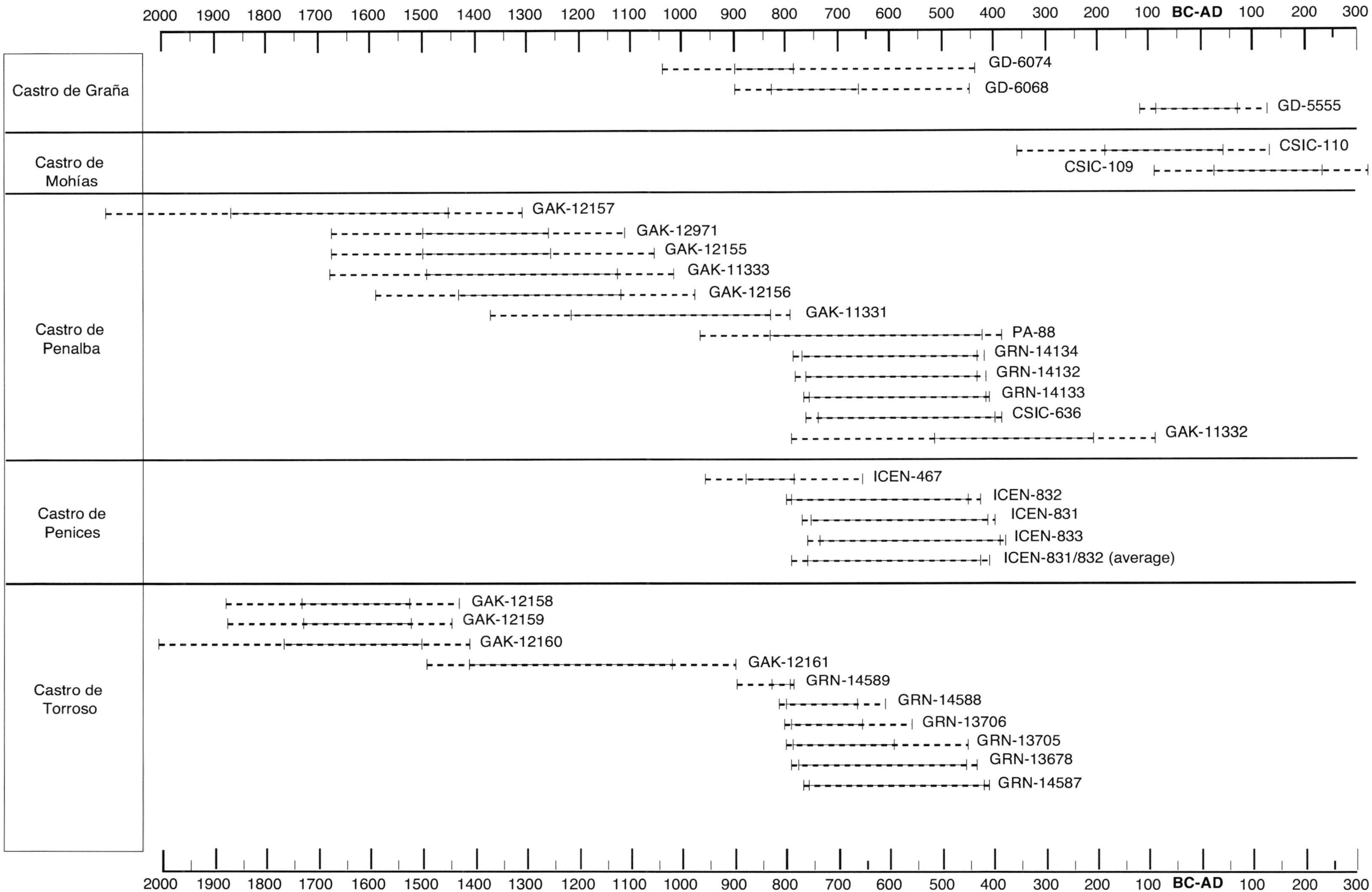
2000
1900
1800
1700
1600
1500
1400
1300
1200
1100
1000
900
800
700
600
500
400
300
200
100
BC-AD
100
200
300
Castro de Graña
GD-6074
GD-6068
GD-5555
Castro de Mohías
CSIC-110
CSIC-109
Castro de Penalba
GAK-12157
GAK-12971
GAK-12155
GAK-11333
GAK-12156
GAK-11331
PA-88
GRN-14134
GRN-14132
GRN-14133
CSIC-636
GAK-11332
Castro de Penices
ICEN-467
ICEN-832
ICEN-831
ICEN-833
ICEN-831/832 (average)
Castro de Torroso
GAK-12158
GAK-12159
GAK-12160
GAK-12161
GRN-14589
GRN-14588
GRN-13706
GRN-13705
GRN-13678
GRN-14587
Figure 2.2

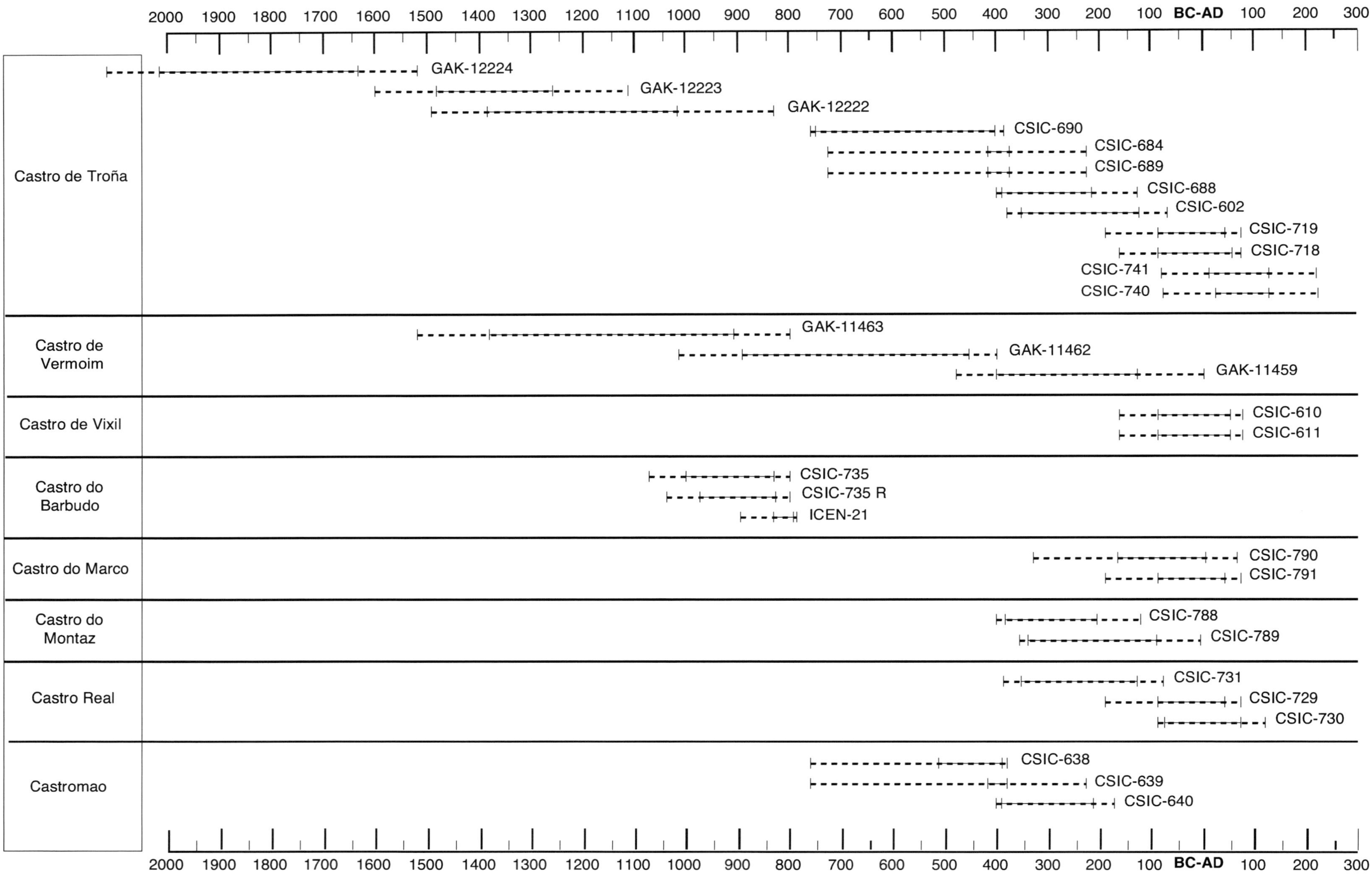
2000 1900 1800 1700 1600 1500 1400 1300 1200 1100 1000 900 800 700 600 500 400 300 200 100 BC-AD 100 200 300
Castro de Troña
GAK-12224
GAK-12223
GAK-12222
CSIC-690
CSIC-684
CSIC-689
CSIC-688
CSIC-602
CSIC-719
CSIC-718
CSIC-741
CSIC-740
Castro de Vermoim
GAK-11463
GAK-11462
GAK-11459
Castro de Vixil
CSIC-610
CSIC-611
Castro do Barbudo
CSIC-735
CSIC-735 R
ICEN-21
Castro do Marco
CSIC-790
CSIC-791
Castro do Montaz
CSIC-788
CSIC-789
Castro Real
CSIC-731
CSIC-729
CSIC-730
Castromao
CSIC-638
CSIC-639
CSIC-640

Figure 2.3

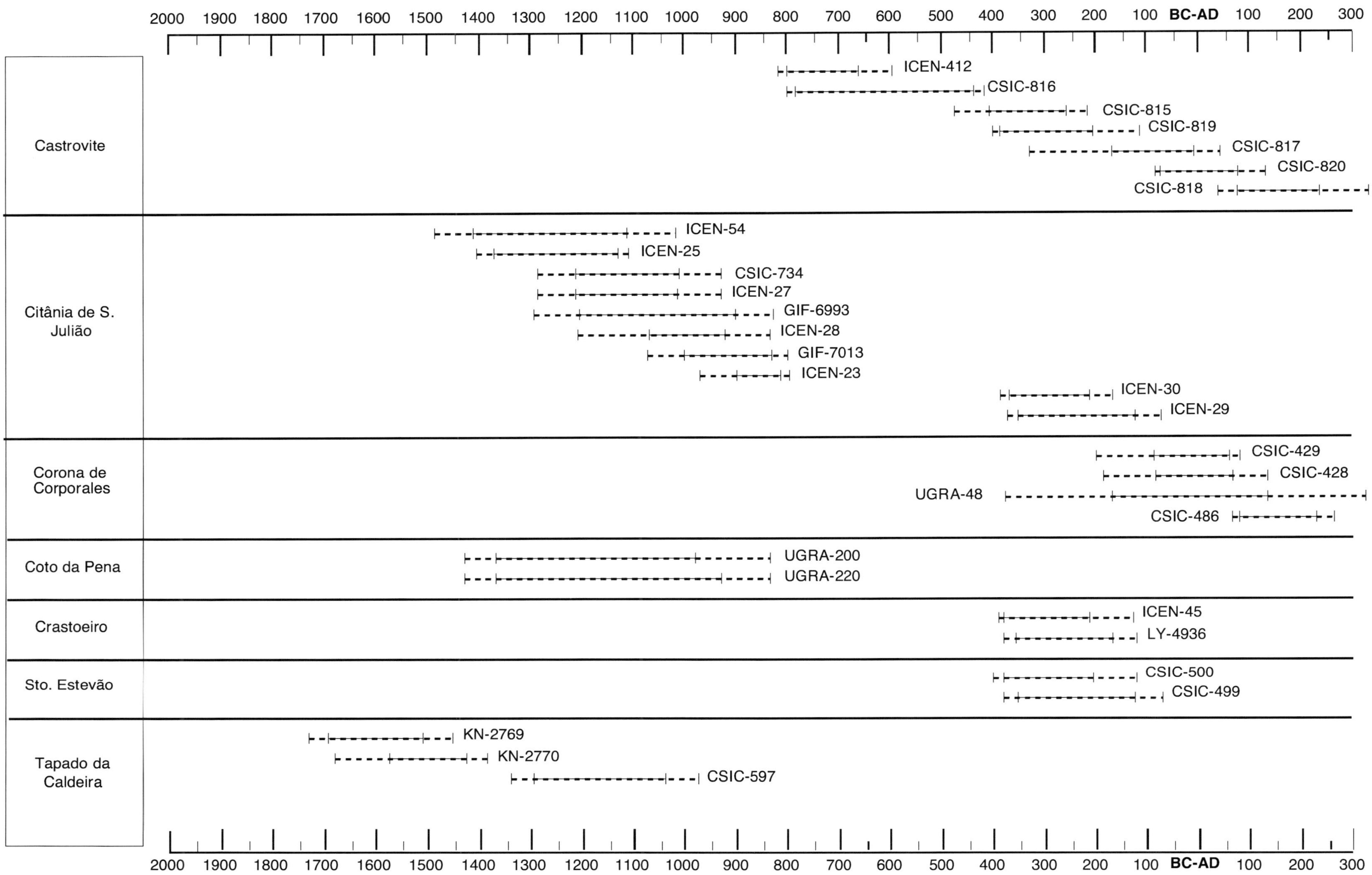

2000 1900 1800 1700 1600 1500 1400 1300 1200 1100 1000 900 800 700 600 500 400 300 200 100 BC-AD 100 200 300
Castrovite
ICEN-412
CSIC-816
CSIC-815
CSIC-819
CSIC-817
CSIC-820
CSIC-818
Citânia de S. Julião
ICEN-54
ICEN-25
CSIC-734
ICEN-27
GIF-6993
ICEN-28
GIF-7013
ICEN-23
ICEN-30
ICEN-29
Corona de Corporales
CSIC-429
CSIC-428
UGRA-48
CSIC-486
Coto da Pena
UGRA-200
UGRA-220
Crastoeiro
ICEN-45
LY-4936
Sto. Estevão
CSIC-500
CSIC-499
Tapado da Caldeira
KN-2769
KN-2770
CSIC-597

Figure 2.4

ILLUSTRATIONS

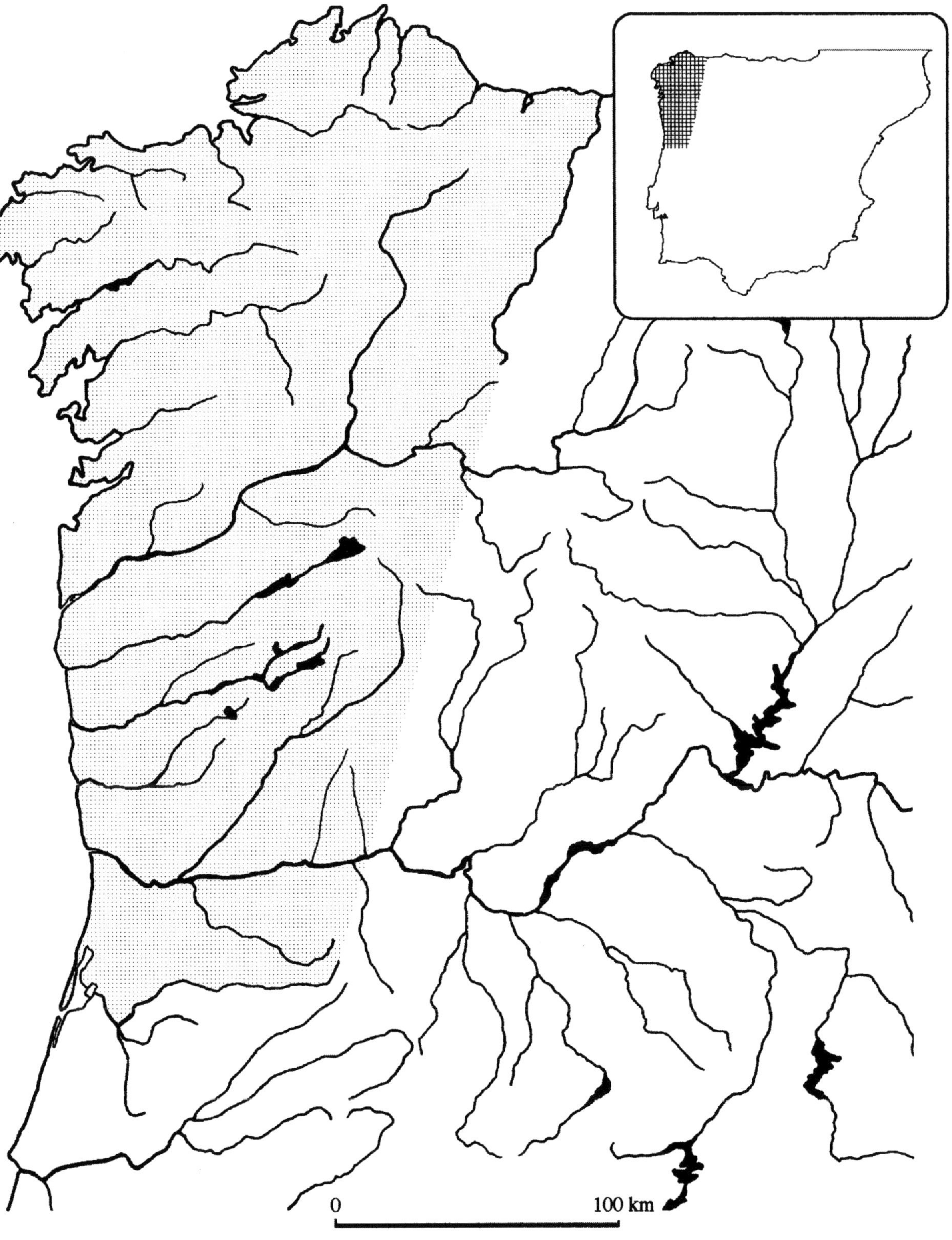

Figure 1

Castros Culture area (after López Cuevillas 1933 and further revisions by various authors)

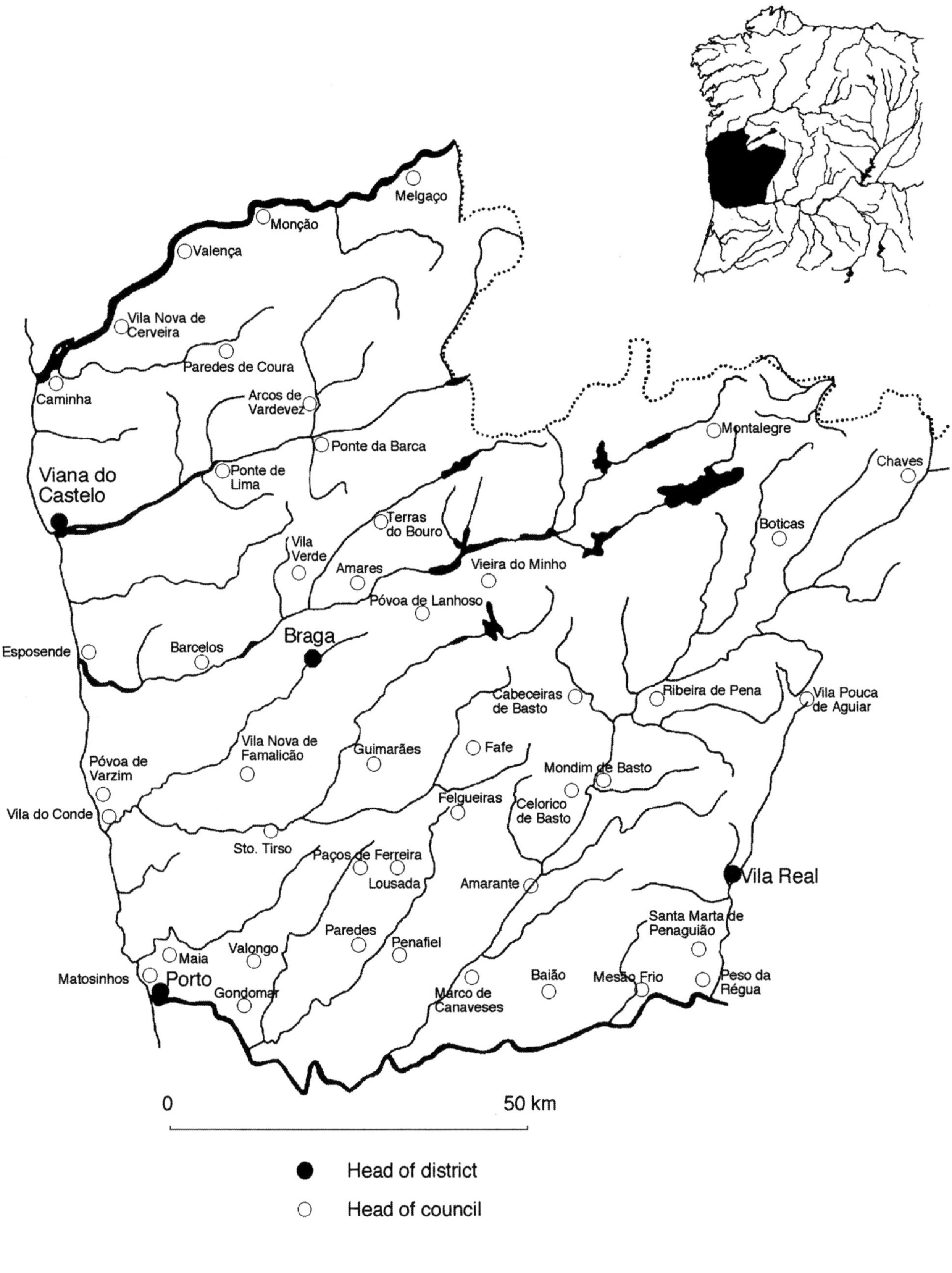

Figure 2 - The region studied showing names of the principal places and administrative districts used in the text

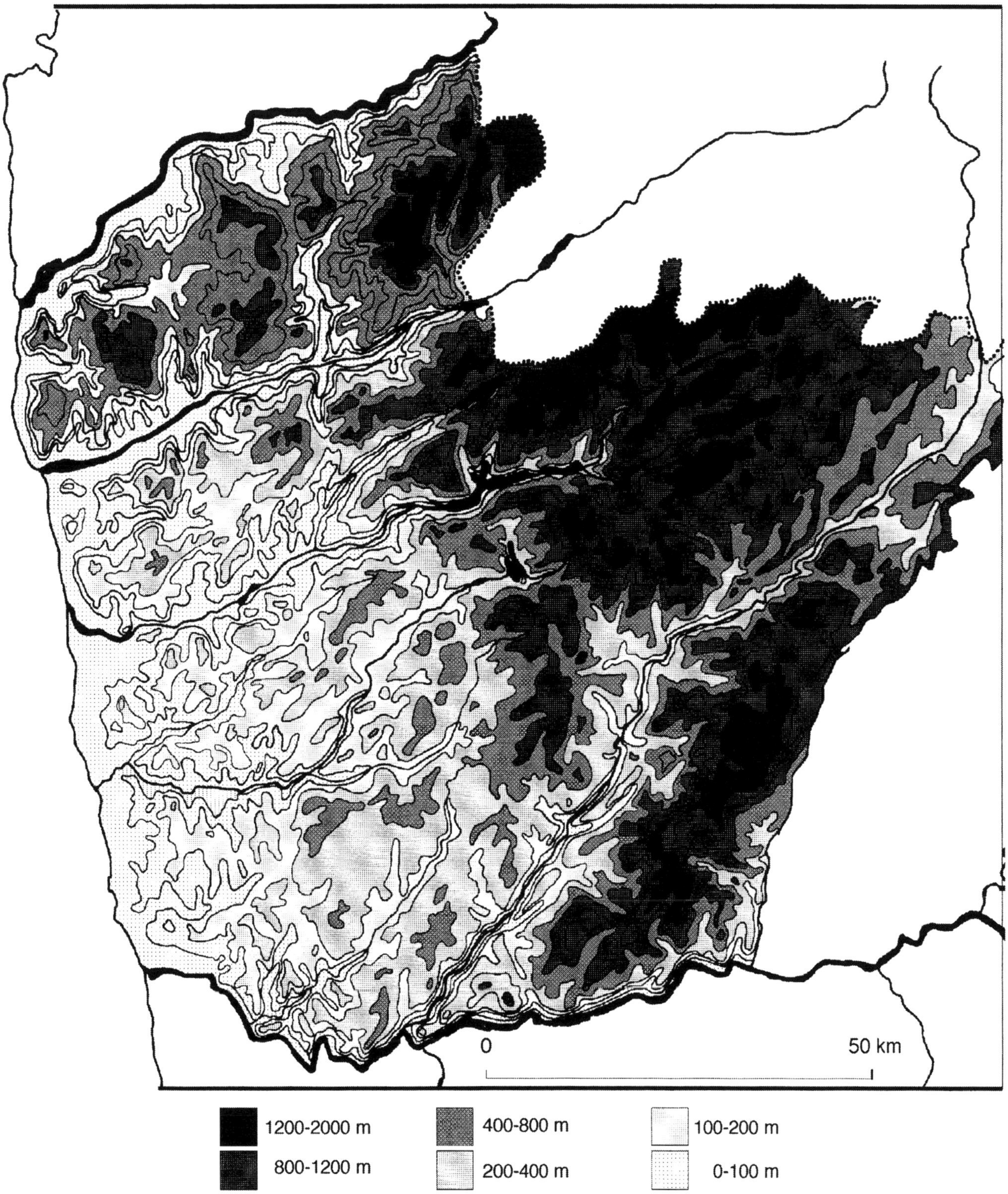

Figure 3 - Orography of the region studied

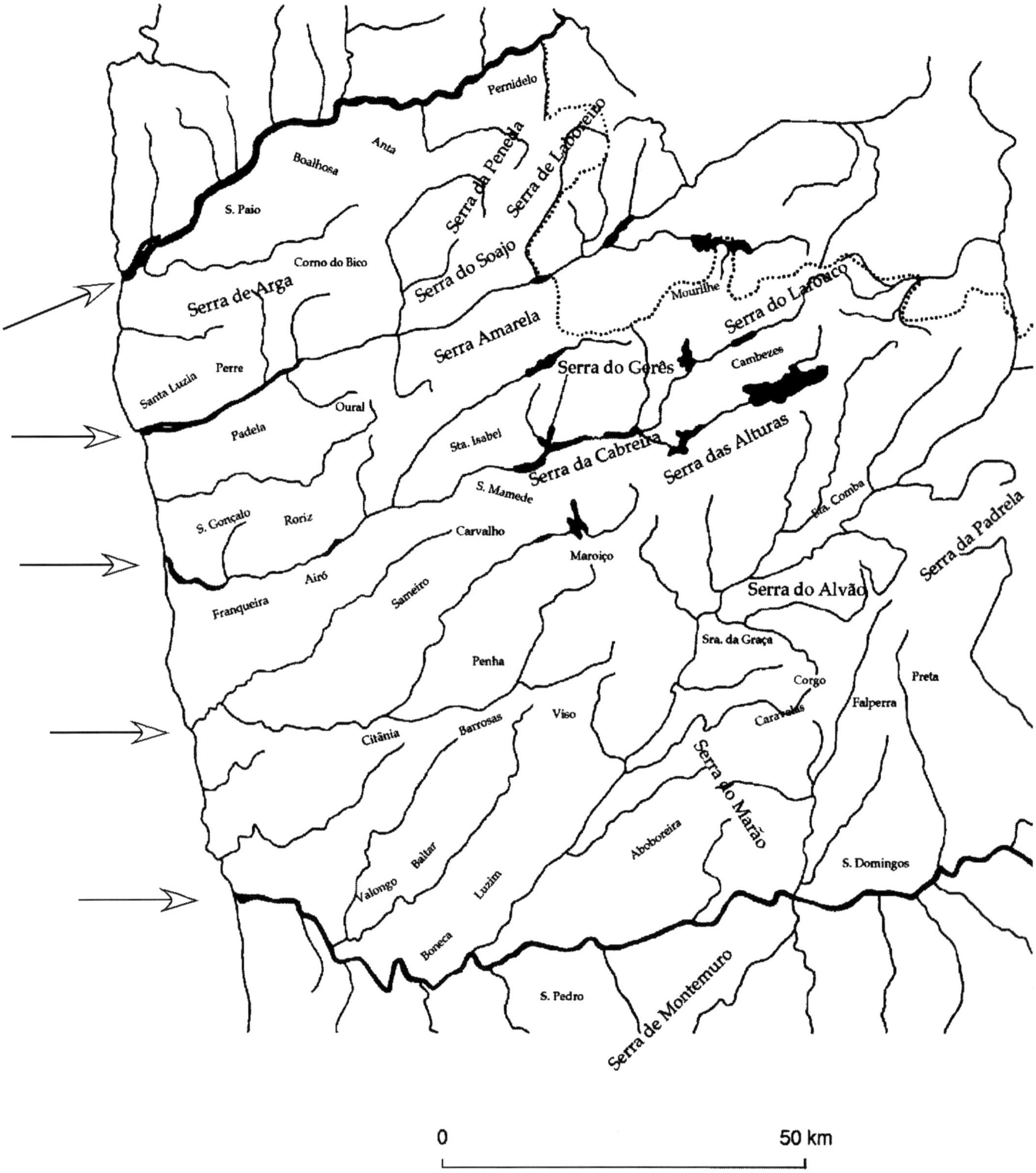

Figure 4 - Mountains, hills and natural routes of hinterland penetration

(after Carta Hipsométrica, Comissão Nacional do Ambiente, 1982)

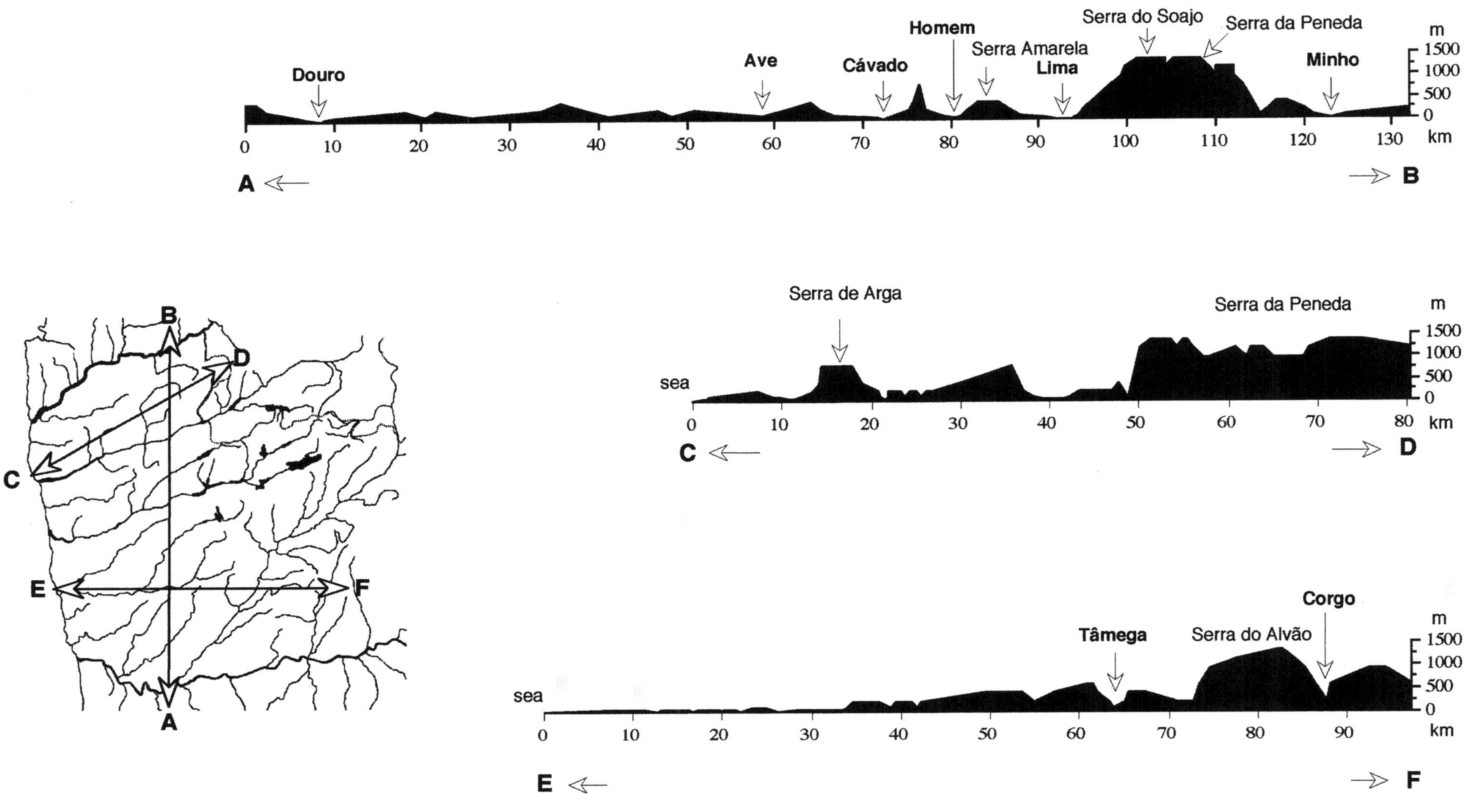

Figure 5 - Cross section of the relief of the nortwest of Portugal.

Names in bold refer to rivers. Schematic output due to scale.

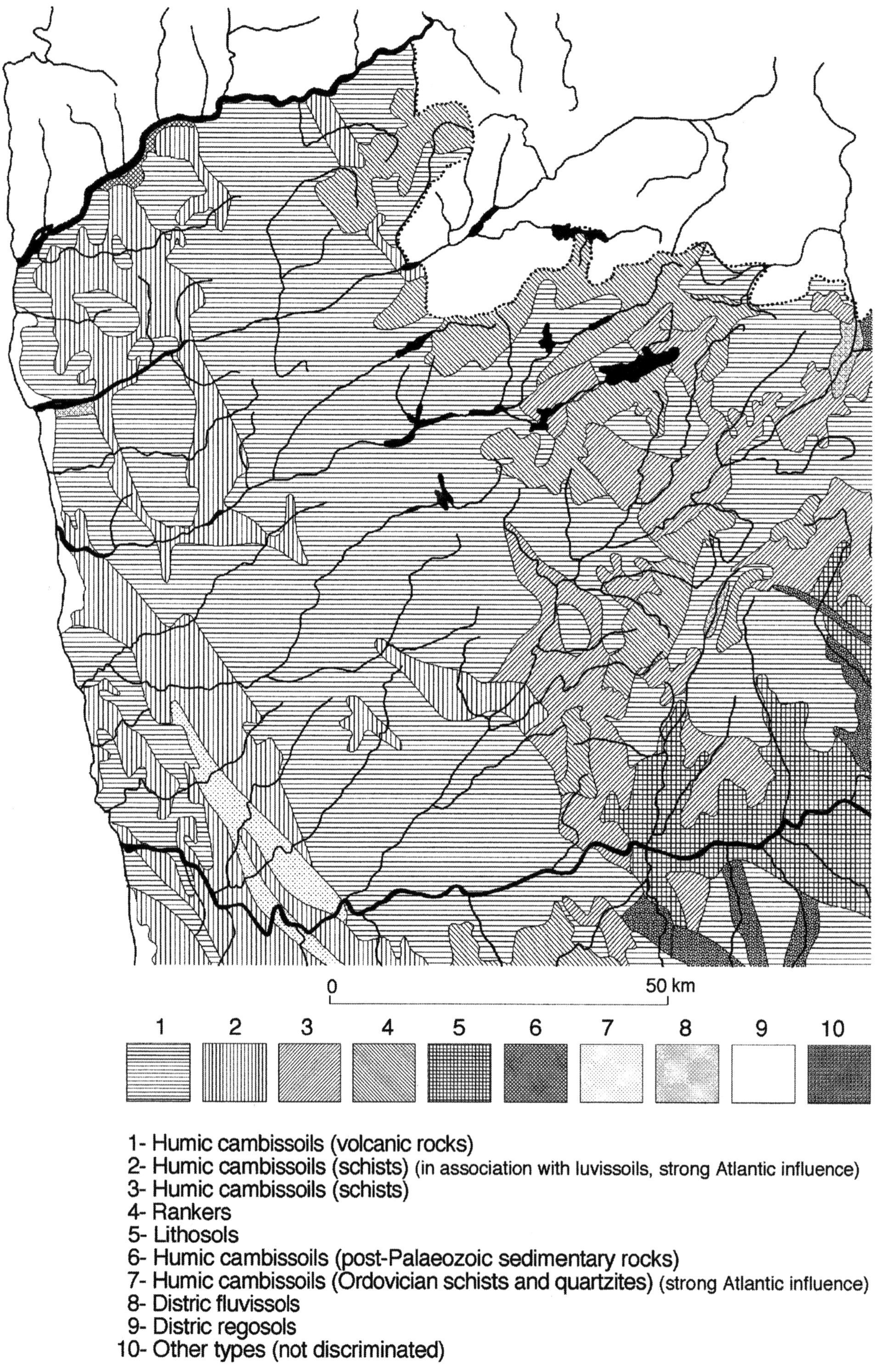

Figure 6 - Chart of types of soil (after Carta dos Solos, Comissão Nacional do Ambiente, 1971).

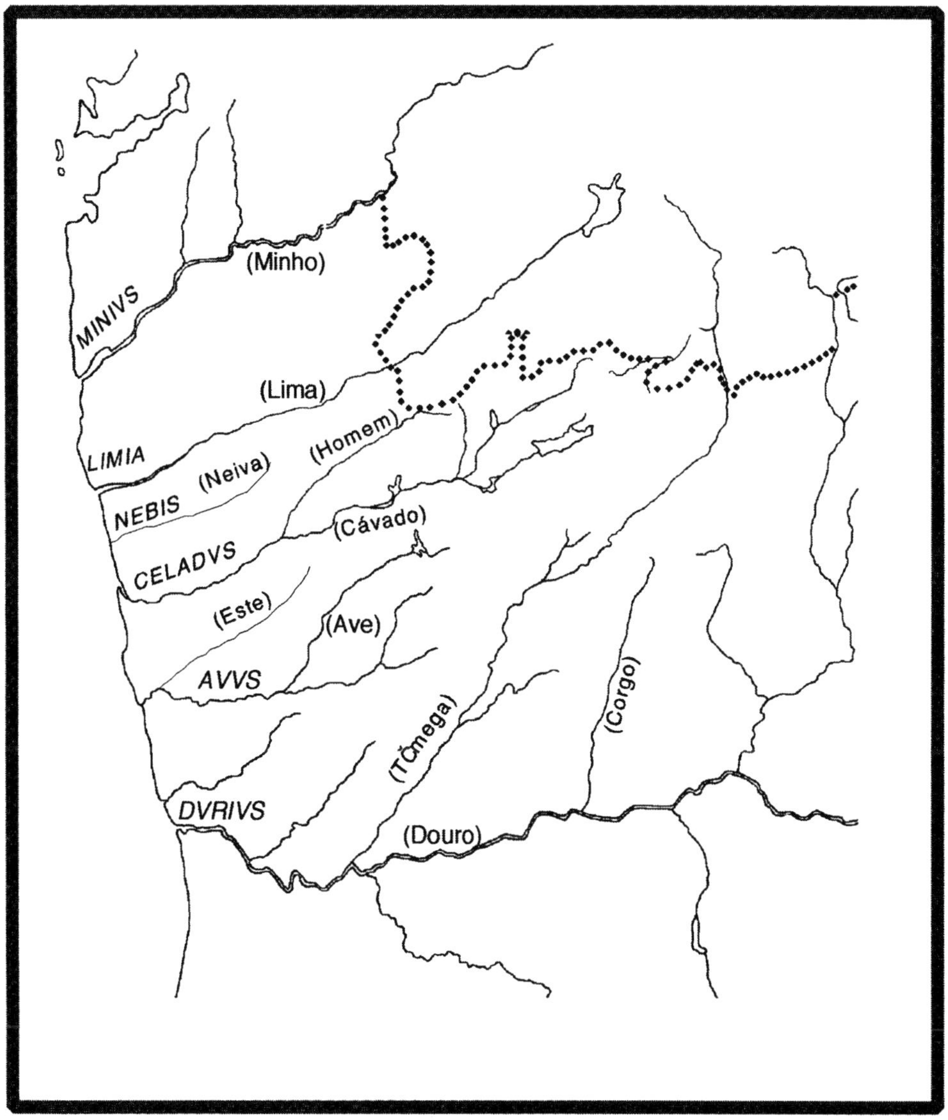

Figure 7 - The main water courses of the northwest of Portugal.

(Current names in brackets, Latin names in italics).

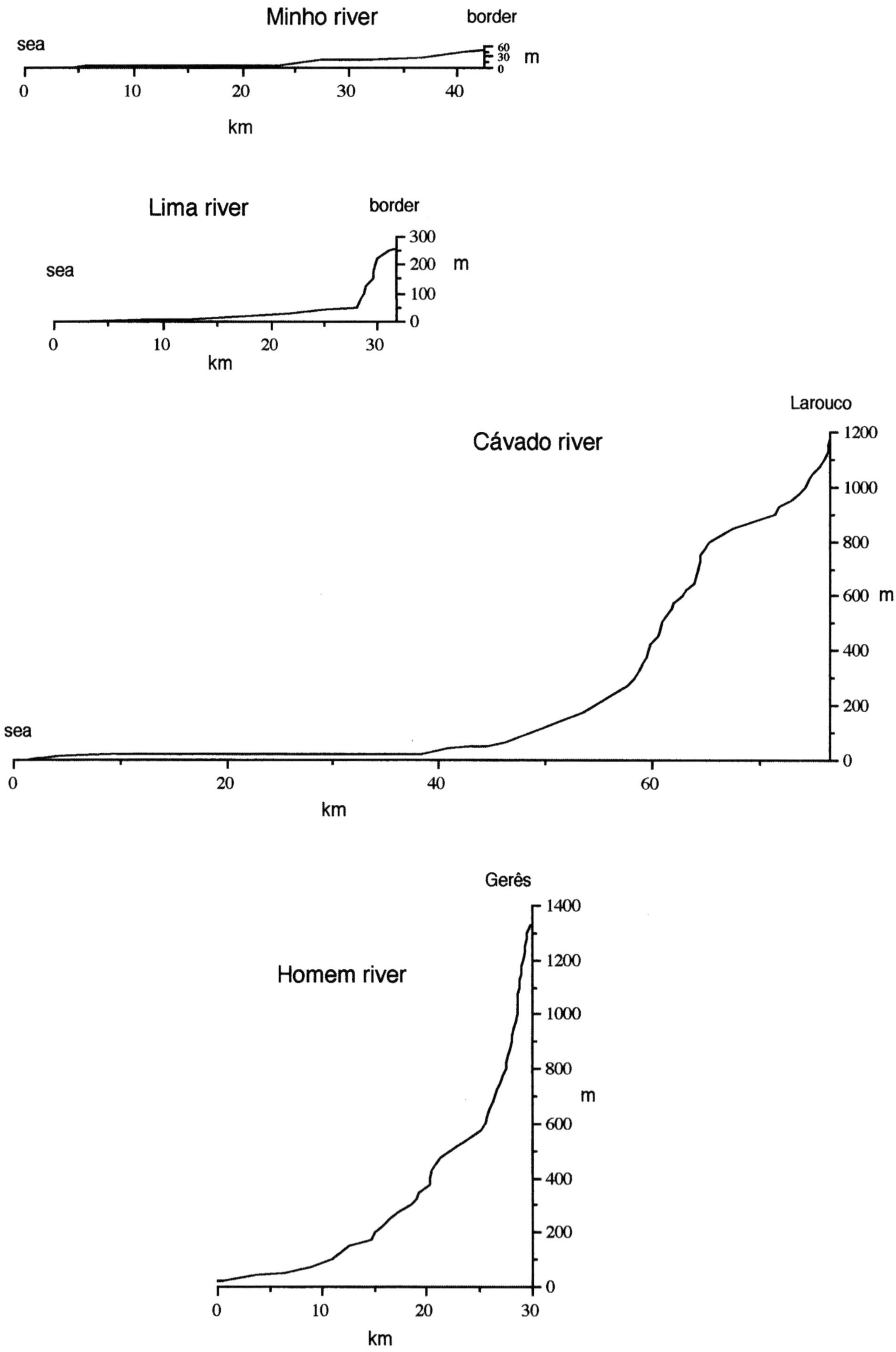

Figure 8.1 - Representation of river gradients in the Portuguese northwest

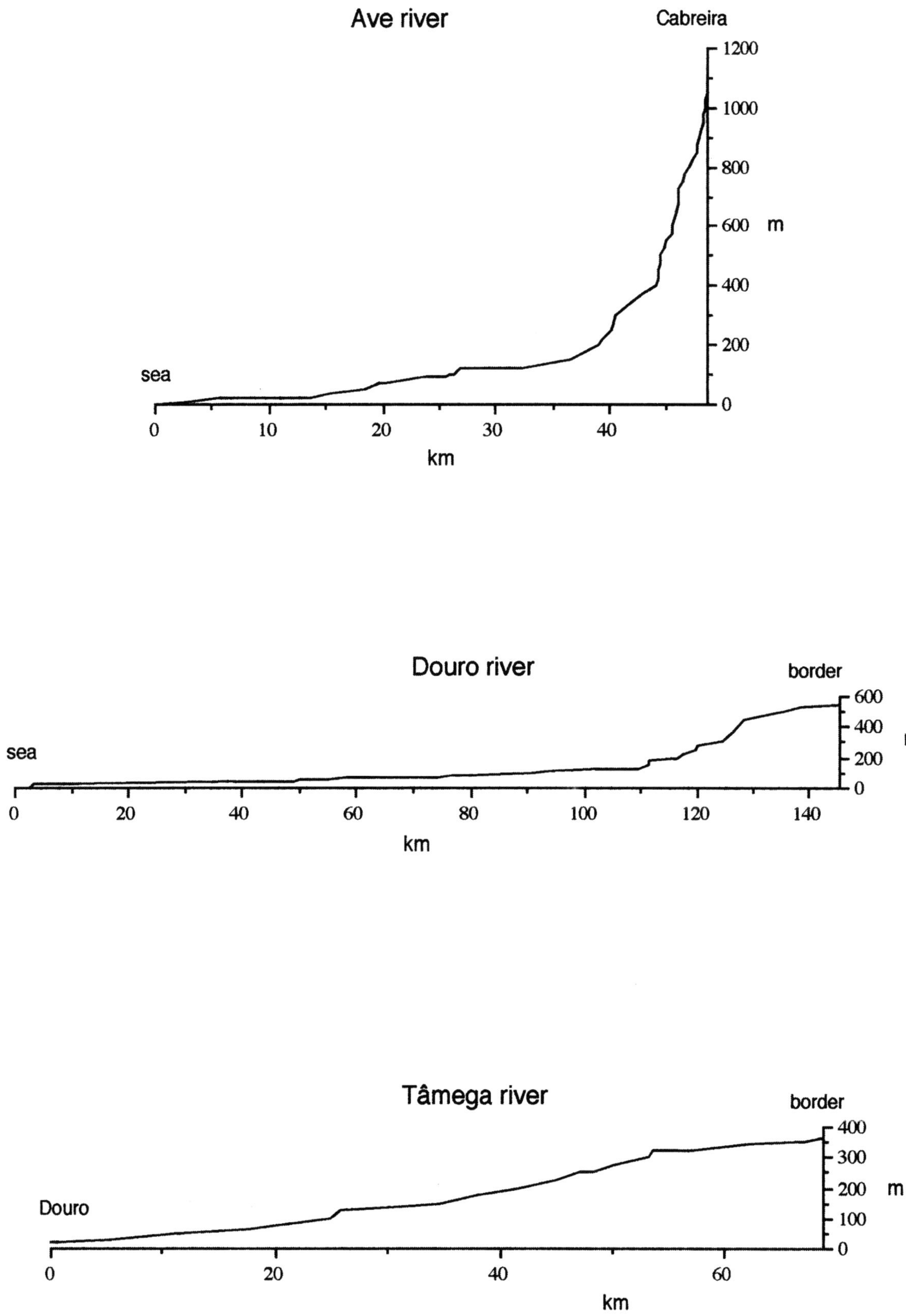

Figure 8.2 - Representation of river gradients in the Portuguese northwest

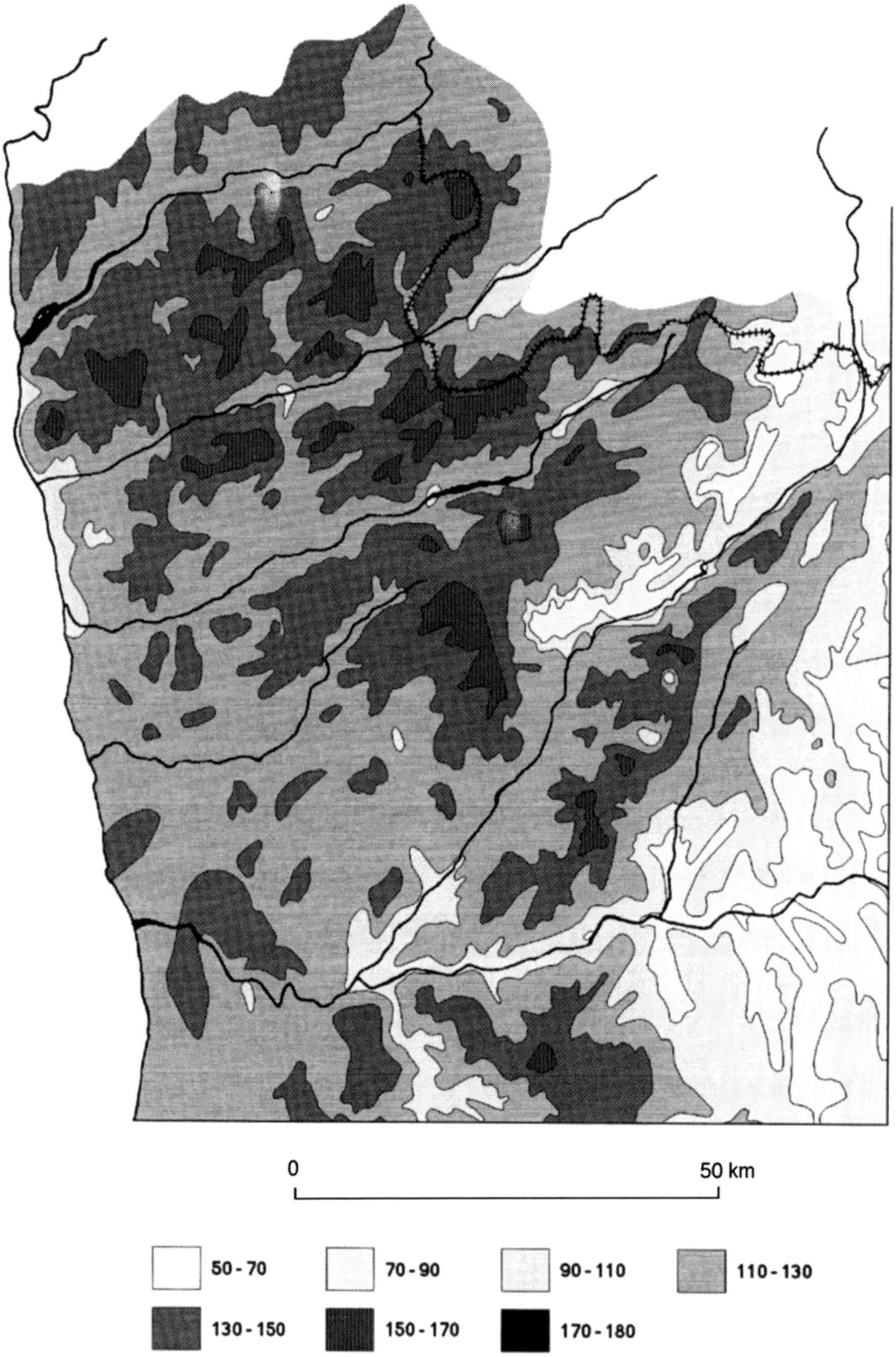

Figure 9 - Chart showing the average number of days of rainfall per annum over the period of (1931-1960). (after Daveau *et al.* 1977, simplified)

species	common name		Late Bronze Age sites		Iron Age sites	
			DEFENDED	OPEN	Pre-Augustean	Augustus-Flavians
Pinus sylvestris	scots pine	%	0.12	-	-	-
Pinus sp.	pine	%	-	-	-	0.01
Taxus baccata	yew	%	2.19	0.08	0.09	-
Quercus (deciduous)	oak	%	45.62	46.8	35.35	30.7
Quercus robur	oak (pedunculate)	%	2.87	3.26	3.02	2.21
Quercus pyrenaica	oak	%	16.06	8.31	-	11.06
Quercus sp.		%	7.49	3.78	4.44	3.78
Corylus avellana	hazel	%	0.24	4.12	1.78	0.96
Crataegus monogyna	hawthorn/ may	%	3.63	0.86	0.62	0.36
Castanea sativa	sweet chestnut	%	1.08	-	-	0.93
Ilex aquifolium	holly	%	-	0.37	-	0.08
Betula pubescens	hairy birch	%	-	0.3	-	0.05
Buxus sempervirens	box	%	-	0.07	-	0.01
Pyrus communis	pear	%	-	0.11	-	0.23
Sorbus aucuparia	rowan	%	0.72	-	-	0.14
Sorbus / Crataegus		%	-	0.41	-	0.18
Sorbus sp.		%	-	0.34	0.18	0.15
Carpinus betulus	hornbeam	%	-	-	-	0.01
Evonymus sp.	spindle	%	-	-	-	0.01
Adenocarpus complicatus		%	0.12	0.15	0.62	0.38
Cytisus sp.	broom	%	-	5.73	-	6.74
Cytisus scoparius	broom	%	0.28	1.98	1.07	0.6
Cytisus laburnum		%	-	-	0.09	0.01
Ulex sp.	gorse/ furze	%	0.32	0.04	-	0.25
Erica cf. cinerea	purple heath	%	-	-	0.18	0.02
Erica tetralix	cross-leaved heath	%	-	-	0.09	2.57
Erica cf. umbellata	heath	%	-	-	-	0.04
Pteridium aquilinum	bracken	%	-	0.15	-	0.01
Fern	fern	%	-	-	-	0.02
Hamilium cf. alyssoides	ploughtail	%	-	-	-	0.15
Leguminosae indet.	legumes	%	11.2	11.61	32.86	12.41
Erica sp.	heath	%	0.12	0.07	1.51	3.57
Cistus sp.	rock-rose	%	-	0.41	-	0.05
Lonicera sp.	honeysuckle	%	-	-	0.09	-
Quercus suber	cork oak	%	0.44	1.01	1.51	2.51
Quercus ilex	holm oak	%	-	0.15	-	0.12
Erica arborea	tree heath	%	-	-	0.98	2.77
Daphne gnidium		%	-	-	-	0.02
Arbutus unedo	strawberry tree	%	-	0.04	-	-
Alnus glutinosa	alder	%	0.12	0.9	3.55	1.8
Frangula alnus	berry bearing alder	%	-	0.04	0.18	0.31
Alnus/Corylus		%	-	-	0.36	0.08
Fraxinus angustifolia	ash	%	0.28	-	0.09	0.66
Fraxinus excelsior	ash	%	-	-	-	0.02
Fraxinus sp.	ash	%	-	0.28	-	0.1
Hedera helix	ivy	%	-	0.22	0.09	2.06
Monocotyledoneae	monocotyledons	%	0.04	-	0.27	0.19
Gramineae	grasses	%	-	-	2.04	-
Populus sp.	poplar	%	-	0.04	0.36	0.06
Salix sp.	sallow/willow	%	0.2	2.02	2.66	1.16
Salix / Populus		%	-	-	0.18	0.03
Salicaceae indet.		%	-	0.11	-	-
Sambucus nigra	elder	%	-	0.97	-	1.18
Sambucus / Hedera		%	-	-	-	0.06
Clematis vitalba	old man's beard	%	0.04	0.97	0.18	0.93
Ficus carica	fig	%	-	-	-	0.05
Prunus avium	wild cherry	%	-	-	-	0.02
Prunus domestica	wild plum	%	-	0.04	-	0.08
Prunus sp.		%	-	-	-	0.01
Juglans regia	walnut	%	-	-	0.09	0.06
Vitis vinifera	grape vine	%	-	-	-	0.05
Vitis sp.	vine	%	-	-	-	0.04
Olea europaea	olive	%	-	-	-	0.05
Rosaceae pomoidea	hawthorn	%	-	0.49	0.71	0.36
Indeterminable		%	6.85	3.78	4.8	7.49

Figure 10 - Chart representing the variation of species in the Ist millennium BC (data from Figueiral 1990)

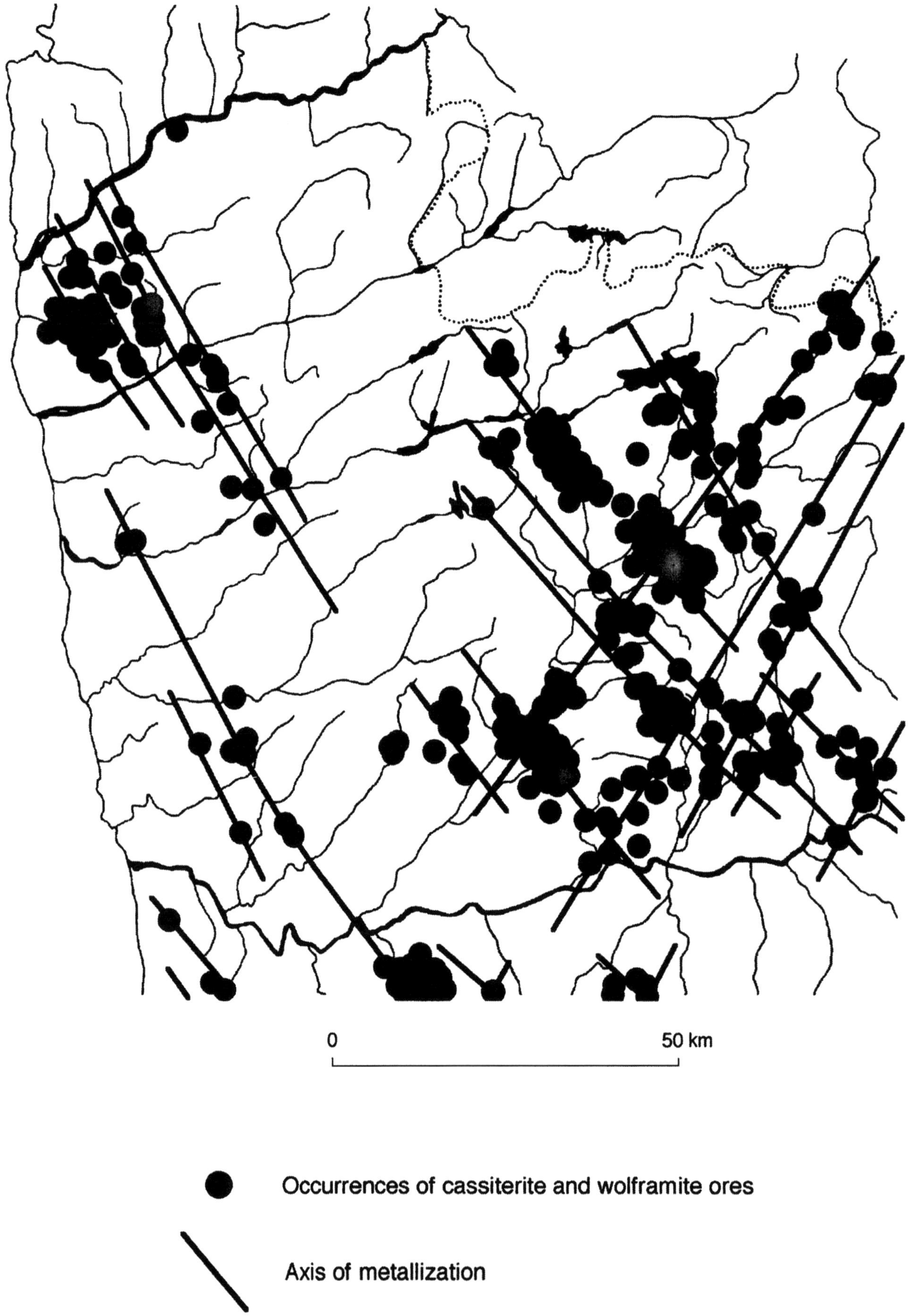

Figure 11 - Occurrences of tin and wolfram ores in the northwest of Portugal
(source Neiva 1944)

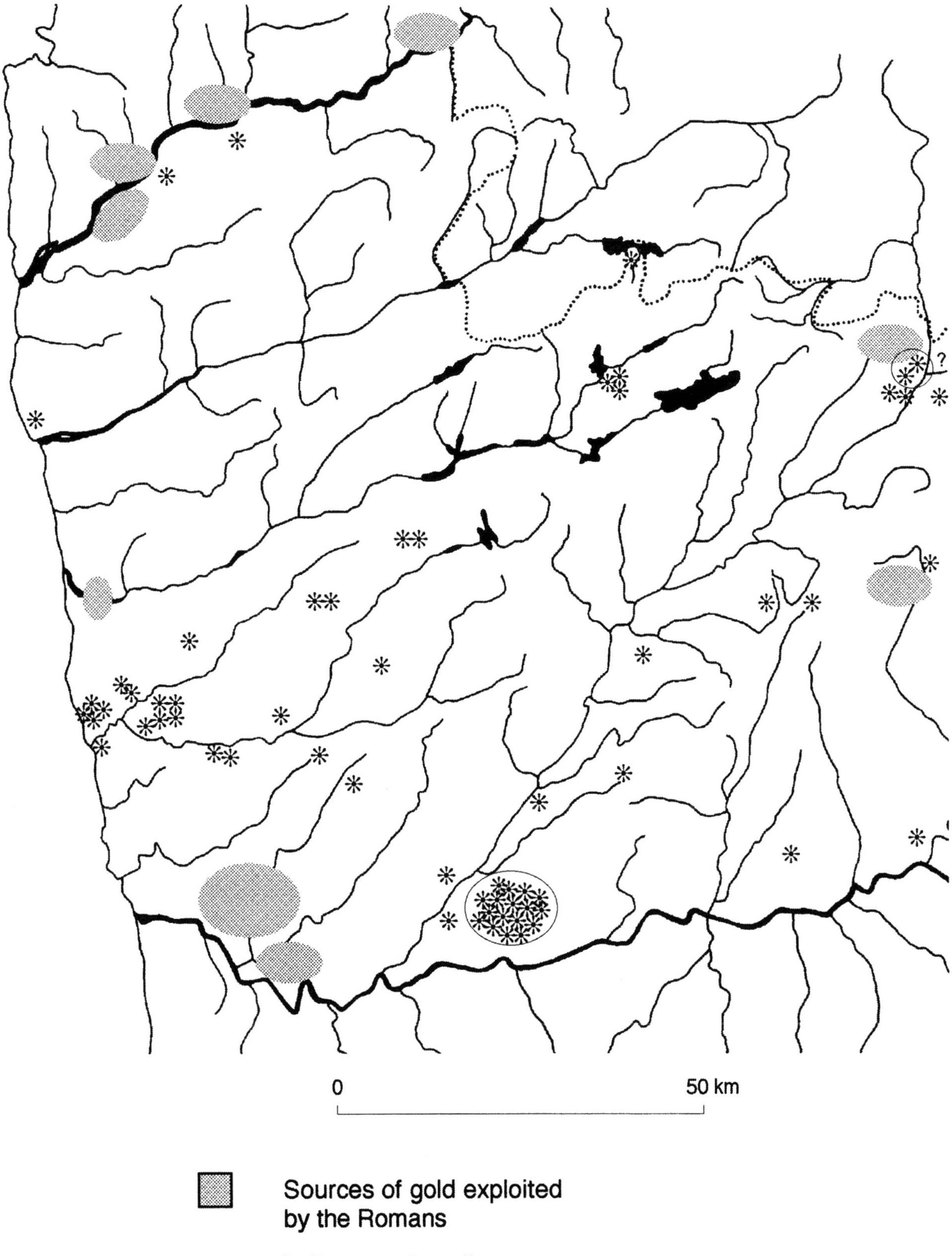

Figure 12 - Evidence for the exploitation of gold sources, and the distribution of indigenous jewellery.

(sources: Alan 1965; Almeida, CAB, 1990; Coffyn1985; Domergue1990, Hartmann 1971; Lopez Cuevillas 1951c and 1953d, Sanchez Palencia 1983; Silva 1986a; Thadeu 1965)

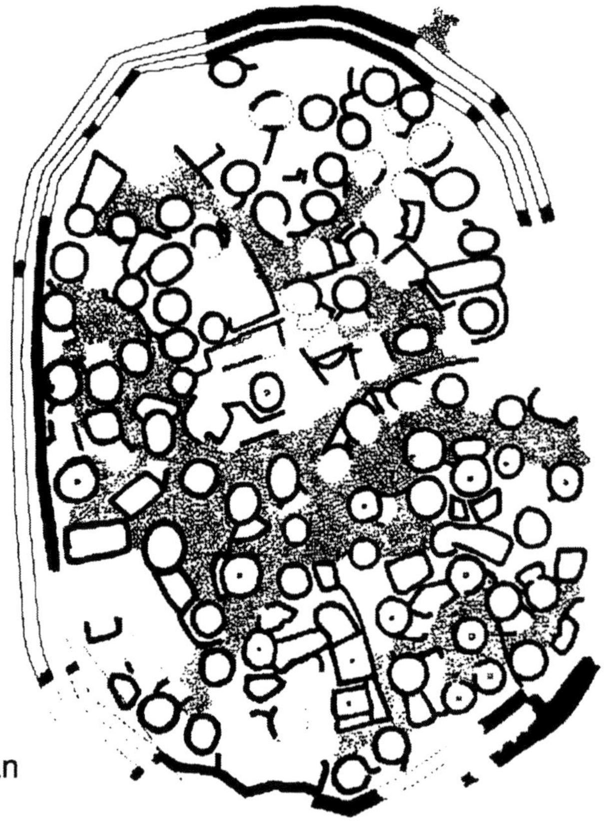

Figure 13 - Two broad types of urban organization of a castro: Pre-Roman (Cividade de Terroso) and under Roman influence (Citânia de Sanfins)
(after Silva 1986a, not to scale)

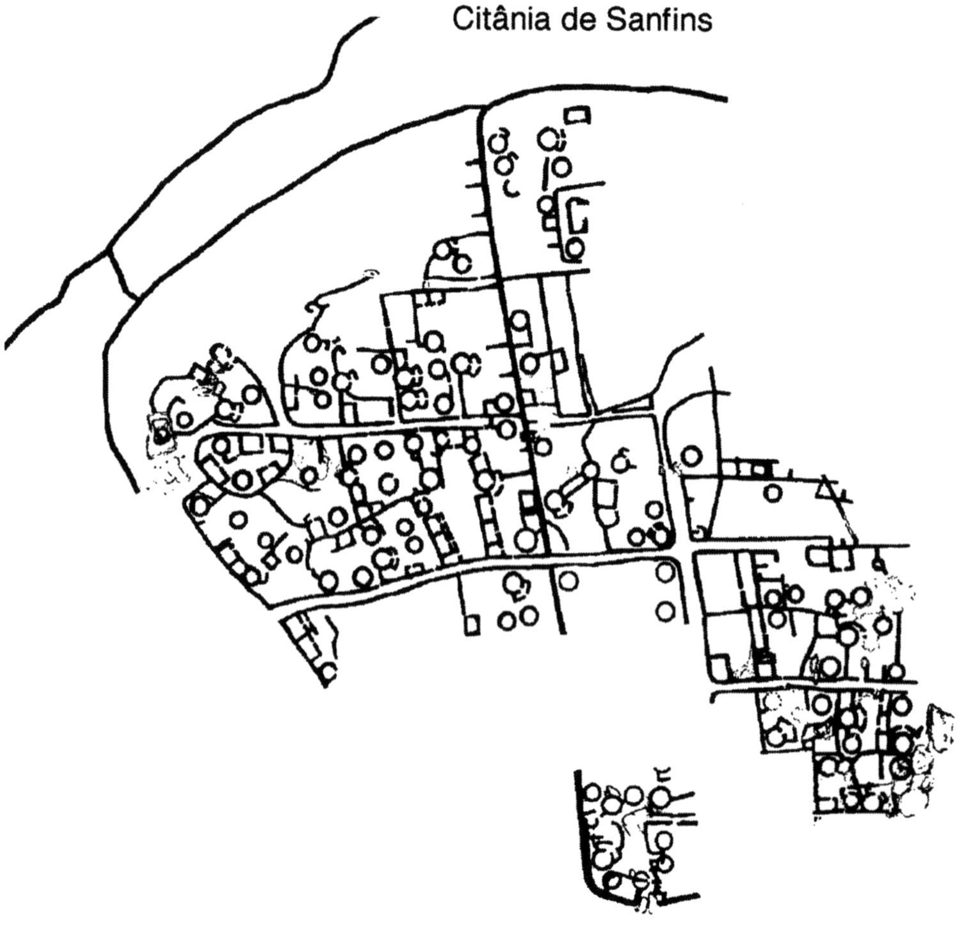

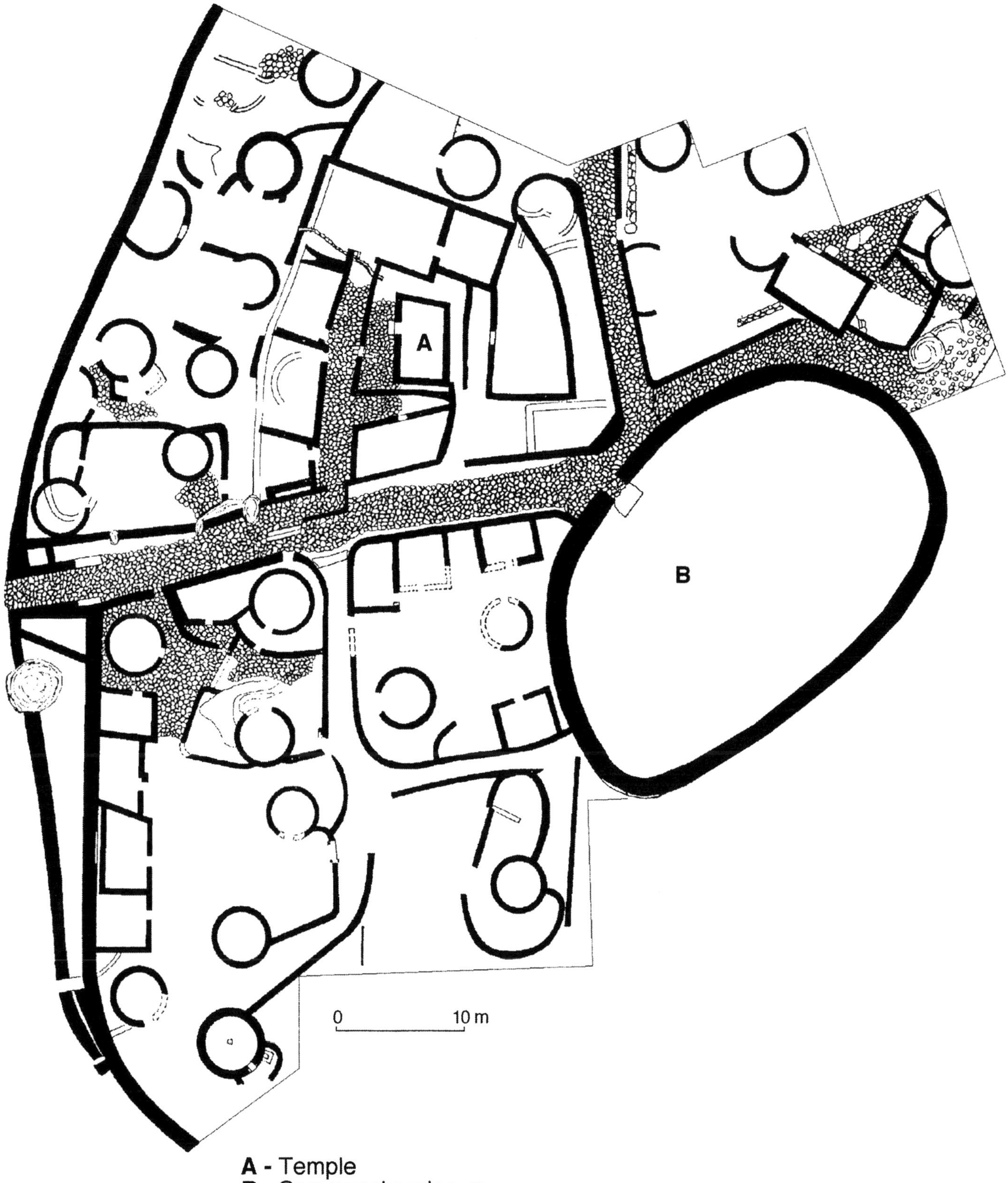

A - Temple
B - Communal enclosure

Figure 14 - Partial plan of Mozinho, Penafiel, showing position of the dwelling enclosures and communal areas.
(after Almeida 1974b, 1977, 1980c and Soeiro 1984, updated)

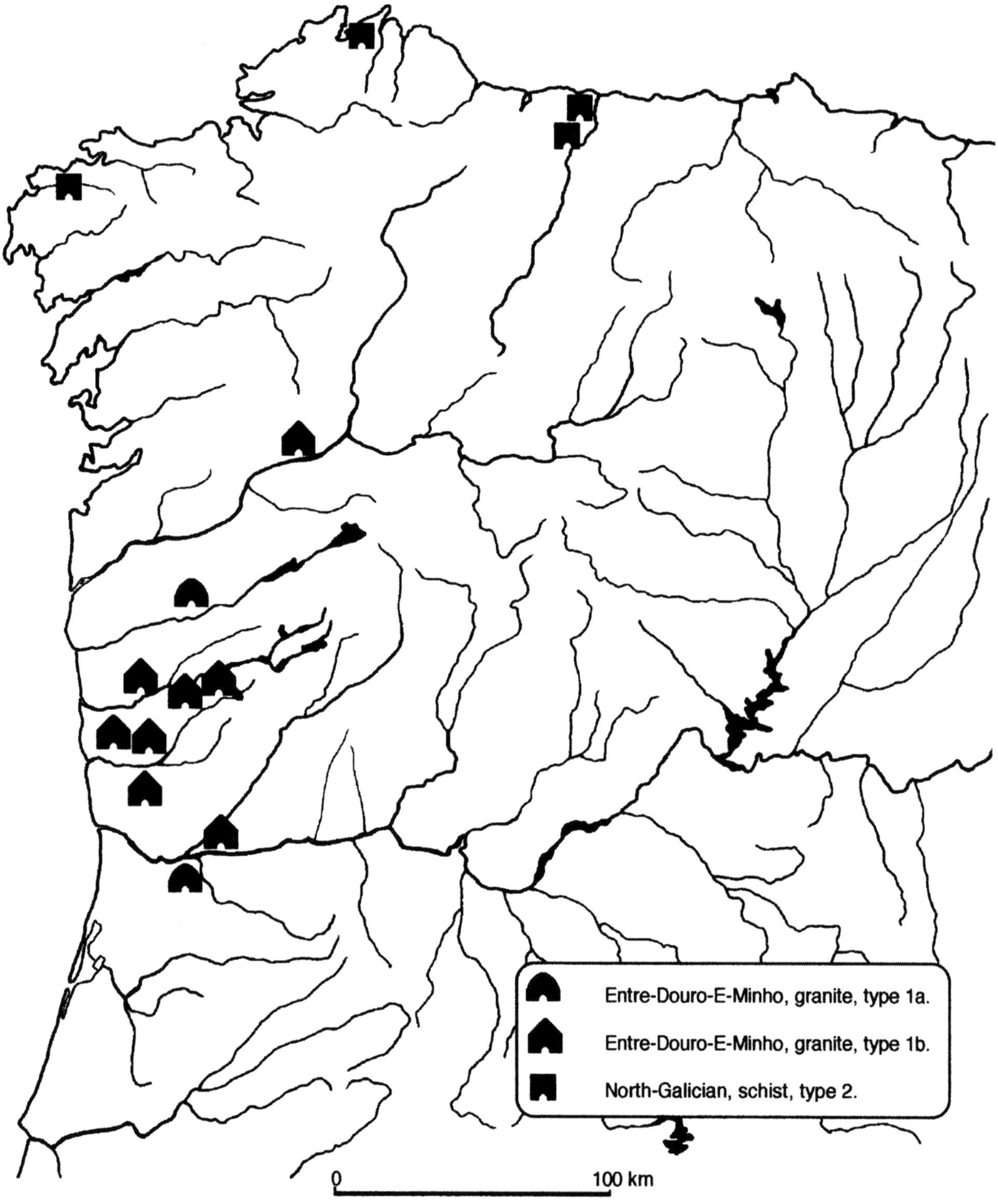

1- Borneiro, Coruña
2- Castro de Prados, Ortigueira, Ortegal
3- Coaña, Navia, Asturias
4- Pendia, Boal, Asturias (2)
5- Aguas Santas, Orense, Galicia
6- Castro de Eiras (?), Arcos de Valdevez
7- Briteiros, Taipas (2)
8- Sabroso, Taipas
9- Galegos, Barcelos
10- Eiras, V.N. Famalicão
11- Saia, Barcelos
12- Sanfins, Paços de Ferreira
13- Freixo, Marco de Canavezes
14- Sardoura, Castelo de Paiva

Figure 15 - The Iron Age bath structures of the Penninsular Northwest.

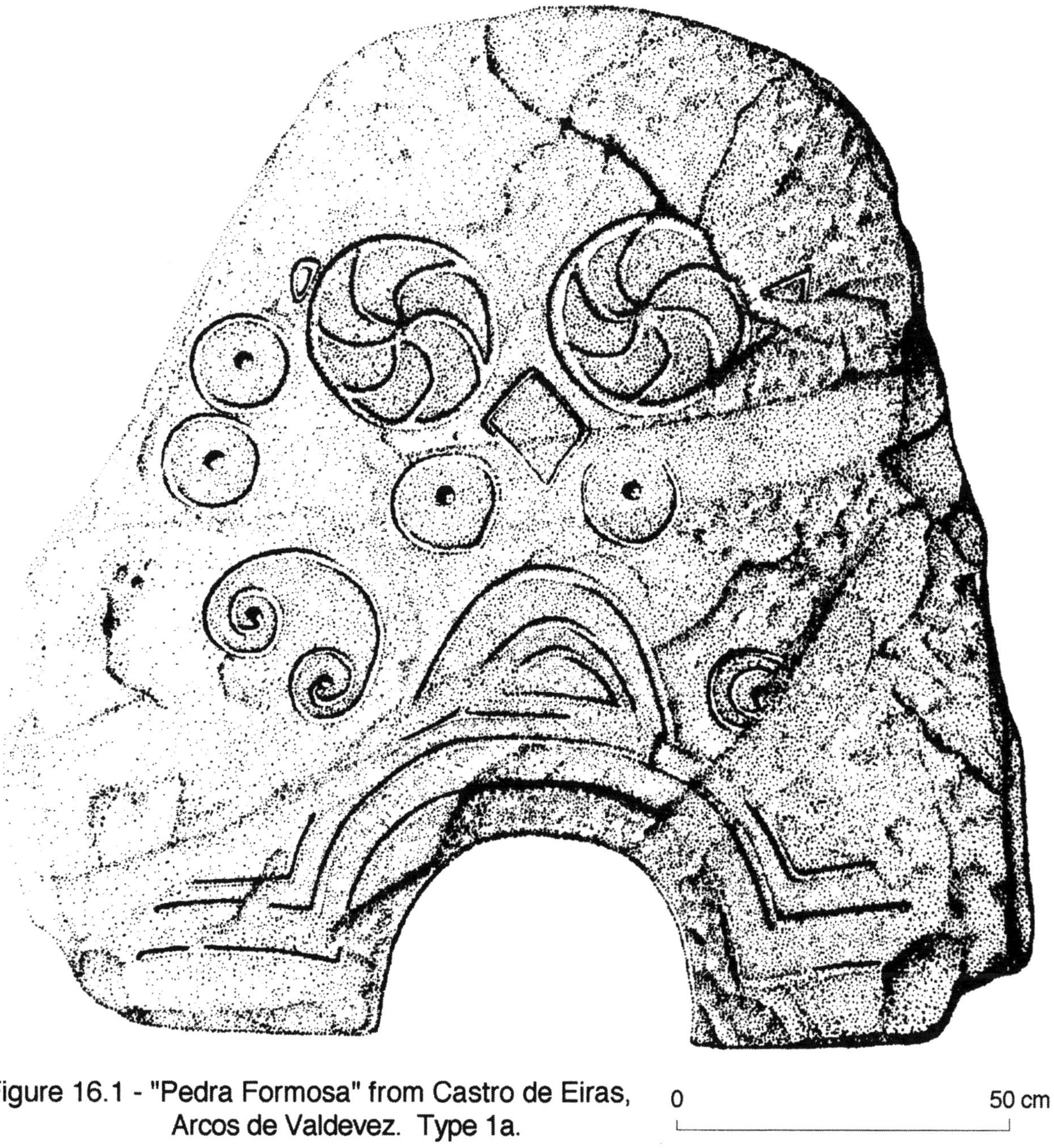

Figure 16.1 - "Pedra Formosa" from Castro de Eiras, Arcos de Valdevez. Type 1a.
(reproduced by kind permission of A. M. Baptista)

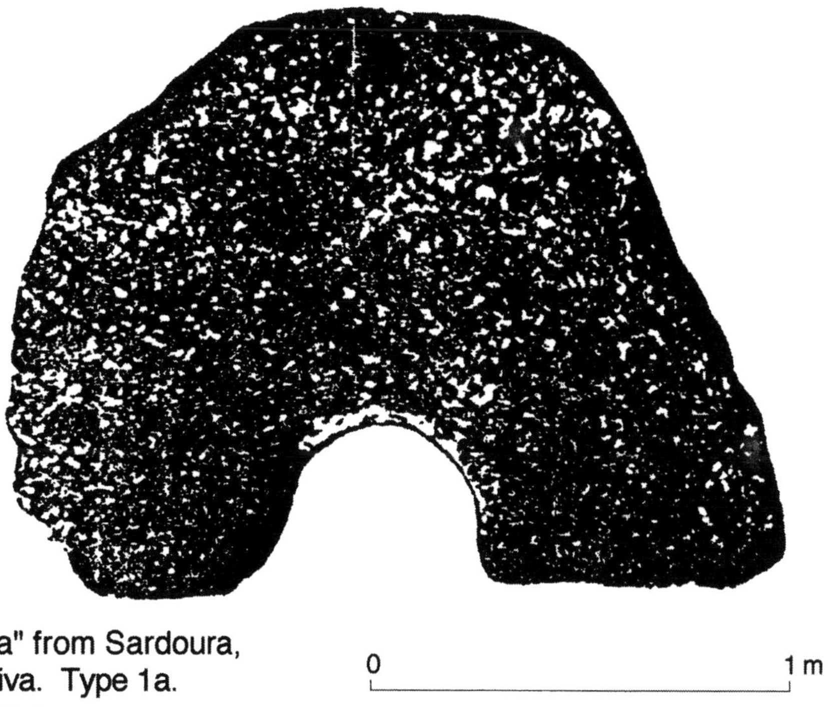

Figure 16.2 - "Pedra Formosa" from Sardoura, Castelo de Paiva. Type 1a.
(after Cardozo 1949c)

Figure 17.1 - "Pedra Formosa" from Castro das Eiras, V.N. Famalicão. Type 1b.
(drawing by Christina Unwin)

Figure 17.2 - "Pedra Formosa" from Citânia de Briteiros, Guimarães. Type 1b.
(not to scale)

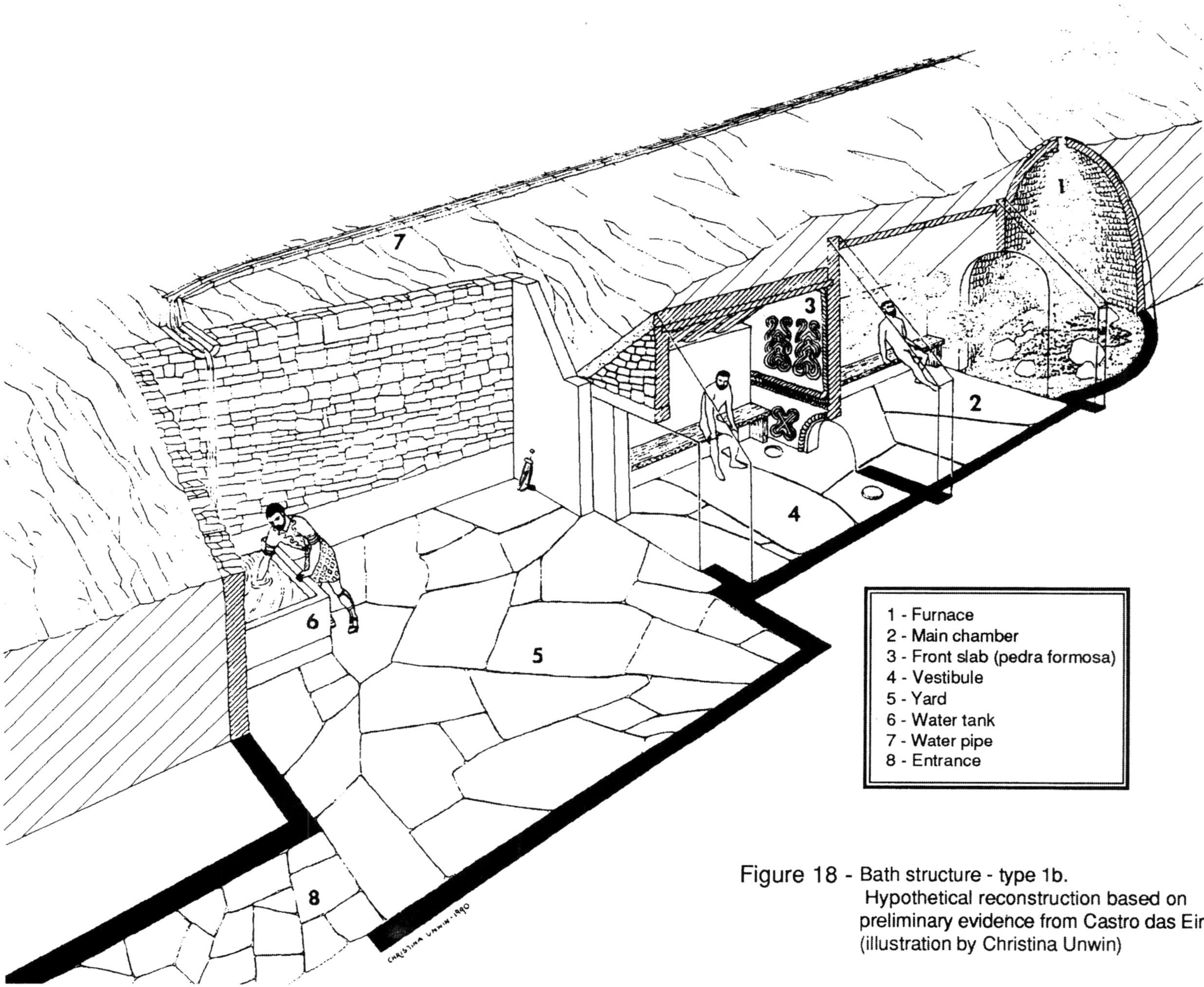

Figure 18 - Bath structure - type 1b.
Hypothetical reconstruction based on preliminary evidence from Castro das Eiras.
(illustration by Christina Unwin)

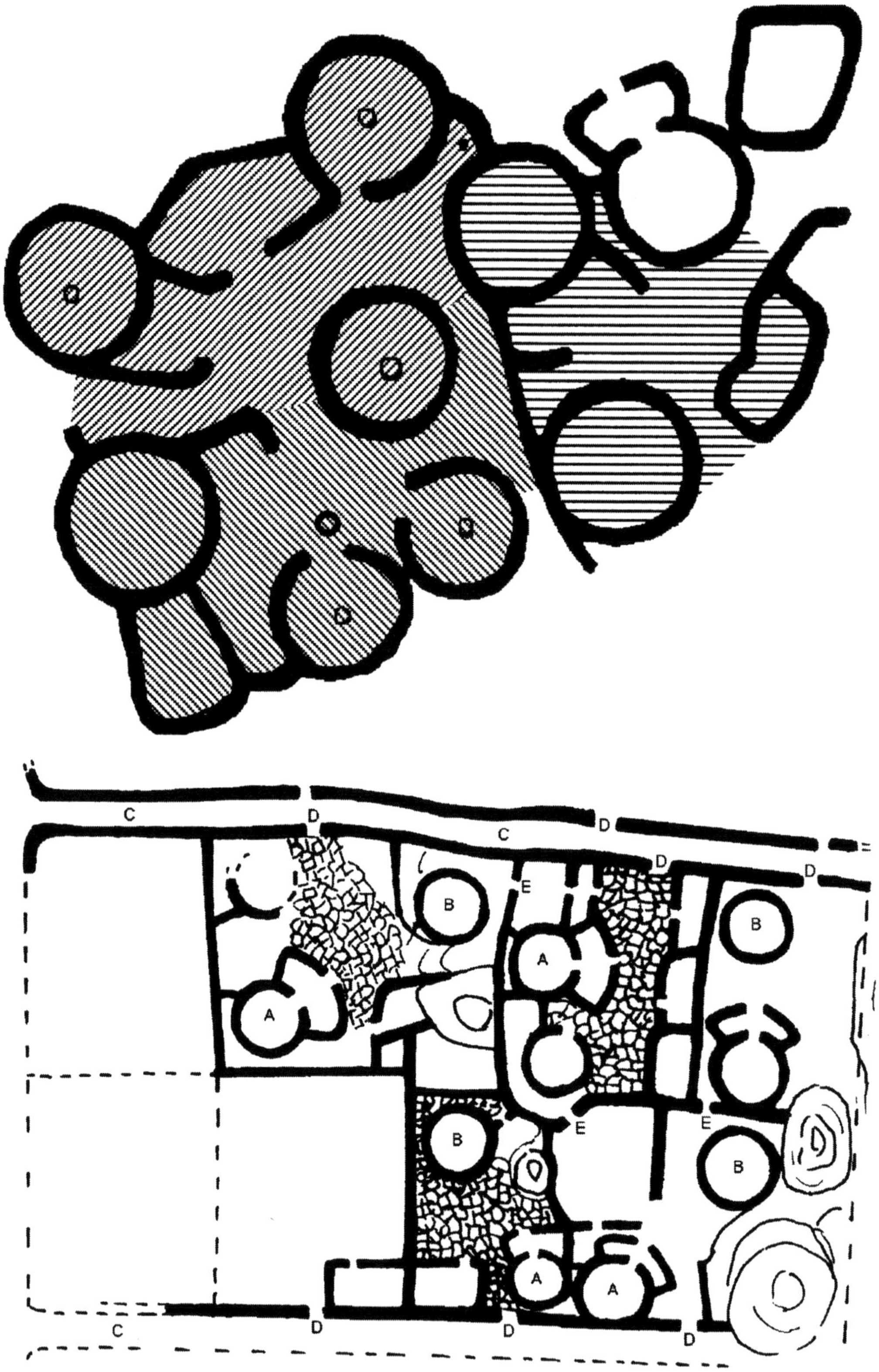

A - House
B - Granary (?)
C - Street
D - Entrance to the dwelling enclosure
E - Communication between enclosures

Figure 19 - Dwelling units from Cividade de Terroso (above), and enclosed dwelling units from Citânia de Sanfins (below).
(sources: Almeida 1984 and Silva 1986a)

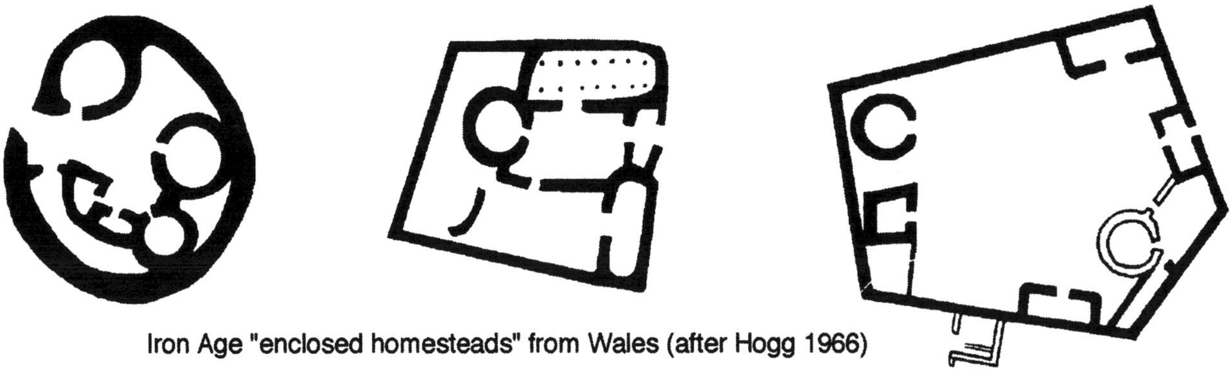

Iron Age "enclosed homesteads" from Wales (after Hogg 1966)

Enclosures from Sanfins (after Romero Masiá 1976)

Enclosure from Santa Luzia
(after Viana 1955a)

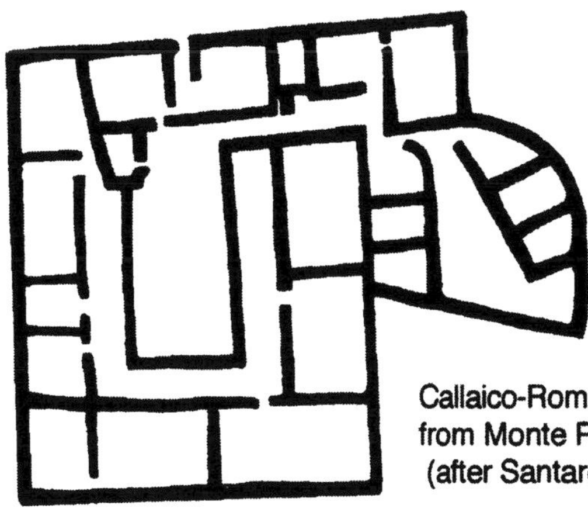

Callaico-Roman *domus*
from Monte Padrão
(after Santarém 1955a)

Figure 20 - Some typical forms of dwelling unit from the north of Portugal, with relevant European examples for comparison. (not to scale)

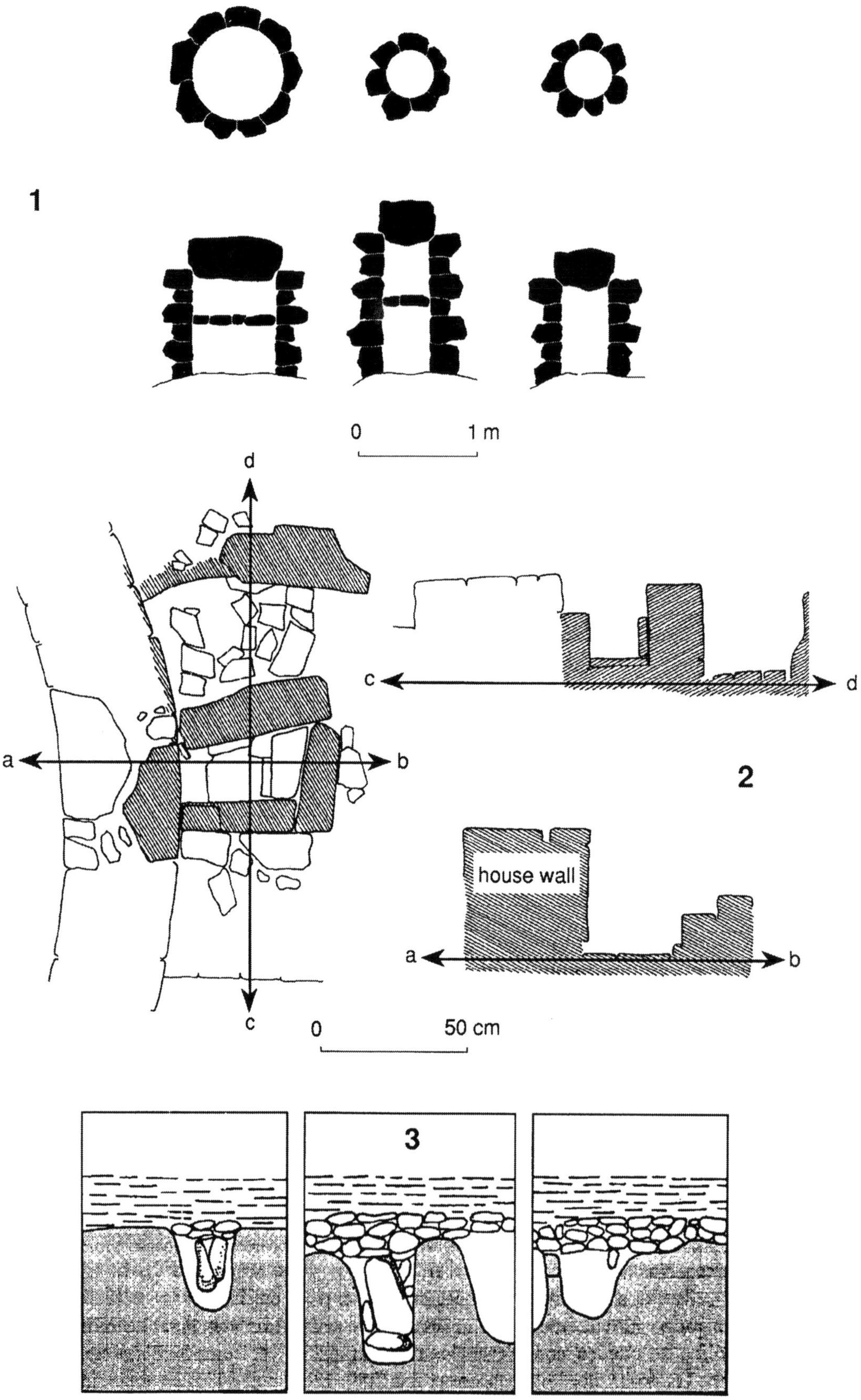

Figure 21 - Iron Age funerary structures

1- Castro de Terroso (after Garcia y Bellido 1966a)

2- Cividade de Âncora (after Silva 1986a)

3- Castro de Meirás (after Romero Masiá 1976)

(Scale not available)

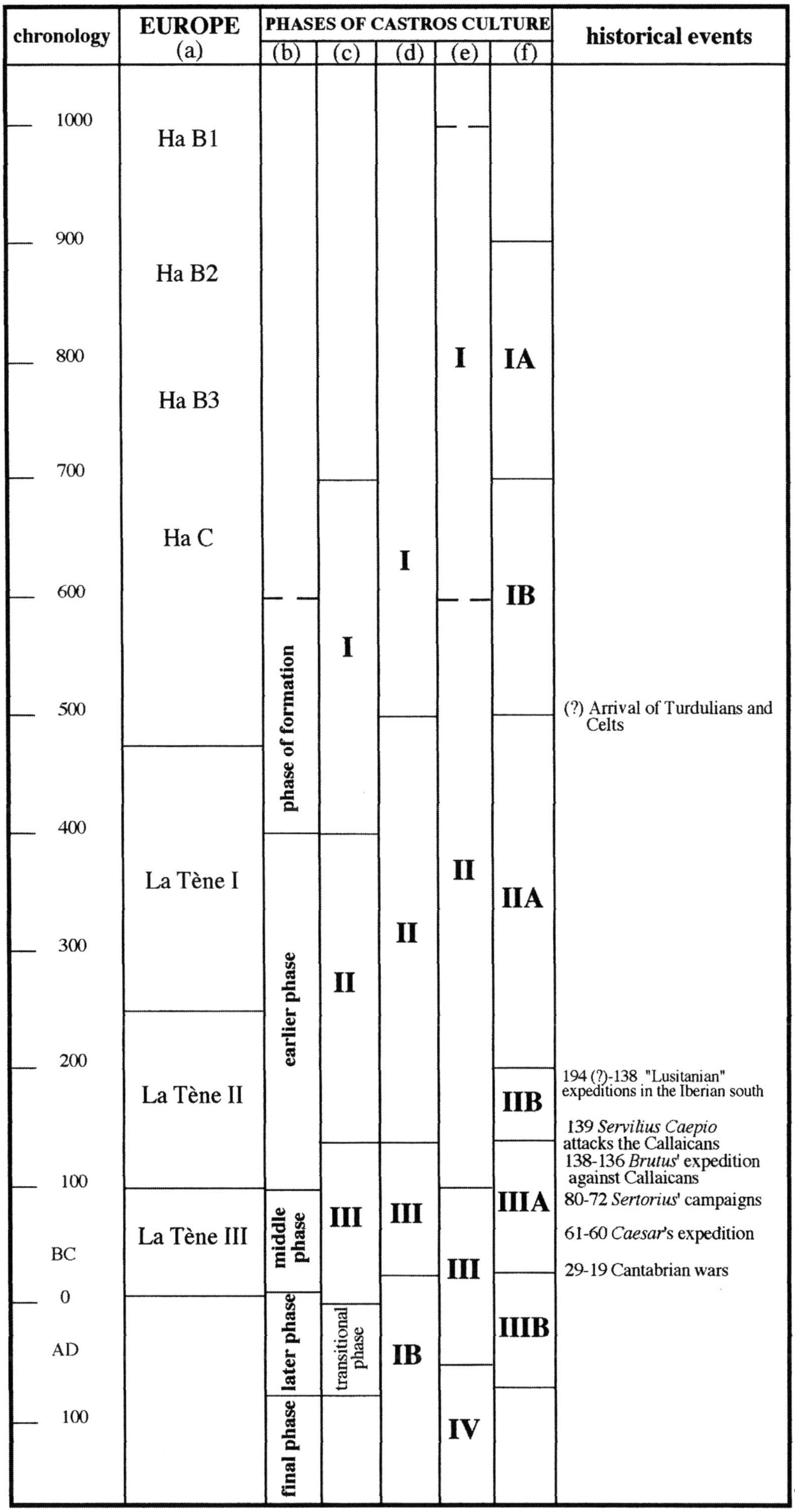

after:
a) Müller-Karpe 1959, Duval 1971, Guilaine 1976, Lumley *et al.* 1976
b) Almeida 1983a.
c) Fariña *et al.* 1983.
d) Maluquer de Motes 1973, 1975a.
e) Martins 1977 (1990).
f) Silva 1986a

Figure 22 - Chronological table showing the relative position of the Castros Culture and its phases in later Prehistoric Europe.

1A - Hilltop, well detached from the massif
1B - Centre or edge of a plateau
1C - Highland spur
1D - Lowland spur over river valley

2A - Higher spur by the sea shore and/or river estuary
2B - Low spur by the sea shore and/or river estuary

3 - River valley

Figure 23 - Patterns of geographical distribution of castros in northern Portugal

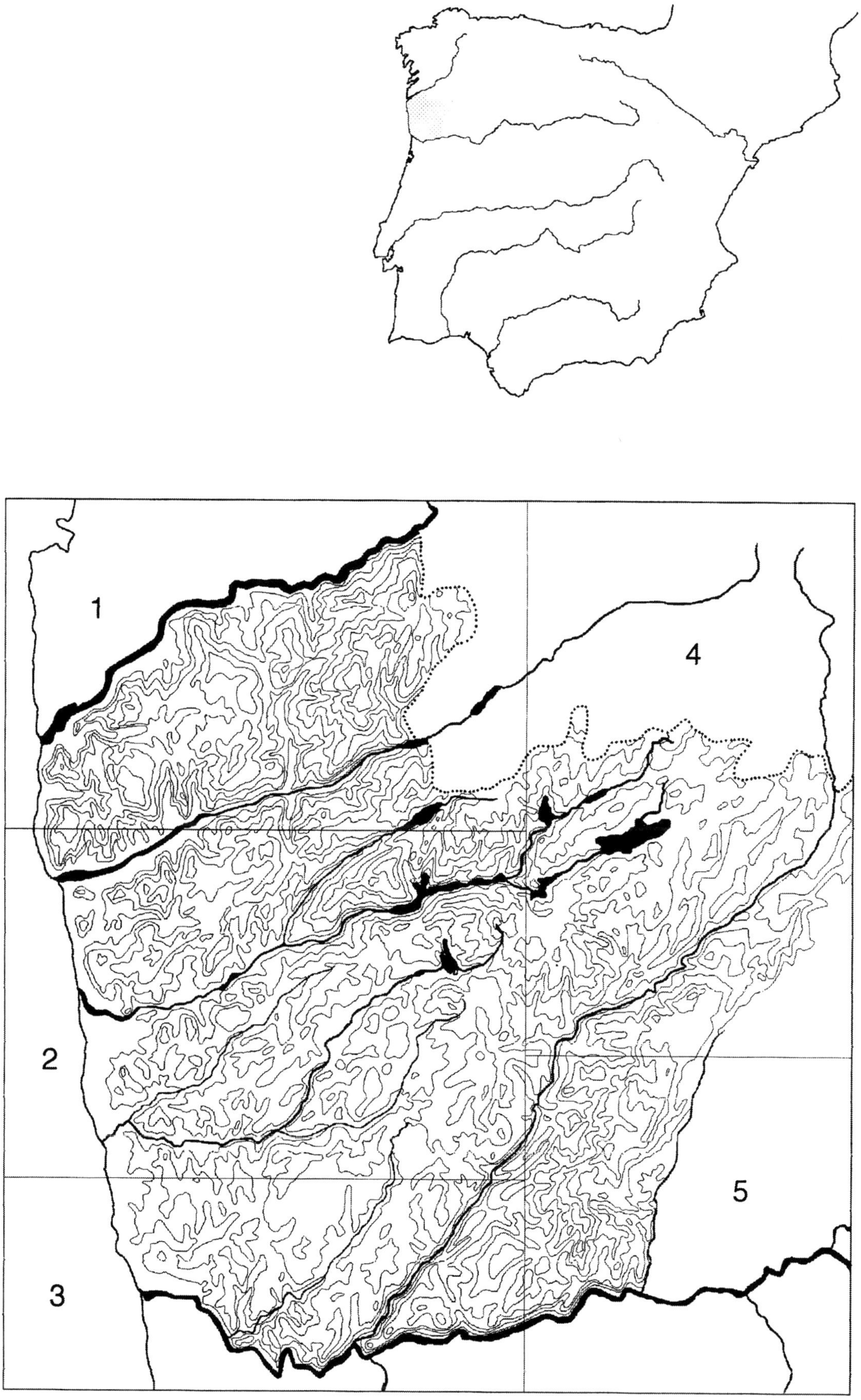

Figure 24 - Key to insets for the location of Castros in the northwest of Portugal.

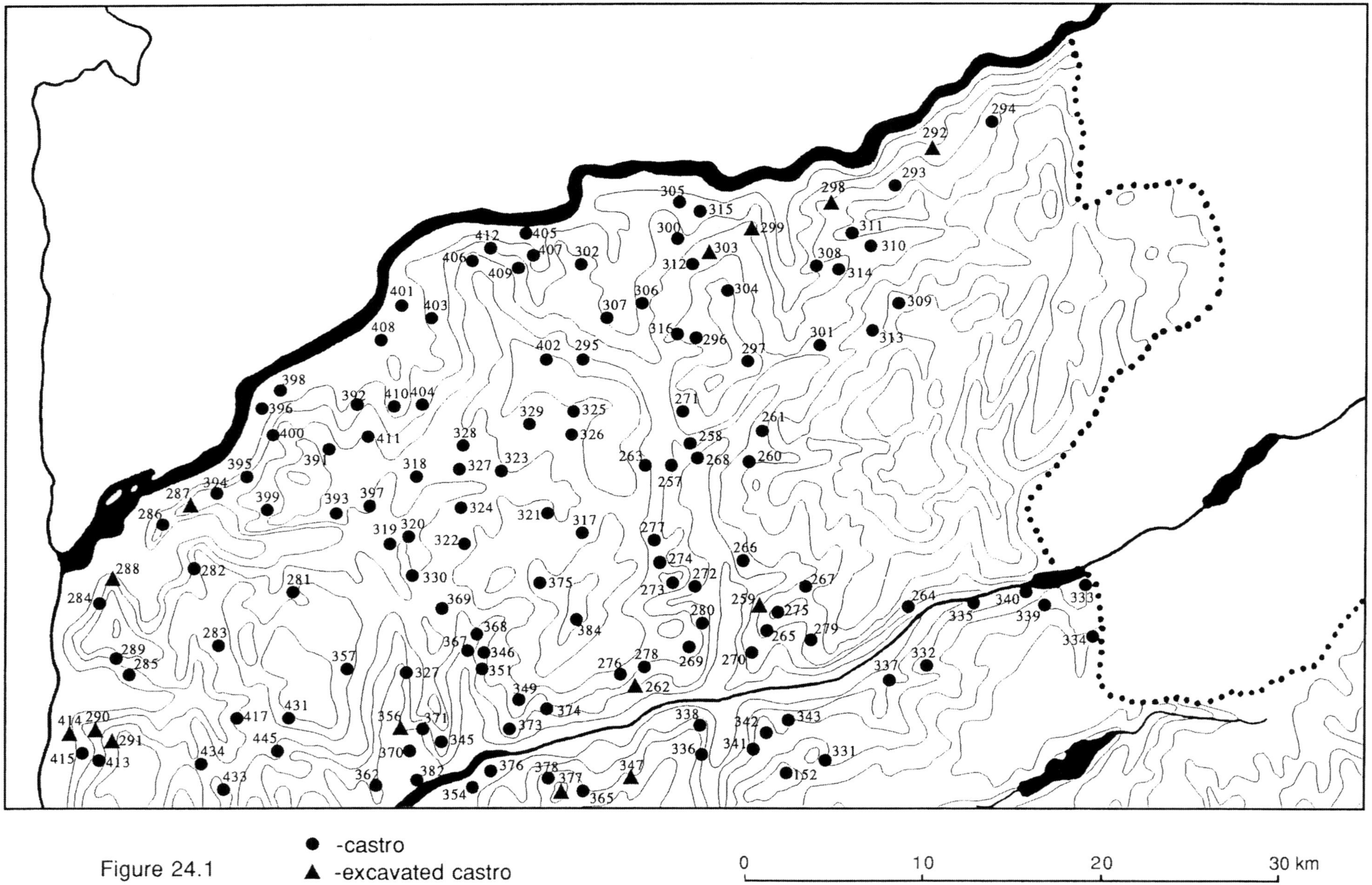

Figure 24.1

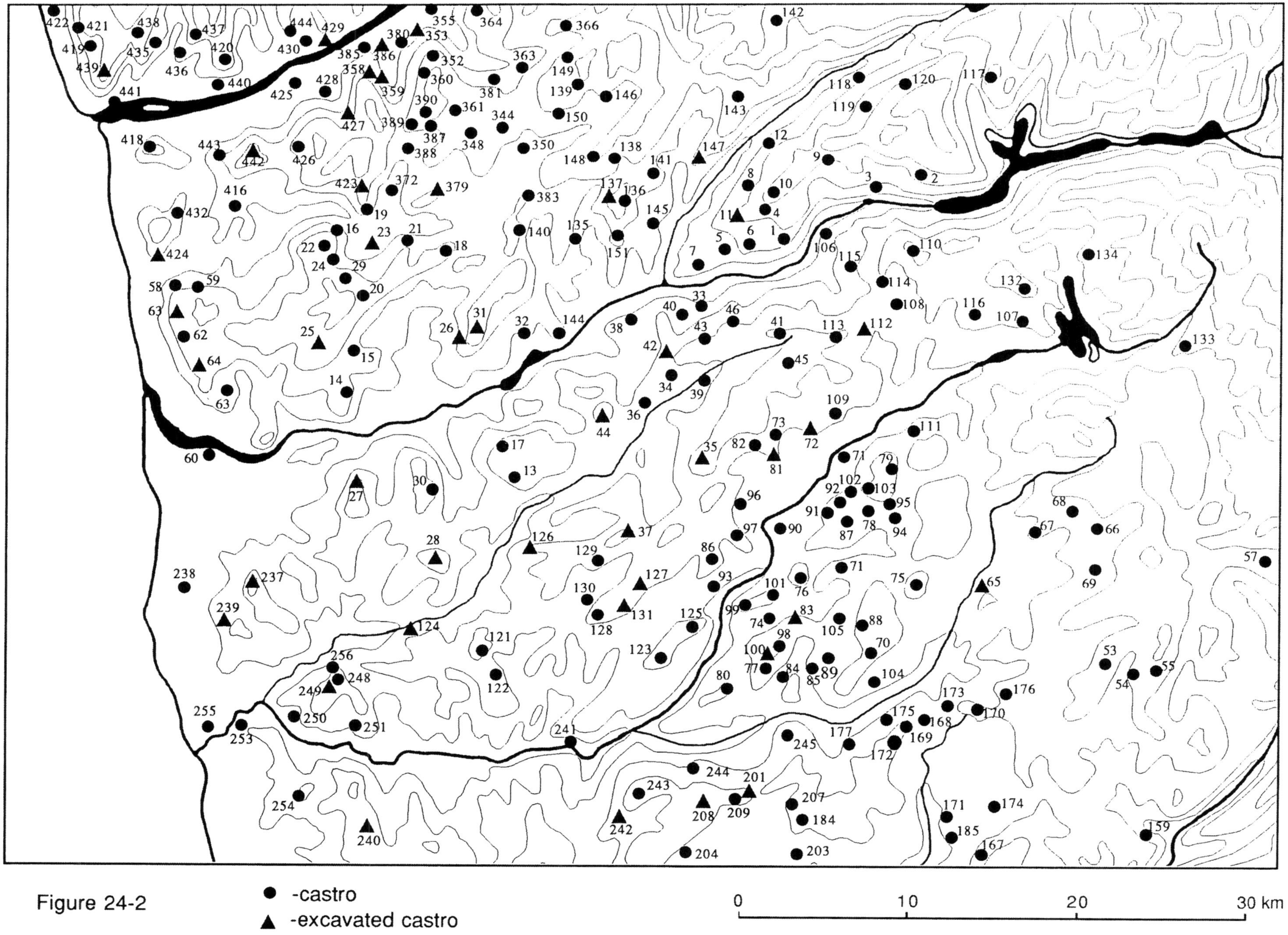

Figure 24-2

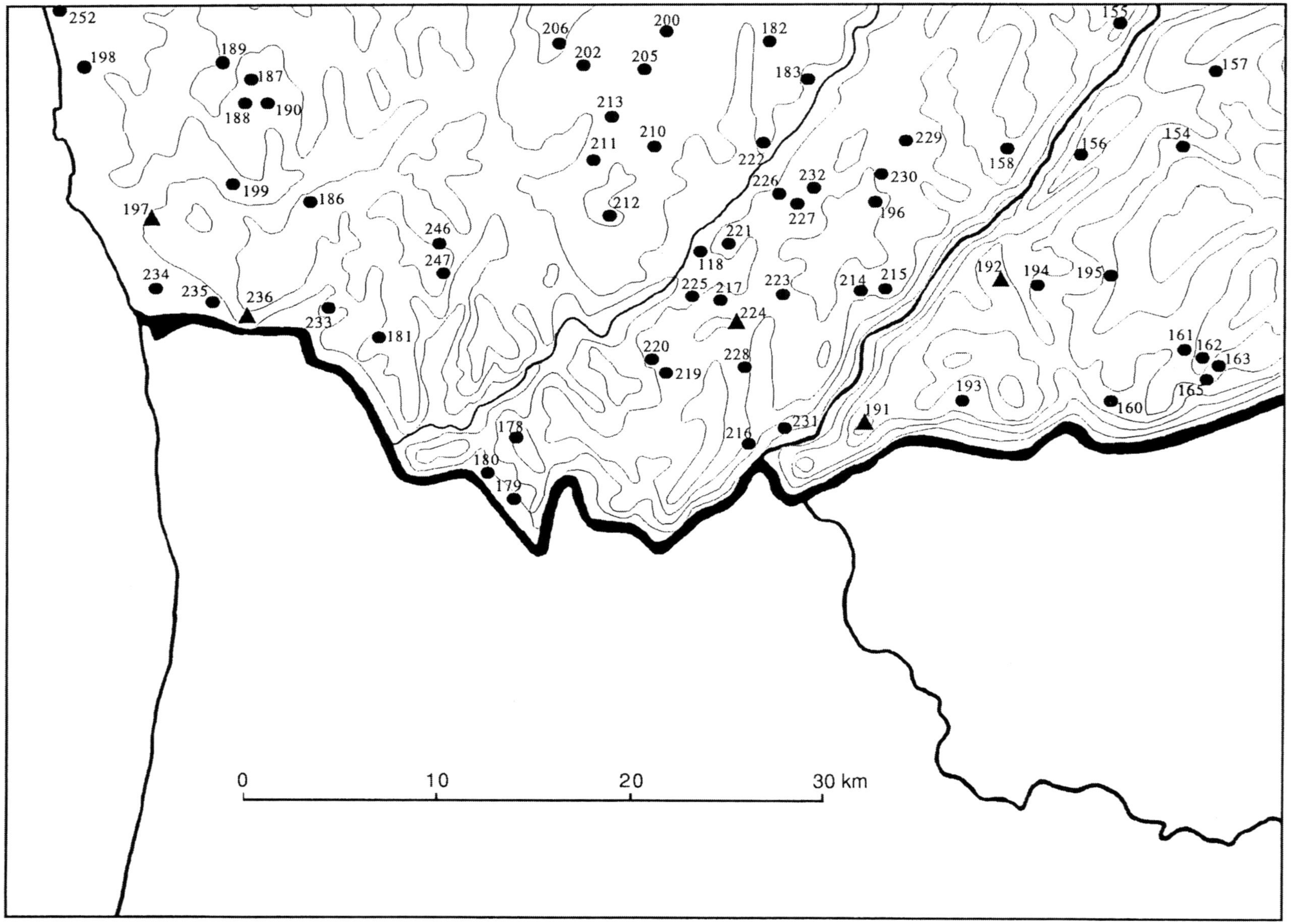

● -castro
▲ -excavated castro

Figure 24.3

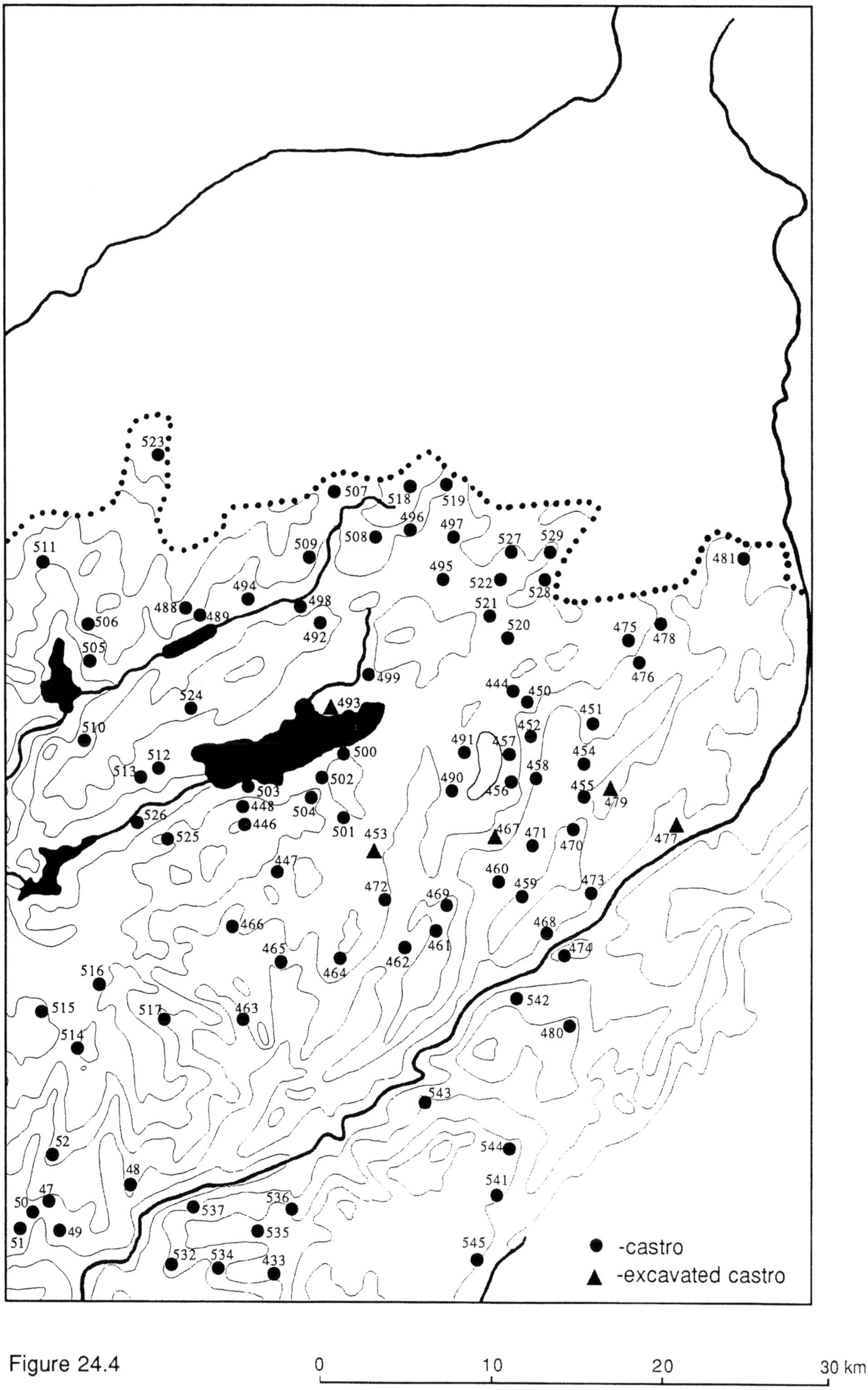

Figure 24.4

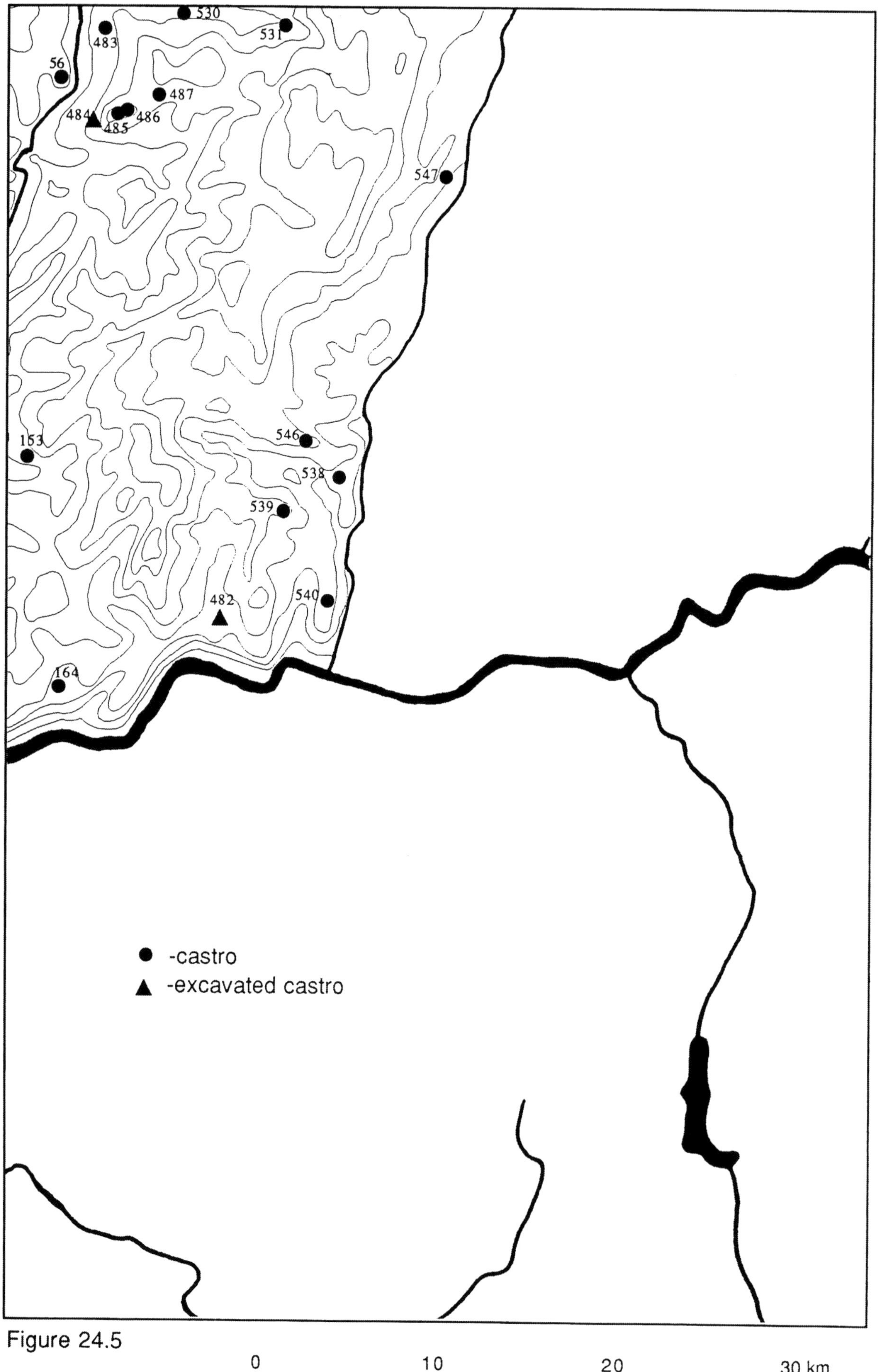
530
531
483
56
487
484
485
486
547
546
153
538
539
540
482
164
-castro
-excavated castro
0
10
20
30 km

Figure 24.5

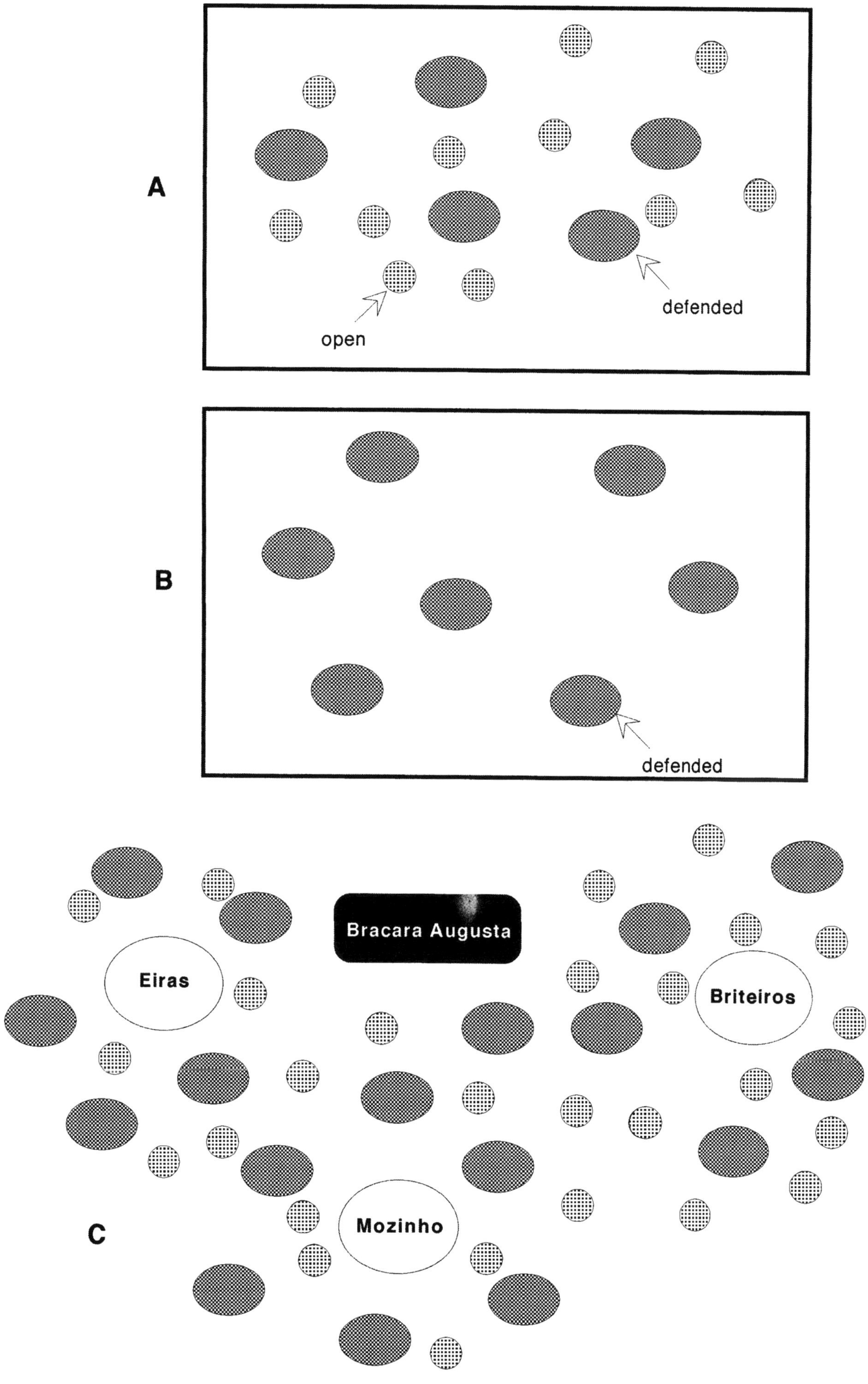

Figure 25 - Model for site hierarchy: A- late Bronze Age, B- second half of the millennium, C- Augustus to Flavian period.

Figure 26 - The hill of Sta. Cristina, Vila Nova de Famalicão.
One particular case of settlement distribution during the late Iron Age

Figure 27.1 - Castelo de Matos, seen from the plateau.

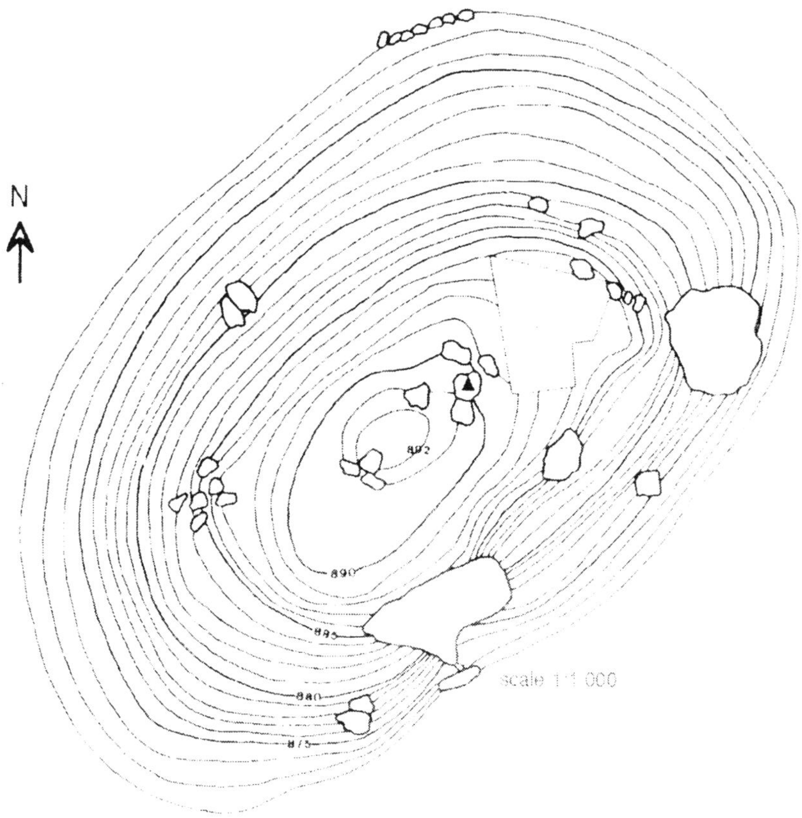

Figure 27.2 - Castelo de Matos. Plan of the hilltop showing position of the excavated area. Scale 1/1.000, contours 1 metre.
(source: Figueiral - Queiroga 1988)

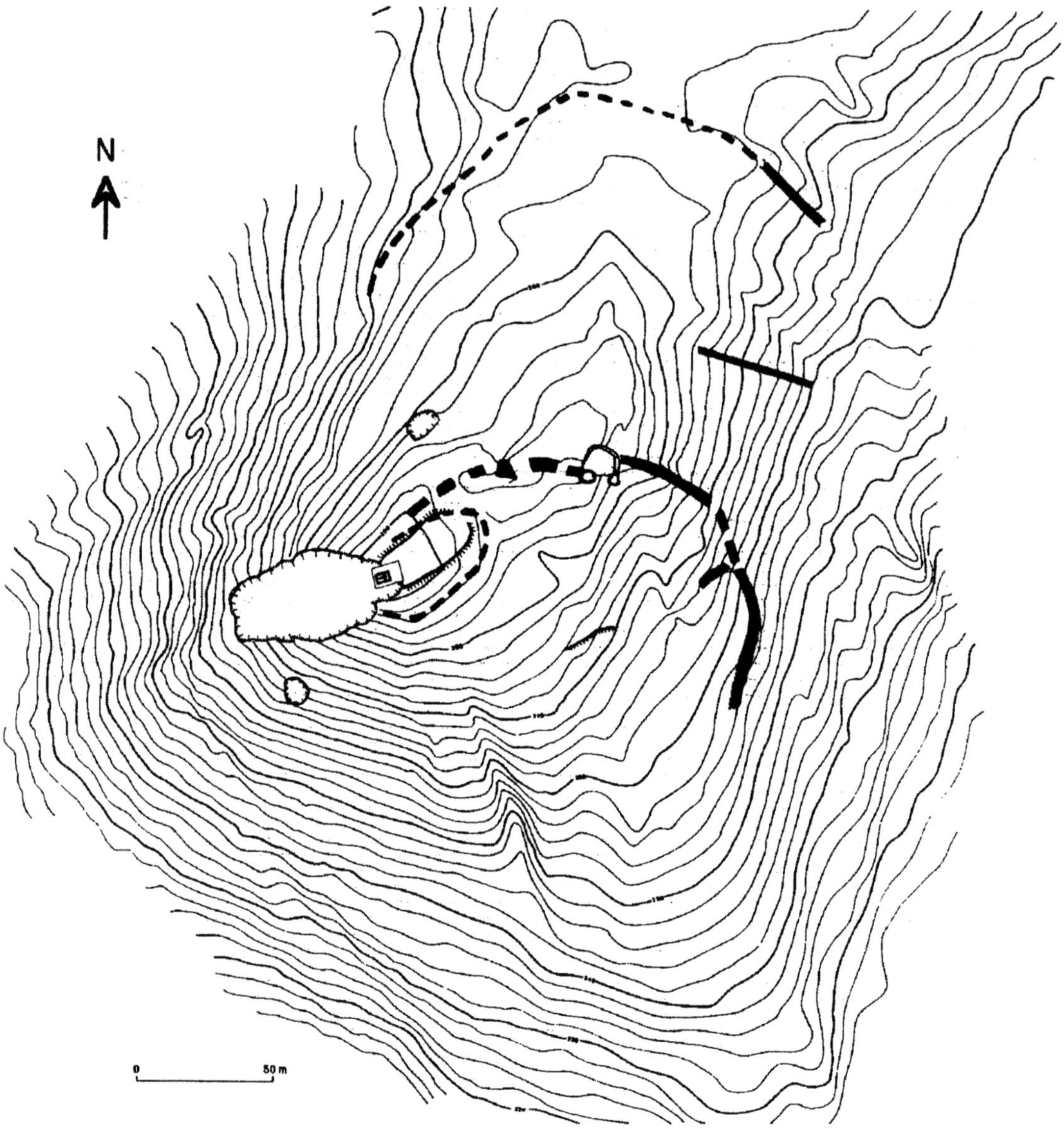

Figure 28.1 - Survey plan of Citânia de S. Julião, Vila Verde.
Thicker lines represent defensive walls of the later phase.
(after Martins 1990)

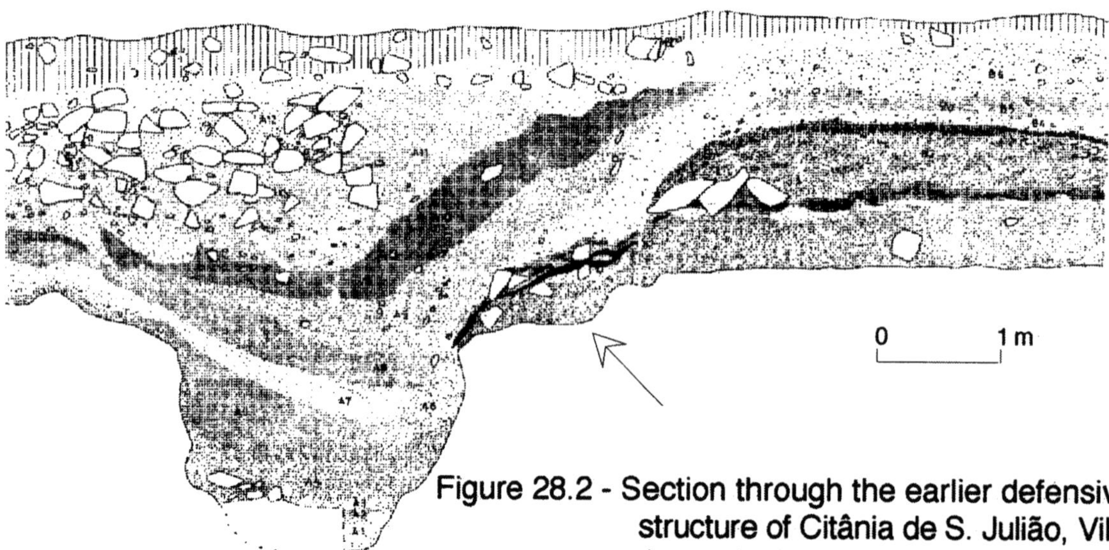

Figure 28.2 - Section through the earlier defensive structure of Citânia de S. Julião, Vila Verde. Arrow indicates the position of a possible enclosing structure, as suggested by stratigraphy. (after Martins 1990)

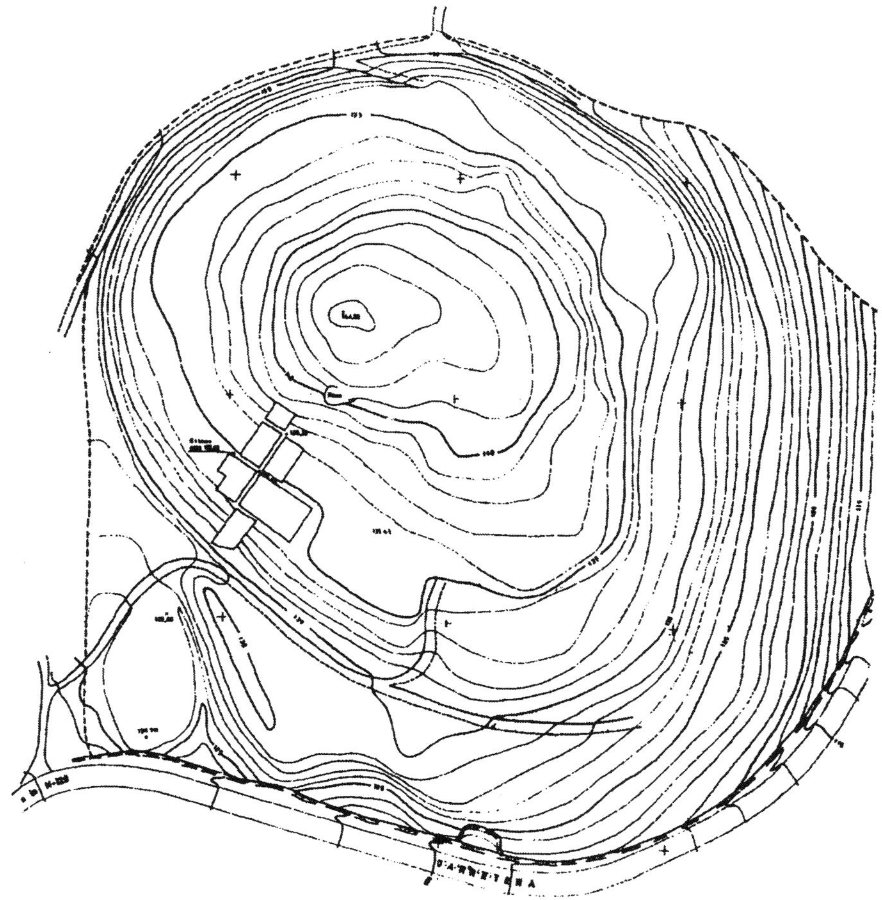

Figure 29 - Plan of Castro de Torroso (scale not available) showing the position of the excavated area, and cross section through the dwelling area and ditch. Characteristics of the stratigraphy contiguous to the ditch suggest the existence of a perishable enclosing structure.
(source Peña Santos 1987)

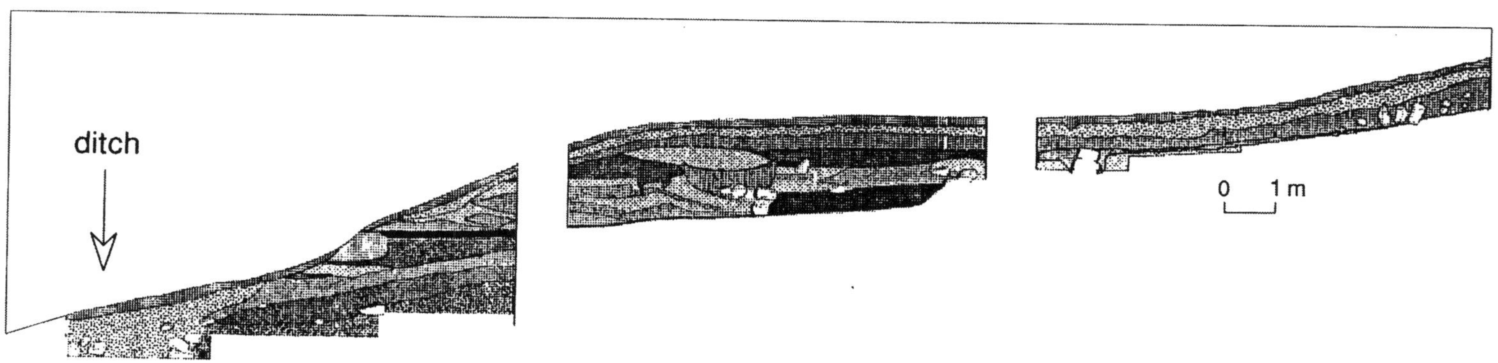

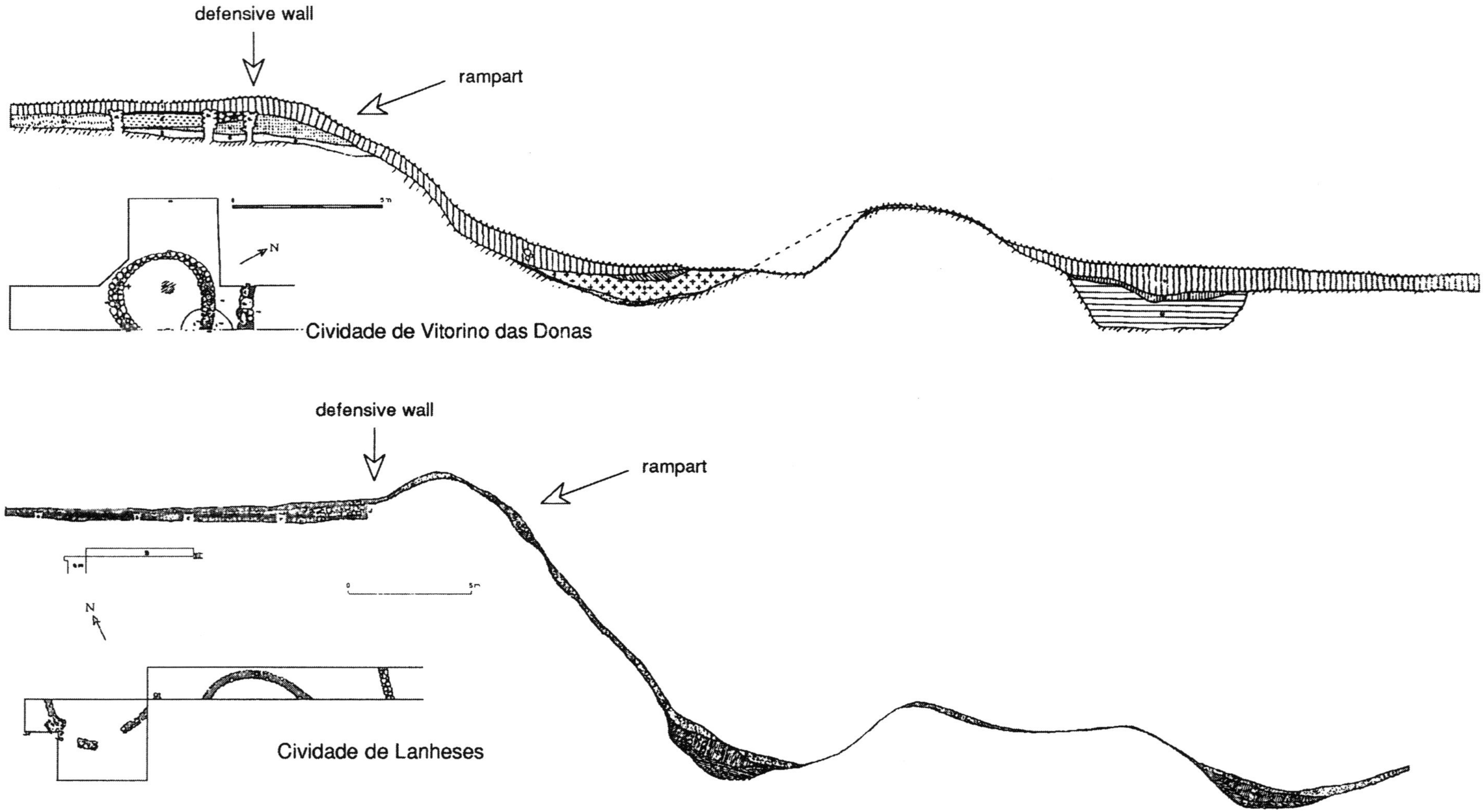

Figure 30 - Cross section through the defensive structures of two castros of type 3
(after Almeida, CAB, 1990, nºs 50 and 125)

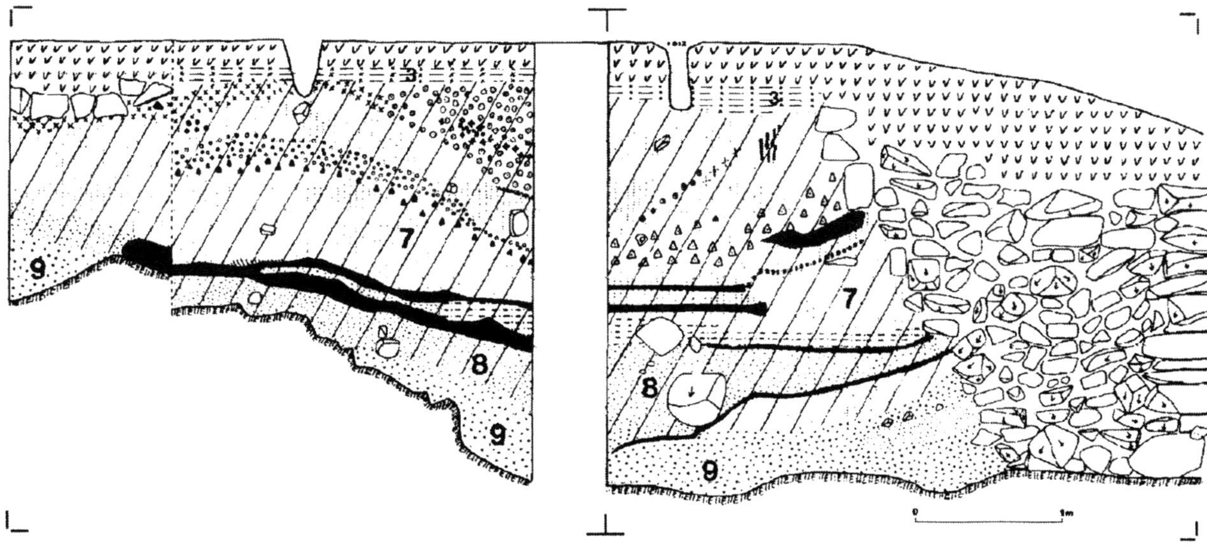

Figure 31.1 - Defensive wall of Castro das Ermidas
(after Queiroga 1985)

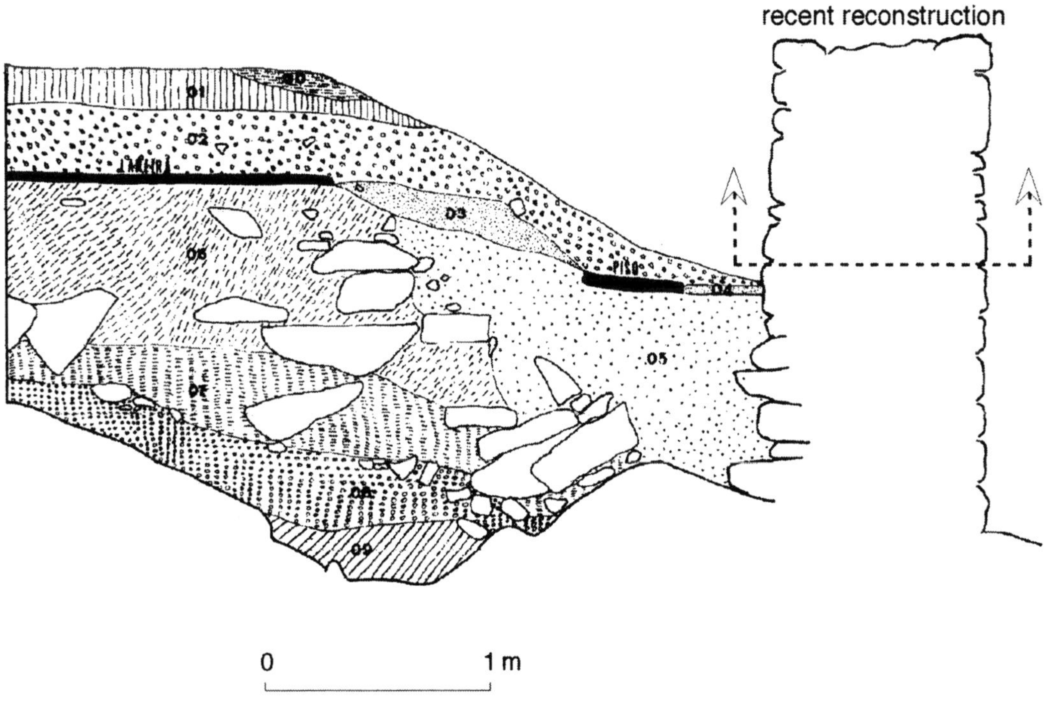

Figure 31. 2 - Inner defensive wall of Citânia de Briteiros
(after Centeno - Silva 1979)

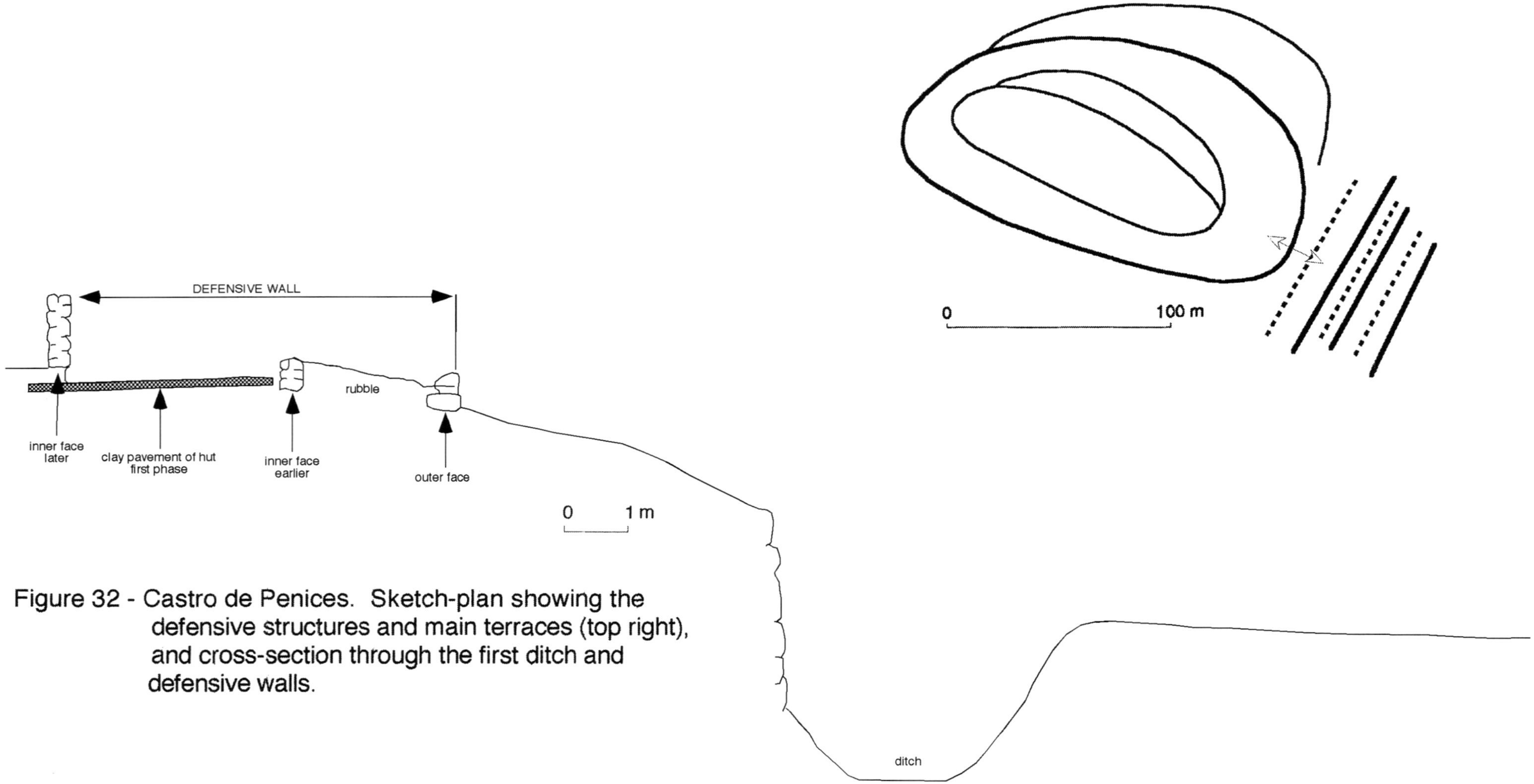

Figure 32 - Castro de Penices. Sketch-plan showing the defensive structures and main terraces (top right), and cross-section through the first ditch and defensive walls.

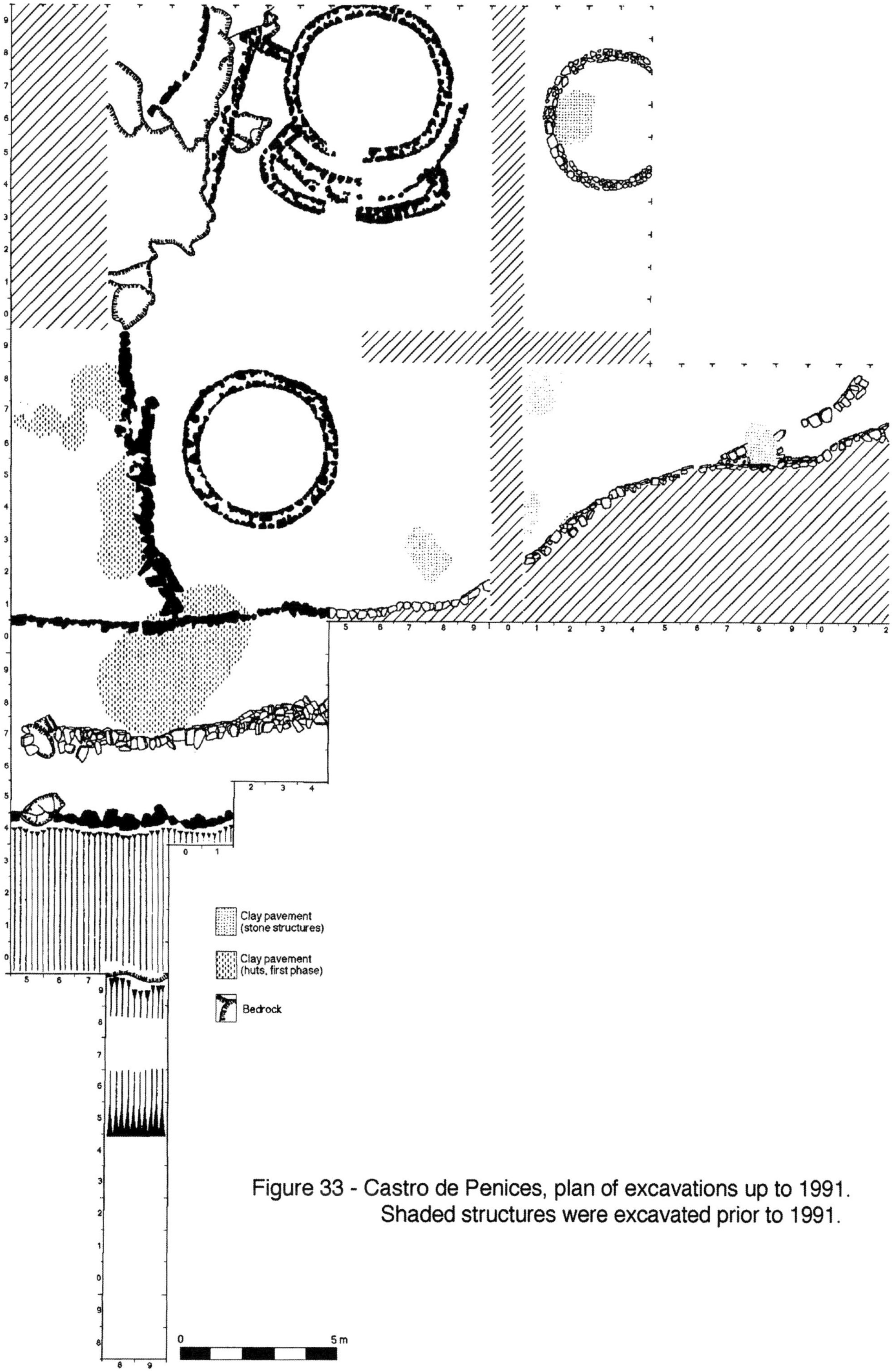

Figure 33 - Castro de Penices, plan of excavations up to 1991.
Shaded structures were excavated prior to 1991.

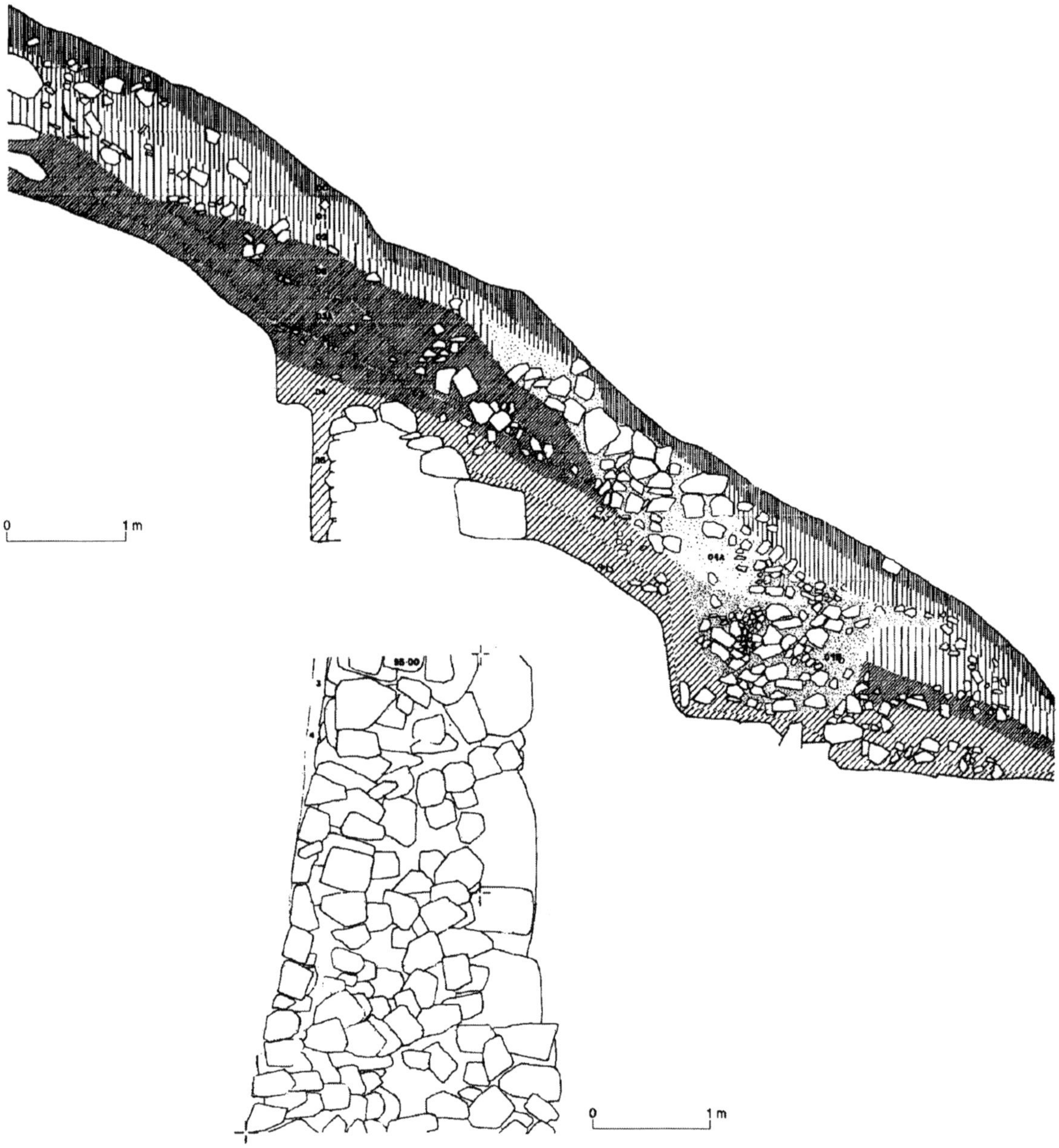

Figure 34.1 - Cross section and plan of the defensive wall of Coto da Pena
(after Silva 1986a)

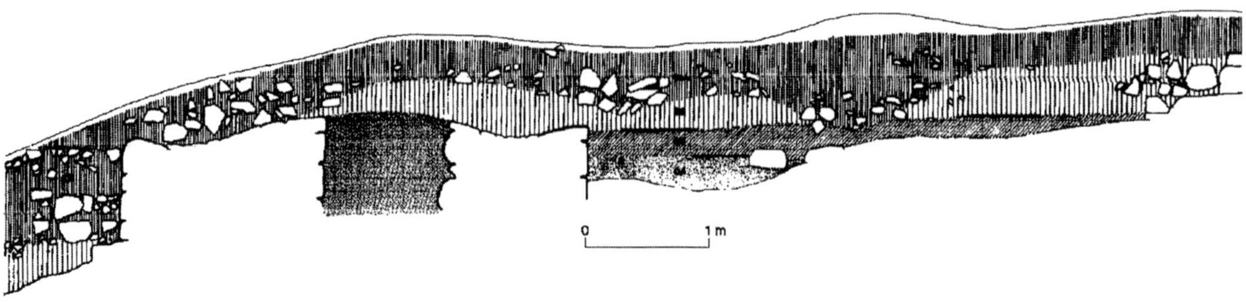

Figure 34.2 - Cross section of the defensive wall of Castro deTerroso
(after Silva 1986a)

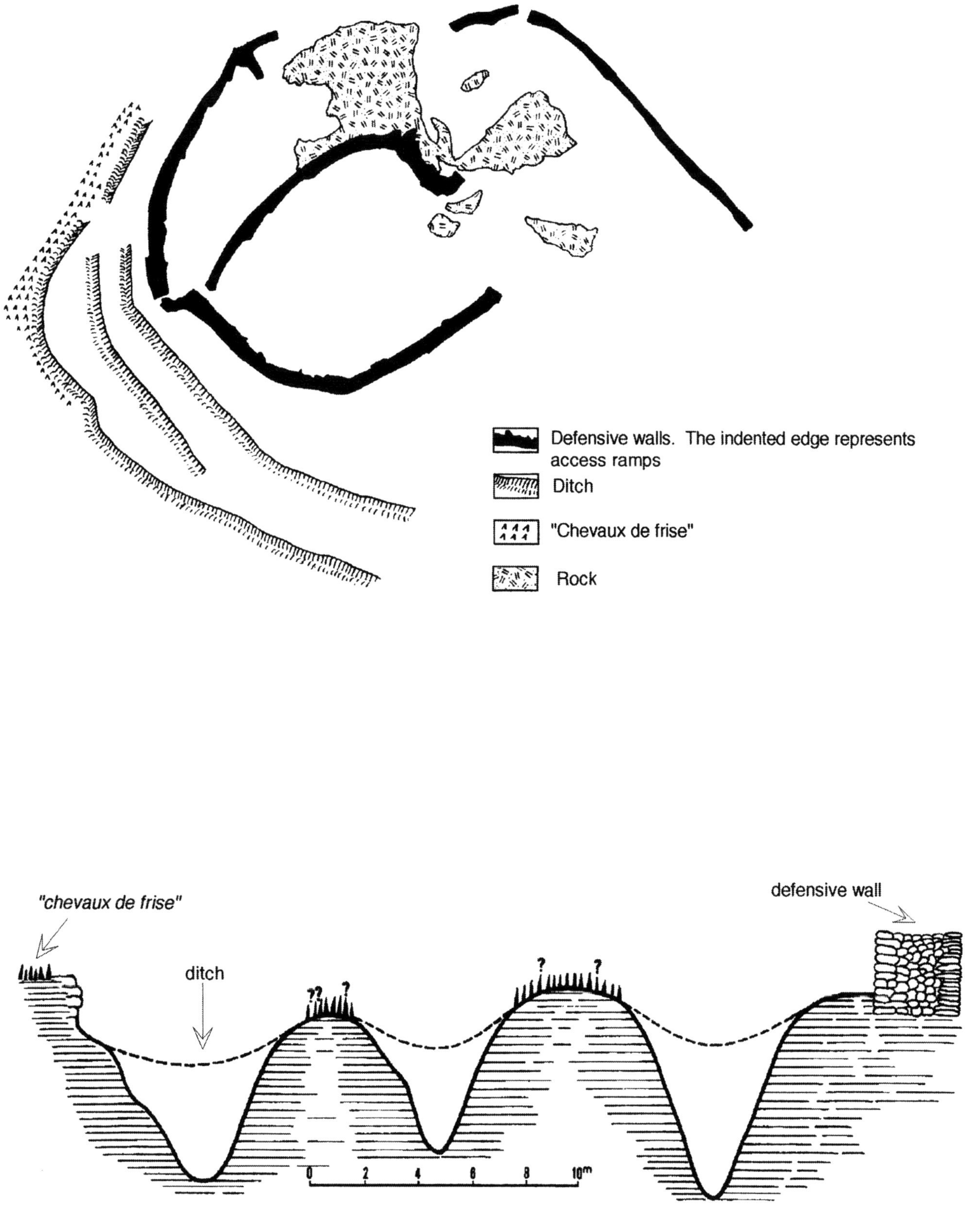

Figure 35 - General plan and western defences of Castro de Carvalhelhos.
(after Santos Júnior 1973)

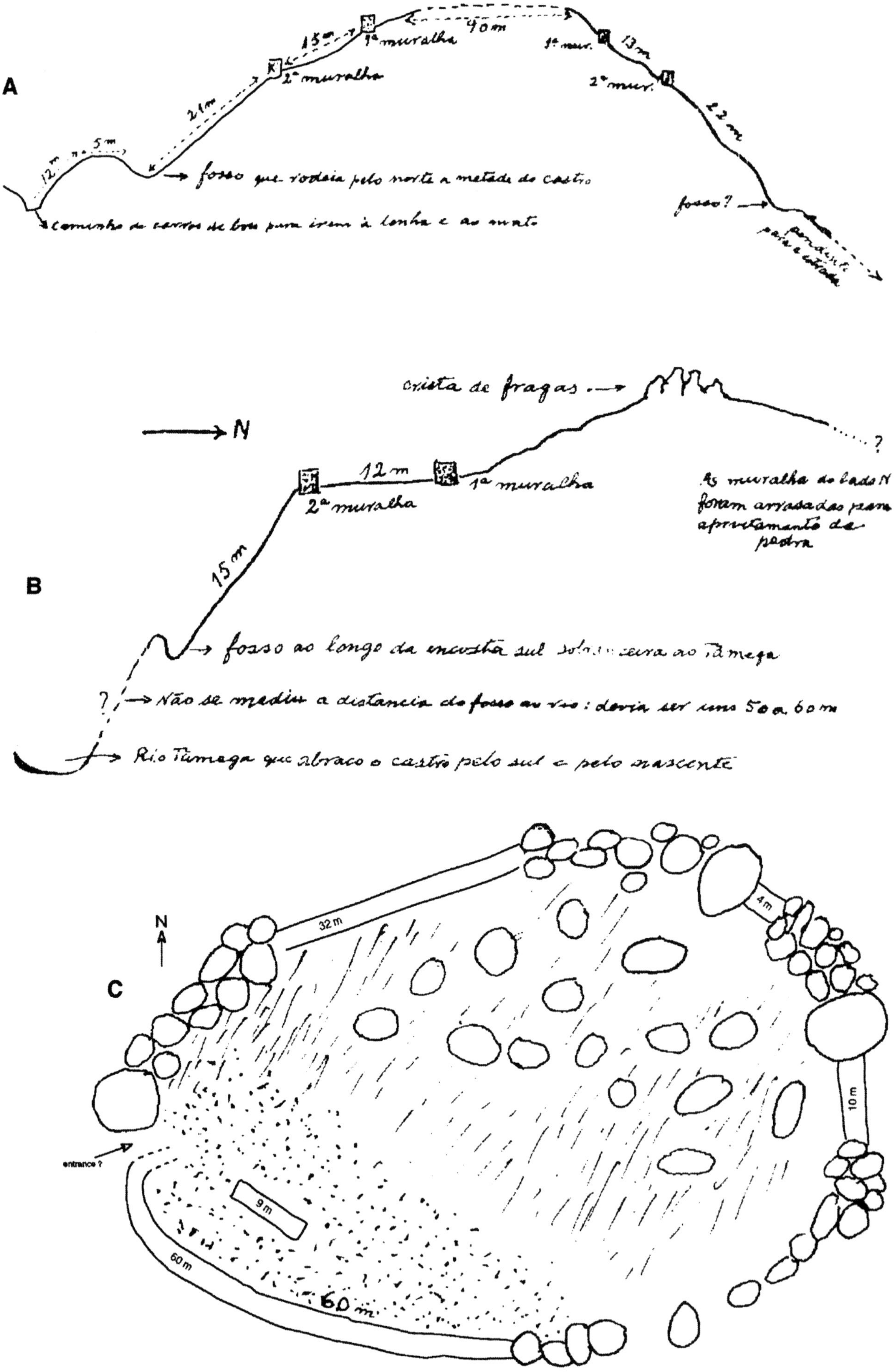

Figure 36 - Draft sections and plan of Castro do Cabeço (A), Castro do Mouril (B), and Castro da Gorda (C), Boticas.
(after field-notes of Miranda Júnior *et al.* 1983)

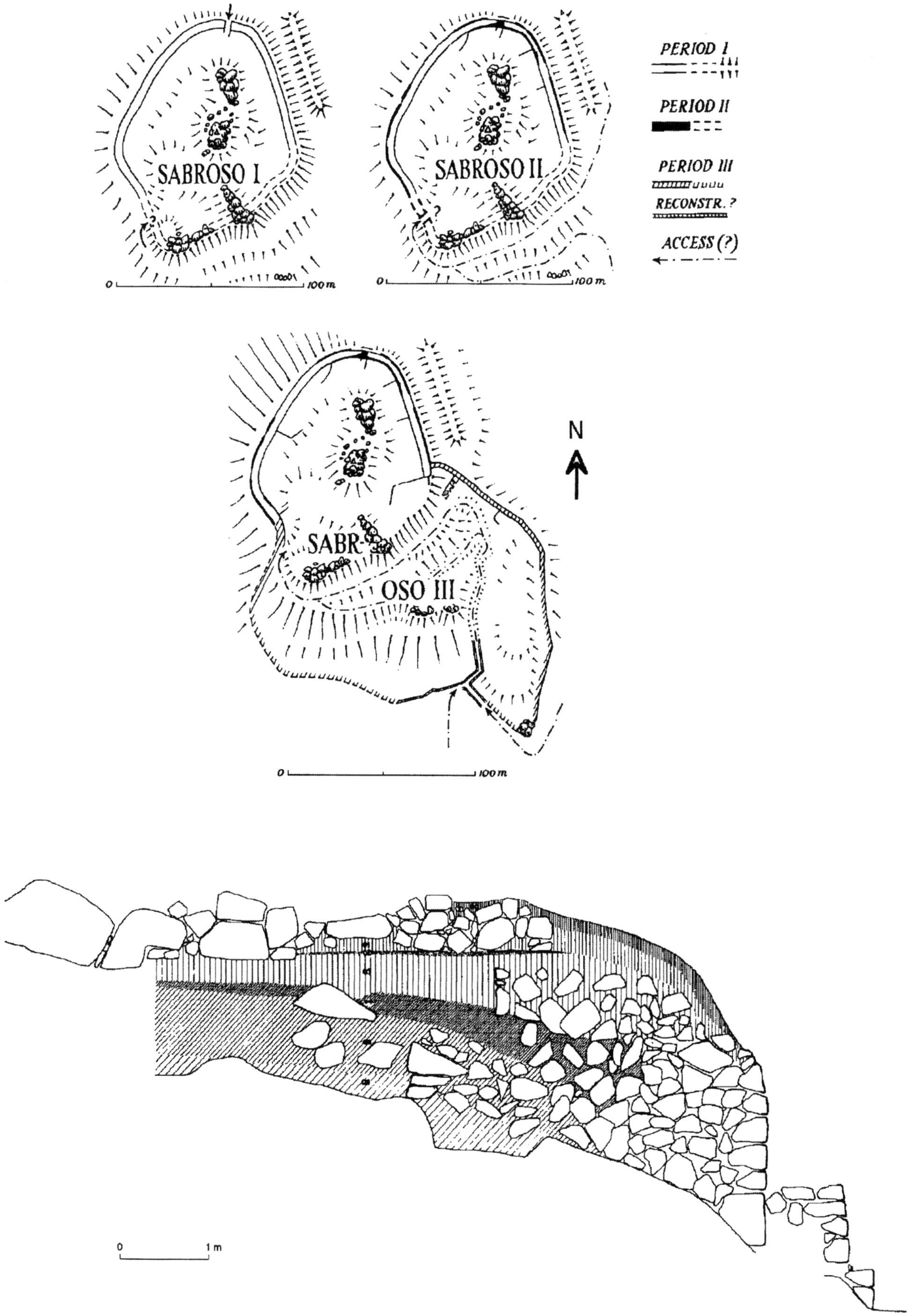

Figure 37 - Fortification phases at Castro de Sabroso, and section through the defensive wall showing the external reinforcement dating to period II.
(after Hawkes 1971 and Soeiro *et al.* 1981, respectively)

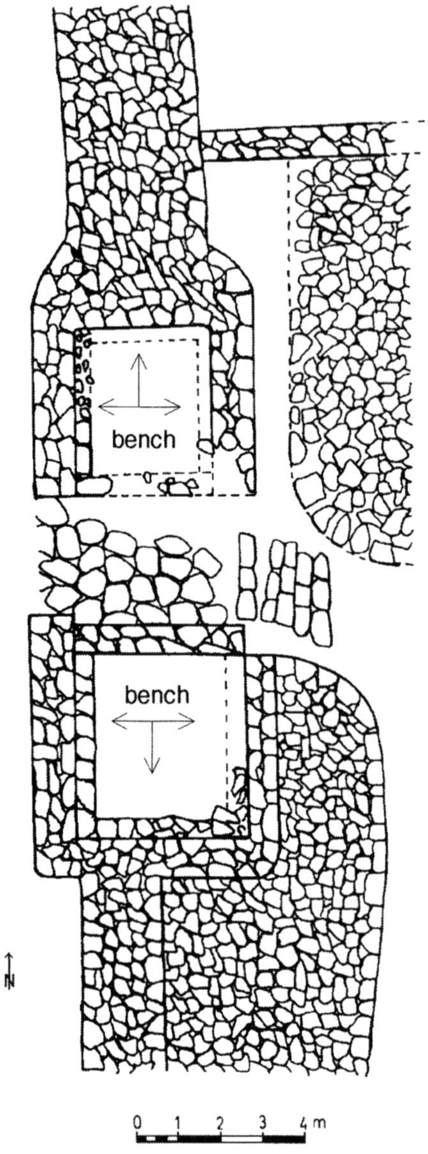

Figure 38 - Eastern gateway of the inner defensive wall of S. Cibrán de Lás. Example of a possible toll-gate.
(after Pérez Outeiriño 1987)

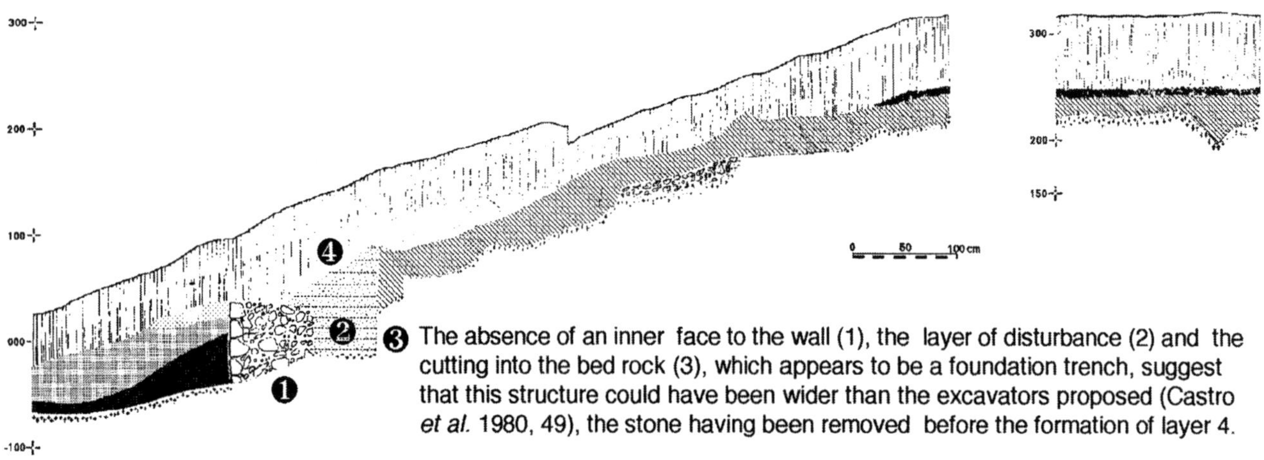

Figure 39 - Section through the defensive wall at Castro Máximo (after Castro *et al.* 1980)

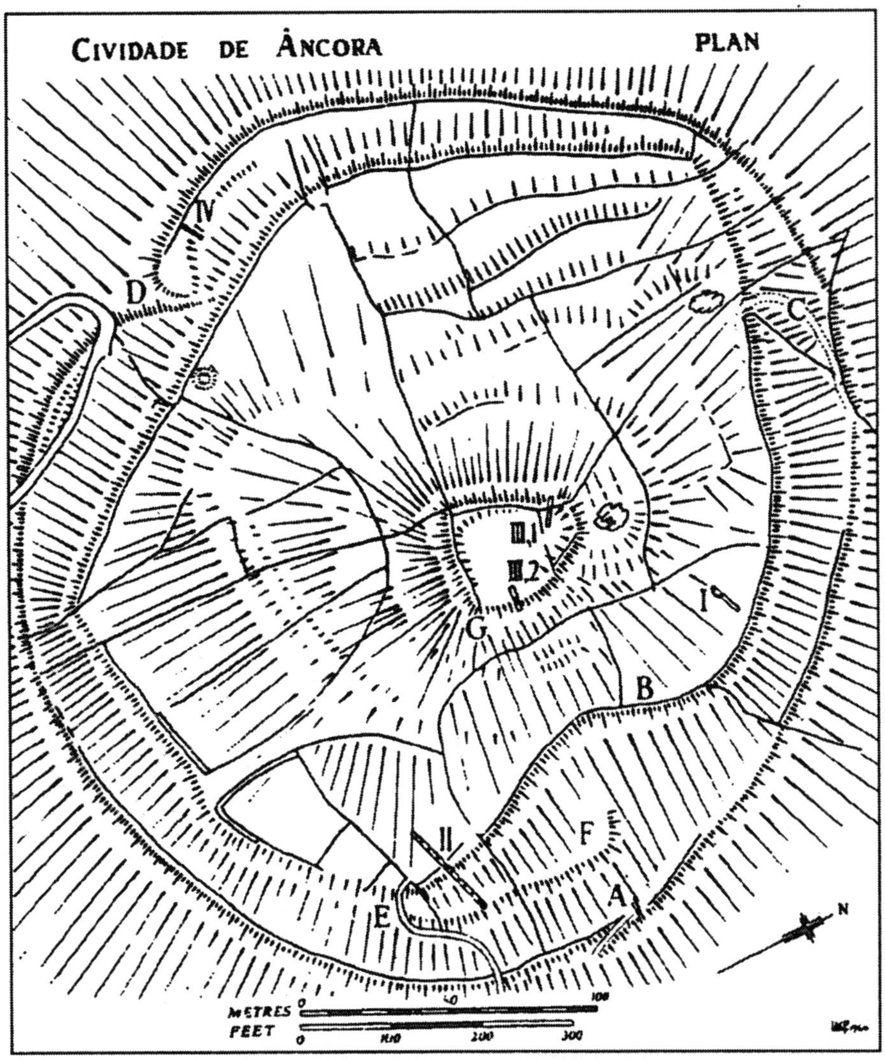

Figure 40 - Plan of Cividade de Âncora, Viana do castelo. (source Hawkes 1984)

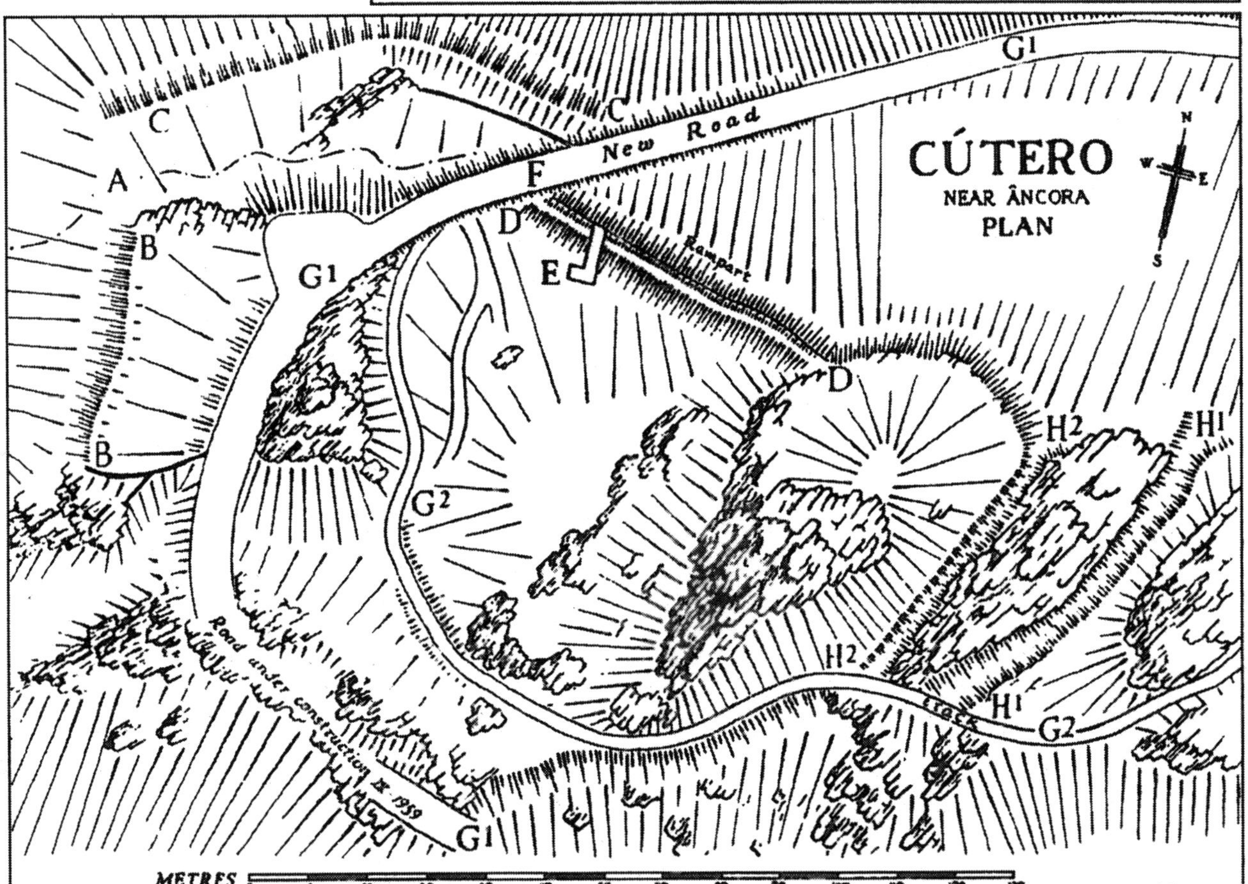

Figure 41 - Plan of Castro do Cútero, Viana do Castelo (source Hawkes 1984)

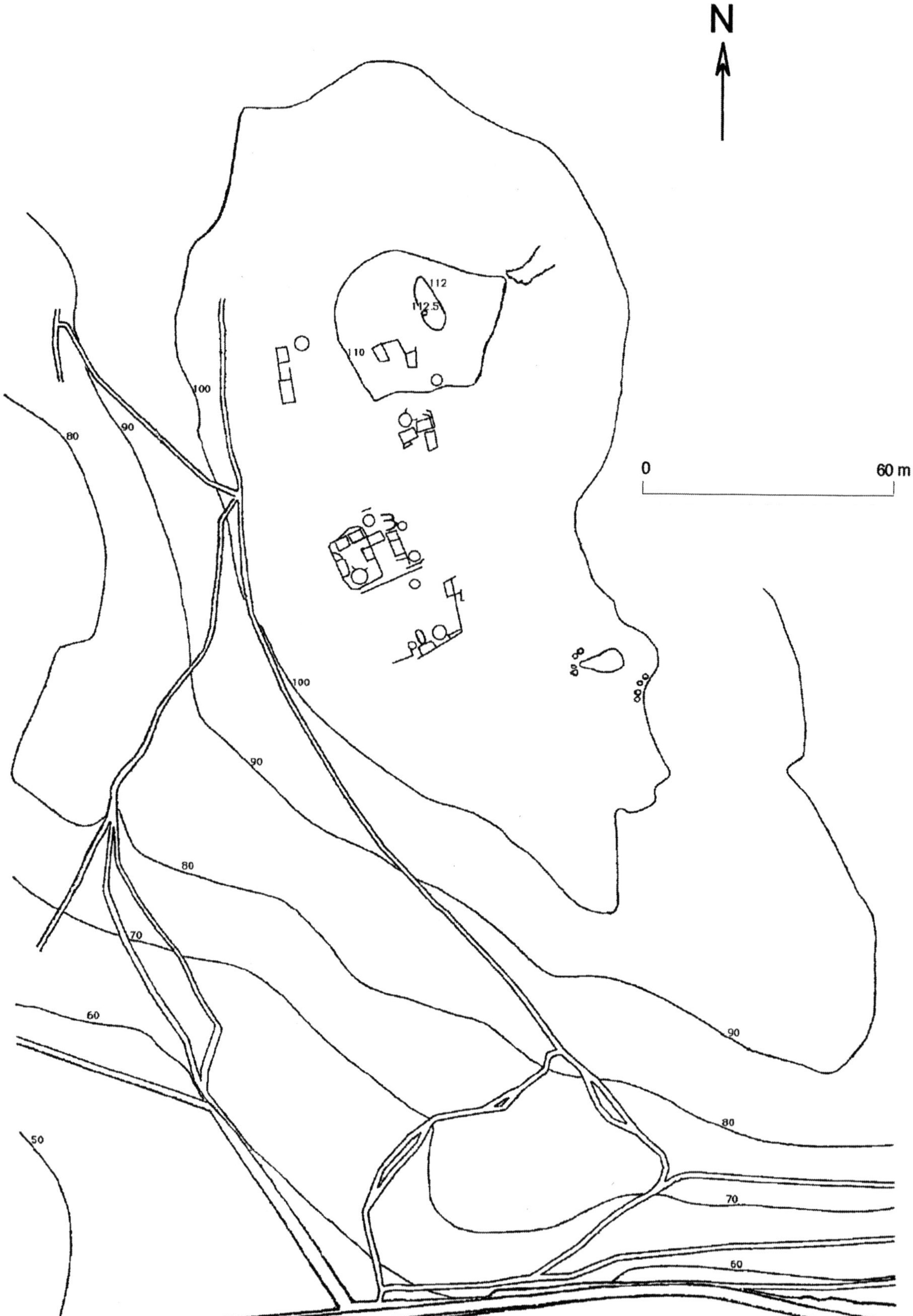

Figure 42 - Survey plan of the Cividade de Bagunte with uncovered Iron Age structures (J.N.E.)

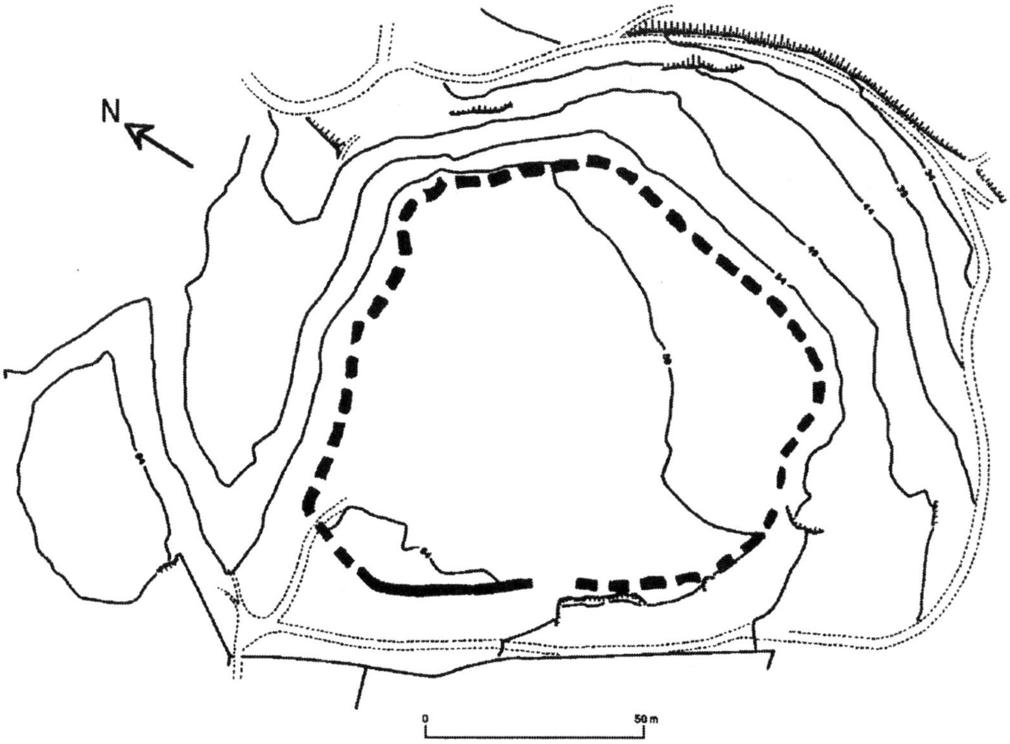

Figure 43 - Survey plan of Castro do Lago, Amares.
Thicker lines represent defensive walls.
(after Martins 1990)

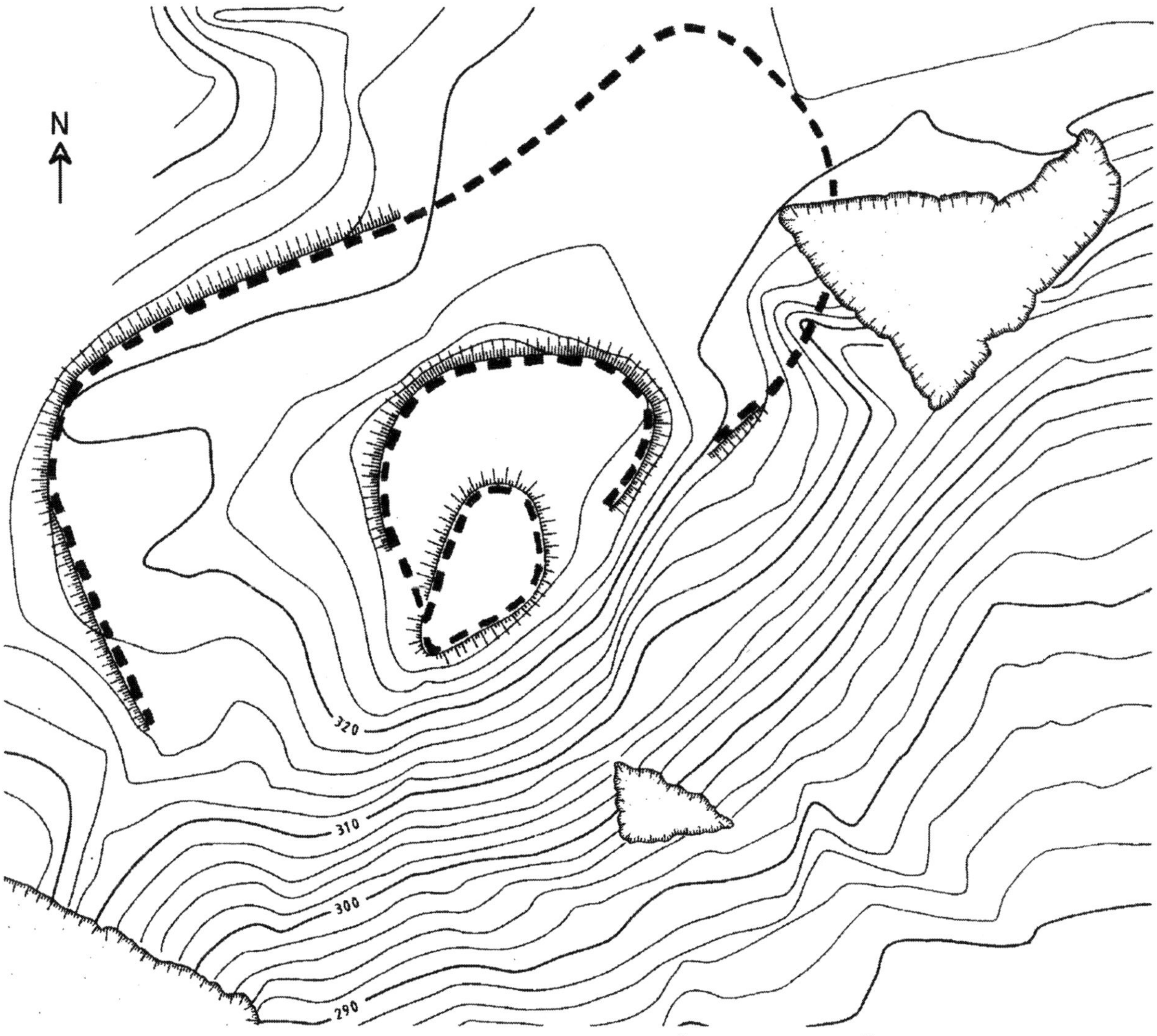

Figure 44 - Survey plan of Castro do Barbudo, Vila Verde.
Thicker lines represent defensive walls.
(after Martins 1990)

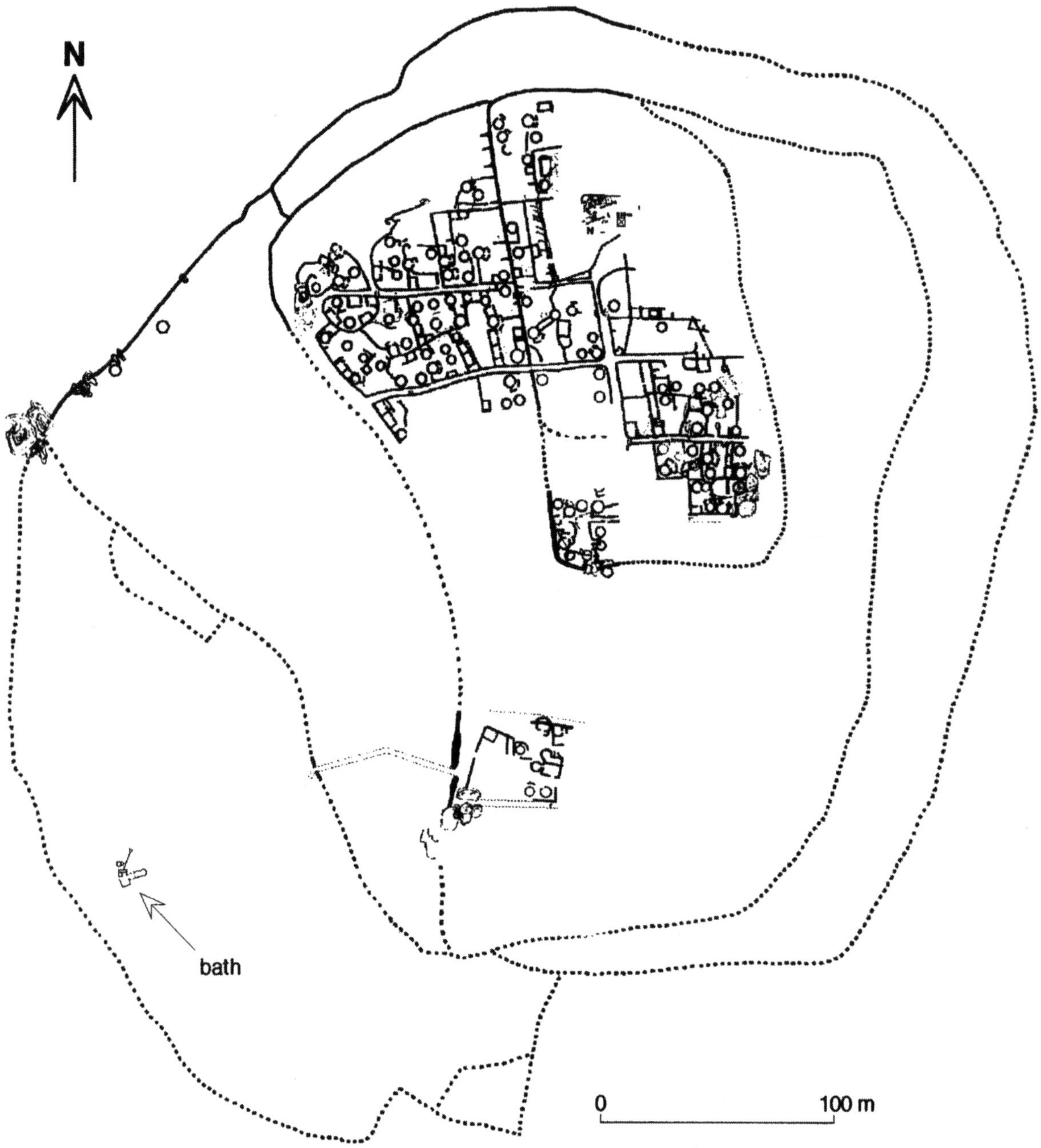

Figure 45 - General plan of Citânia de Sanfins. Dotted lines represent the possible alignment of the defensive walls.
(after Paço 1968b and Silva 1986a, revised)

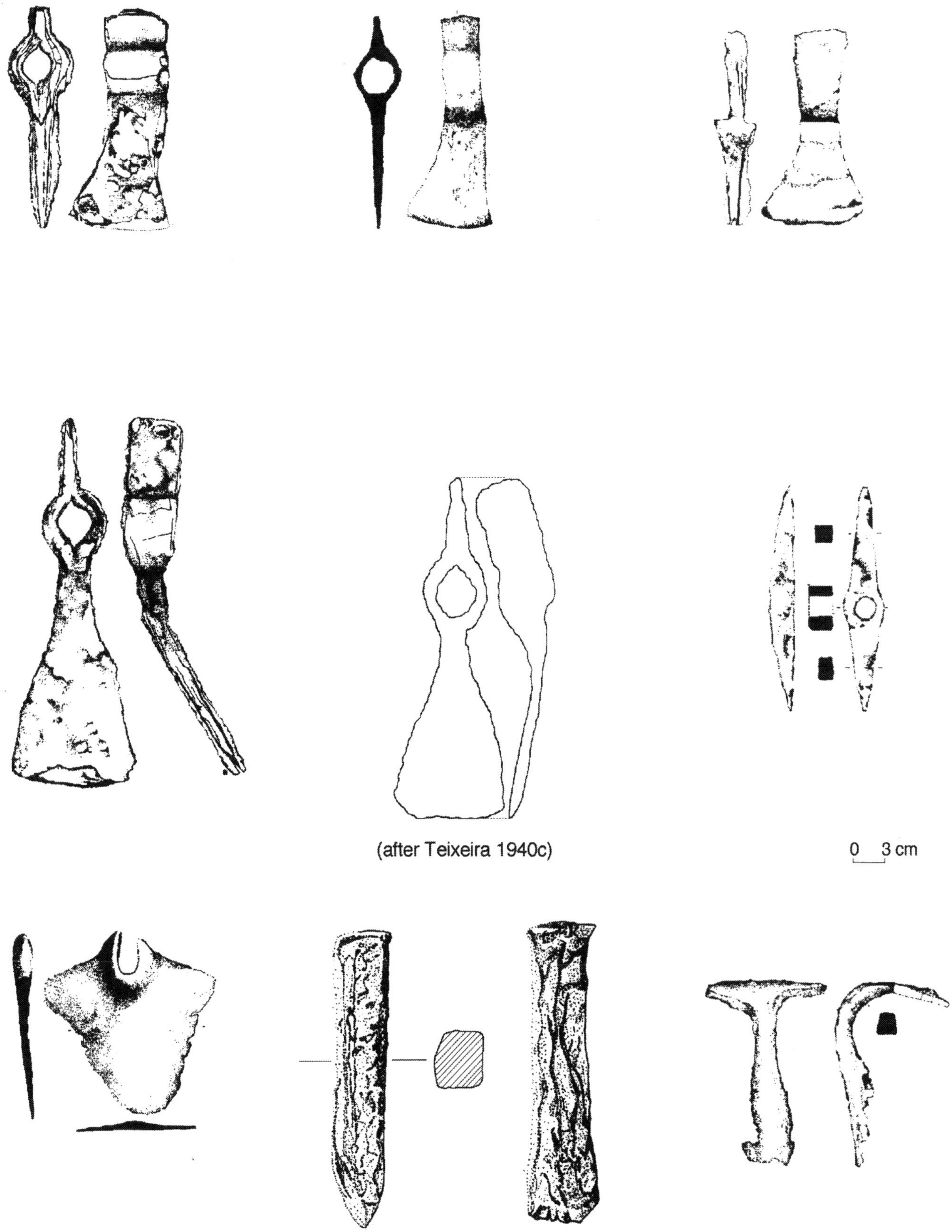

Figure 46 - Implements for agriculture and crafts
(source, Silva 1986a otherwise indicated)

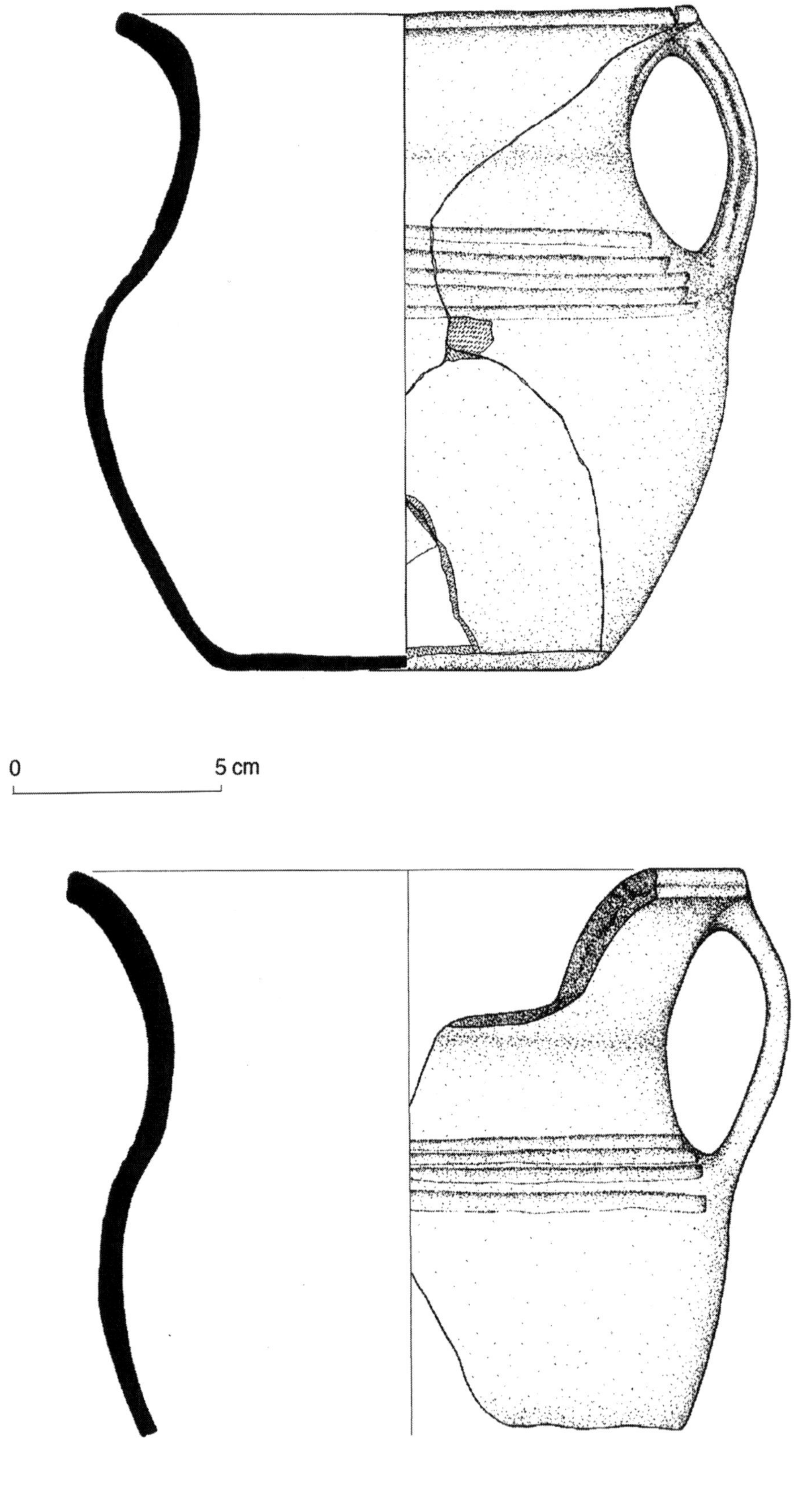

Figure 47 - Two vessels from Castro de Penices dating to the time of Augustus-Tiberius. An example of the mass production of pottery from the change of millennium onwards.

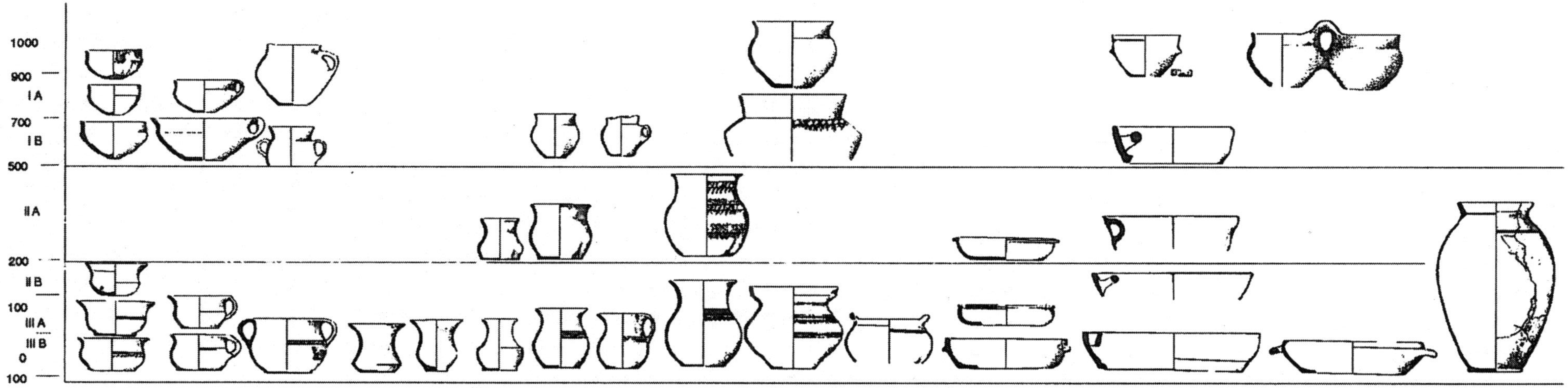

A - Development of pottery form according to phase and function. (Source Silva 1986a for phasing; and 1991, also exhibition brochure)

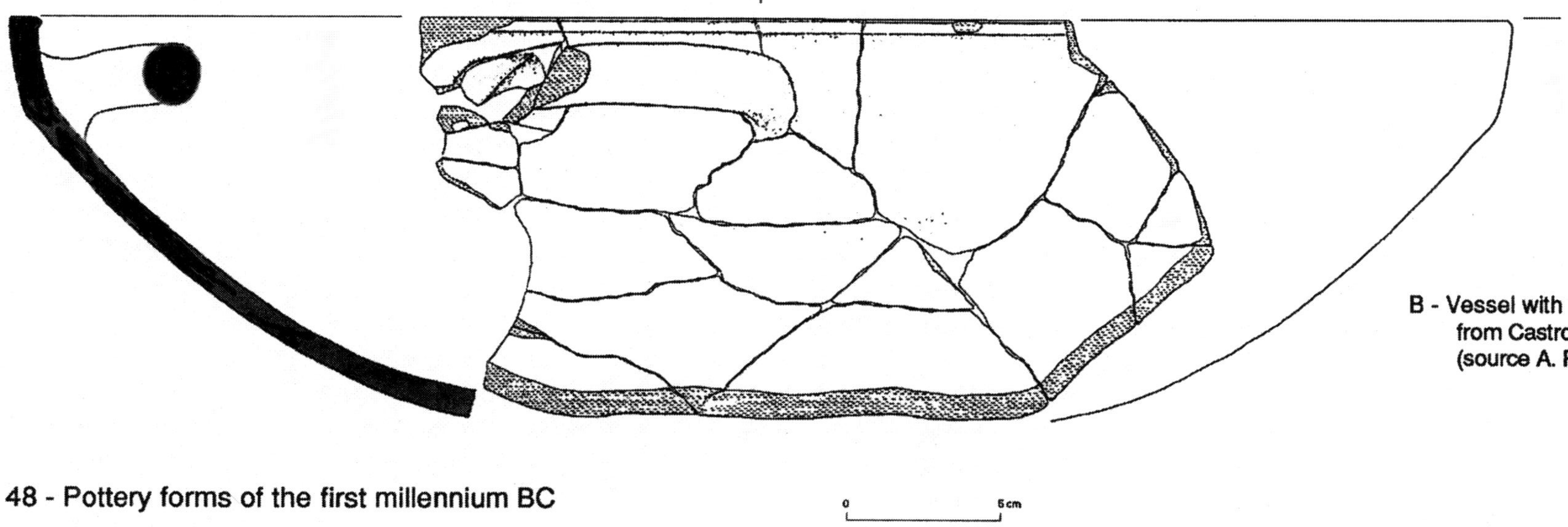

B - Vessel with interior handles, from Castro do Crastoeiro. (source A. P. Dinis)

Figure 48 - Pottery forms of the first millennium BC

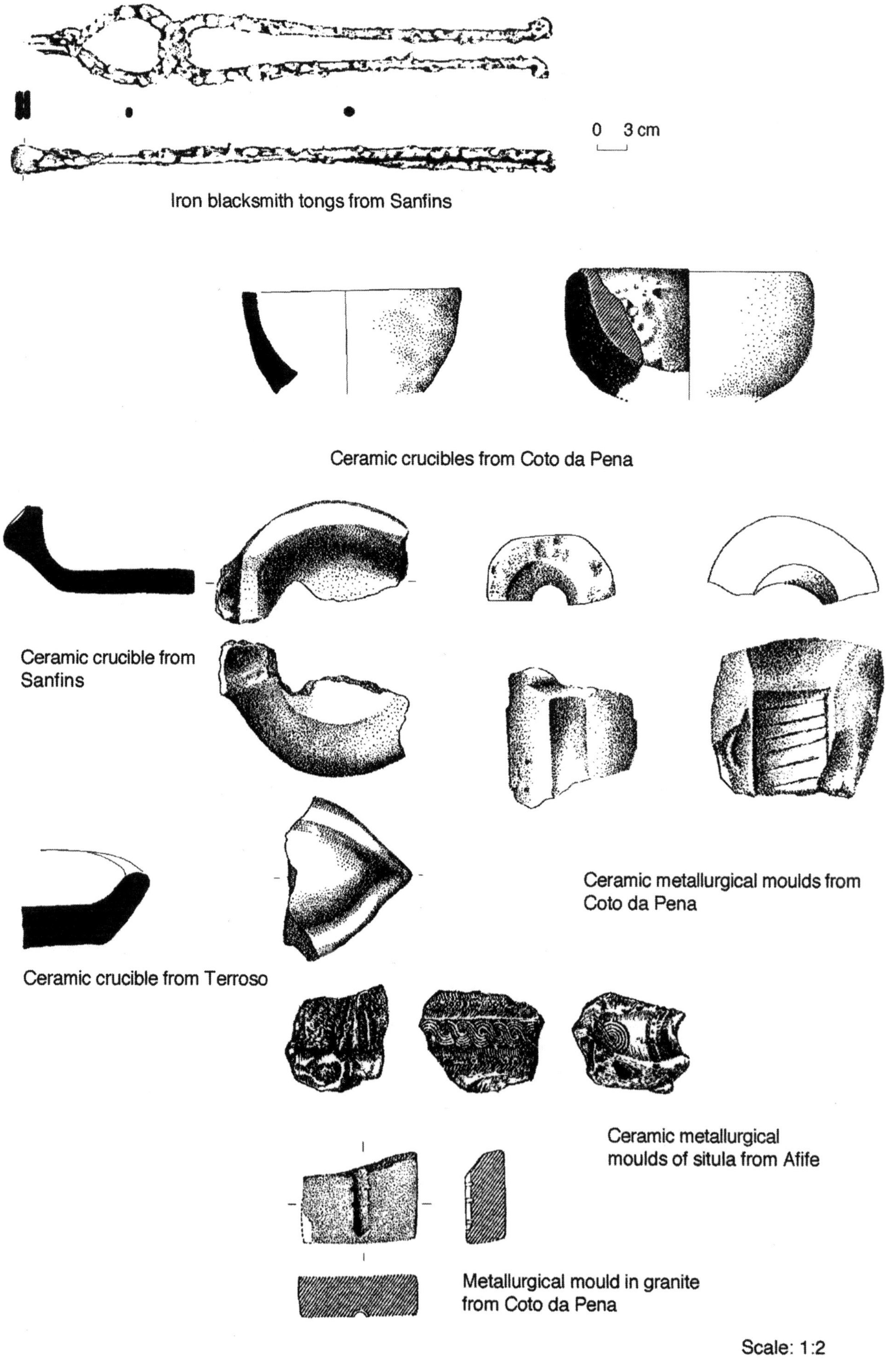

Figure 49 - Evidence for metallurgical activities
(source Silva 1986a)

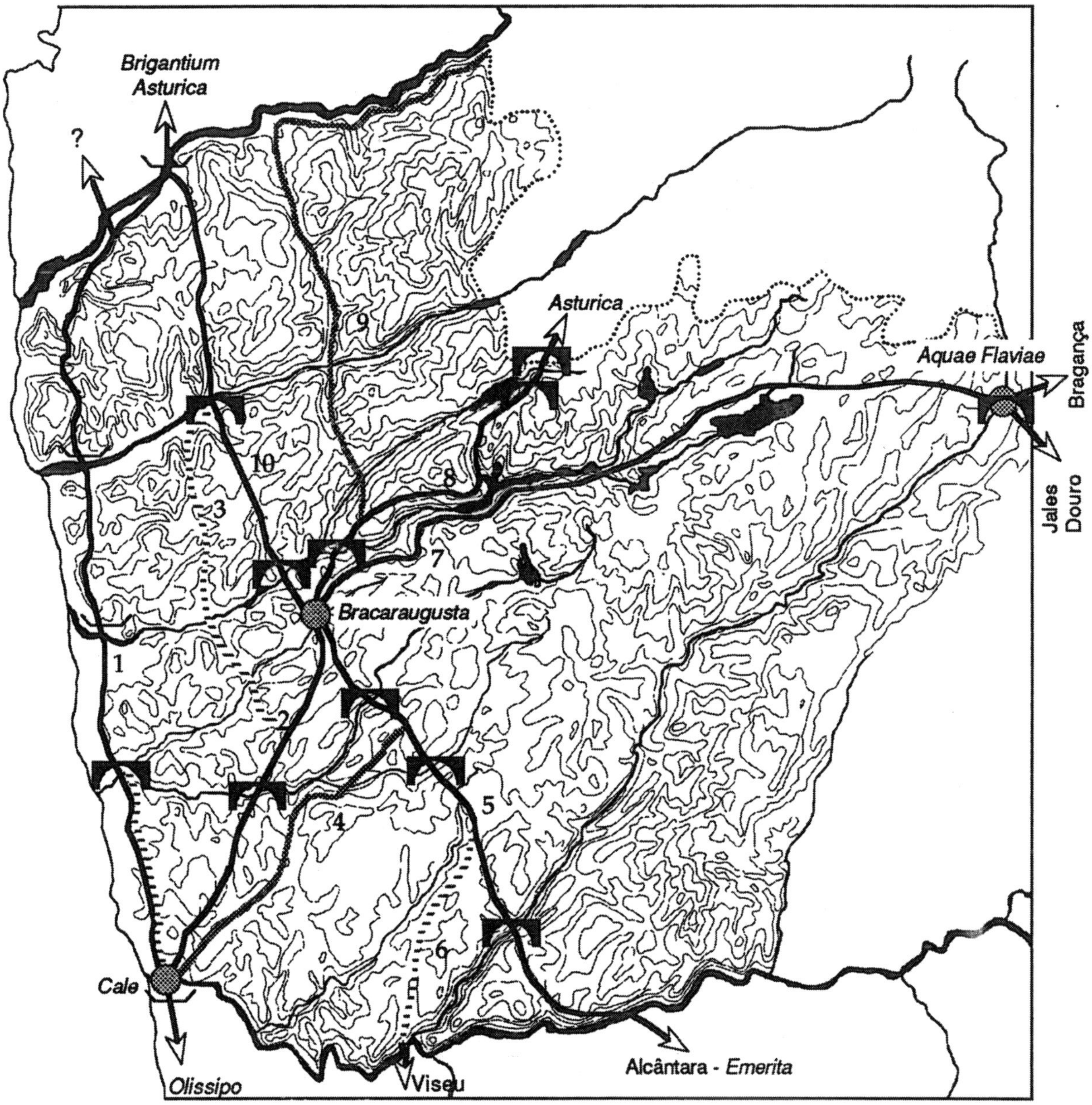

Roman roads referred to by the *Itinerary* of Antoninus or documented by archaeological evidence

Road referred to as *antiqua* in the XIIth century documents. May possibly follow an earlier road.

Medieval road, which may follow earlier Roman road

Bridge

Barge

Figure 50 - Evidence for Roman roads. (after personal notes kindly provided by Prof. Almeida, Almeida, CAB, 1979, Naveiro López 1991, Tranoy 1981)

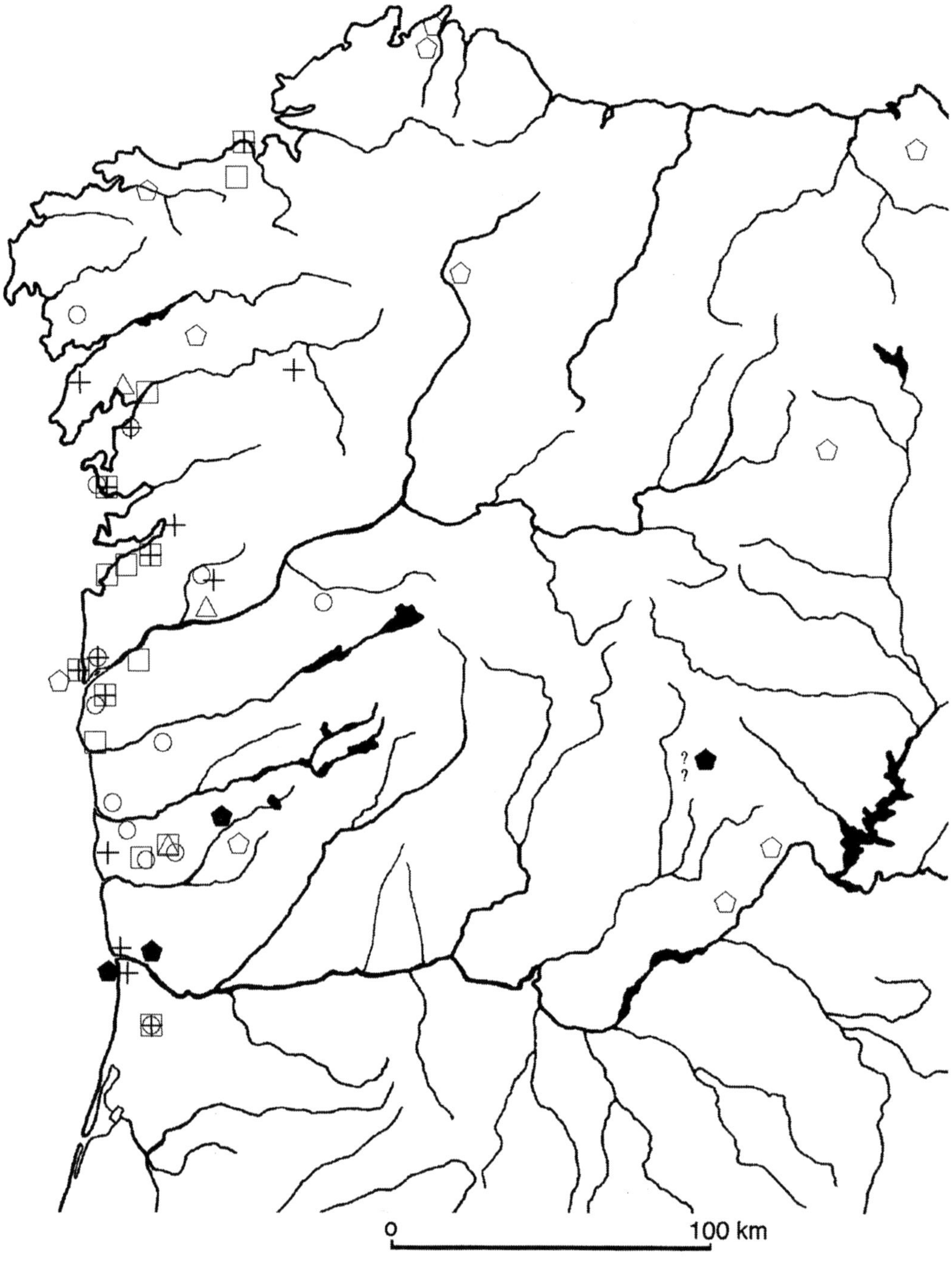

○ Greek pottery

+ Punic-type pottery

□ Campanian ware

△ Sand-core glass

⬠ Ibero-Punic coins

⬟ Greek coins

A Coruña (Galicia)
A Lanzada (Pontevedra, Galicia)
Astorga (León)
Carballo (A Coruña, Galicia)
Bares (A Coruña, Galicia)
Castelo de Faria (Barcelos, Portugal)
Castelo de Gaia (Gaia, Portugal)
Castro da Forca (Pontevedra, Galicia))
Castro das Ermidas (V.N. Famalicão, Portugal)
Castro de Alobre (Pontevedra, Galicia)
Castro de Baroña (A Coruña, Galicia)
Castro de Elviña (A Coruña, Galicia)
Castro de Fozara (Pontevedra, Galicia)
Castro de Penices (V.N. Famalicão, Portugal)
Castro de Recarea (A Coruña, Galicia)
Castro de Romariz (Vila da Feira, Portugal)
Castro de Sabroso (Taipas, Portugal)
Castro de S. Lourenço (Esposende, Portugal)
Castro de Sto. Estevão (Ponte de Lima, Portugal)
Castro de Troña (Pontevedra, Galicia)
Castro de Vigo (Pontevedra, Galicia)
Castrovite (A Coruña, Galicia)
Castro Pedra da Moura (Pontevedra, Galicia)
Castro Verde (Pontevedra, Galicia)
Castromao (Ourense, Galicia)
Citânia de Sta. Luzia (Viana do Castelo, Portugal)
Cividade de Terroso (Póvoa de Varzim, Portugal)
Coto da Pena (Caminha, Portugal)
Gondomar (Portugal)
Guimarães (Portugal)
Isla de Toralla (Pontevedra, Galicia)
Lovelhe (V. N. Cerveira, Portugal)
Lugo (Galicia)
Mañónn (A Coruña, Galicia)
Miranda do Douro (Portugal)
Morro da Sé (Porto, Portugal)
O Neixón (A Coruña, Galicia)
Oviedo (Asturias)
Peredo da Bemposta (Portugal)
Santa Trega (Pontevedra, Galicia)
Santiago de Compostela (A Coruña, Galicia)
Serra do Pilar (V.N. de Gaia, Portugal)
Torres do Oeste (Pontevedra, Galicia)

Figure 51 - Mediterranean imports (2nd half of the 1st mill. BC) and list of locations. (after various sources)

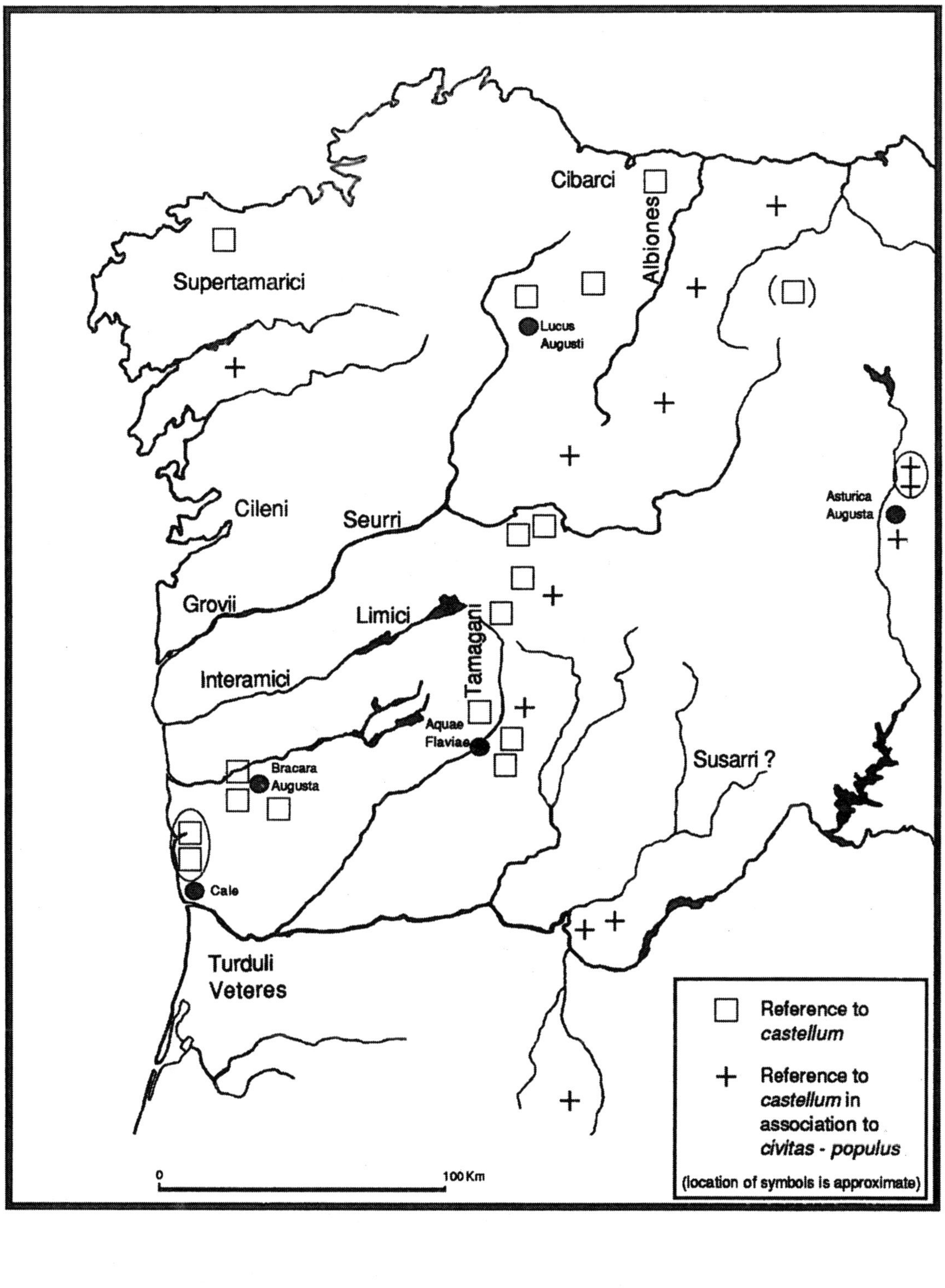

Figure 52 - References to *castella* in epigraphic sources in the Northwest.
(after Albertos Firmat 1975, 1977; Le Roux - Tranoy 1984; Pereira - Santos 1980; Santos Yanguas 1985)

Figure 53 - *Tesserae hospitales* from Monte Murado, V.N. Gaia (not to scale)
(after Silva 1983b)

Figure 54 - *Tabula patronatus* from La Coruña
(after Pereira Menaut 1985-86)

Figure 55 - Reconstruction of the dress of a Callaican warrior.
(ill. by Christina Unwin)

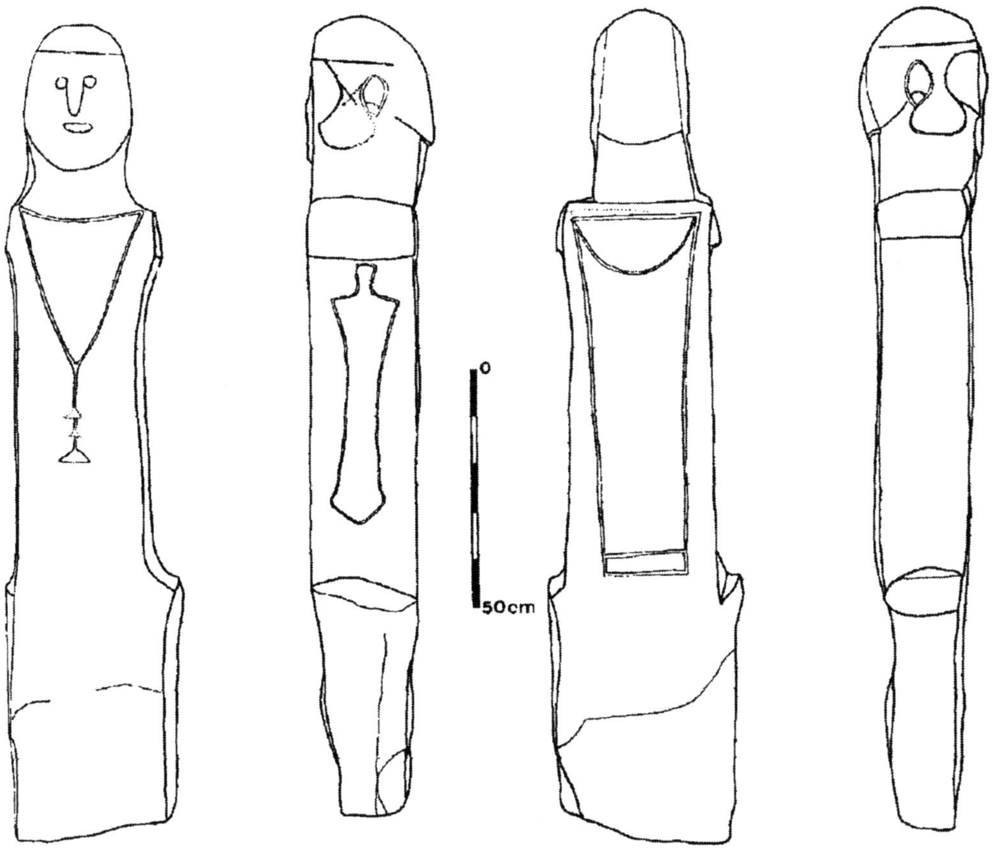

Figure 56.1 - Statue from S. João de Ver.
(source Jorge-Jorge 1983)

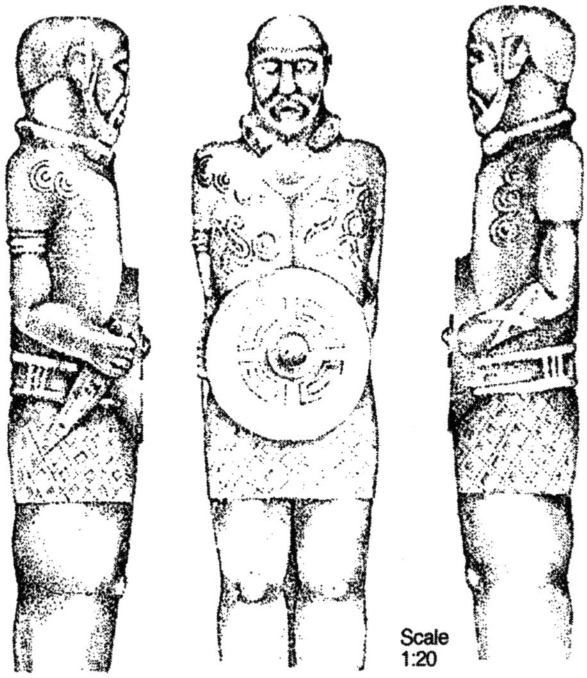

Figure 56.2 - Statue from Lezenho
(source Silva 1986)

Figure 56.3 - Statue from S. Julião, Vila Verde.
(source Martins - Silva 1984)

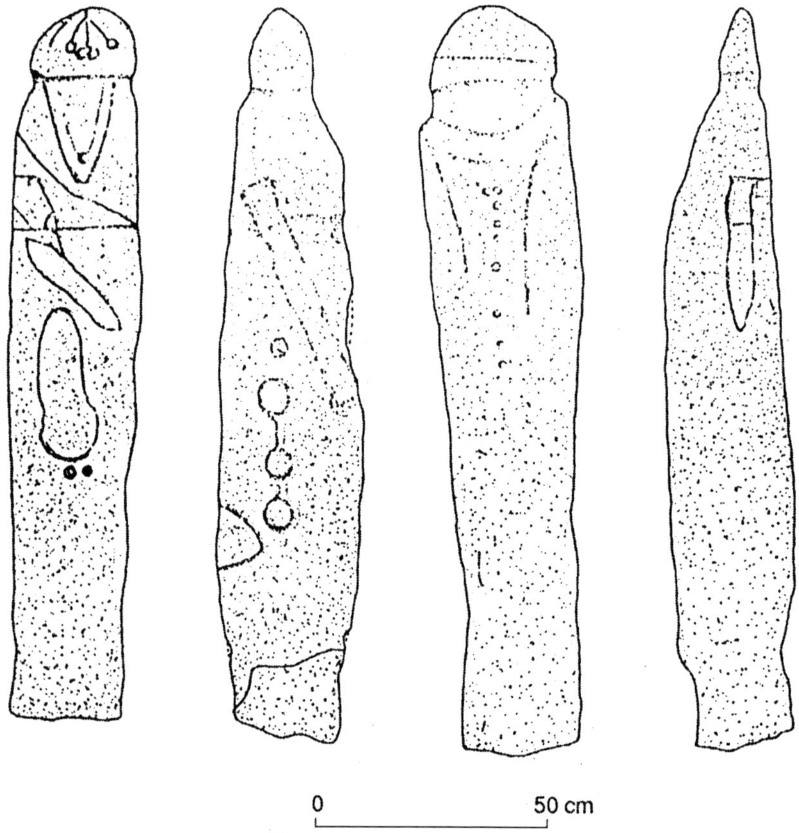

Figure 57.1 - Bronze Age statue from the Tâmega river bed, Chaves.
(after Jorge - Almeida 1980)

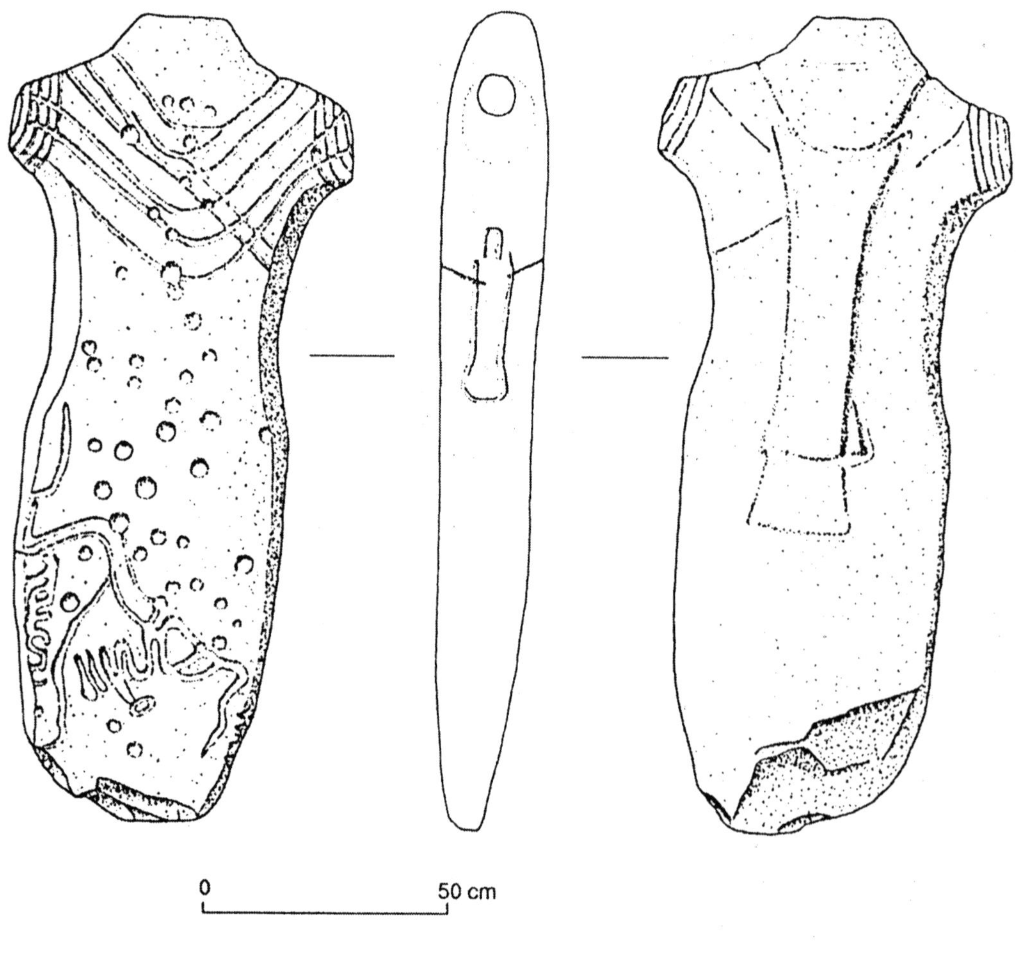

Figure 57.2 - Bronze Age statue from Faiões, Chaves. (after Almeida - Jorge 1979)

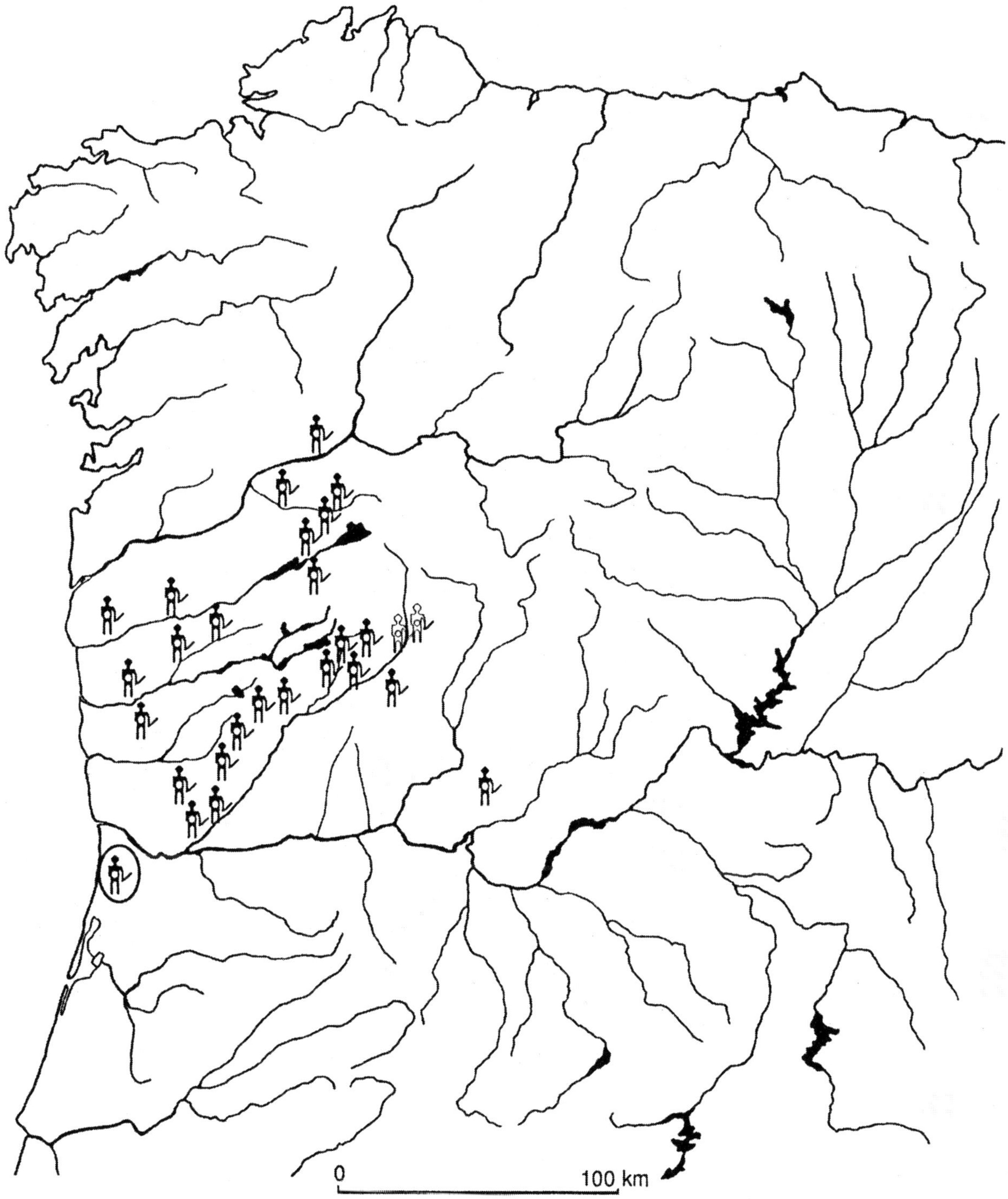

Figure 58 - Distribution of the statues representing warriors.
(after various sources)

Warrior statue (Iron Age)
Warrior statue (Bronze Age)

4-Outeiro de Lezenho, Campos, Boticas
1-Castro de Rubiás, Celanova, Ourense
1-Castro de Santa Adega, Villamarín, Ourense
1-Castro de Santo Ovídio, Fafe
1-Castro do Mau Vizinho, Cendufe, Arcos de Valdevez
2-Cibdá de Armeá, (Sta. Mariña das Aguas Santas), Allariz, Ourense
1-Cidadelhe, Britelo, Ponte da Barca
1-Citânia de Roriz, Roriz/Oliveira, Barcelos
1-Citânia de S. Julião, Ponte/Coucieiro, Vila Verde
1-Citânia de Sanfins, Paços de Ferreira
1-Leiro, Ribadavia, Ourense
1-Meixedo, Viana do Castelo
1-Midões, Barcelos
1-Monte do Crasto, Capeludos, Vila Pouca de Aguiar
2-Mozinho, Oldrões, Penafiel
1-Refojos de Basto, Cabeceiras de Basto
1-S. João de Ver, Feira (enclosed by oval)
1-S. Jorge de Vizela, Guimarães
1-Sta. Comba, Refojos de Basto, Cabeceiras de Basto
1-Sra. dos Anúncios, Vilarelhos, Alfândega da Fé
1-Vilar de Bárrio, Allariz, Ourense
2-Faiões and Tâmega river bed, Chaves

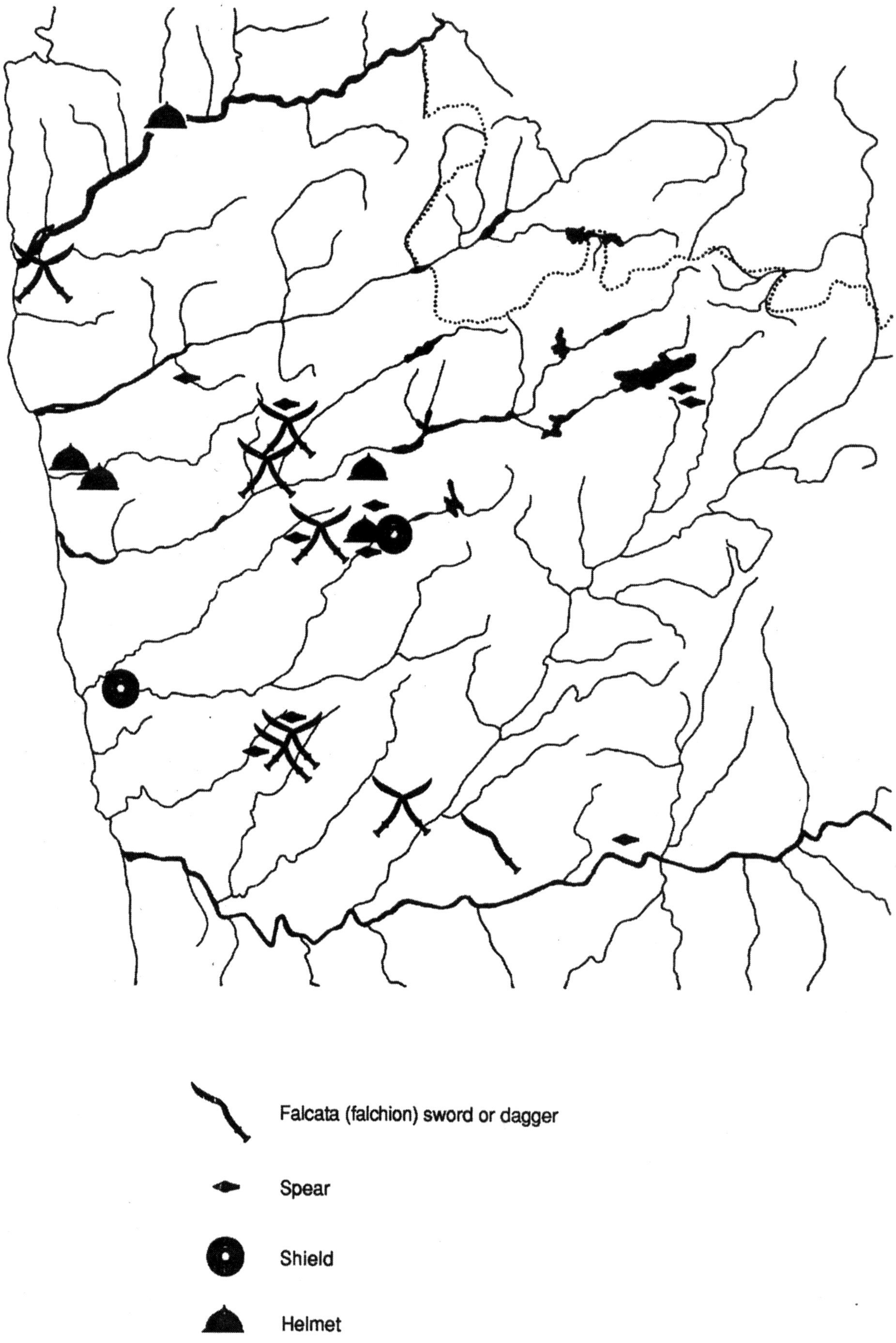

Figure 59 - Iron Age war equipment in the northwest of Portugal
(after various sources)

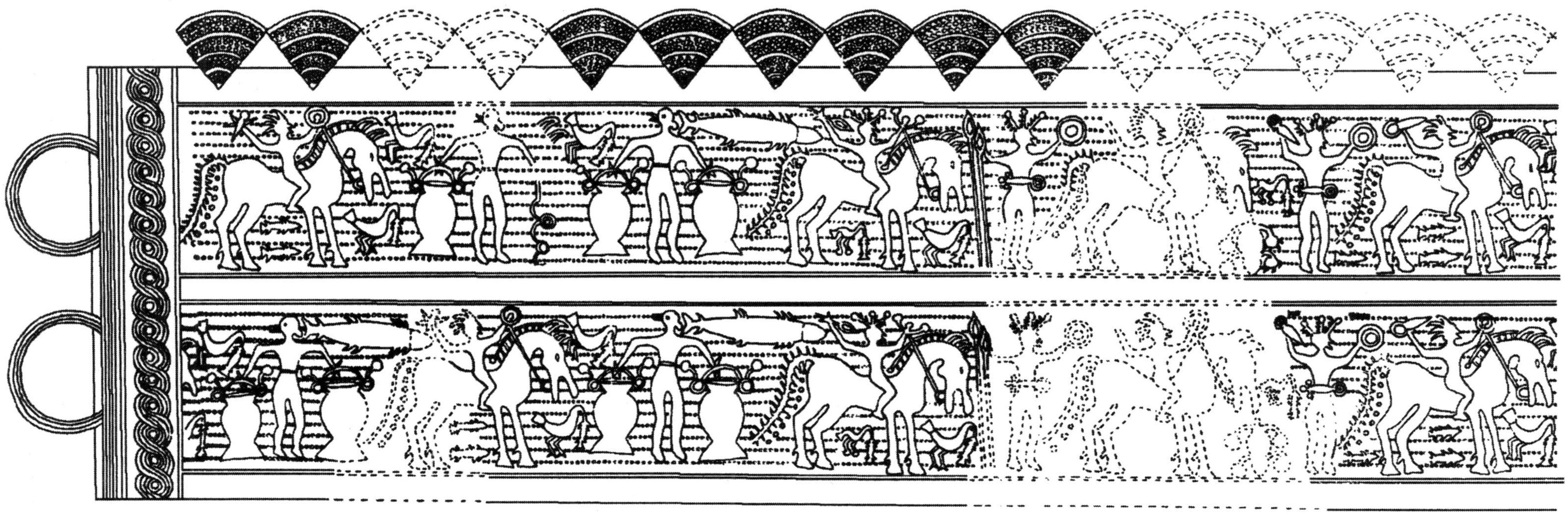

Figure 60 - Fragments of the diadem from Ribadeo (or San Martín de Oscos).
(reconstruction by López Monteagudo 1977)

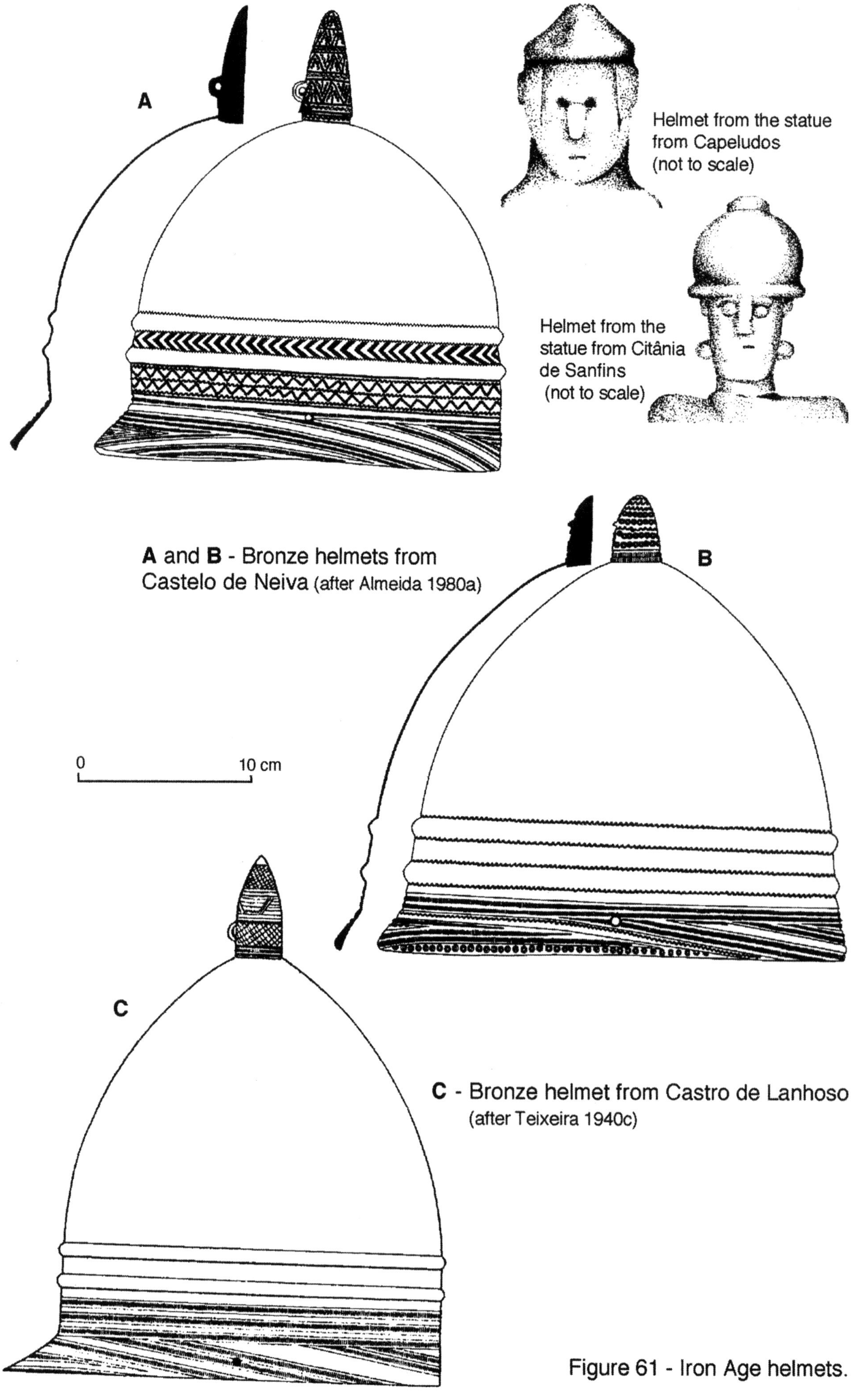

Figure 61 - Iron Age helmets.

Weapons represented on one coin of the "caetra" mint (not to scale)

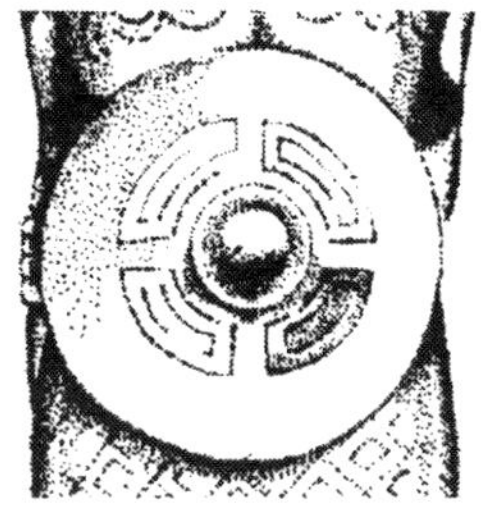

Caetra from one of the statues of Outeiro de Lezenho, Boticas (not to scale)

La Tène- type shield boss from Castro de Alvarelhos (after Soeiro 1980)

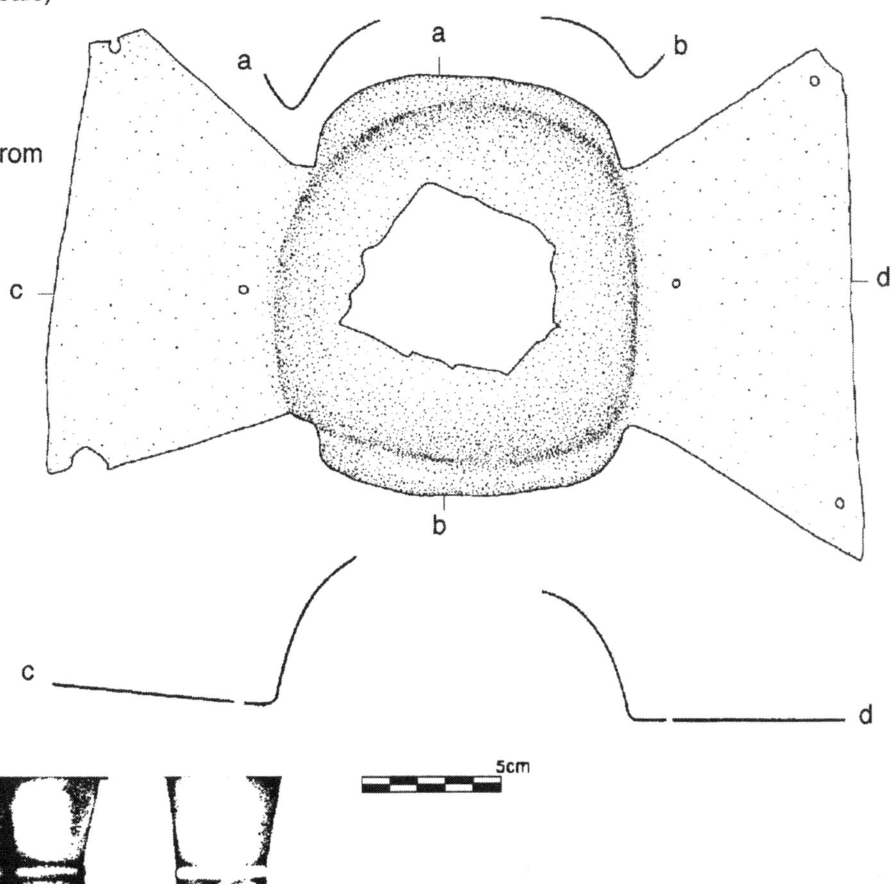

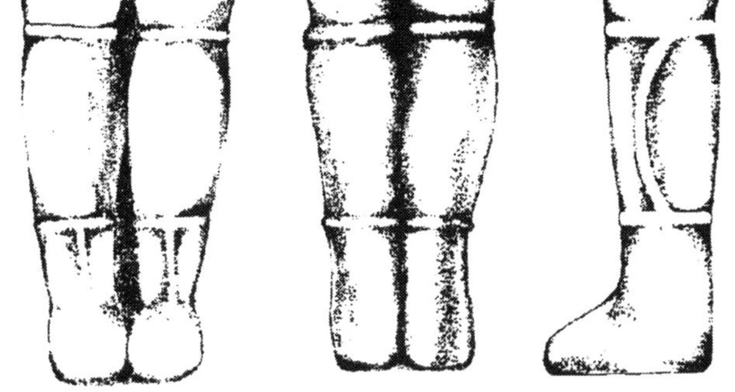

Greaves depicted on the statue from Santa Comba de Basto (not to scale)

Figure 62 - Iron Age protective gear

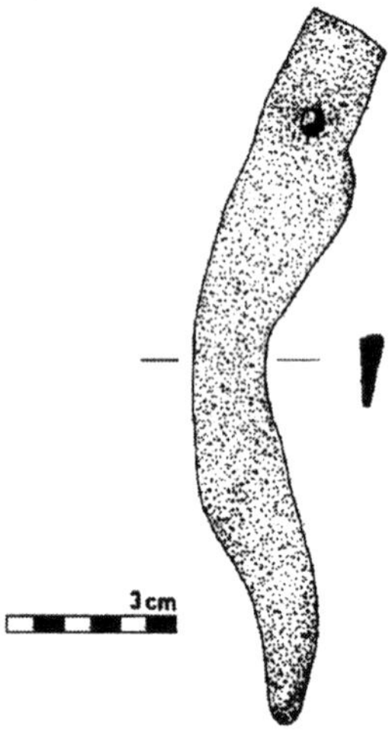

Falcata-type dagger from Castro do Cruito (after Pereira - González 1988)

Daggers depicted on the statues from Sto. Ovídio and S. Julião (not to scale)

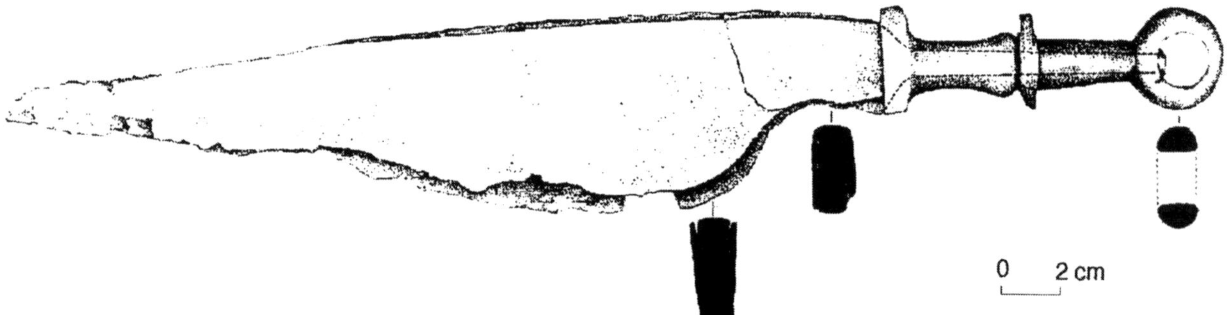

Dagger / sword from S. Julião (after Silva 1986a)

The Iberian falcata (after Sandars 1913) (not to scale)

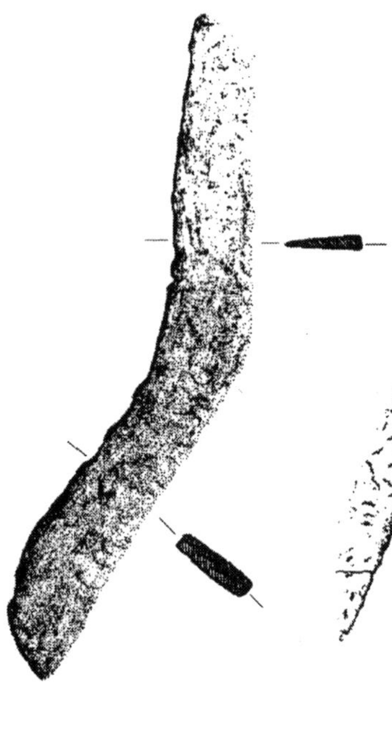

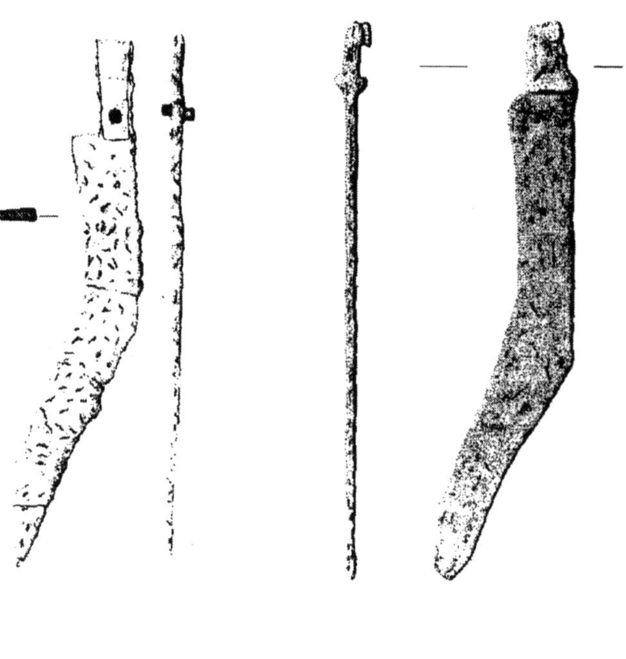

Falcata - type swords from Citânia de S. Julião. (after Martins 1988b) (not to scale)

Figure 63 - Iron Age swords and daggers

Iron arrow-head (possibly Medieval)
from Castro de Lanhoso
(after Teixeira 1940c)
(scale not known)

scale 1:2

Iron spear head from Mozinho
(after Almeida 1975)

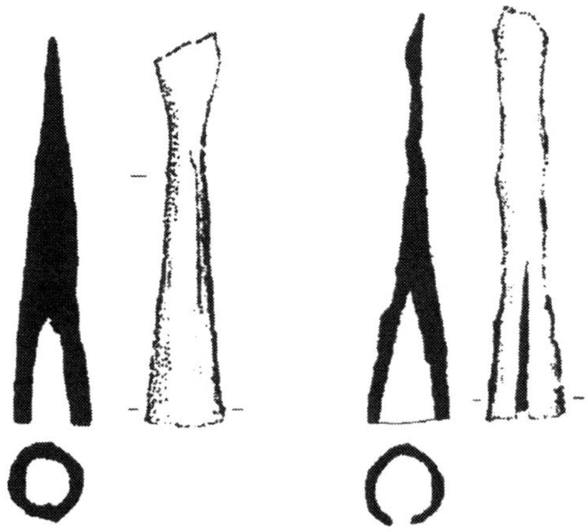

scale 1:3

Iron spear heads from Carvalhelhos
(after Silva 1986a)

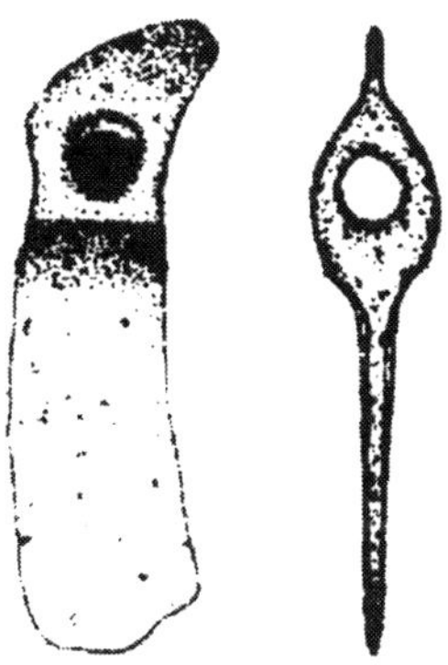

Bronze axe from Castro de S. Simão,
Cossourado. (actual size)
(after Almeida, CAB, 1984)

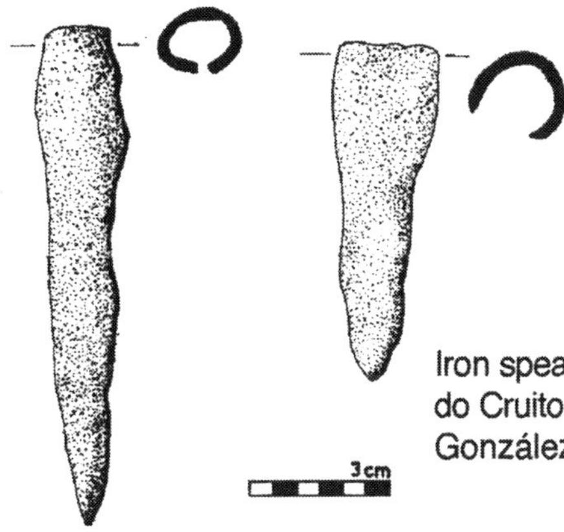

Iron spear heads from Castro
do Cruito (after Pereira -
González 1988)

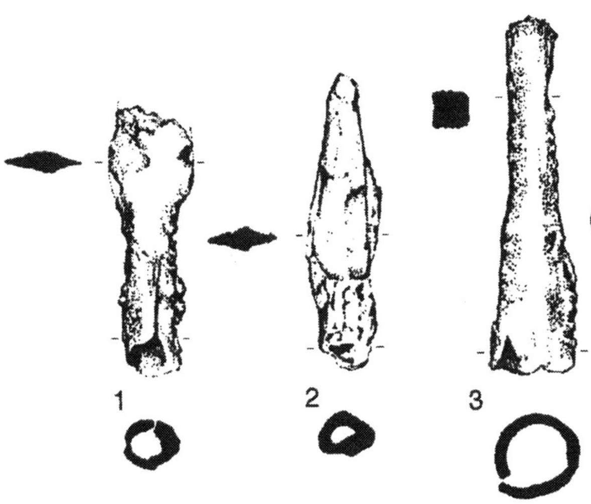

scale 1:3
(after Silva 1986a)

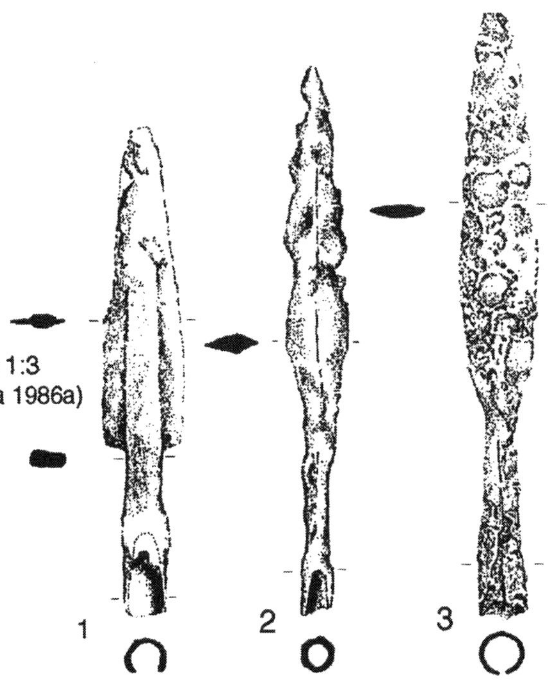

Iron spear heads from Briteiros(1),
Cidadelhe(2), S. Julião (3)

Iron spear heads from Facha (1),
Briteiros (2), Castro Máximo (3)

Figure 64 - Iron Age thrusting weapons

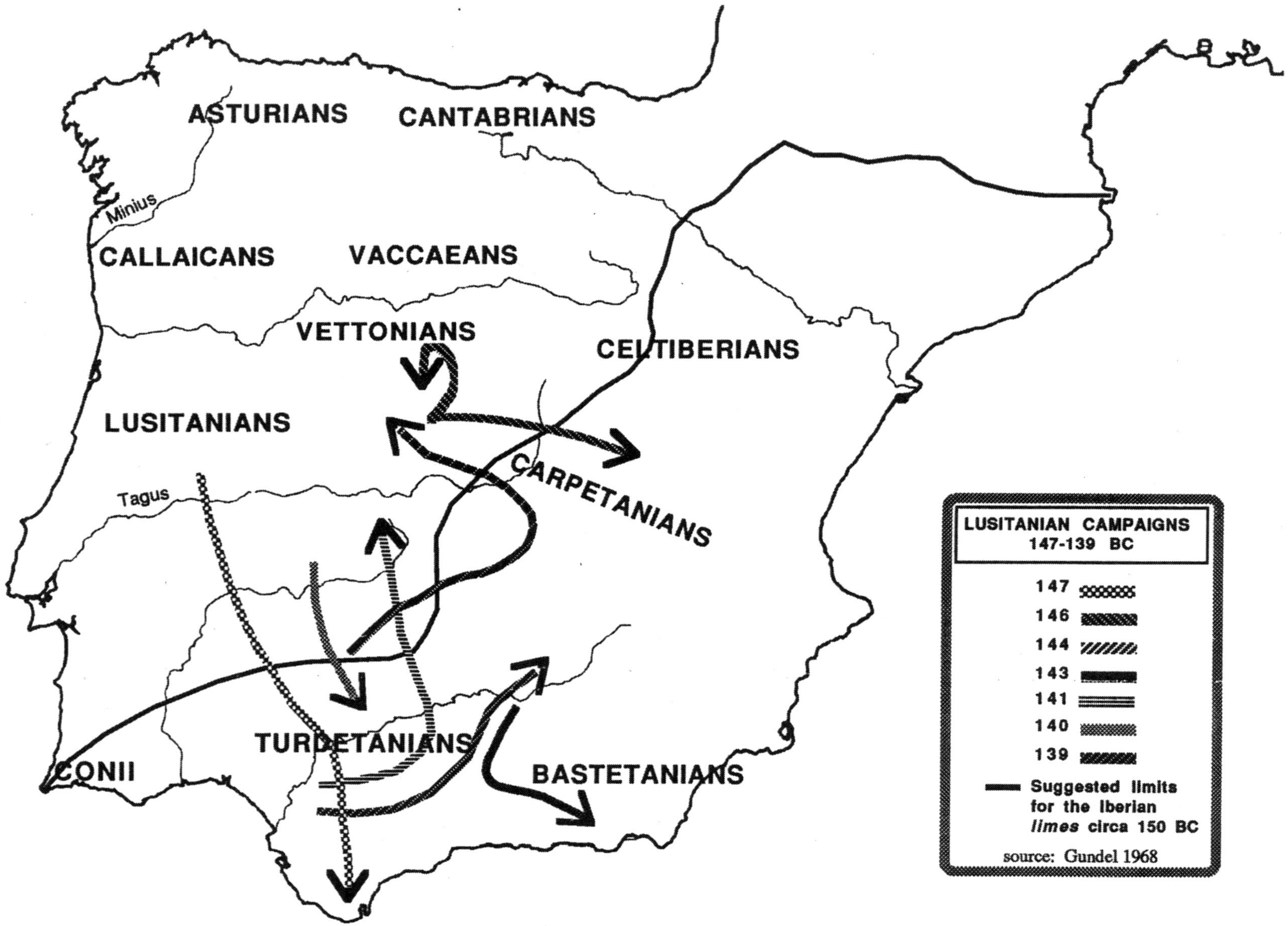

Figure 65 - Development of the "Lusitanian wars" with location of the Iberian peoples.

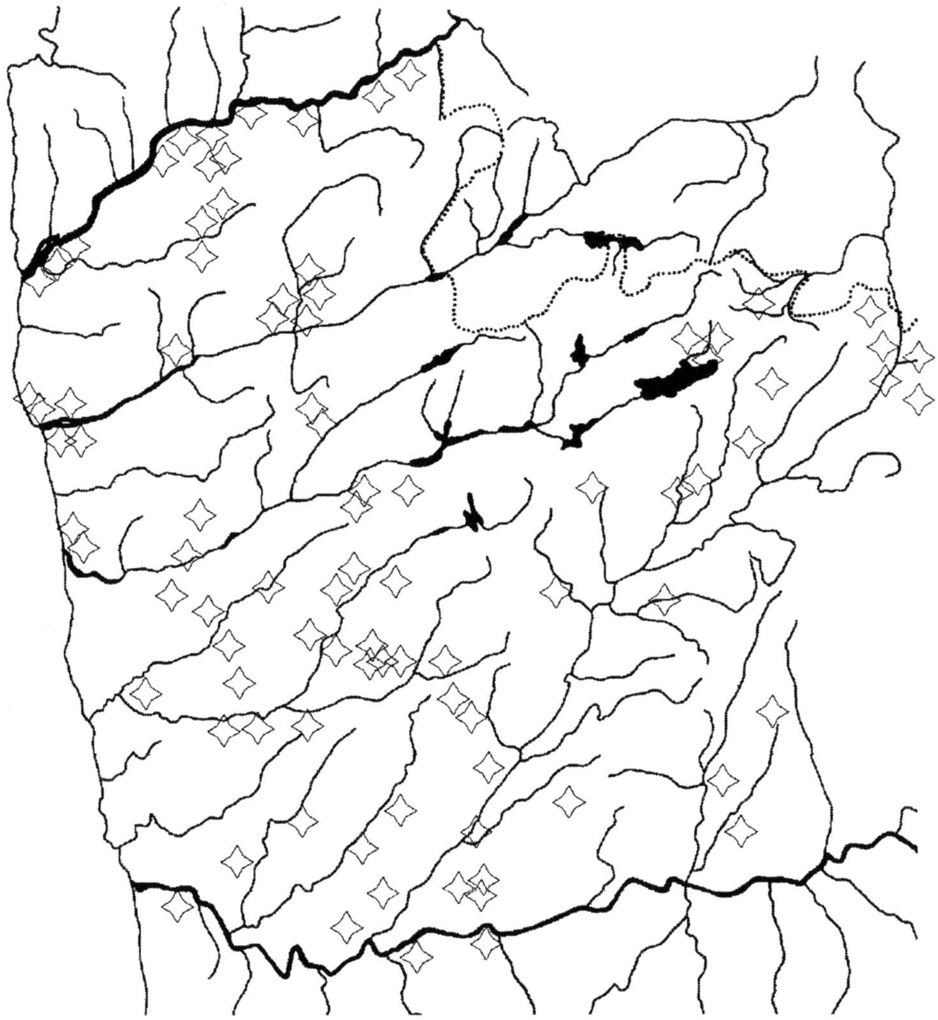

Figure 66 - Distribution of Late Bronze Age hoards and stray finds.
(after Kalb 1980a, revised)

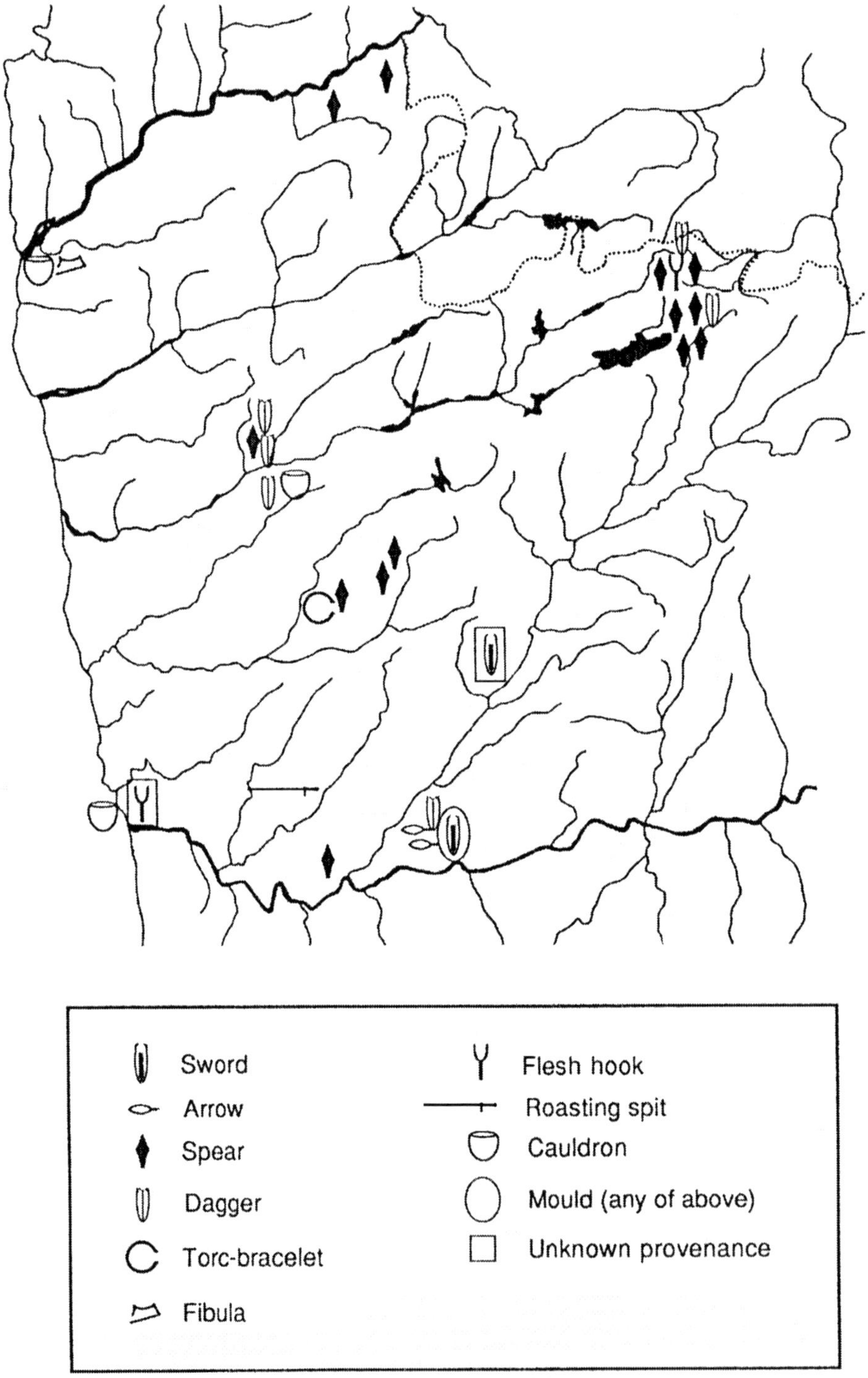

Figure 67 - Distribution of Late Bronze Age war and display equipment, and feasting utensils.
(after Coffyn 1985, Silva 1986a, and other sources)